Dynamical Systems and Automatic Control

Prentice Hall International
Series in Systems and Control Engineering

M. J. Grimble, Series Editor

BAGCHI, A., *Optimal Control of Stochastic Systems*
BENNETT, S., *Real-time Computer Control: An introduction*
BITMEAD, R. R., GEVERS, M. and WERTZ, V., *Adaptive Optimal Control*
BUTLER, H., *Model Reference Adaptive Control*
COOK, P. A., *Nonlinear Dynamical Systems*
ISERMANN, R., LACHMANN, K. H. and MATKO, D., *Adaptive Control Systems*
KUCERA, V., *Analysis and Design of Discrete Linear Control Systems*
LUNZE, J., *Feedback Control of Large-scale Systems*
MATKO, D., KARBA, R. and ZUPANCIC, B., *Simulation and Modelling of Continuous Systems: A case study approach*
MCLEAN, D., *Automatic Flight Control Systems*
OLSSON, G. and PIANI, G., *Computer Systems for Automation and Control*
PARKS, P. C. and HAHN, V., *Stability Theory*
PATTON, R., CLARK, R. N. and FRANK, P. M. (editors), *Fault Diagnosis in Dynamic Systems*
PETKOV, P. H., CHRISTOV, N. D. and KONSTANTINOV, M. M., *Computational Methods for Linear Control Systems*
SODERSTROM, T. and STOICA, P., *System Identification*
SOETERBOEK, A. R. M., *Predictive Control: A unified approach*
STOORVOGEL, A., *The H_∞ Control Problem*
WATANABE, K., *Adaptive Estimation and Control*
WILLIAMSON, D., *Digital Control and Instrumentation*

Dynamical Systems and Automatic Control

J. L. Martins de Carvalho

Prentice Hall
New York London Toronto Tokyo Singapore Sydney

First published 1993 by
Prentice Hall International (UK) Limited
Campus 400, Maylands Avenue
Hemel Hempstead
Hertfordshire, HP2 7EZ
A divison of
Simon & Schuster International Group

Typeset in 10 pt Times
by Columns Design and Production Services Ltd, Reading

Printed and bound in Great Britain
by Redwood Books, Trowbridge, Wiltshire

Library of Congress Cataloging-in-Publication Data

Carvalho, J. L. Martins de.
Dynamical systems and automatic control/J. L. Martins de Carvalho.
p. cm. – (Prentice-Hall international series in systems and control engineering)
Includes bibliographical references and index.
ISBN 0–13–221755–4
1. Automatic control. 2. Dynamics. I. Title. II. Series.
TJ213.C2933 1993
629.8–dc20 93–3829
CIP

British Library Cataloguing in Publication Data

A catalogue record for this book is available from
the British Library

ISBN 0–13–221755–4

1 2 3 4 5 97 96 95 94 93

To the memory of my grandfather, Joaquim

Contents

Preface

This text is the result of courses on automatic control and linear systems, given by the author at the Engineering Faculty of Oporto University (FEUP) for the last ten years. It is addressed not only to students in the fields of electrical, mechanical, aeronautical and chemical engineering, but also to scientists and professionals. These will find it particularly useful for self-study and specific training courses in industry.

The background knowledge required to use this book is an elementary knowledge of differential equations and general physics. A working knowledge of Laplace transforms and matrix manipulation is also assumed; for the reader's convenience, these are reviewed in the appendices, to the depth they are used in the book.

The book covers modelling, analysis and design of both continuous-time and computer-controlled systems. Mathematical models of physical systems, consisting of sets of coupled ordinary differential equations, are required by a vast class of controller design methods. Modelling of control system components, and simple plants, is reviewed in Chapter 2.

Mathematical models tell us a great deal about the limitations and achievable performance of a well-designed control system. In this respect, the role of system zeros is particularly significant and it is analyzed in several parts of the book. The ubiquitous PID (proportional, integral and derivative) controller is treated in detail, and a number of recent developments, in connection with the simultaneous satisfaction of tracking and regulation objectives, are presented.

Chapter 8 addresses the state-space and the algebraic design methods. In the second part of the chapter, the reader has the opportunity to look back from the vantage point of the algebraic framework at the design methods presented earlier, and easily assess how far each can go in terms of (achievable) performance.

The mathematical setting to treat computer-controlled systems is established in Chapter 9. With the exception of the frequency domain, all the methods presented in earlier chapters, for analog controller design, are applicable in the discrete-time case, with obvious modifications. Consequently, they are only briefly surveyed in Chapter 10. In this chapter we concentrate mainly on the differences and special issues, such as the alias phenomenon, block diagram

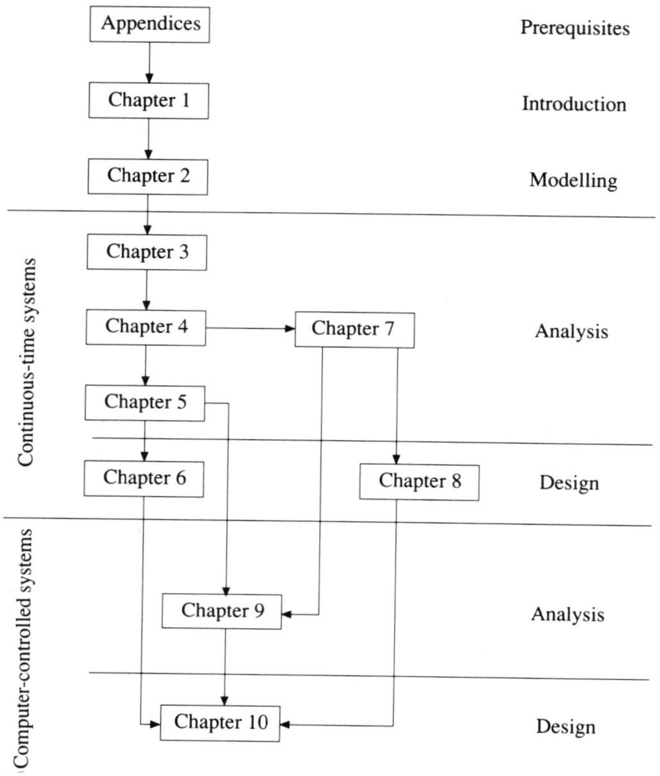

Figure 1 Structure of the book and reader's guide.

algebra of computer-controlled systems, choice of sampling frequency and frequency warping.

The book can be used in several ways, as depicted in Fig. 1. The material represents the core of two, one-semester courses on feedback control as they are lectured by the author at FEUP. Chapters 1–6 constitute the first course; the material on state-space analysis can be omitted without loss of continuity if the reader is not familiar with matrix algebra.

Although Chapters 7–10 make up the above-mentioned second course, other combinations are possible. Chapters 7 and 8 can be used as a quarter course in more advanced design methods; Chapters 9 and 10 constitute a follow-up course on computer-controlled systems, with or without the inclusion of the state-space material.

Exercises are given throughout the book. They are intended to help the reader to master the subject and also as a form of continuous self-assessment.

Despite the fact that the entire text can be followed without access to a computer, the use of one is strongly recommended. The full power of the methods presented can only be assessed through examples which are no longer amenable to hand calculations. Nowadays, personal computers constitute an indispensable tool, given their computing power and the availability of affordable and compatible software for a multitude of applications, namely for the analysis and design of control systems.

J. L. Martins de Carvalho

Acknowledgements

The author would like to thank colleagues for valuable comments and suggestions on various aspects of the book. Thanks are also due to his students for their useful feedback and for proofreading many parts of the manuscript.

The author wishes to express his gratitude to Teresa Costa, André Catarino, Pedro Garrido, Joaquim Godinho, Rodrigo Torres and João Resende for executing the simulations and for drafting many of the figures.

J. L. Martins de Carvalho
18/1/93

1

Introduction

1.1 DYNAMICS OF FEEDBACK SYSTEMS

In this book we shall be concerned with dynamical systems and ways of making them perform so as to achieve the desired result. A system which is unable instantaneously to transfer input variations to its output is called a dynamical system. Any form of physical system with internal energy storage, e.g. electrical, thermal, mechanical, is a dynamical system and this explains why understanding such systems is so important in engineering.

Let us start by analyzing an example from everyday life – the heating of a room. This can be regarded as a dynamical system whose input is heat and whose output is the room temperature. As the heater is switched on, the input changes instantaneously, but the *output* (temperature) rises gradually because of the thermal inertia. When the heat input becomes equal to the heat losses, e.g. through the walls and windows, the temperature will eventually stabilize. But how can we control the process?

One solution is the use of a *predetermined program*. Such a program could be based on a typical profile of the outside temperature throughout the day. The heat supply required could then be computed as a function of this information and the desired room temperature. Another possibility is simply to use a feedback control system with a thermostat!

Between these two approaches there are striking differences. The former, an example of *open-loop* control, has the disadvantage of being unable to cope with the effect of *disturbances*. Unexpected variations of outside temperature, changes in wind speed, the opening and closing of doors and windows, fluctuations in the number of people in the room, inaccuracies in the room model used in the computation of fuel settings (calibration), etc. will produce deviations from the desired temperature, and our control scheme will be totally insensitive to them. This scheme is shown in Fig. 1.1.

In contrast, the thermostat keeps the temperature close to the desired value, in spite of the above-mentioned disturbances and without resorting to an environmental model. It continuously compares the room temperature with the desired value and uses the result of this comparison to determine the fuel supply.

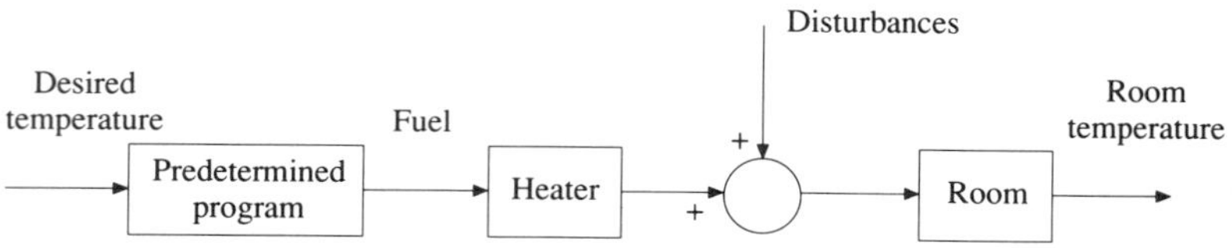

Figure 1.1 Open-loop temperature control system.

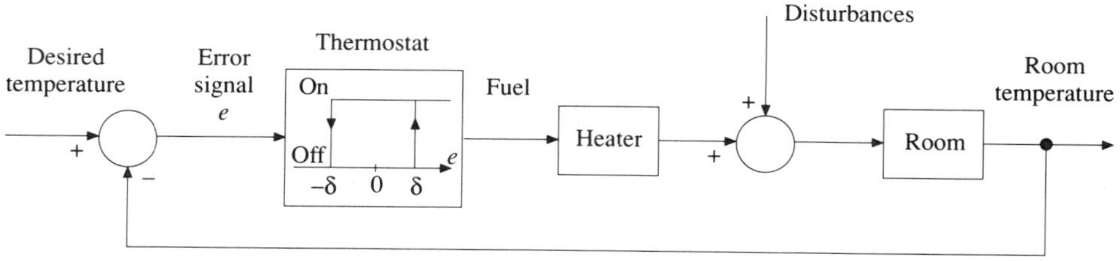

Figure 1.2 Feedback temperature control system.

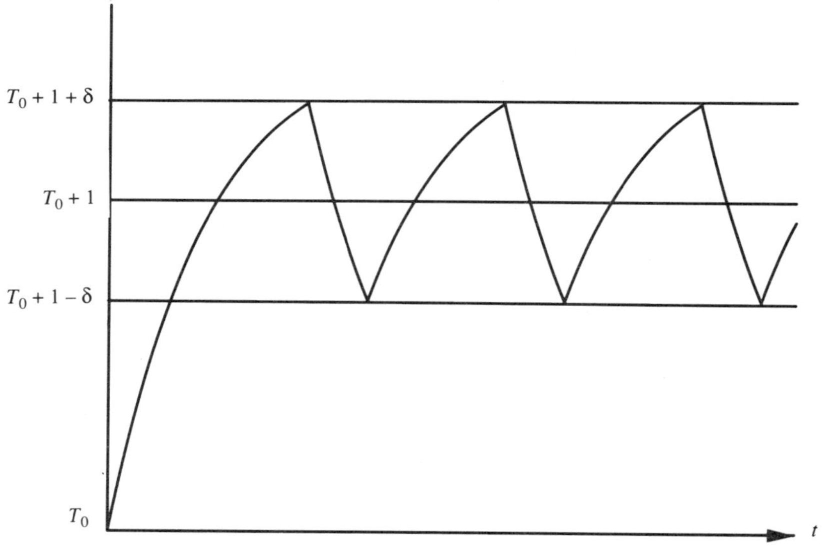

Figure 1.3 Evolution of room temperature following a unit-step change in the reference signal of the system in Fig. 1.2.

This approach is an example of feedback control and is shown in Fig. 1.2. The evolution of the temperature after a sharp change in thermostat setting at time $t=0$ is represented in Fig. 1.3.

Once the temperature reaches the new value it starts oscillating with an amplitude that depends on δ, the thermostat differential gap or hysteresis. In a real thermostat, δ is always greater than 0 and therefore such a system is unable to hold the temperature at a constant value. If we reduce δ the amplitude of oscillation will decrease. However, this has an adverse effect on the life span of the thermostat because it increases the frequency of the switching.

The essential difference between open loop control and the feedback system lies in the ability of the latter to sense the evolution of the controlled variable (room temperature) and to respond with corrective action in the case of deviation from the desired value. In contrast, the open loop scheme always acts in the same fashion no matter how great the variations.

Almost all aspects of life at either or both microscopic and macroscopic levels are controlled by this very simple principle of feedback. Microbes, the human body and human societies all possess self-regulating powers which emanate from a large number of interconnected feedback systems. It is therefore surprising that, although the application of the principle dates back more than 2000 years (Mayr, 1970), only recently, i.e. in the 1930s, have humans begun to explore its possibilities.

To summarize: if we could predict everything exactly, we could do without feedback control systems, since their key feature is their ability to cope with *uncertainty*. This uncertainty can refer to the plant (room temperature) which is being controlled. For example, if the same thermostatic control system is implemented in another room with different characteristics it is most likely that it will also function there satisfactorily. The uncertainty can also refer to the environment, e.g. we are unable to forecast exactly the outside temperature and the wind gusts. Therefore the feedback gives a system the ability to *adapt* to environmental changes. The stereotype configuration of a feedback control system is shown in Fig. 1.4. If the reference is normally held constant, we say we have a regulator. If it is varying, as, for example, in an aircraft following radar, it is called a *servo system*.

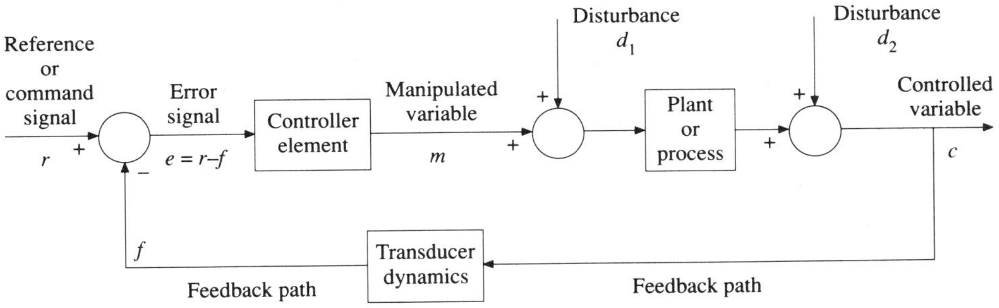

Figure 1.4 Archetypal configuration of a feedback control system.

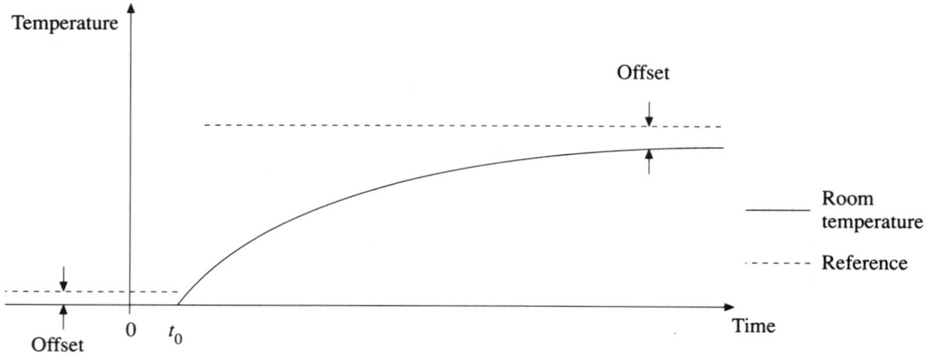

Figure 1.5 Evolution of room temperature following a step change in reference at time t_0 using a proportional controller instead of a thermostat, in the system of Fig. 1.2.

The *reactive* nature of feedback control precludes the possibility of anticipating the effect of disturbances entering the system, i.e. it has to wait until the effect of a disturbance shows up on the controlled variable. This is unsatisfactory in situations where disturbances can be measured before entering the system and it is possible to take steps to counteract their effects. This is the underlying purpose of *feedforward* control which we shall discuss later.

The discussion so far may have led the reader to conclude that a simple feedback loop is all we need to achieve a satisfactory control system, i.e. one that is both *accurate* and *stable*. Unfortunately, in the majority of cases, a lot more has to be done before reaching this goal. In the remainder of this chapter we shall try to get the gist of the main problems that arise in the design of a feedback control loop.

Let us start by analyzing the problem of accuracy. Inspection of Fig. 1.3 immediately reveals that the thermostatically controlled system, although stable (whatever this means), is unable to control the temperature within an interval amplitude 2δ. Because δ cannot be made arbitrarily small we must resort to other types of controllers if we wish to achieve a more accurate system. Assume, then, that we have replaced the thermostat by a *proportional controller* – a device whose output m is proportional to its input, i.e. $m=Ke$. In the absence of disturbances we can now hold the temperature constant but at the expense of some error as shown in Fig. 1.5 – although this error can be made arbitrarily small by a large enough K. Therefore in this case a high gain feedback system is required to obtain an accurate and stable regulator.

> **Exercise 1.1** Explain why in Fig. 1.5 an increase in reference gives rise to an increase in offset (difference between desired and actual temperatures).

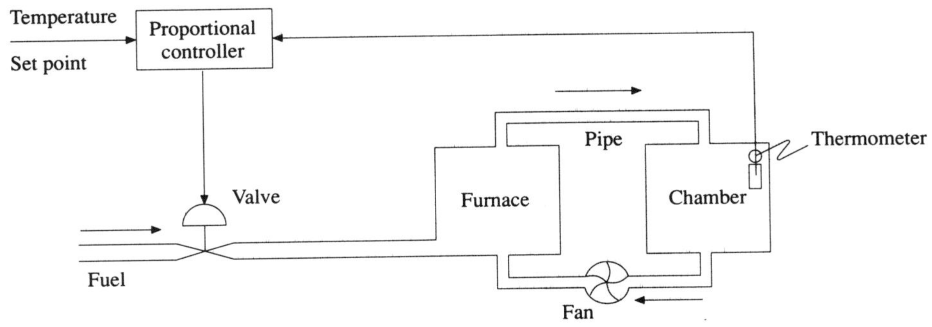

Figure 1.6 Temperature control system of a plant with a time delay.

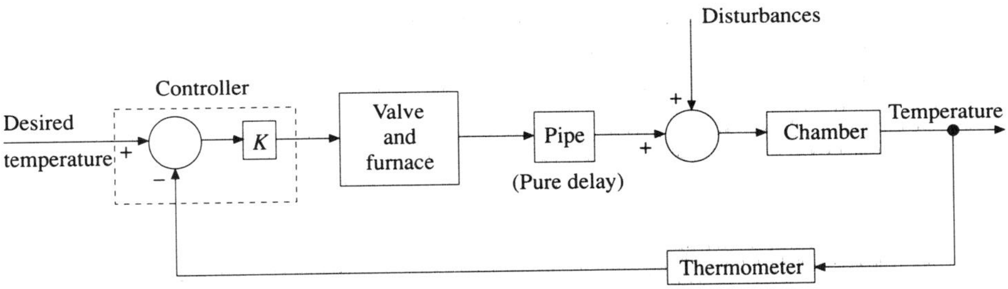

Figure 1.7 Block diagram of the system of Fig. 1.6.

This may be true for the example which we are discussing, but in general large gains have an adverse effect on the system stability, especially when regulation or measurement is *slow* or the plant under control is *non-minimum* phase. We have all come across such systems in our daily lives. For example, if we can instantaneously adjust the level of lighting in a room (with a dimmer switch), why can we not similarly instantaneously regulate the temperature of the water in our shower?

In an attempt to gain some insight into these problems let us take a closer look at the transient behaviour of the feedback control system of two non-minimum phase plants. The first is depicted in Figs 1.6 and 1.7. Its purpose is to hold the chamber temperature at a prescribed value. In this system heat is pumped into the chamber by means of a hot air stream, forced to circulate with an adjustable speed v by means of a blower. Because of the finite air velocity, any change in the temperature of the furnace will be detected by the thermometer in the chamber only after a time l/v, where l is the length of the pipe. If, for

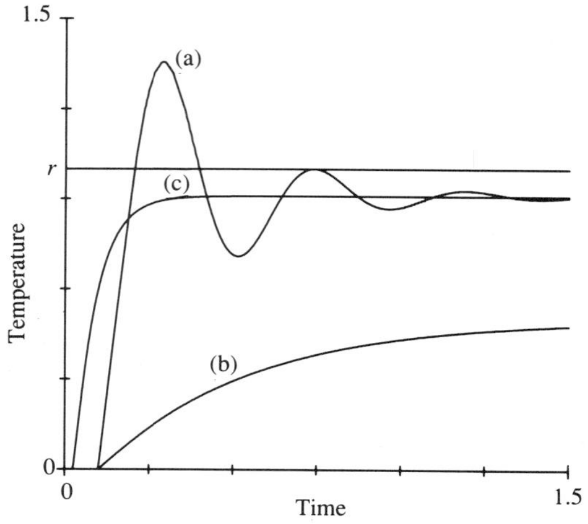

Figure 1.8 Variation of chamber temperature for the system in Fig. 1.7 following an increase r in reference temperature for the following values: (a) K=10 and delay =0.1; (b) K=1 and delay =0.1; (c) K=10 and delay =0.025.

example, the reference is incremented by a value r at time 0 the error will also increase by the same amount and remain constant (assuming no disturbances) until time l/v when it starts to decrease. Eventually it will change sign at some time t_c. When this occurs the controller immediately starts reducing the heat input to the furnace. However, the air entering the chamber until time (t_c+l/v) will contain more heat than necessary. When the newly adjusted heat supply eventually arrives the temperature is already above the desired value. Consequently the same pattern now repeats but in the opposite direction and so on. The temperature will then fluctuate with a period of approximately 4 l/v as shown in Fig. 1.8(a).

With lower values of K ('cautious control') we can avoid oscillations – but at the expense of *accuracy* and *speed of response*. This is illustrated in Fig. 1.8(b) where the temperature converges very slowly (and monotonically) to a lower value. In an attempt to achieve a response which is both accurate and fast, the gain K of Fig. 1.8(a) was maintained and the air speed increased four times. This resulted in a drastic improvement as shown in Fig. 1.8(c). (The plots in Fig. 1.8 were obtained by digital simulation with the following parameters: unit transfer function for the valve, furnace and thermometer; chamber with transfer function $1/(s+1)$.) Notice the absence of response until time equals the delay and the same *offset* in (a) and (c).

Exercise 1.2 Give examples of feedback control systems and show how time delays, if any, can give rise to oscillatory behaviour.

We have seen that *time delays* (also known as transport delays or dead times) can cause undesirable effects in a feedback system. But such effects are not exclusive to engineering systems – they can also have political implications. For example, the function of a government is to ensure that the life of a country remains balanced. We can, therefore, compare it to a regulating system with negative feedback. But because it has such large delays it is difficult to control. It can take months, even years, before a government measure takes effect, and measurement of the effect is slow as its performance can only be assessed after data have been collected and processed. Obviously such a system is potentially unstable; even more so nowadays than in the past. The media and the speed of communication allow the public to express dissatisfaction with the measure and demand corrective action before it has had a chance to prove its worth. Under such pressure the government may be obliged to push further in the direction of improvement and large oscillations may result – perhaps this is one of the reasons why governments change so frequently these days!

The second non-minimum phase plant to be considered is one of the *inverse response* type. These plants have a peculiar form of behaviour – the transient response starts out in the opposite direction from the input. A typical example of such a system is the hydraulic turbine shown in Fig. 1.9. When subjected to a sudden increase in flow rate, there is a momentary decrease in the efficiency of the blades due to a sudden loss in torque caused by the turbulence. It is not

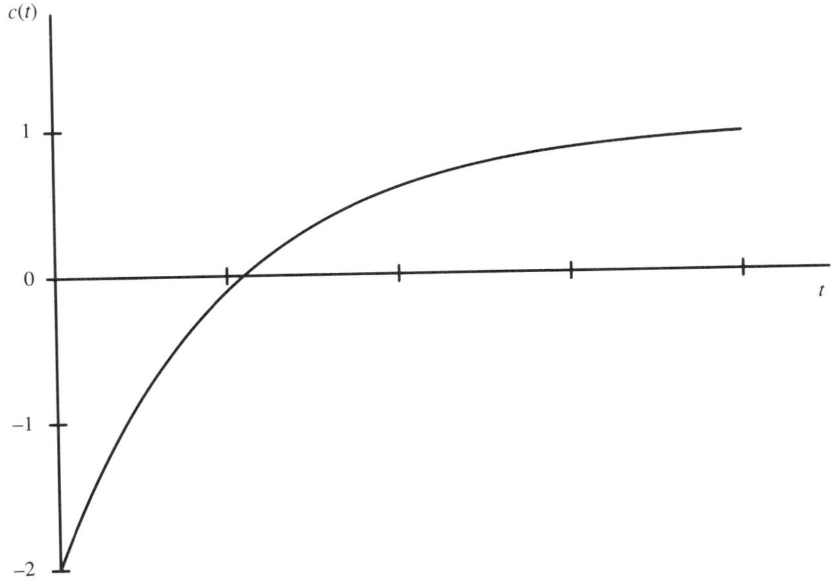

Figure 1.9 Response of an hydraulic turbine (with transfer function $(1-2s)/(1+s)$) to a positive unit step increase in flow rate.

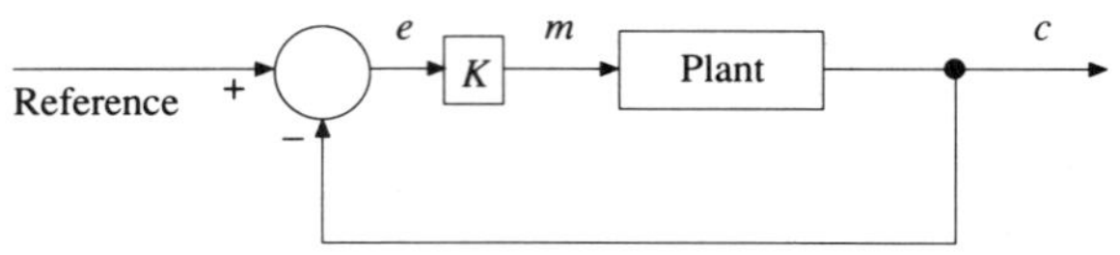

Figure 1.10 Feedback control loop for the plant in Fig. 1.9.

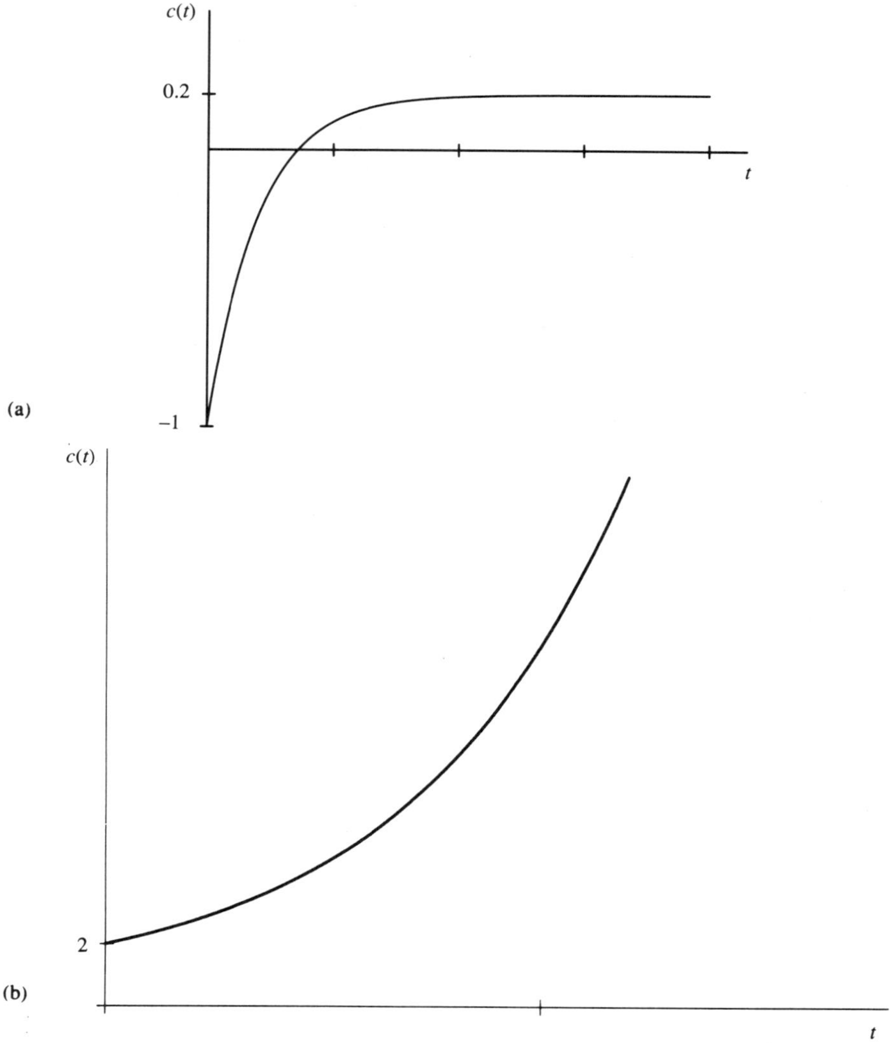

Figure 1.11 Response of system of Fig. 1.10 to a unit-step change in reference signal with: (a) $K=1/4$; (b) $K=1$.

difficult to predict the effect of inserting such a plant in a feedback control loop as shown in Fig. 1.10. The plant's response in the opposite direction to the input 'confuses' the controller because it does not realize that it is unwise for this type of plant to increase by too much its output m immediately after an error increase. Therefore, to avoid instability the gain K must be kept small (see Fig. 1.11). In this case it would have been much better if the controller had remained idle for a time before making a decision. We shall see later that, under computer control, this can be achieved by selecting a sufficiently large sampling period provided the response is zero at time $t=0$. Then the computer does not 'see' the inverse behaviour, thus giving, in principle, a better performance than an analog controller.

Inverse response plants and plants with pure transport delays are known as non-minimum phase because they introduce too much phase shift compared with other plants with similar gains. Loosely speaking we can say that for such plants, the gain at very low frequencies (DC gain) is the negative of the gain at some high frequency. As a result, when they are inserted in a feedback loop, with *negative* feedback, they behave like an ordinary plant under *positive* feedback – an unsafe type of feedback. For example, if we have a sinusoidal reference signal in the negative feedback system of Fig. 1.4 and the plant introduces a phase shift equal to half the period, the error becomes the sum of two sinusoids in phase and as a result the error will build up indefinitely. In Chapter 5 we shall analyze this behaviour in detail.

1.2 SCOPE OF THE BOOK

The book covers modelling, analysis and design of both continuous time and computer-controlled systems, and is structured as shown in Fig. 1 on p. xiv. From what we have discussed so far we can see that although feedback combined with simple on–off or proportional controllers has been perhaps the greatest step forward so far in automatic control, the performance of a feedback system is very dependent on the dynamic characteristics of the plant. Fortunately, with more information on the plant its performance can be greatly improved. Such information can be in the form of a mathematical model, and this is why we dedicate Chapter 2 to this problem. In Chapter 2, we model control system components and elementary processes, on the principle of conservation or continuity of mass, energy and momentum.

Chapters 3–5 cover the time and frequency domain analysis of open- and closed-loop systems, described by transfer functions, and pave the way for the design of controllers by the so-called classical methods, i.e. PID, feedforward, root-locus and frequency-domain.

In Chapter 3, the state-space representation is also introduced and related with the transfer function concept. In this fashion, the generating mechanisms of the solution of ordinary differential equations, the mode and the stability concepts arise naturally. The effect of the zero on the transient response is also analyzed.

The feedback concept is dealt with at length in Chapter 4. A feedback system, no matter how complicated it might look, can always be reduced to a simple standard form; this is why we spend some time with block diagram reduction before moving into the wonders of feedback. After the steady-state analysis, we present the root-locus method which constitutes an elegant way to express the richness of the dynamics of a feedback system.The PID controller is introduced here. Although this might seem a little premature, we believe that from the moment the reader becomes familiar with root-locus ideas, namely the role of the zeros as loci attractors, and the ability of integrators to eliminate steady-state errors, the PID controller appears as a natural corollary.

Chapter 5 presents the reader with a competing, though less intuitive, tool for system analysis: the frequency-domain. While root-locus analysis is impaired by system order – the information content for the human designer gradually decreases as the system order increases – higher-order systems constitute no obstacle for frequency-domain methods. The reason is that transfer functions (poles and zeros), being a parametric representation, require the explicit knowledge of the system order. The frequency-domain representation, being a non-parametric representation, is free from this burden, and this is why it is favoured when we are dealing with higher-order systems. This chapter starts by introducing the frequency response concept, and then describes several forms of representing it, namely by using polar and Bode plots. Frequency-domain is also endowed with powerful analysis tools, such as the Nyquist stability criterion, which also constitutes an example of how fundamental mathematical results can have important applications in engineering. The Nyquist criterion is presented in the second part of Chapter 5.

Chapter 6 deals with controller design by the so-called classical methods – namely root-locus and frequency-domain, which are trial-and-error methods. An in-depth analysis of the state–space representation, is the topic of Chapter 7. Considerable time is spent with modal analysis, and concepts such as controllability, observability and stabilizability, in order to pave the way for Kalmans canonical decomposition theorem, and the state-space design methods of the following chapter.

In contrast with Chapter 6, the design methods presented in Chapter 8, state-space and algebraic design, are analytical methods. They yield systematic design procedures and insight into the problem. Unfortunately they are very sensitive to modelling errors, which is a disadvantage when compared with the classical methods of chapter 6.

Chapters 9 and 10 deal with (digital) computer-controlled systems. In Chapter 9 the independent time variable belongs to a countable set, i.e. the system is seen as from the computer. It is shown that most of the theory developed in Chapters 3–5 applies, *mutatis mutandis*, to this new domain. However, things get more complex as we consider, simultaneously, discrete-time and continuous-time signals, as happens when we have continuous-time specifications and a digital controller. New and puzzling problems arise in this situation. In Chapter 10 we concentrate on these issues and present a survey of digital controller design methods.

REFERENCE

Mayr, O. (1970) 'The origins of feedback control', *Scientific American*, **223**(4), 110–18, October.

2

Mathematical Models

Frequently novice students in control experience difficulties when facing dynamical phenomena outside their narrow field of specialization. This is not surprising given the multidisciplinary nature of control. Our experience has shown that one way to overcome these difficulties is to present the student with a fair number of simple examples, from a variety of fields, and to encourage him or her to develop a 'feeling' for such models by simulating them on a computer for several values of the parameters. Such a task can easily be accomplished by most of the present-day home microcomputers. Furthermore, as a consequence of the decrease in prices of these machines and the simultaneous increase of their computing power, software libraries for scientific use are easily available.

2.1 MODELS OF PHYSICAL SYSTEMS

In this chapter we model continuous-time dynamical systems by means of ordinary linear differential equations. If the model one ends up with is a partial differential equation we usually discretize it in its spatial variable to obtain a much easier, although less detailed, description consisting of a system of ordinary differential equations, i.e. a *lumped* parameter model. In essence an ordinary differential equation model identifies integrators with independent energy stores in the system.

Integrators are the building blocks of differential equations; an nth order ordinary linear differential equation:

$$\mathrm{d}^n y/\mathrm{d}t^n + a_1\, \mathrm{d}^{n-1}y/\mathrm{d}t^{n-1} + \ldots + a_n y = b_0\, \mathrm{d}^n u/\mathrm{d}t^n + b_1\, \mathrm{d}^{n-1}u/\mathrm{d}t^{n-1} + \ldots + b_n u \qquad (2.1)$$

requires n integrators as shown in Fig. 2.1. From this diagram it is now apparent that if $b_0=0$, as happens in the majority of cases, the system is not only unable to transfer instantaneously variations from input u to its output y but also that sharp input variations will be smoothed out by the successive integrations. Such a diagram follows immediately from equation (2.1) as shown in Chapter 3.

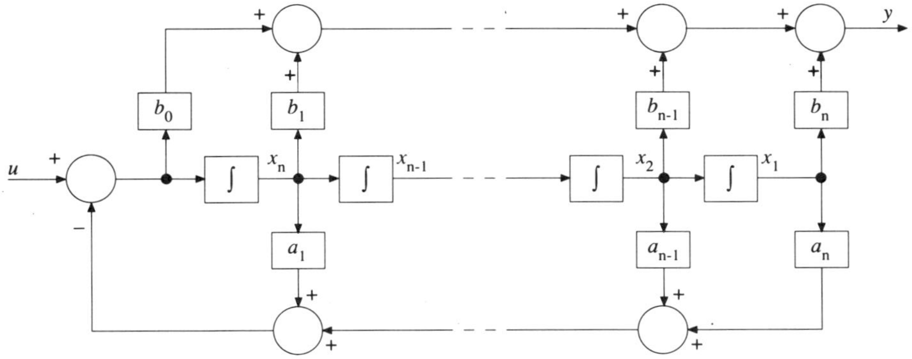

Figure 2.1 Block diagram for the *n*th-order linear ordinary differential equation (2.1): controllable canonical form.

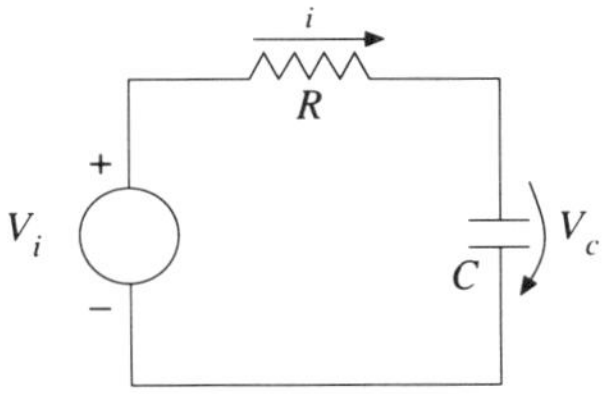

Figure 2.2 An RC circuit.

The way a system deviates from a desired equilibrium point or trajectory is a particular concern for control. Such deviations may be due to disturbances or the inability of the system to cope with fast and prescribed variations. If the non-linearities present are not too strong, a linear model describing these deviations may be satisfactory. This model will then provide us with a set of valuable information, known as the state of the system whose values are required for determining, for example, a control strategy that (hopefully) will bring the system back to the desired operating point or trajectory. Most of the models described below constitute a set of archetype examples of control system components and elementary processes which act as building blocks for more complex systems. The underlying modelling principle is conservation or continuity of mass, energy and momentum. Theoretical modelling is also of importance in methods for diagnosing process faults based on process parameters (Isermann, 1984). It is therefore highly desirable to have models in which the parameters are expressed in terms of process coefficients whose changes may indicate process faults. For a general discussion on principles of model building we refer the reader to Richards (1979) and Fasol and Jörgl (1980). A deep discussion on modelling and sensitivity to model errors can be found in Wierzbicki (1984).

2.1.1 RC Circuit

The circuit illustrated in Fig. 2.2 is one of the simplest examples of dynamic systems and it is also a useful analog in the study of other physical systems. The energy stored by the capacitor is $1/2CV_C^2$. Therefore the voltage across its terminals cannot change abruptly, because the laws of physics do not allow energy transfers in zero time. In fact the constitutive relation for a (linear) capacitor is

$$V_C(t) = 1/C \int_0^t i(t')\, dt' + V_C(0) \tag{2.2}$$

where $V_C(0)$ denotes the voltage of the capacitor when we started counting the time. Therefore $V_C(t)$ will always be, for practical purposes, a *continuous* function of t.

From (2.2) and Kirchhoff's voltage law a mathematical model for this circuit follows immediately:

$$V_i(t) = Ri(t) + 1/C \int_0^t i(t')\, dt'$$

assuming $V_C(0)=0$, where V_i is the applied voltage.

Because we rather prefer to work with differential equations we can apply instead Kirchhoff's current law and get

$$(V_i - V_C)/R = C\, dV_C/dt$$

where $C\, dV_C/dt$ is the current through the capacitor. Writing this equation in a more familiar form

$$RC\, dV_C/dt + V_C = V_i \tag{2.3}$$

we see that our circuit is modelled by a first-order ordinary differential equation.

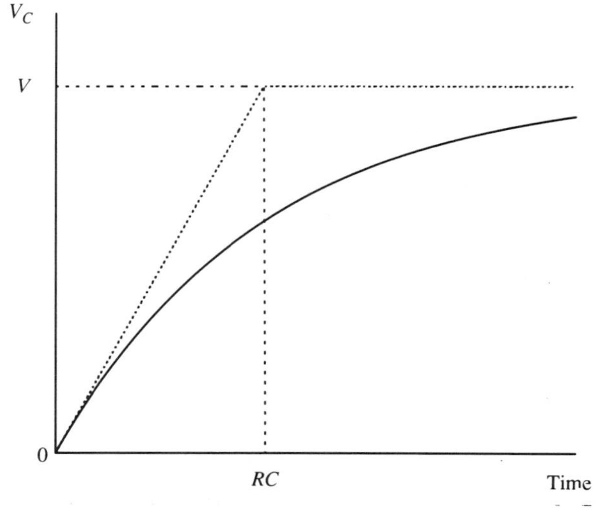

Figure 2.3 Response of the RC circuit to a step increase in V_i of magnitude V.

If V_i changes abruptly, at time $t=0$, from 0 to a value V then V_C will evolve according to

$$V_C = V(1 - e^{-t/RC})$$

which is depicted in Fig. 2.3. RC always has the units of time and is called the circuit *time constant*. Notice that the response immediately after the change in V_i behaves approximately like a straight line with slope V/RC. Therefore the larger R or C, the slower the response.

Exercise 2.1 Let $R = 10\ \text{k}\Omega$ and $C = 1\ \mu\text{F}$ in the circuit of Fig. 2.2. If a unit step voltage is applied at time $t=0$, and $V_C(0)=0$, what is the value of V_C at time $t=10^{-3}$ s? And at time $t=RC$? How can we use the last result to compute RC experimentally?

2.1.2 Mercury Thermometer

The dynamic behaviour of a thermometer can also be described by a linear first-order differential equation, if we neglect the heat capacity of the glass walls and assume that the mercury is at a uniform temperature, which we denote by $\theta_m(t)$. Now assume that, at time $t=0$, the thermometer is suddenly immersed in a fluid at a temperature θ_0. The heat flow rate q (amount of heat per unit of time into the thermometer) is given by

$$q = (\theta_0 - \theta_m(t))/R$$

where R is the resistance to heat flow between glass and fluid. On the other hand, the increment in heat stored by the mercury as the result of a temperature change from $\theta_m(0)$ to $\theta_m(t)$ is

$$C(\theta_m(t) - \theta_m(0))$$

where C denotes the thermal capacitance of the mercury; recall that C is the amount of heat required to produce a unit increase in temperature. The heat balance equation tells us that the input flow equals the accumulation, because no heat is being given away by the thermometer. Therefore

$$(\theta_0 - \theta_m(t))/R = C\ d/dt\ (\theta_m(t) - \theta_m(0))$$

or equivalently

$$RC\ d\theta_m/dt + \theta_m = \theta_0 \qquad (2.4)$$

A comparison of the above equation with (2.3) shows that RC is the time constant of the thermometer; it is therefore a measure of the time it takes to adjust to a new temperature. Figure 2.2 is also a useful analog of our thermometer where V_i stands for θ_0 and V_C for θ_m, R is the resistance to heat flow and C the capacity for heat storage. A more accurate description of the thermometer can be obtained by including the effects of the thermal capacitance of the walls. Heat is first transferred from the fluid to the glass and from there into the mercury. A diagram of the thermometer together with the temperature profile and electrical analog are depicted in Fig. 2.4: C_g is the glass thermal capacitance assumed to be at a uniform temperature θ_g.

Exercise 2.2 (a) Show that the equation describing the system in Fig. 2.4 is

$$C_g C_m \, d^2\theta_m/dt^2 + (C_g/R_{gm} + C_m/R_{fg} + C_m/R_{gm}) \, d\theta_m/dt + 1/(R_{fg}R_{gm})\theta_m$$
$$= 1/(R_{fg}R_{gm}) \, \theta_0$$

(b) Find a state-space representation for this equation (cf. with (2.2)).
The evolution of θ_m following a unit step change in θ_0, at time t=0, is shown in Fig. 2.5. Notice that the response is slower in comparison with Fig. 2.3 and that the tangent to the curve at the origin is now zero.

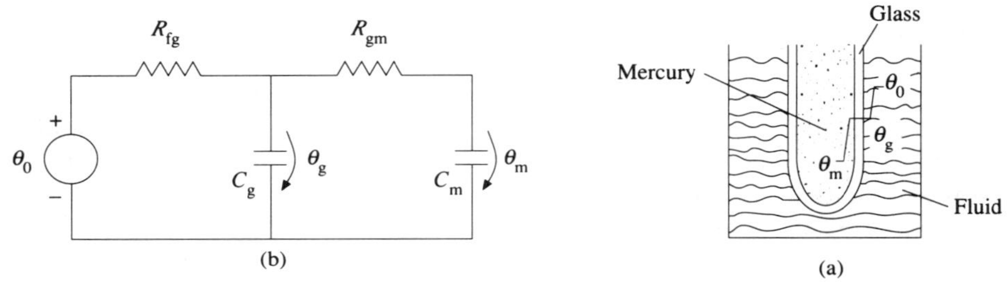

Figure 2.4 Thermometer with non-negligible glass thermal capacitance: (a) temperature profile; (b) electrical analog.

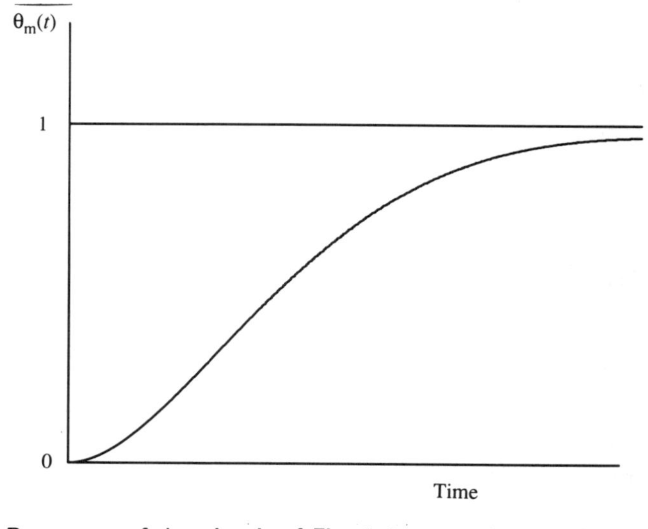

Figure 2.5 Response of the circuit of Fig. 2.4 to a unit-step change in θ_0.

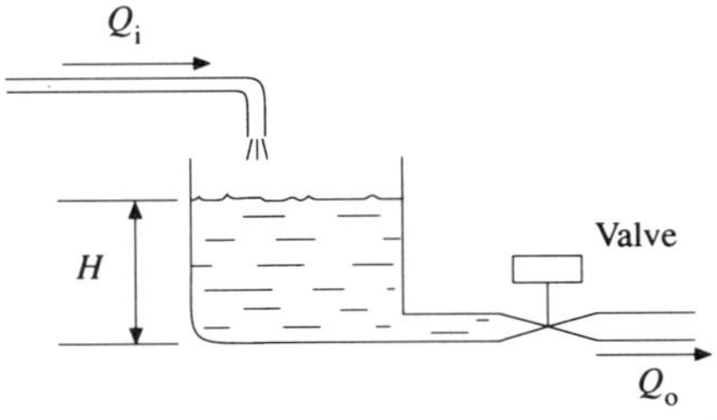

Figure 2.6 Liquid-level system.

2.1.3 Liquid-Level Systems

Assume a tank with constant cross sectional area A as shown in Fig. 2.6. The law of mass conservation tells us that

$$A\, dH/dt = Q_i - Q_0 \tag{2.5}$$

i.e. in a small time interval the difference between the inflow and outflow equals the additional amount stored. It is also known that the flow of water through a restriction is usually a function of the square root of the pressure drop across it; for the case of a discharge to atmospheric pressure, the pressure drop is obviously proportional to H. We then have

$$Q_0 = K_t \sqrt{H} \tag{2.6}$$

Notice that we are assuming a constant and small valve opening. More rigorously, we should have also considered the pressure drop in the discharge line which, for turbulent flow, is approximately proportional to the square of the flow. Substituting (2.6) in (2.5) and rearranging terms, we get the following model:

$$A\, dH/dt + K_t \sqrt{H} = Q_i \tag{2.7}$$

which is a *non-linear* differential equation. Obviously linear equations are easier to solve and we must attempt some simplification. Fortunately in most situations we wish to keep H constant, say equal to $\bar{H}$, and if the control system is working properly only *small* deviations h from the *steady-state* value $\bar{H}$ will be allowed. We can then linearize (2.7) around the operating point $\bar{H}$. Let $\bar{Q}_i$ and $\bar{Q}_0$ denote the steady-state values of the in- and outflows respectively, which are obviously equal.

If Q_i is suddenly increased by a constant small amount q_i , Q_0 will gradually increase towards $\bar{Q}_i + q_i$; expressing the variables in (2.5) in terms of their steady-state components and deviations we get

$$A\, d/dt\, [\bar{H} + h] = (\bar{Q}_i + q_i) - (\bar{Q}_0 + q_0)$$

which is equivalent to

$$A\, dh/dt = q_i - q_0 \tag{2.8}$$

Also

$$q_0 = K_t/(2\sqrt{\bar{H}})\cdot h + O(h^2) \tag{2.9}$$

Substituting (2.9) in (2.8) we arrive at

$$(2A\sqrt{\bar{H}}/K_t)\, dq_0/dt + q_0 = q_i \tag{2.10}$$

which is the desired linear equation.

Let us now analyze how (2.9) was derived from (2.6). Expanding $\sqrt{H}$ in a Taylor series around $\bar{H}$ we get

$$\sqrt{(\bar{H}+h)} = \sqrt{\bar{H}} + h/(2\sqrt{\bar{H}}) - h^2/8\, (1/\sqrt{\bar{H}})^3 + \ldots$$

Neglecting non-linear terms, which is acceptable if h is *small*, and substituting in (2.6) we get

$$Q_0 = \bar{Q}_o + q_0 = K_t\, (\sqrt{\bar{H}} + 1/2\; h/\sqrt{\bar{H}})$$

But

$$\bar{Q}_0 = K_t\sqrt{\bar{H}}$$

Then (2.9) follows.

At this point we may question the usefulness of these equations if K_t is unknown. In fact it can be computed by experimental means simply by plotting, for each value of Q_i within the desired range, the corresponding stabilizing value of H. The curve will look like the one shown in Fig. 2.7. We can then compute the slope of the tangent at the operating point $\bar{H}$, which is equal to $K_t/(2\sqrt{\bar{H}})$.

Equation (2.10) is of the same type as equation (2.3). Therefore if an equilibrium is disturbed by an input step increase q_i we know that q_0 will evolve towards the new equilibrium value according to

$$q_0(t) = q_i(1 - \exp(-K_t t/(2A\sqrt{\bar{H}})) \tag{2.11}$$

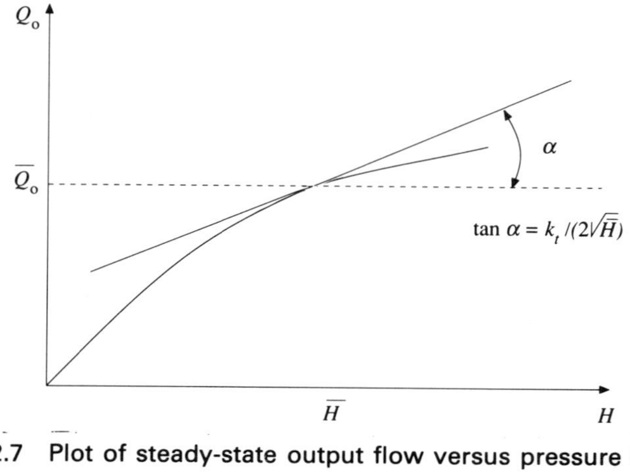

Figure 2.7 Plot of steady-state output flow versus pressure head.

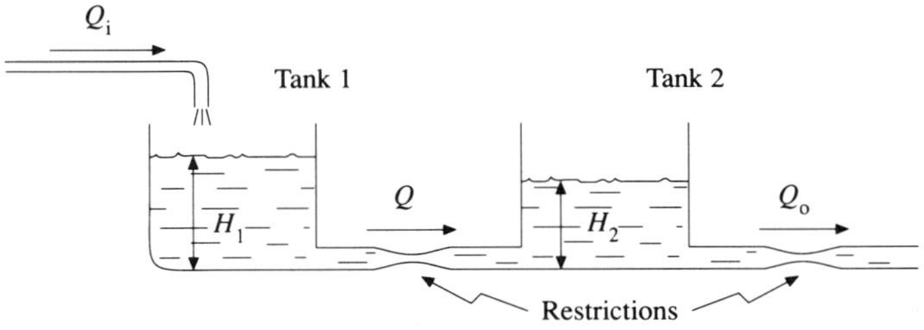

Figure 2.8 System for exercise 2.4.

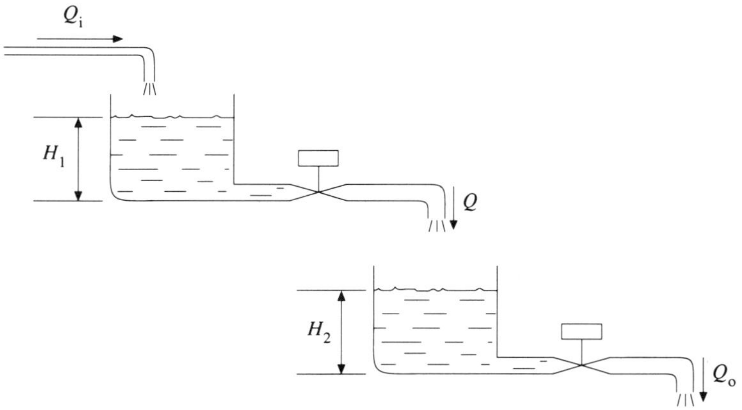

Figure 2.9 System for exercise 2.5.

whose graph is similar to Fig. 2.3. This means, for example, that Q_0 is unable to follow fast variations in Q_i; as a result the tank can be used to 'filter' flow fluctuations allowing a steadier output flow Q_0 in the presence of a fluctuating input Q_i. The greater the system time constant, the better the filtering action. In our case the time constant is $2A\sqrt{\bar{H}}/K_t$ which shows that the larger the cross-sectional area and the pressure head, the better the 'filtering', as expected. From (2.9) we can conclude that $2\sqrt{\bar{H}}/K_t$ is the *dynamic* resistance of the tank. In general, resistance is defined as the rate of change of driving force (effort) with flow. Here A is the tank 'capacity'.

Exercise 2.3 Show that the tank in Fig. 2.6 behaves like an integrator when the output valve is closed.

Exercise 2.4 Establish the equations for the two interacting tanks shown in Fig. 2.8 assuming that the flow Q through the restriction is laminar and therefore obeys the law $Q=K\Delta P$, where ΔP is the pressure drop across the restriction. Also neglect the frictional pressure drop in the discharge lines.

Exercise 2.5 Derive the equations for the system of Fig. 2.9 (non-interacting tanks) assuming law (2.6) for both output flows.

Notice that in exercise 2.5 the dynamics of the two tanks can be studied *separately* in contrast with the system of Fig. 2.8 where the flow Q depends on H_1 and H_2, i.e. the tanks *interact.*

Exercise 2.6 Assume the tank of Fig. 2.6 with an area of 0.1 m^2 and a steady discharge rate of 500 litres/hour for a depth of 1.2 m. If the input flow rate is suddenly increased to 550 litres/hour, compute:
(a) The new steady-state value for the depth.
(b) The tank time constant.
(c) The time it takes the output flow to reach 517 litres/hour.
Answers: (a) 1.45 m; (b) 28 m in 48 s; (c) 12 min.

2.1.4 Manometer

Another interesting example, particularly for chemical engineers, is the manometer shown in Fig. 2.10. In order to establish its mathematical model let:

g be gravity acceleration
V be volume of fluid
A be cross-sectional area
P be applied pressure
M be mass of fluid
R be frictional resistance

In the following derivation we shall use a force balance equation, namely that the product of the acceleration of the fluid (assumed incompressible) by its mass equals the sum of the forces acting on it. Let us consider that the wall exerts a friction on the fluid which is proportional to its velocity, giving rise to a pressure drop 1/2 $R\, dH/dt$; we must also bear in mind that, as the fluid rises, its own weight starts to counteract the applied pressure exerting a force $g\, H\, A\, M/V$. The dynamic equation then becomes

$$M\, d^2H/dt^2 = 2A\,(P - (M/V)\, gH) - RA\, dH/dt$$

or equivalently

$$M/A\, d^2H/dt^2 + R\, dH/dt + 2(Mg/V)\, H = 2P \tag{2.12}$$

which is a second-order linear differential equation.

The reader may have already observed that when a sudden presure is applied to one of the sides of the tube the fluid in the manometer *oscillates*. This is a characteristic not found in first-order differential equations. Any system that stores only one type of energy is unable to produce oscillations. However, the manometer stores two types of energy: kinetic and potential. With reference to

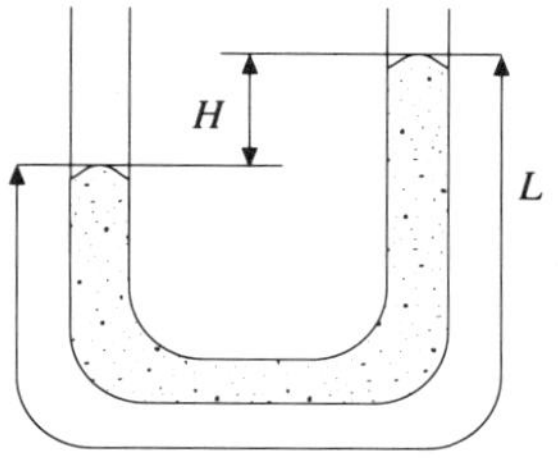

Figure 2.10 A manometer.

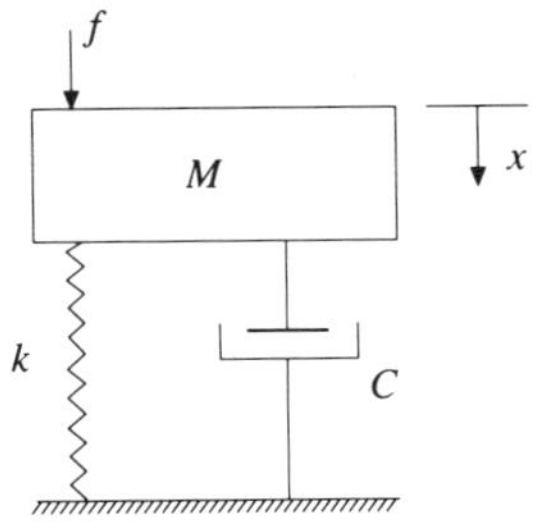

Figure 2.11 Mass–spring–damper system.

Fig. 2.10, when H reaches its maximum the energy in the system is entirely potential (H is stationary); when the system passes through its equilibrium position the energy, measured from the equilibrium state of rest, is entirely kinetic. The same type of reasoning can be applied to a mass hanging from a spring or a simple pendulum, for example.

2.1.5 Mass–Spring–Damper System

This system is illustrated in Fig. 2.11. It is made from a mass M, attached to a linear spring and a linear damper. The spring produces a reaction force proportional to displacement and the damper reacts against variations in displacement.

The mathematical model for this system can be easily derived by noticing that the algebraic sum of the forces acting on the mass equals the product of the mass times its acceleration. Since f is an (externally) applied force, the system will react against variations induced by this force; therefore

$$f - C\frac{\mathrm{d}x}{\mathrm{d}t} - kx = M\frac{\mathrm{d}^2x}{\mathrm{d}t^2} \tag{2.13}$$

where x measures the deviation from the equilibrium state of rest; ($-C\ dx/dt$) and ($-kx$) are reaction forces exerted by the damper (viscous friction) and by the spring, respectively.

Exercise 2.7 If the spring is pulled and then released the system starts oscillating. Assuming negligible friction, i.e. $c=0$ show:

(a) $x(t)=A\cos\omega t$, with $\omega^2=k/M$ (notice that the frequency of oscillation is independent of the amplitude);
(b) the mass passes through its equilibrium position with velocity ωA;
(c) the potential energy, when x is maximum, is given by $1/2\ kA^2$;
(d) the expression of the total energy at $x=0$ and $x=A$.

Can you derive the frequency of oscillation from the previous result?

Exercise 2.8 For a simple pendulum of length l and mass M derive the linearized equation of motion and then show that the frequency of oscillation is given by $\omega=\sqrt{(g/l)}$, therefore independent of the amplitude of oscillation and the mass of the pendulum.

2.1.6 Mechanical Accelerometer

Such a device is based on the principle that all bodies react against changes in speed. We all experience this principle in a train or car when it pulls up or starts. A mechanical accelerometer is represented in Fig. 2.12: x is the displacement of the case with respect to an inertial frame and y is the position of the mass relative to the accelerometer case, such that $y=0$ when the entire system is at rest. Equating the forces acting on the mass to its *inertial* acceleration we get

$$M\ d^2/dt^2\ (y - x) = -ky - c\ dy/dt$$

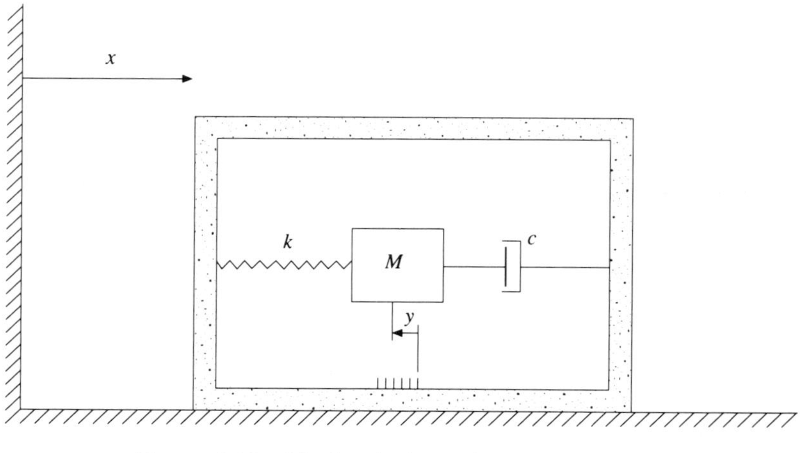

Figure 2.12 Mechanical accelerometer.

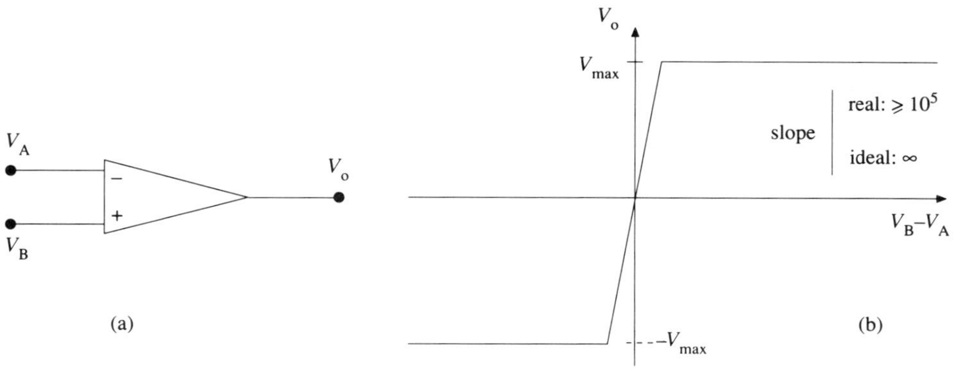

Figure 2.13 Operational amplifier symbol (a) and input–output characteristic (b).

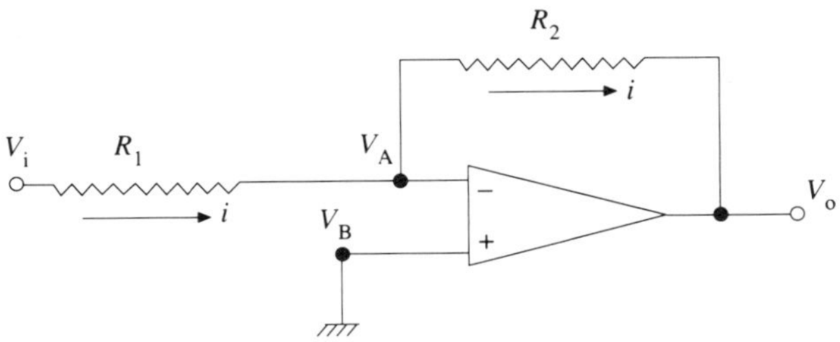

Figure 2.14 Ideal inverting amplifier.

or equivalently

$$d^2y/dt^2 + (c/M)\,dy/dt + (k/M)\,y = d^2x/dt^2 \tag{2.14}$$

If the case is being submitted to a constant acceleration, i.e. d^2x/dt^2 is constant, then y tends to $M/k\ d^2x/dt^2$. This means that, when all transients have died out, the mass will remain displaced from its rest position at a distance *proportional* to the acceleration of the case; such distance is independent of the damping coefficient c and the bigger the mass M, the larger the displacement, as expected.

2.1.7 Operational Amplifier

The standard symbol for an operational amplifier and input–output characteristics are shown in Fig. 2.13. The usefulness of an operational amplifier derives from its high gain, typically around 10^5, which permits the implementation of an endless number of useful devices when used with a feedback configuration. In open loop it has limited use; in fact since V_A, V_B and V_o are signals of the same order of magnitude, such a high gain makes the amplifier behave as a two state output device, i.e. $V_o = \pm V_{max}$ if used in open loop. An ideal operational amplifier is assumed to have infinite gain, no current through the input terminals, i.e. infinite input impedance and zero output impedance. The former implies $V_A = V_B$ and input zero current at the two terminals, and the latter that the device allows whatever current is required at the output terminal without changing V_o. Operational amplifiers play an important role in the development of circuits for control applications and we discuss some of them next. Figure 2.14 represents one of the simplest possible set-ups: the inverting amplifier.

Assuming an ideal amplifier, we have the same current through R_1 and R_2 as the input current at operational amplifier terminal (−) is zero and $V_A = V_B = 0$ as the gain of the operational amplifier is infinite. Therefore $V_o = (-R_2/R_1)\ V_i$. The circuit input impedance is now R_1.

As we have seen in Chapter 1 an important signal in any feedback control system is the *error* which is the difference between the actual and the desired output signal. The circuit shown in Fig. 2.15 produces a signal which is proportional to the difference in inputs as the following simple analysis reveals:

$$V_o = (i_2 - i_1)R_2$$

$$V_2 - V_1 = (i_2 - i_1)R_1$$

therefore

$$V_o = (V_2 - V_1)R_2/R_1$$

We are already aware of the importance of the integrator in the study of dynamical systems. An integrator can be easily implemented with an operational amplifier as Fig. 2.16 shows. Neglecting the input current of the amplifier we have

$$V_i/R = -C\ dV_o/dt$$

or equivalently

$$V_o(t) = -1/RC \int_o^t V_i(t')\ dt' + V_o(0) \tag{2.15}$$

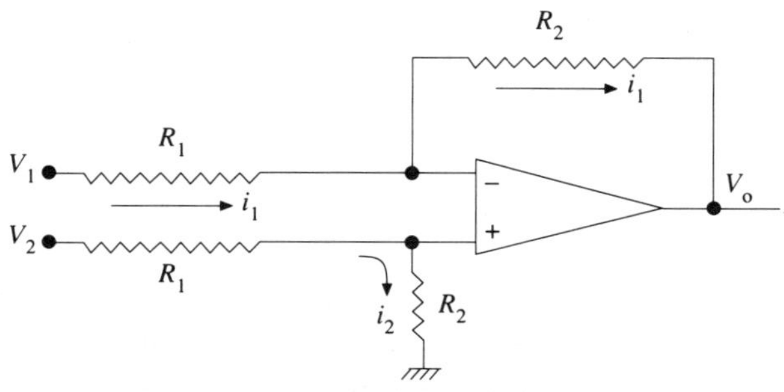

Figure 2.15 Differential amplifier.

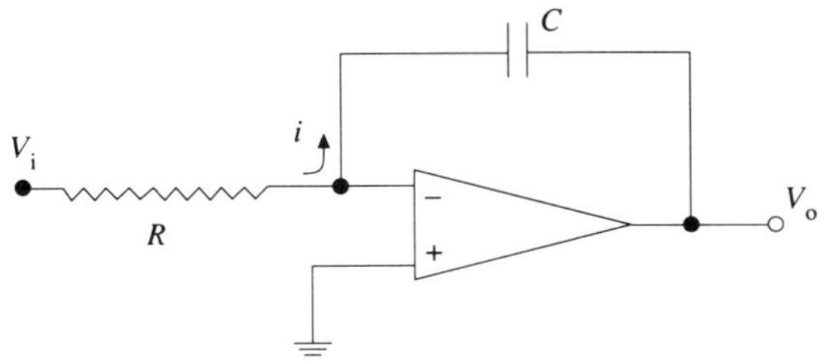

Figure 2.16 Integrator.

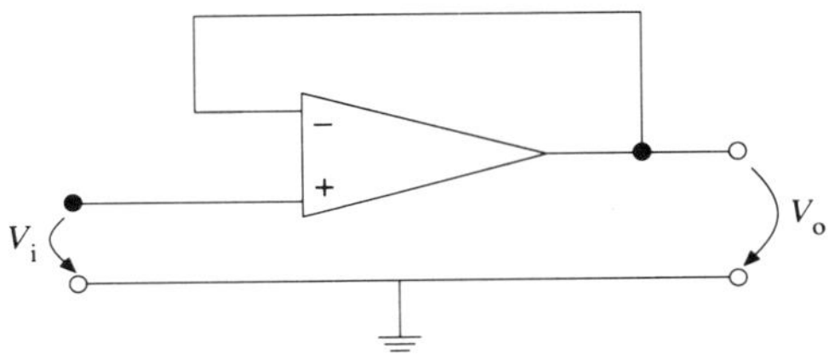

Figure 2.17 Voltage follower.

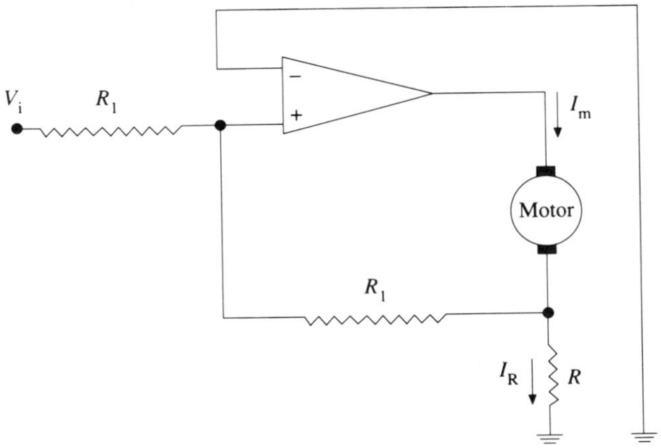

Figue 2.18 Torque control system.

Obviously V_o will not increase for ever; it is bounded above by the operational amplifier voltage supply.

The reader may have noticed that the feedback element is always connected to the negative input V_A of the operational amplifier. If, for an ideal amplifier, this is irrelevant it makes a lot of difference when non-ideal dynamic effects begin to show up; in fact feedback through the (+) input terminal may lead to

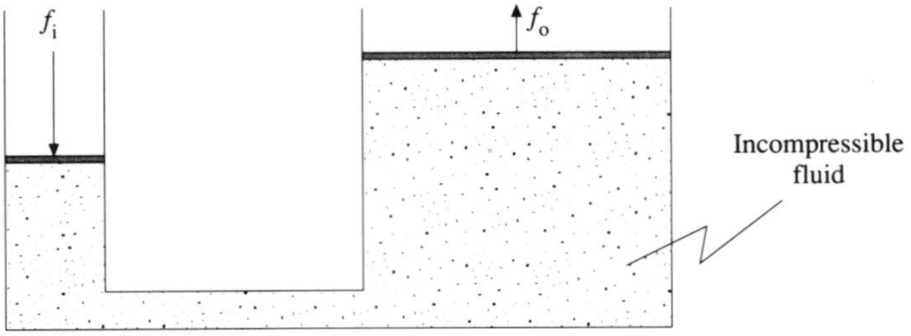

Figure 2.19 A simple hydraulic amplifier.

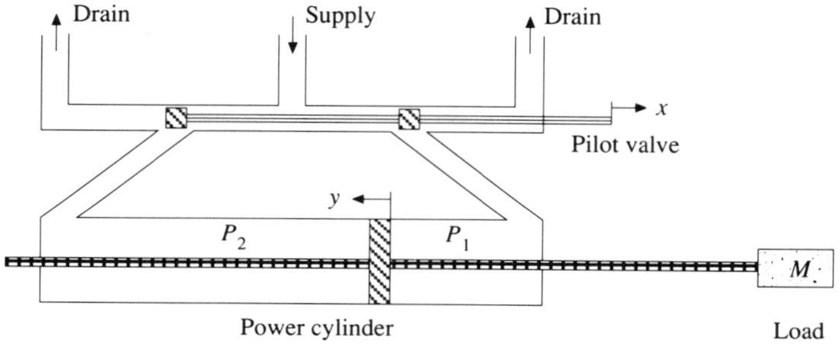

Figure 2.20 Hydraulic amplifier.

instability as will be seen later. Sometimes this effect is deliberately used to produce an oscillator.

Exercise 2.9 Determine the relation between V_o and V_i and the input impedance for the circuit in Fig. 2.17.

Exercise 2.10 In DC permanent magnet motors torque is proportional to motor current I_m. Therefore a constant current must be provided to the motor if the torque is to be kept constant. Show that a possible set-up to achieve this is the circuit illustrated in Fig. 2.18 where $R_1 >> R$ and $V_i = -I_d R$, I_d being the desired value for I_m.
Answer: $I_m = I_d(1 - R/R_1) \approx I_d$

2.1.8 Hydraulic Amplifier

A simple hydraulic amplifier is shown in Fig. 2.19. It converts a force f_i applied to the forcing piston into a larger force f_o on the working piston. Assuming an incompressible fluid we have that the hydraulic pressure p is the same on both pistons; therefore

$$p = f_i/A_i = f_o/A_o$$

where A_i and A_o are the areas of the pistons. As a result f_i is amplified by a factor A_o/A_i. If a supply of oil under constant pressure is available, we can implement a more ingenious amplifier capable of producing controlled displacements of heavy loads with virtually no effort from the operator. See Fig. 2.20. In fact if the pilot valve is moved to the right then oil under pressure is admitted to the right hand side of the power cylinder and the load begins to move to the left. The oil on the other side of the cylinder goes into a sump where it is pumped back into the system. Notice that the net effect of the oil pressure on the pilot valve is null which means it can be displaced with negligible force.

In the derivation of the mathematical model we shall assume that x takes only small values and therefore we can apply the formula of flow of fluid through a restriction which is given by

$$q = K A \sqrt{\Delta P} \tag{2.16}$$

where q is the volume flow rate, A is the orifice area, ΔP is the pressure drop across the orifice and K is a constant depending on the type of fluid and orifice (cf. (2.7)). If the load effect is negligible we can regard ΔP as constant and q will be a function of A (or x) only. The load effect is of an inertial nature: the mass M will react against any variations of speed of the power piston; this reaction translates into a variation of the pressure P_1 and therefore ΔP. For example, a positive increment in x will produce a positive increase in P_1 in an attempt to reduce the flow of fluid into the power cylinder. These variations of P_1 are called the load or back pressure and we will designate them by Δ_{pl}. Assuming the supply and drain pressures are constant, changes in ΔP will depend *only* on Δ_{pl}. A linear approximation of equation (2.16) around $x = x_o$, $\Delta P = \Delta P_o$ is given by

$$\Delta q = (\partial q/\partial x)_{\Delta P = \Delta P_o} \Delta x + (\partial q/\partial \Delta P)_{x = x_o} \Delta_{pl} \tag{2.17}$$

or

$$\Delta q = K_1 \Delta x - K_2 \Delta_{pl} \tag{2.18}$$

with $K_1>0$ and $K_2>0$. The minus sign arises because a positive increment in P_1, i.e. $\Delta_{pl}>0$ causes a decrease in flow. K_1 and K_2 can be computed from the operating curves for the hydraulic amplifier (See exercise 2.11). Denoting by S the cross-sectional area of the power piston we have

$$\Delta q = S\, \mathrm{d}y/\mathrm{d}t$$

and

$$\Delta_{pl} S = M\, \mathrm{d}^2y/\mathrm{d}t^2$$

The above equations combined with (2.18) finally yield

$$K_2 M/S\, \mathrm{d}^2y/\mathrm{d}t^2 + S\, \mathrm{d}y/\mathrm{d}t = K_1 \Delta x \tag{2.19}$$

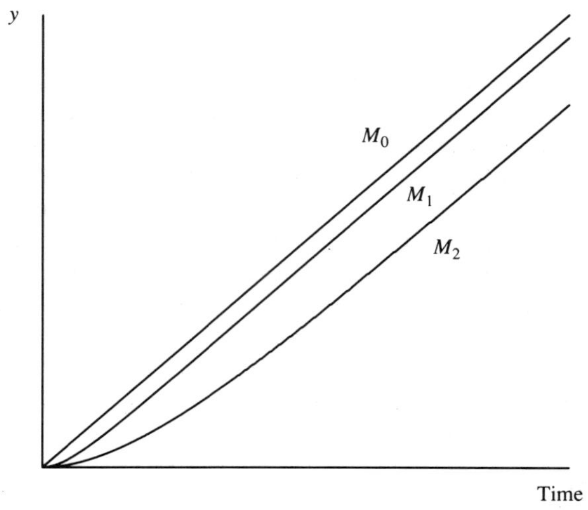

Figure 2.21 Displacement of the power cylinder of Fig. 2.20 following a step change in *x* at time t=0, for several values of M: M_0=0 , 0<M_1<M_2.

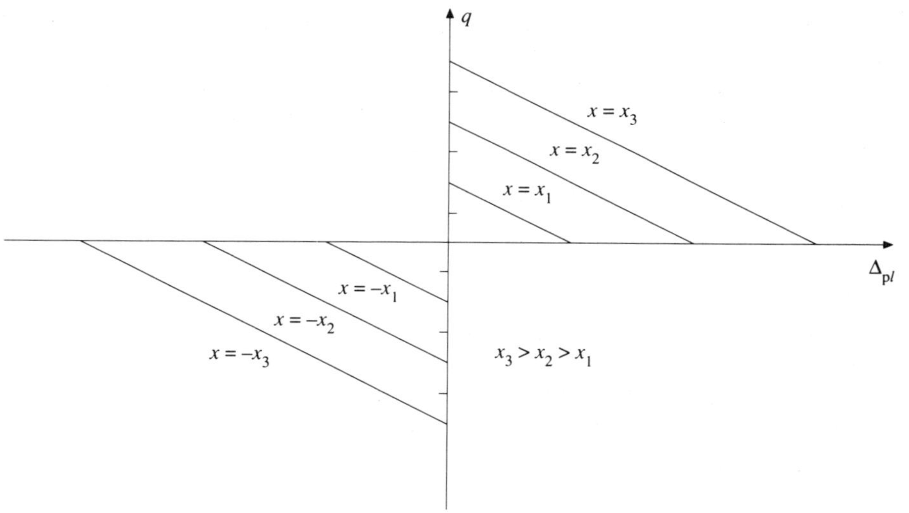

Figure 2.22 Linearized characteristic curves for a hydraulic amplifier. Along each line *x* is constant.

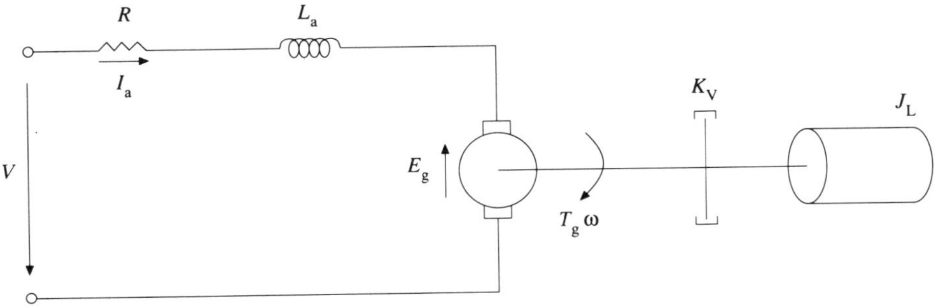

Figure 2.23 Permanent magnet DC motor and load.

From the previous equation we conclude that if there are no load effects, i.e. $M=0$, then for x constant, say $x=\bar{x}$, the amplifier behaves like a pure integrator producing a constant speed $K_1\bar{x}/S$, of the power piston. If M is not negligible then the velocity of the piston will increase gradually until eventually it reaches the speed $K_1\bar{x}/S$. Such behaviour is illustrated in Fig. 2.21.

Exercise 2.11 Figure 2.22 shows the (linearized) characteristic curves for an hydraulic amplifier. Show how K_1 and K_2 in equation (2.18) can be computed from them.

2.1.9 Permanent Magnet DC Motor

This type of motor is particularly suited for servo-systems, because of its linear torque–speed characteristics and high accelerating torques, besides being smaller and lighter than other types of motors, for a given output power. The motor and load equivalent circuit is shown in Fig. 2.23 where

L_a is the armature winding inductance
I_a is the armature winding current
T_g is the torque generated by the motor
E_g is the back electromotive force (back e.m.f.). E_g is an internally generated voltage proportional to the motor velocity ω, i.e. $E_g=K_g\,\omega$.

From Fig. 2.23 we can write the electrical equation of the motor

$$V = L_a\, dI_a/dt + RI_a + K_g\,\omega \tag{2.20}$$

The electromechanical equation is simply

$$T_g = K_T\, I_a \tag{2.21}$$

because the magnetic field is constant. The generated torque T_g is opposed by a

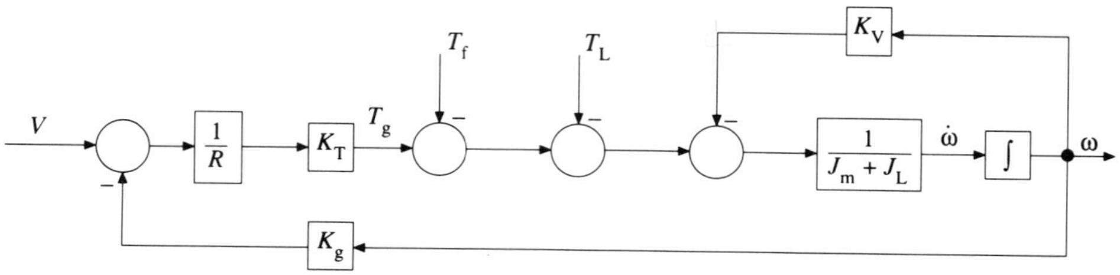

Figure 2.24 Block diagram of the motor of Fig. 2.23 neglecting the armature inductance.

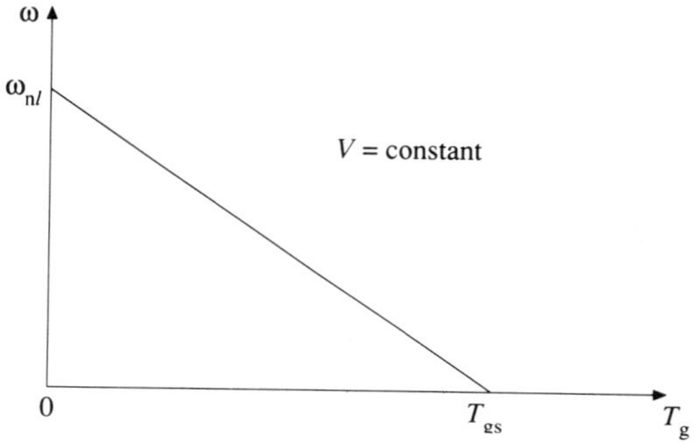

Figure 2.25 Permanent magnet DC motor speed–torque curve.

variety of torques; some originate inside the motor, namely T_f (constant friction torque) and $J_m\, d\omega/dt$ originated by the motor moment of inertia J_m. Others are produced by the load such as:

$J_L\, d\omega/dt$: load inertial torque
$K_v\, \omega$: load viscous friction torque
T_L: constant load opposing torque

Equating all these torques we get the mechanical equation of the motor

$$T_g = (J_m + J_L)\, d\omega/dt + K_v\, \omega + T_f + T_L \tag{2.22}$$

Equations (2.20), (2.21) and (2.22) are the dynamic equations of the motor. Combining them we get the block diagram of Fig. 2.24. From this diagram it is apparent why the back e.m.f. helps to desensitize motor speed against load variations. In fact when the speed decreases, the motor generated torque increases; without internal feedback (K_g=0) the generated torque would remain unchanged. The curve relating T_g and the *steady-state* speed is shown in Fig. 2.25. It can be obtained by a combination of (2.20) and (2.21) setting $dI_a/dt=0$:

$$V = (R/K_t)T_g + K_g\, \omega \tag{2.23}$$

If $\omega=0$ then $T_g=T_{gs}=V\,K_t/R$ which is the torque that the motor produces when it is switched on. It is known as the accelerating torque or torque at stall. ω_{nl} is the 'no load velocity'.

Exercise 2.12 For a motor with $K_T=0.1$ Nm/A, $K_g=10$ V/k r.p.m. and $R=1.3\ \Omega$, compute the no load speed and the torque at stall for $V=10$, 15 and 20 volts.

2.1.10 Motors and Loads

Often the load is not directly coupled to the motor shaft. In such cases it is convenient to compute an equivalent system with a load having the same angular movement of the motor in order to apply equation (2.22). Such transformation relies on the *coupling ratio* between motor and load, i.e. the angular movement of the motor compared to the movement of the load. Next we show how to compute the moment of inertia of the load as felt by the motor shaft, which we will designate by J_{LS}, in a number of situations.

Let us begin with the gear coupled system shown in Fig. 2.26. Denote by ϕ_m and ϕ_L the shaft angular displacement of the motor and load respectively. Then $\omega_m=d\phi_m/dt$ and $\omega_L=d\phi_L/dt$. For a system in which the load is rotated the coupling ratio is commonly known as *gear ratio* and we shall denote it by N. Then

$$N = \phi_m/\phi_L = \omega_m/\omega_L \tag{2.24}$$

The moment of inertia, J_{LS}, of the equivalent system without gearing must be such that the kinetic energy in both systems is the same, i.e.

$$1/2\ J_{LS}\ \omega_m^2 = 1/2\ J_L\ \omega_L^2$$

therefore

$$J_{LS} = J_L/N^2 \tag{2.25}$$

In other words, a load with moment of inertia J_L/N^2, directly coupled to the motor shaft, will produce the same transient behaviour.

Exercise 2.13 If the load on Fig. 2.26 also has viscous friction, K_v, producing a resisting torque $K_v\ \omega_L$, then show that this will give rise, on the equivalent systems without gearing, to a viscous friction torque $(K_v/N^2)\ \omega_m$. *Hint*: notice that the power dissipation, which is given by $K_v\ \omega_L^2$, must be the same in both systems.

We now consider a situation where a load with mass M is linearly translated as shown in Fig. 2.27. If r is the radius of the pulley and x the linear displacement of the load we have $x=\phi_m r$. Equating the kinetic energies as above we get

$$1/2\ J_{LS}\ \omega_m^2 = 1/2\ M\ (dx/dt)^2$$

and therefore

$$J_{LS} = M\,r^2 \tag{2.26}$$

If we assume a constant opposing force F from the belt and neglect friction torques, the mechanical equation for this system becomes

$$T_g = (J_m + M r^2)\, d\omega_m/dt + Fr \qquad (2.27)$$

Notice that Fr is a constant opposing torque on the motor shaft and that the coupling ratio is now $1/r$. The latter expresses the motor rotation for one unit of linear motion and is therefore expressed in rad/m or rad/in.

Exercise 2.14 For the worm screw drive system with 'pitch' P represented in Fig. 2.28 compute J_{LS}. Pitch is the number of revolutions per unit of linear motion.

Moments of inertia are responsible for most of the power consumption during transient periods. For a desired velocity profile the larger the moment of inertia the larger must be the torque supplied by the motor which, for a permanent magnet DC motor, is proportional to armature current. If the motor is subject to frequent accelerations and decelerations, as in incremental motion control (a control system which drives the load from one position to another by small steps) armature heat dissipation can be a problem. It can be shown that such energy dissipation can be minimized by an appropriate choice of the coupling ratio making the moment of inertia of the load as felt by the motor shaft approximately equal to the motor moment of inertia. This result is known as *inertia match*.

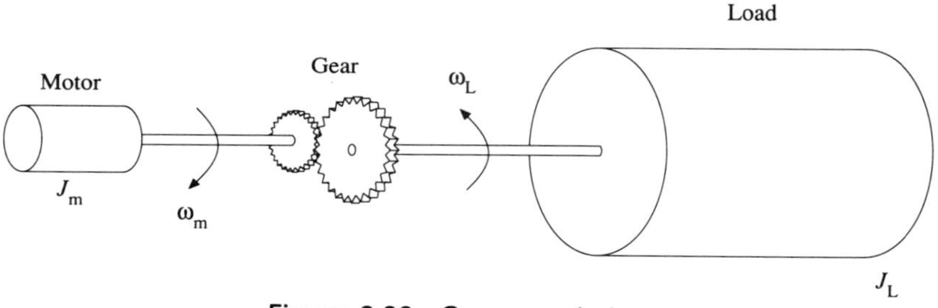

Figure 2.26 Gear coupled system.

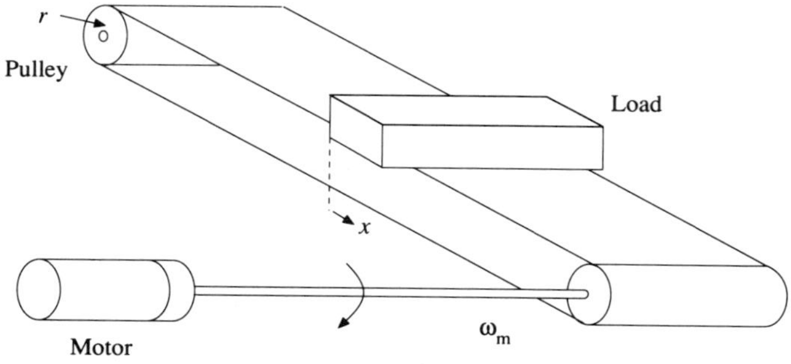

Figure 2.27 Belt-pulley drive system.

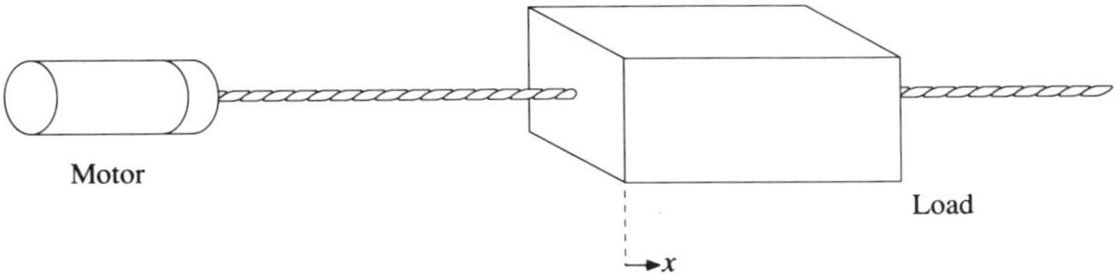

Figure 2.28 Worm screw drive system.

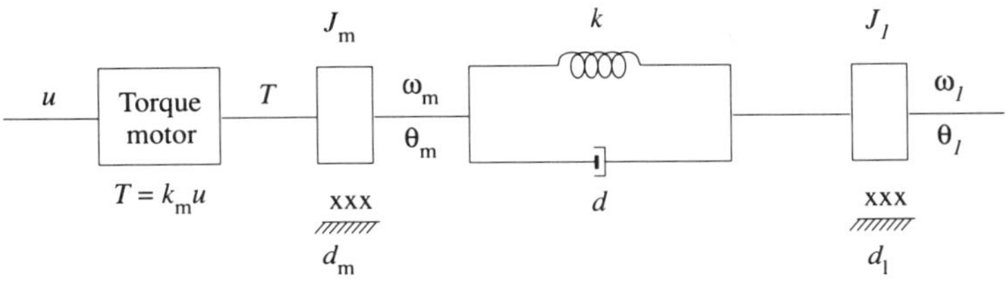

Figure 2.29 Torque motor coupled to an inertial load via an elastic shaft.

There are situations where the dynamics of the motor can be neglected. This happens, for example, in the case of torque motors (DC motors with current feedback); their torque, T, can be regarded as proportional to input voltage u, when used in high performance robot drives, in the presence of lightly damped torsional modes in gears (and harmonic drives). These raise severe control problems, as discussed later in the book. Let us then model the set-up as in Fig. 2.29, showing such a motor coupled to an inertial load, via an elastic shaft.

Equating torques, we get the following system of equations

$$\begin{cases} J_m \ddot{\theta}_m + d_m \dot{\theta}_m + d(\dot{\theta}_m - \dot{\theta}_l) + k(\theta_m - \theta_l) = T = K_m u \\ J_l\ddot{\theta}_l + d_l \dot{\theta}_l + d(\dot{\theta}_l - \dot{\theta}_m) + k(\theta_l - \theta_m) = 0 \end{cases} \tag{2.28}$$

It is interesting to note that, in the absence of elasticity (d=0 abd $k=\infty$), the equations become

$$\begin{cases} \theta_m = \theta_l \\ (J_l + J_m) \ddot{\theta}_m + (d_l + d_m) \dot{\theta}_m = T \end{cases}$$

as expected.

2.1.11 Distributed Parameter System

The models encountered so far have been ordinary differential equations which are examples of *lumped parameter* or macroscopic system descriptions. These are the simplest and the most general models of dynamic systems. In the next levels of complexity we have the maximum gradient and multiple gradient descriptions. We can take the detail even further by modelling at the microscopic or molecular levels. However as the level of detail (and accuracy) increases we not only get more complex models but also restrict their generality. Therefore a compromise is always necessary and that depends on the purpose of the model. As an example, let us assume that we wish a mathematical description of the temperature within the pipe linking the furnace and the chamber of the plant shown in Fig. 1.6. Such a model has to describe the temperature T at any *time* t and at any *point* in the pipe, i.e. $T=T(t,d)$, $0 \leq d \leq l$, where d denotes the distance from the furnace and l the pipe length. Because we now have two independent variables the possibility of modelling by an ordinary differential equation is ruled out. The model can be derived from simple heat balance continuity equation for the fluid on an element of pipe of length Δd. Neglecting the heat capacity of the pipe and assuming the fluid incompressible and no heat losses though the pipe wall this equation becomes:

accumulation of heat within the element = heat entering the element − heat leaving the element

Then in time Δt we can write

$$\Delta_d\ (\Pi D^2/4)\ \rho c_p\ \partial T/\partial t\ \Delta t = \rho\ (\Pi D^2/4)\ v c_p\ T\,\Delta_t$$
$$- \rho\ (\Pi D^2/4)\ v c_p\ (T + \partial T/\partial d\ \Delta d)\Delta t$$

where

v is fluid velocity
c_p is specific heat capacity
D is the internal diameter of pipe
ρ is fluid density

Dividing by Δt and taking limits we get

$$v\ \partial T/\partial d + \partial T/\partial t = 0 \qquad (2.29)$$

This is a partial differential equation in two independent variables. Partial differential equations are in general more difficult to solve than ordinary ones. The solution of the latter is a finite dimensional problem – besides the system input all that is required for their solution is a finite set of initial conditions which is equivalent to a point in $\mathbb{R}^n$. On the contrary a partial differential equation requires not a point but a function – a boundary condition – in order to find the solution. In (2.29) this would correspond, for example, to the knowledge of the temperature distribution at time $t=0$, i.e. the knowledge of the function $T(d)=T(0,d)$, $d \in [0,l]$. Because the space of functions is infinite dimensional we no longer have a finite dimensional problem.

This is why partial differential equations cannot be represented in state-space form. Fortunately for most of our purposes the simplest type of lumped parameter description, like (2.29), will do.

Exercise 2.15 Sketch in [3] the evolution of T assuming $T(0,d)$ known and constant and $T(t,0)=t$ for $t,d \geqslant 0$.

2.1.12 Discrete-time Model

Very often the variables under study are not continuously monitored but observed at a defined sequence of (equally spaced) time points. The underlying phenomenon must then be described by a *difference equation*. Such an equation relates the value of the variable at instant k (k an integer) to values at other neighbouring instants. We shall look at these equations in some detail when studying digital control systems. Difference equations are particularly well suited for digital computers because they allow the solutions to be computed recursively, as the next example shows.

Imagine two enemy squadrons, S_1 and S_2, engaged in an air battle. Let h_i, $i=1,2$, denote the hitting power of each aircraft in squadron S_i, and $n_i(k)$ the number of its aircraft at time k. The hitting power of an aircraft is defined as the number of enemy units it can shoot down per unit of time. Assuming the battle starts at time 0, we then have, for $k=0,1,2,\ldots$, the following system of *difference equations*:

$$\begin{cases} n_1(k+1) - n_1(k) = -h_2 n_2(k) \\ n_2(k+1) - n_2(k) = -h_1 n_1(k) \end{cases}$$

or in matrix form

$$\boldsymbol{n}(k+1) = \mathbf{A}\,\mathbf{n}(k) \qquad (2.30)$$

where

$$\boldsymbol{n}(k) = (n_1(k); n_2(k))^{\mathrm{T}}$$

and

$$\boldsymbol{A} = \begin{bmatrix} 1 & -h_2 \\ -h_1 & 1 \end{bmatrix}$$

If, for example, $h_1=0.4$, $h_2=0.1$ and $n(0)=(50;90)^{\mathrm{T}}$ the solution of (2.30) can be computed recursively leading to

$$\boldsymbol{n}(1) = \begin{pmatrix} 41 \\ 70 \end{pmatrix}; \ \boldsymbol{n}(2) = \begin{pmatrix} 34 \\ 54 \end{pmatrix}; \ \boldsymbol{n}(3) = \begin{pmatrix} 29 \\ 40 \end{pmatrix};$$

$$\boldsymbol{n}(4) = \begin{pmatrix} 25 \\ 29 \end{pmatrix}; \ \boldsymbol{n}(5) = \begin{pmatrix} 22 \\ 19 \end{pmatrix}$$

which suggests squadron 1 will eventually win. In general, the final outcome will

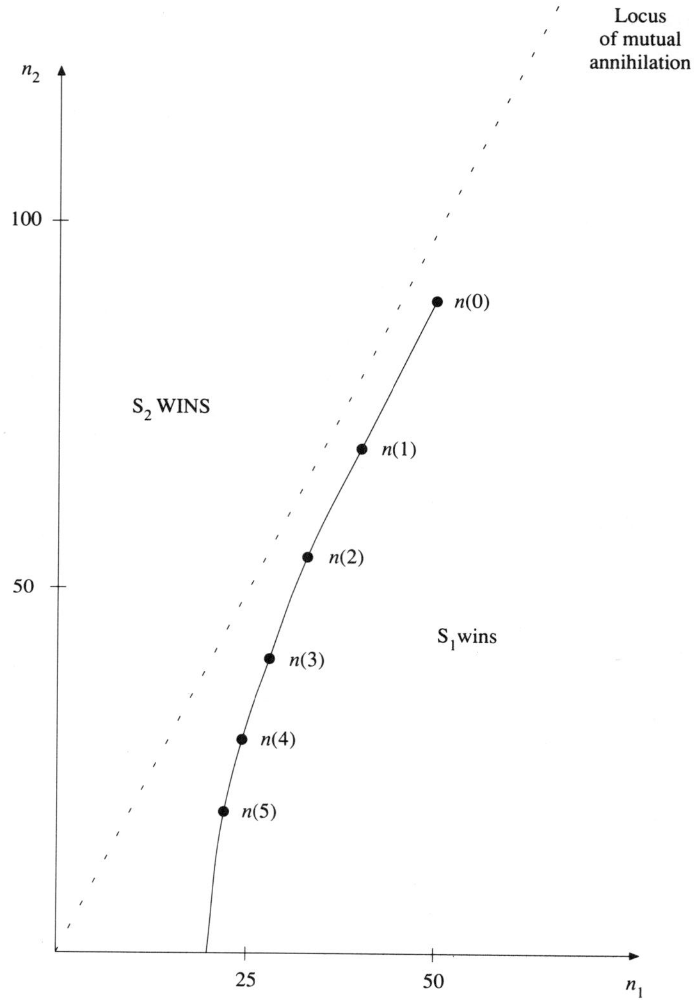

Figure 2.30 Evolution of $\mathbf{n}(k)$ for $\mathbf{n}(0)=(50;90)^T$, $h_1=0.4$ and $h_2=0.1$ in equation (2.30). The line joining the origin with the point $(50;90)^T$ is an eigenvector of matrix **A**.

depend both on the hitting power and the size of each squadron. Figure 2.30 shows how the battle will evolve as a function of these parameters. Model (2.30) is known as the Lanchester Model of Warfare.

Exercise 2.16 Compute $\boldsymbol{n}(k)$, $k=0,1,\ldots,5$, for $\boldsymbol{n}(0)=(50;100)^T$ and $\boldsymbol{n}(0)=(40;100)^T$ and compare the results with Fig. 2.30.

2.2 TIME-SCALING

In practice, it is often desirable to change the time-scale of the model, in order to speed up or slow down its simulation on a computer. Consider, for example, the system described by the equation

$$\frac{d^3y}{dt^3} + 3\frac{d^2y}{dt^2} + 2\frac{dy}{dt} + 0.1y = 0.1\frac{d^2u}{dt^2} + 0.3\frac{du}{dt} + 0.2u \tag{2.31}$$

Assume we wish to select a time-scale, where the system looks ten times faster. If t_n denotes the new time-scale, then

$$t = 10t_n \tag{2.32}$$

This means that after one new time unit, the system reaches the same value as ten old time units later.

The derivatives, with respect to the new time-scale, become:

$$\frac{dy}{dt} = \frac{dy}{dt_n}\frac{dt_n}{dt} = \frac{dy}{dt_n} \times 0.1$$

$$\frac{d^2y}{dt^2} = \frac{d}{dt}(0.1\frac{dy}{dt_n}) = \frac{d}{dt_n}(0.1\frac{dy}{dt_n})\frac{dt_n}{dt} = (0.1)^2\frac{d^2y}{dt_n^2}$$

$$\frac{d^3y}{dt^3} = (0.1)^3\frac{d^3y}{dt_n^3}$$

In the new time-scale, the differential equation (2.31) becomes

$$\frac{d^3y}{dt_n^3} + 30\frac{d^2y}{dt_n^2} + 200\frac{dy}{dt_n} + 100y = \frac{d^2u}{dt_n} + 30\frac{du}{dt_n} + 200u \tag{2.33}$$

Let us illustrate this result with the spring–mass–damper system of Fig. 2.11, which is described by equation (2.13). If we wish to make our system look ten times faster, then we change the time-scale as in (2.31), obtaining a new equation

$$\frac{M}{100}\frac{d^2x}{dt_n^2} + \frac{c}{10}\frac{dx}{dt_n} + kx = f \tag{2.34}$$

This result has a nice physical interpretation: reducing the mass and the damping by factors of 100 and 10 respectively, we get a tenfold increase in system speed. It is shown later in the book that the effect of such time-scaling (2.31) is to increase the magnitude of the system poles and zeros (cf. section 3.1), by a factor of 10. In fact, the poles of the transfer function $Y(s)/U(s)$ are -2.047, -0.8990, -0.05435 for equation (2.31) and -20.47, -8.8990, -0.5435 for equation (2.33).

REFERENCES

Fasol, K.H. and H.P. Jörgl (1980) 'Principles of model building and identification', *Automatica*, **16**, 505–18.

Isermann, R. (1984) 'Process fault detection based on modelling and estimation methods – a survey', *Automatica*, **20**(4), 387–404.

Richards, R.J. (1979) *An Introduction to Dynamics and Control*, Longman.

Wierzbicki, A. (1984) *Models and Sensitivity of Control Systems*, Elsevier.

3

Time-domain Analysis of Continuous-time Open-loop Systems

In this chapter we will study the behaviour of systems described by ordinary linear constant coefficient differential equations, i.e. equations of the type

$$\mathrm{d}^n y/\mathrm{d}t^n + a_1\,\mathrm{d}^{n-1}y/\mathrm{d}t^{n-1} + \ldots + a_{n-1}\,\mathrm{d}y/\mathrm{d}t + a_n y = b_0\mathrm{d}^m u/\mathrm{d}t^m + b_1\,\mathrm{d}^{m-1}u/\mathrm{d}t^{m-1} + \ldots + b_{m-1}\,\mathrm{d}u/\mathrm{d}t + b_m u \qquad (3.1)$$

where $m \leqslant n$, which have already been introduced in Chapter 2. Restricting ourselves to (finite dimensional) linear systems allows us to keep the mathematical treatment at an elementary level and capitalize on geometrical intuition. Input–output and state-space representations will be discussed, with an emphasis on the former.

3.1 TRANSFER FUNCTIONS AND STATE-SPACE REPRESENTATIONS

The solution of the above equation can easily be computed by means of the Laplace Transform. By transforming differential equations into algebraic ones, the Laplace Transform not only simplifies the computation of solutions but also provides a systematic way of studying their properties as a function of the coefficients of the equation.

For example, if we take the Laplace Transform of both sides of the equation

$$f(t) = M\,\mathrm{d}^2x/\mathrm{d}t^2 + c\,\mathrm{d}x/\mathrm{d}t + kx$$

which models the spring–mass–damper system of Fig. 2.11, we get

$$F(s) = (Ms^2 + cs + k)X(s) - M(sx(0) + \dot{x}(0)) - cx(0)$$

where $\dot{x}=\mathrm{d}x/\mathrm{d}t$. The time solution, $x(t)$, is then the inverse Laplace transform of $X(s)$ where

$$X(s) = \frac{1}{Ms^2 + cs + k}F(s) + \frac{M(sx(0) + \dot{x}(0)) + cx(0)}{Ms^2 + cs + k} \qquad (3.2)$$

Notice that the response is the sum of two terms: the first one, a function of the input only, is the forced response, while the second term, which depends only upon the initial conditions, is the free response.

If the initial conditions are zero, we have

$$\frac{X(s)}{F(s)} = \frac{1}{Ms^2 + cs + k} \tag{3.3}$$

Such a ratio, which is independent of the system inputs and initial value of the state, and hence a characteristic of the system, is called the Transfer Function. Notice that although we assumed zero initial conditions we have lost no information because the differential equation describing the system can be reconstructed from the transfer functions simply by replacing $s^l X(s)$ by $d^l x/dt^l$, l=0,1,2. For the general equation (3.1) the transfer function $W(s)$ is then

$$\frac{Y(s)}{U(s)} = \frac{b_0 s^m + b_1 s^{m-1} + \ldots + b_m}{s^n + a_1 s^{n-1} + \ldots + a_n} \tag{3.4}$$

This enables us to regard differential equations of type (3.1) as linear operators in the space of Laplace transforms of functions of a real variable. Assuming zero initial conditions, the solution $Y(s)$ is simply obtained by multiplying the Laplace transform of the input $U(s)$ by the transfer function $W(s)$.

3.1.1 Proper, Strictly Proper, Biproper and Improper Rational Transfer Functions

Given a transfer function $W(s)$ as in (3.4) we say it is a rational transfer function because both the numerator and denominator are polynomials. The roots of the numerator are called the *zeros* of the transfer function; the roots of the denominator are known as the transfer function *poles*. If $m>n$, $W(s)$ is called an *improper* transfer function; if $m \leq n$, $W(s)$ is a *proper* transfer function; if $m<n$, $W(s)$ is a *strictly proper* transfer function. When $m=n$, the transfer function is known as *biproper*, because its inverse is also proper.

Exercise 3.1 Show that when $m=n$ (3.4) can be written in the form

$$W(s) = b_0 + \frac{\alpha_1 s^{n-1} + \ldots + \alpha_{n-1}s + \alpha_n}{s^n + a_1 s^{n-1} + \ldots + a_{n-1}s + a_n}$$

and compute $\alpha_1, \ldots, \alpha_n$.
Answer: $\alpha_1 = b_1 - b_0 a_1; \ldots; \alpha_n = b_n - b_0 a_n$.

Let us now attempt the computation of a transfer function, in a more complex situation, such as that of Fig. 2.29. Laplace transforming both sides of equation (2.28) we get

$$[J_m s^2 + (d_m + d)s + k]\theta_m(s) - (ds + k)\theta_l(s) = K_m U(s) = T(s)$$

$$[J_l s^2 + (d_l + d)s + k]\,\theta_l(s) - (ds + k)\,\theta m(s) = 0 \qquad (3.5)$$

By eliminating θ_l or θ_m we can compute $\Omega_m(s)/T(s)$ or $\Omega_l(s)/T(s)$, where $s\theta_m = \Omega_m$ and $s\theta_l = \Omega_l$. Then

$$\frac{\Omega_m(s)}{T(s)} = \frac{J_l s^2 + (d_l + d)s + k}{s^3(J_m J_l) + s^2[J_l d_m + J_m d_l + (J_m + J_l)d] + s[d_m d_l + d(d_m + d_l) + k(J_m + J_l)] + k(d_m + d_l)} \qquad (3.6)$$

Because

$$\theta_l(s) = \frac{(ds + k)}{J_l s^2 + (d_l + d)\,s + k} = \theta_m(s) \qquad (3.7)$$

we also have

$$\frac{\Omega_l(s)}{T(s)} = \frac{ds + k}{s^3(J_l J_m) + s^2[J_l d_m + J_m d_l + (J_m + J_l)d] + s[d_l d_m + d(d_l + d_m) + k(J_m + J_l)] + k(d_l + d_m)} \qquad (3.8)$$

From these expressions a lot of useful information about the dynamic behaviour of the system can be easily obtained:

1. In the case of rigid coupling ($k=\infty$), and $d=0$, we have

$$\frac{\Omega_2}{T} = \frac{\Omega_1}{T} = \frac{1}{s(J_l + J_m) + (d_1 + d_2)} \qquad (3.9)$$

as expected.

2. Assuming a constant input torque, $T(s) = 1/s$, we conclude from the final and initial value theorems of Laplace transform theory:
   (a) The final speed is only determined by the damping coefficients d_l and d_m; since

$$\lim_{t\to\infty} \omega_1(t) = \lim_{t\to\infty} \omega_2(t) = \frac{1}{d_l + d_m} \qquad (3.10)$$

   if $T(s) = 1/s$.
   (b) The speed of response is mainly determined by the system inertias, since the first non-zero derivatives at time $t=0^+$, are

$$\frac{d^3\theta_l}{dt^3} = \frac{d}{J_l J_m} \text{ and } \frac{d^2\theta_m}{dt^2} = \frac{1}{J_m} \qquad (3.11)$$

   for a unit step input torque. This means θ_m will start 'faster' than θ_l, because

$$\theta_m - \theta_l = \frac{s(J_l s + d_l)}{J_l s^2 + (d_l + d)s + k}\,\theta_m \qquad (3.12)$$

We also conclude that, when ω_m is constant ($\theta_m(s) = \bar{\omega}_m/s^2$), θ_l will lag behind θ_m by

$$\frac{d_l}{k}\bar{\omega}_m$$

Let us address now the proof of equivalence between (3.1) (or (2.1)) and the simulation diagram of Fig. 2.1. If we denote by x_1 the output of a system with input u and transfer function

$$\frac{1}{s^n + a_1 s^{n-1} + \ldots + a_{n-1}s + a_n}$$

then $Y(s)$ in (3.4) is given by

$$Y(s) = b_0 s^m X_1(s) + b_1 s^{m-1} X_1(s) + \ldots + b_{m-1} s X_1(s) + b_m X_1(s)$$

Bearing in mind that

$$X_2(s) = sX_1(s),\ X_3(s) = s^2 X_1(s),\ \ldots,\ X_m(s) = s^{m-1} X_1(s)$$

the diagram of Fig. 2.1 follows.

The mathematical relations characterizing Fig. 2.1 can also be written as

$$\begin{aligned}
&\mathrm{d}x_1/\mathrm{d}t = x_2\\
&\quad\vdots\\
&\mathrm{d}\,x_{n-1}/\mathrm{d}t = x_n\\
&\mathrm{d}x_n/\mathrm{d}t = -a_1 x_n - a_2 x_{n-1} - \ldots - a_{n-1} x_2 - a_n x_1 + u\\
&y = b_n x_1 + b_{n-1} x_2 + \ldots + b_1 x_n + b_0\,[u - a_1 x_n - a_2\, x_{n-1} - \ldots - a_n\, x_1]
\end{aligned}$$

or in a more compact form

$$\begin{cases} \dot{\boldsymbol{x}} = \boldsymbol{A}\boldsymbol{x} + \boldsymbol{b}u \\ y = \boldsymbol{c}\boldsymbol{x} + b_0 u \end{cases} \tag{3.13}$$

where

$$\boldsymbol{A} = \begin{bmatrix} 0 & 1 & 0 & \ldots & 0 \\ 0 & 0 & 1 & \ldots & 0 \\ \cdot & \cdot & \cdot & & \cdot \\ \cdot & \cdot & \cdot & & \cdot \\ \cdot & \cdot & \cdot & & \cdot \\ 0 & 0 & 0 & \ldots & 1 \\ -a_n & -a_{n-1} & -a_{n-2} & \ldots & -a_1 \end{bmatrix};\ \boldsymbol{b} = \begin{bmatrix} 0 \\ 0 \\ \cdot \\ \cdot \\ \cdot \\ 0 \\ 1 \end{bmatrix};$$

$$c^{\mathrm{T}} = \begin{bmatrix} b_n - a_n b_0 \\ b_{n-1} - a_{n-1} b_0 \\ \cdot \\ \cdot \\ \cdot \\ b_1 - a_1 b_0 \end{bmatrix} ; \; \boldsymbol{x} = \begin{bmatrix} x_1 \\ x_2 \\ \cdot \\ \cdot \\ \cdot \\ x_n \end{bmatrix} ; \; \dot{\boldsymbol{x}} = \begin{bmatrix} \mathrm{d}x_1/\mathrm{d}t \\ \mathrm{d}x_2/\mathrm{d}t \\ \cdot \\ \cdot \\ \cdot \\ \mathrm{d}x_n/\mathrm{d}t \end{bmatrix}$$

The system of equations (3.13) is called a *state-space representation* of equation (2.1). The controllable canonical form will be seen later.

An alternative dynamic diagram for the linear differential equation (2.1) is shown in Fig. 3.2. In fact, if we integrate n times both sides of this equation we get

$$y + a_1\int y + a_2\int\int y + \ldots + a_n\int \ldots \int y = b_0 u + b_1\int u + b_2\int\int u + \ldots + b_n\int \ldots \int u \tag{3.14}$$

or equivalently

$$y = b_0 u + \int(b_1 u - a_1 y) + \ldots + \int \ldots \int(b_n u - a_n y) \tag{3.15}$$

and the diagram of Fig. 3.1 follows immediately.

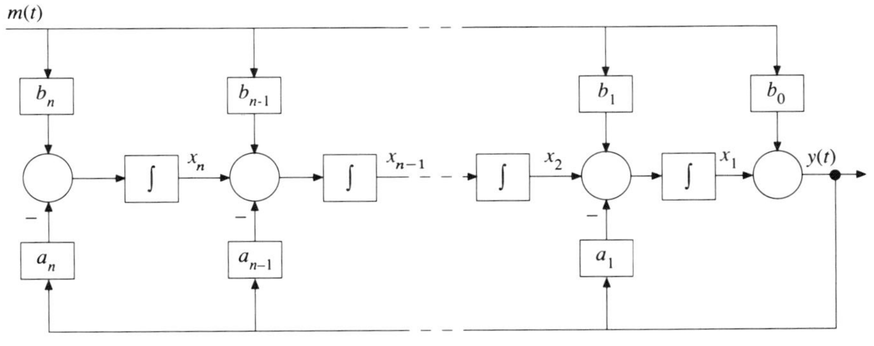

Figure 3.1 Block diagram for the nth-order linear ordinary differential equation (2.1): observable canonical form.

From Fig. 3.1, we obtain the following state-space representation of equation (2.1):

$$\begin{bmatrix} \dot{x}_1 \\ \dot{x}_2 \\ \cdot \\ \cdot \\ \cdot \\ \dot{x}_{n-1} \\ \dot{x}_n \end{bmatrix} = \begin{bmatrix} -a_1 & 1\,0 \ldots 0 \\ -a_2 & 0\,1 \ldots 0 \\ \cdot & \cdot\;\cdot \\ \cdot & \cdot\;\cdot \\ \cdot & \cdot\;\cdot \\ -a_{n-1} & 0\,0 \ldots 1 \\ -a_n & 0\,0 \ldots 0 \end{bmatrix} \begin{bmatrix} x_1 \\ x_2 \\ \cdot \\ \cdot \\ \cdot \\ x_{n-1} \\ x_n \end{bmatrix} + \begin{bmatrix} b_1-a_1b_0 \\ b_2-a_2b_0 \\ \cdot \\ \cdot \\ \cdot \\ b_{n-1}-a_{n-1}b_0 \\ b_n-a_nb_0 \end{bmatrix} u$$

$$y = (1,0, \ldots, 0) \begin{bmatrix} x_1 \\ x_2 \\ \cdot \\ \cdot \\ \cdot \\ x_n \end{bmatrix} + b_0u$$

which is an observable canonical form, as shown later in the book.

From Fig. 2.1 we also conclude that to compute the output after a time $t \geqslant t_0$, knowledge of u(t), $t \geqslant t_0$ is not enough; the *state* of each integrator at time t_0 is also required. Such information is condensed in the vector $\boldsymbol{x}$ – the state vector – which acts as the *memory* of the system. Also note that the evolution of the components of $\boldsymbol{x}$ is independent of $b_0,b_1,\ldots,b_n$. Therefore the system's dynamical properties (stability, for example) only depend upon $a_1,a_2,\ldots,a_n$. Obviously when u is a function of y, as happens under feedback, the dynamic properties also become dependent upon $b_0,\ldots,b_n$. $\boldsymbol{A}$ is known as the *system matrix*.

For a state-space representation

$$\begin{aligned} \dot{\boldsymbol{x}} &= \boldsymbol{Ax} + \boldsymbol{B}u \\ y &= \boldsymbol{Cx} + Du \end{aligned} \tag{3.16}$$

when u and y are scalars, the transfer function is

$$\frac{Y(s)}{U(s)} = \boldsymbol{C}(s\boldsymbol{I}_n-\boldsymbol{A})^{-1}\,\boldsymbol{B} + D \tag{3.17}$$

$\boldsymbol{I}_n$ denotes the $n \times n$ identity matrix. The proof of this result is straightforward: by taking the Laplace transform of both equations in (3.16) and eliminating $\boldsymbol{X}(s)$ we get (3.17). Recall that the Laplace transform of a vector of time functions

$$\boldsymbol{x}(t) = (x_1(t),x_2(t), \ldots, x_n(t))^{\mathrm{T}}$$

is defined as the vector of Laplace transforms of its components, i.e.

$$\begin{aligned} \mathcal{L}(\boldsymbol{x}(t)) &= (\mathcal{L}(x_1(t)),\mathcal{L}(x_2(t)),\ldots,\mathcal{L}(x_n(t))^{\mathrm{T}} \\ &= \boldsymbol{X}(s) \end{aligned}$$

Analogously we have

$$\mathcal{L}[\mathrm{d}/\mathrm{d}t(\boldsymbol{x}(t))] = (sX_1(s) - x_1(0), sX_2(s) - x_2(0), \ldots, sX_n(s) - x_n(0))^{\mathrm{T}}$$
$$= s\boldsymbol{X}(s) - \boldsymbol{x}(0)$$

Exercise 3.2 Show result (3.17).

Because the determinant of $(s\boldsymbol{I}_n - \boldsymbol{A})$ divides all the entries of the inverse of $(s\boldsymbol{I}_n - \boldsymbol{A})$ we can see the denominator of the transfer function is a factor of the characteristic polynomial of the system matrix $\boldsymbol{A}$. Or stated in other words: the poles of the transfer function are eigenvalues of the matrix $\boldsymbol{A}$. Recall that the determinant of $(s\boldsymbol{I}_n - \boldsymbol{A})$ is a polynomial of degree n in s, known as the characteristic polynomial of $\boldsymbol{A}$, whose roots are the eigenvalues of $\boldsymbol{A}$.

The state-space representation (controllable canonical form) has the important property that the characteristic polynomial $P(s)$ of the system matrix $\boldsymbol{A}$ can be obtained by simple inspection as follows.

Fact The characteristic polynomial $P(s)$ of the 'companion' matrix

$$\begin{bmatrix} 0 & 1 & 0 & \ldots & 0 \\ 0 & 0 & 1 & \ldots & \cdot \\ \cdot & \cdot & \cdot & \cdot & \cdot \\ \cdot & \cdot & \cdot & \ldots & 0 \\ 0 & 0 & 0 & \ldots & 1 \\ -a_n & -a_{n-1} & -a_{n-2} & \ldots & -a_1 \end{bmatrix}$$

is

$$P(s) = s^n + a_1 s^{n-1} + \ldots + a_{n-1}s + a_n$$

Exercise 3.3 Prove the above statement. (*Hint*: basically the proof consists of expanding the determinant of $(s\boldsymbol{I}_n - \boldsymbol{A})$ along the last row.)

Another useful representation for the transfer function of (3.16) is

$$\frac{Y(s)}{U(s)} = \frac{\det \left[\begin{array}{c|c} s\boldsymbol{I}_n - \boldsymbol{A} & -\boldsymbol{B} \\ \hline \boldsymbol{C} & D \end{array}\right]}{\det\,[s\boldsymbol{I}_n - \boldsymbol{A}]} \tag{3.18}$$

This is an immediate consequence of the following identity, for the determinant of a block matrix:

$$\det \left[\begin{array}{c|c} \boldsymbol{M} & \boldsymbol{Q} \\ \hline \boldsymbol{P} & N \end{array}\right] = \det \boldsymbol{M} \det\,[N - \boldsymbol{P}\boldsymbol{M}^{-1}\boldsymbol{Q}] \tag{3.19}$$

In fact, we can then write

$$\det \left[\begin{array}{c|c} s\boldsymbol{I}_n - \boldsymbol{A} & -\boldsymbol{B} \\ \hline \boldsymbol{C} & D \end{array}\right] = \det\,[s\boldsymbol{I}_n - \boldsymbol{A}]\,(D + \boldsymbol{C}\,[s\boldsymbol{I}_n - \boldsymbol{A}]^{-1}\boldsymbol{B})$$

which is just (3.18).

The transfer function concept is easily generalized to the multivariable case. If, in (3.16), $u \in \mathbb{R}^p$ and $y \in \mathbb{R}^q$, we can write

$$Y(s) = \{\boldsymbol{C}(s\boldsymbol{I}_n - \boldsymbol{A})^{-1} \boldsymbol{B} + D\} \; U(s)$$

The matrix in brackets has p columns and q rows and is called the *transfer matrix*. The (i,j)th element of this matrix is the transfer function between the ith output and the jth input when all the other inputs are zero.

Exercise 3.4 Prove the identity

$$\det \begin{bmatrix} M & Q \\ P & N \end{bmatrix} = \det M \det [N - PM^{-1}Q]$$

Hint: Recall that

$$\begin{bmatrix} M & Q \\ P & N \end{bmatrix} = \begin{bmatrix} I & 0 \\ PM^{-1} & I \end{bmatrix} \begin{bmatrix} M & 0 \\ 0 & N-PM^{-1}Q \end{bmatrix} \begin{bmatrix} I & M^{-1}Q \\ 0 & I \end{bmatrix}$$

3.2 TRANSFER FUNCTION AND CONVOLUTION INTEGRAL

The description of a system by means of its transfer function is called a 'parametric' description because the transfer function is specified by a finite set of numbers; these can be determined either by physical considerations, as in Chapter 2, or by a variety of techniques under the heading of system identification. An excellent review of classical techniques can be found in Rake (1980).

Another type of description is the 'non-parametric'. The impulse response of a system falls into this category. By impulse response, we mean the time response $w(t)$, $t \in \mathbb{R}$, of a system in the quiescent state to the unit (Dirac) impulse, $\delta(t)$ applied at time $t=0$. The importance of this response stems from the fact that its Laplace transform is equal to the system transfer function (recall that $\mathcal{L}(\delta(t))=1$). Therefore the impulse response provides another characterization of the system.

From the convolution theorem of Laplace transform theory we have that the response $y(t)$ to an input $u(t)$ applied at time $t=0$ is given by

$$y(t) = \int_0^t w(t - t')\, u(t')\, \mathrm{d}t' \tag{3.20}$$

for non-negative values of t. Obviously $y(t)=0$ for negative t. Although (3.20) is not very attractive for hand calculations, it is particularly suited for simulating the system on a digital computer; besides $w(t)$ is relatively straightforward to obtain. Experimental methods specially designed for the determination of the impulse response can be found in Godfrey (1980). Figure 3.2 provides an interpretation of (3.20); from there the following conclusions can be drawn:

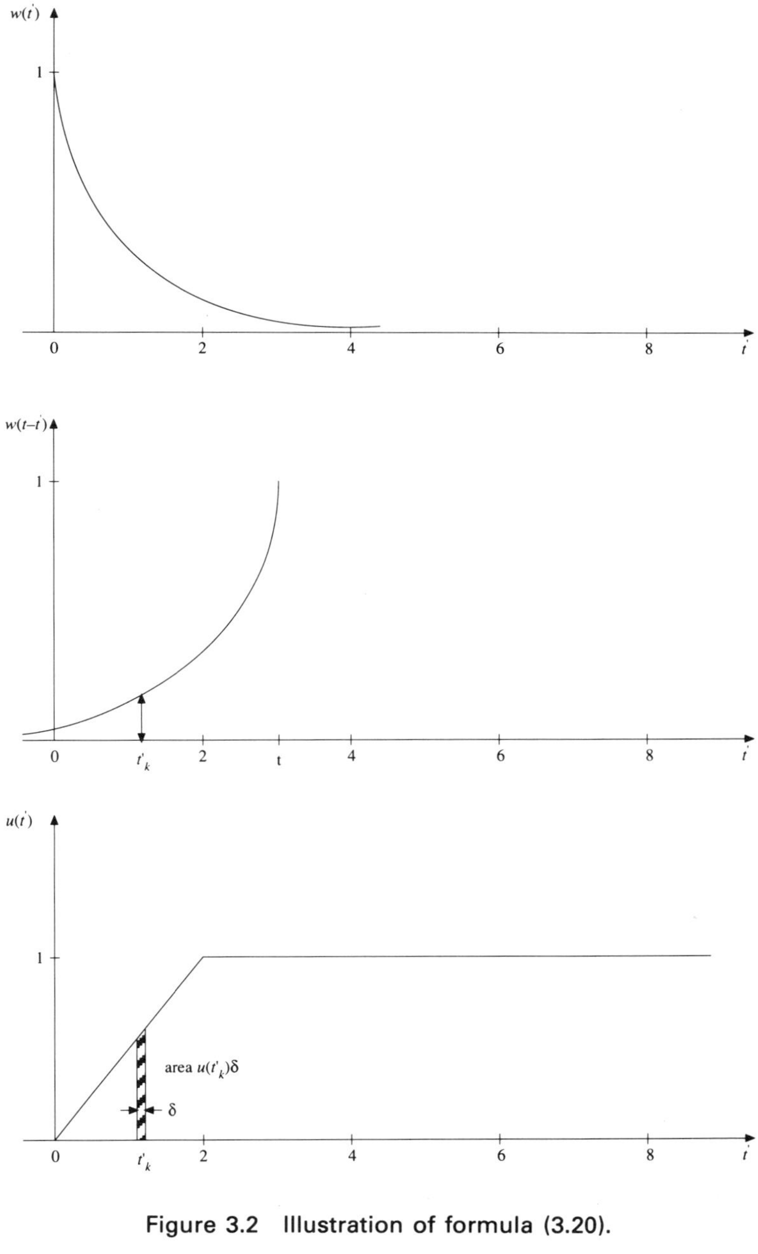

Figure 3.2 Illustration of formula (3.20).

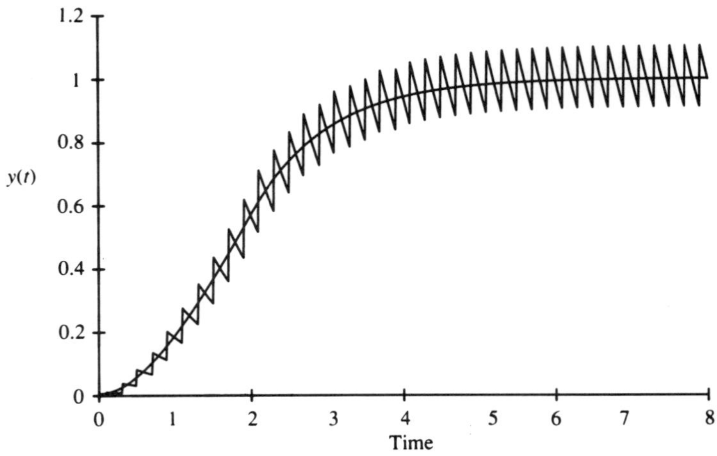

Figure 3.3 Exact response of the system in Fig. 3.1 and an approximation calculated by means of formula (3.21) with $\delta=0.2$.

1. The system is 'causal', i.e. the response at time t does not depend on future values of the input.
2. The system has 'memory' because the response at time t depends on past input values.
3. The memory of the system is a 'forgetting factor' weighting past input values, as shown in Fig. 3.2. The input value at time $t'_k<t$ contributes towards the response at time t weighted by the factor $w_k=w(t-t'_k)$; furthermore, as we go back in time the weighting factor goes to zero. In fact if we approximate (3.20) by

$$\sum_{k=1}^{n} w(t - t'_k)u(t'_k)\delta; \; \delta = t/n \tag{3.21}$$

we see how the input value at time t'_k contributes to the final response. In Fig. 3.3 the exact response to the input $u(t)$, as in Fig. 3.2, is compared with an approximation obtained via (3.21) with $\delta=0.2$.

For pratical purposes we may regard $w(t)=0$ for $t>t_s$, where t_s is the *settling time*. Its definition will be given later in the book.

Exercise 3.5

(a) Show the linearity of the convolution integral.

(b) Compare the responses of the system of Fig. 3.2 to the inputs $u_1=\sin(\omega_1 t)$ and $u_2=\sin(\omega_2 t)$ assuming ω_1 is very small and $\omega_2>>\omega_1$.

Hint: if ω_2 is very large, the function ω will be almost constant during a period of u_2. Then the positive and negative areas under u_2 will almost cancel the effect of each other in (3.20), the result being a (sinusoidal) output with very small amplitude. Conversely if the period of u_1 is much larger than the settling time, the output will exhibit a large amplitude.

3.3 TRANSIENT RESPONSE ANALYSIS OF FIRST- AND SECOND-ORDER SYSTEMS

In this section we study the dynamic behaviour of first- and second-order systems. Their importance stems not only from the fact that they are the building blocks for the study of higher-order systems but, above all, because they constitute good working approximations in the majority of cases.

Because we cannot test the response of a system to all the types of inputs it might be subjected to during its life span, we shall only analyze its transient response to the unit impulse, unit step and unit ramp inputs. Experience has shown that these functions simulate satisfactorily the most common situations that occur in practice, namely, the unit impulse for sudden and undesirable disturbances, the unit step for sharp set-point changes that occur very infrequently and the unit ramp for tracking actions. These test signals can be obtained from a single source by successive integration as shown in Fig. 3.4.

In principle, it would suffice to study the response to one of these signals. The system being linear and time invariant implies that if a signal u produces a response y then the signal $\mathrm{d}u/\mathrm{d}t$ will produce a response $\mathrm{d}y/\mathrm{d}t$; similarly, input $\int u$ will produce the response $\int y$.

Exercise 3.6 Prove the above statement.
Hint: use the transfer function and recall that $sU(s)=\mathscr{L}[\mathrm{d}u/\mathrm{d}t]$ if $U(s)=\mathscr{L}(u(t))$.

3.3.1 First-order Systems

These systems are described by the linear differential equation

$$\tau \frac{\mathrm{d}y}{\mathrm{d}t} + y = Ku \tag{3.22}$$

or equivalently by the transfer function

$$\frac{Y(s)}{U(s)} = \frac{K}{(\tau s + 1)} \tag{3.23}$$

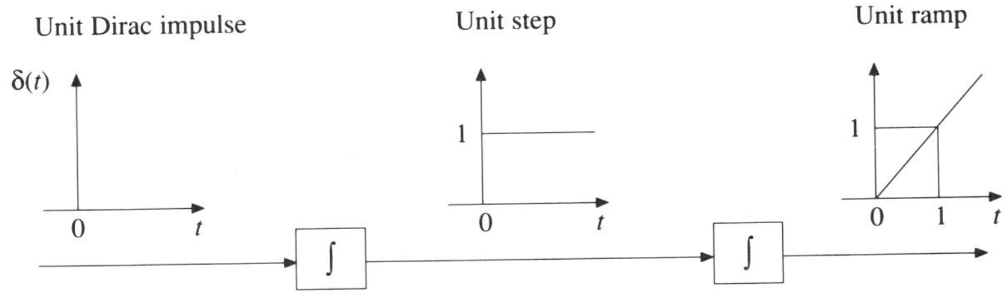

Figure 3.4 Generation of the unit step and ramp from the unit impulse.

The RC circuit modelled by equation (2.3) falls into this category. τ is the system time constant.

3.3.1.1 Impulse response

Because the Laplace transform of the Dirac δ function is 1, we have

$$Y(s) = \frac{K}{\tau s + 1} = \frac{K}{\tau}\frac{1}{(s + 1/\tau)}$$

From the Laplace transform Table A.1 in Appendix A we get

$$y(t) = \frac{K}{\tau}\exp(-\,t/\tau) \tag{3.24}$$

which is plotted in Fig. 3.5. Notice the discontinuity of the response at time $t=0$. This is a characteristic of first-order systems. As will be seen shortly, the impulse response of second-order systems does not exhibit such a discontinuity. Think of the mass–spring–damper system illustrated in Fig. 2.11: a non-zero mass cannot 'jump' from one place to another in zero time. The figure also shows that after time $t=\tau$ the output is approximately 37% of its maximum value and at time 3τ it has been reduced to 5%.

One can therefore compute τ experimentally using this knowledge. A more accurate way of determining τ is to compute the slope of the straight line representing $\ln y$ as a function of t, which is $-1/\tau$.

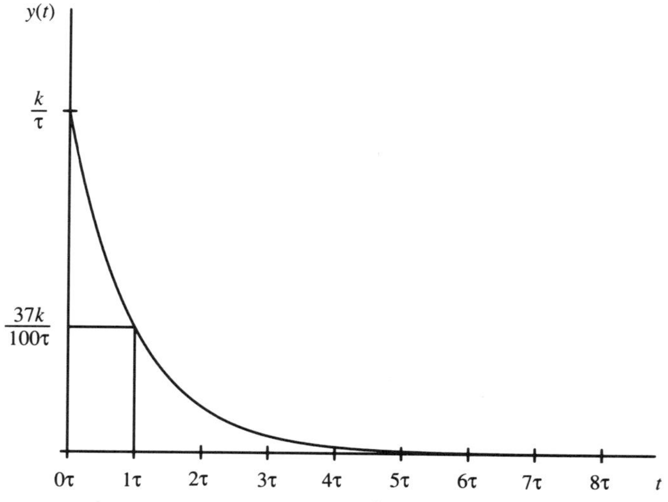

Figure 3.5 Impulse response of system (3.22).

Exercise 3.7 Prove the above statement.
Answer: $\ln y = \ln (K/\tau) - t/\tau$.

3.3.1.2 Step response

In this case, the Laplace transform of the response is

$$Y(s) = \frac{K}{s(\tau s + 1)} \tag{3.25}$$

where $1/s$ is the Laplace transform of the unit step. By a partial fraction decomposition we get

$$Y(s) = K\left(\frac{1}{s} - \frac{\tau}{\tau s + 1}\right) \tag{3.26}$$

or equivalently

$$y(t) = K(1 - \exp(-t/\tau)) \tag{3.27}$$

This function is plotted in Fig. 3.6. Notice that the slope of the tangent at the origin is K/τ. As expected (3.27) is the integral of (3.24). Compare this result with the step response of the *RC* circuit illustrated in Fig. 2.3. From (3.27) we can see, after a time equal to the time constant, i.e. $t=\tau$, the response is 0.632 of its final value; after four time constants, $t=4\tau$, the response is $0.982K$.

Exercise 3.8 Show that τ can be determined experimentally from:
(a) The above remarks.
(b) The slope of the tangent to the response curve at time $t=0$.
(c) The plot of $\ln(K-y(t))$.

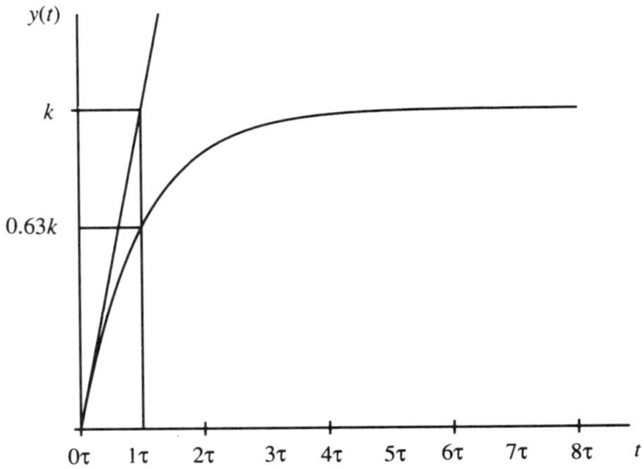

Figure 3.6 Step response of system (3.22).

In practice we cannot wait an infinite amount of time for the response to reach its final value. Therefore it is generally accepted that the steady-state has been reached once the response remains within an interval of $\pm 1.8\%$ of its final value. Such an instant is called the '*settling time*' and in our example it is equal to four time constants.

3.3.1.3 Ramp response

Assume the system input is now the unit ramp, i.e.

$$u(t) = 0, \; t<0$$

$$u(t) = t, \; t \geqslant 0$$

whose Laplace transform is $1/s^2$. Thus

$$Y(s) = \frac{K}{s^2(\tau s + 1)} \tag{3.28}$$

The calculation of the partial fraction expansion of (3.28) is slightly more complicated than (3.26) given the presence of a multiple pole; (3.28) can be expanded as

$$\frac{Y(s)}{K} = \frac{K_1}{s^2} + \frac{K_2}{s} + \frac{K_3}{(\tau s + 1)} \tag{3.29}$$

where K_1 and K_3 are given by

$$K_1 = \lim_{s \to 0} s^2 Y(s)/K = 1$$

and

$$K_3 = \lim_{s \to -1/\tau} (\tau s + 1) Y(s)/K = \tau^2$$

The computation of K_2 is a little more complex because we cannot use

$$\lim_{s \to 0} s Y(s)/K = \lim_{s \to 0} \left(\frac{K_1}{s} + K_2 + \frac{sK_3}{(\tau s + 1)} \right)$$

which leads to $\infty = \infty + K_2 + 0$. To avoid this indetermination we must use instead

$$\lim_{s \to 0} [\mathrm{d}/\mathrm{d}s(s^2 Y(s)/K)] = \lim_{s \to 0} \frac{\mathrm{d}}{\mathrm{d}s} \left(K_1 + sK_2 + \frac{s^2 K_3}{(\tau s + 1)} \right)$$

leading to $K_2 = -\tau$. Therefore

$$y(t) = K(t - \tau + \tau \exp(-t/\tau)) \tag{3.30}$$

for $t \geqslant 0$ which is plotted in Fig. 3.7.

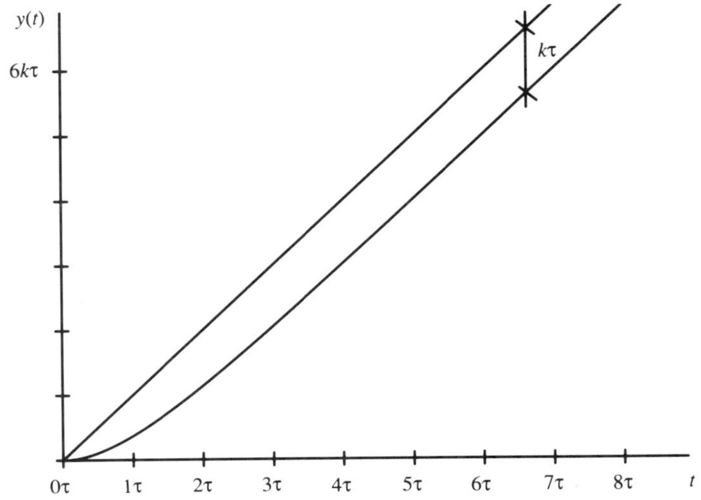

Figure 3.7 Response of system (3.22) to the unit ramp.

As $t \to \infty$ the response approaches a straight line that follows the line $y = Kt$ at a (constant) distance equal to $K\tau$. The smaller the time constant τ, the faster this convergence. Again we find that the response in Fig. 3.6 is the derivative of $y(t)$ in Fig. 3.7 as expected.

3.3.2 Second-order Systems

This type of system exhibits a much richer dynamic behaviour. The standard form of its equation is

$$\frac{d^2y}{dt^2} + 2\zeta\omega_n \frac{dy}{dt} + \omega_n^2 y = \omega_n^2 u \tag{3.31}$$

The spring–mass–damper system described in (2.14) falls into this category. The transfer function corresponding to (3.31) is

$$\frac{Y(s)}{U(s)} = \frac{\omega_n^2}{s^2 + 2\,\zeta\,\omega_n s + \omega_n^2} \tag{3.32}$$

The response of this system can be divided into three categories, depending on the value of the parameter ζ, the damping ratio, being as follows:

underdamped when $0 \leq \zeta < 1$
critically damped for $\zeta = 1$
overdamped when $\zeta > 1$

We now analyze the unit step response of (3.32) for each of the above cases. We have

$$Y(s) = \frac{\omega_n^2}{s(s^2 + 2\zeta\omega_n s + \omega_n^2)} \tag{3.33}$$

The inversion of this transform can be obtained directly from the tables, respectively yielding for:

1. $0 \leq \zeta < 1$

$$y(t) = 1 - \frac{\exp(-\zeta\omega_n t)}{(\surd(1 - \zeta^2)} \sin(\surd(1 - \zeta^2)\ \omega_n t + \arccos \zeta) \tag{3.34}$$

2. $\zeta = 1$

$$y(t) = 1 - \exp(-\omega_n t)\ (1 + \omega_n t) \tag{3.35}$$

3. $\zeta > 1$

$$y(t) = 1 + \frac{1}{2\surd(\zeta^2 - 1)} \left[\frac{\exp - (\zeta + \surd(\zeta^2 - 1))\omega_n t}{\zeta + \surd(\zeta^2 - 1)} - \frac{\exp - (\zeta - \surd(\zeta^2 - 1))\omega_n t}{\zeta - \surd(\zeta^2 - 1)} \right] \tag{3.36}$$

These results are plotted in Fig. 3.8. Notice the independent variable is dimensionless. In fact, by plotting the result as a function of $\omega_n t$, instead of t, we avoid specifying the value of ω_n, achieving thereby a general representation with a single graph. This figure reveals two distinct patterns of behaviour: an oscillatory response for $\zeta < 1$ and a monotonically increasing response for $\zeta \geq 1$. Although the

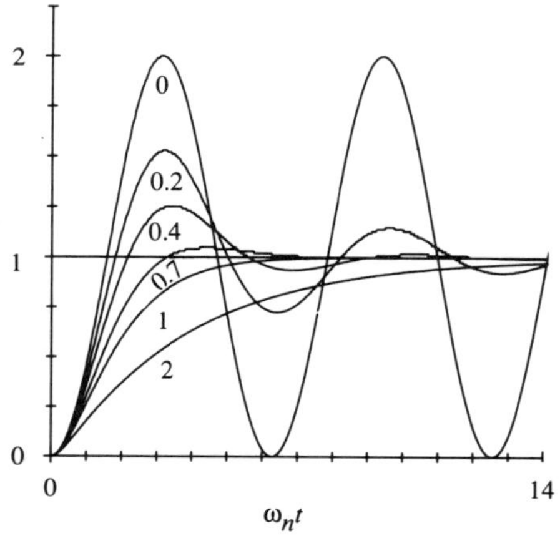

Figure 3.8 Unit step response of second order system (3.32), for $\zeta \in (0, 0.2, 0.4, 0.7, 1, 2)$.

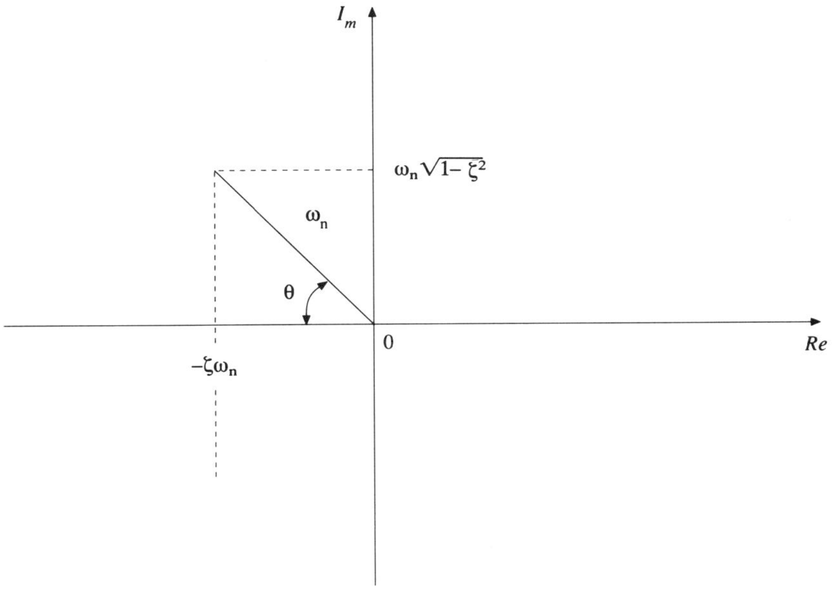

Figure 3.9 Poles of transfer function (3.32) when $0 \leqslant \zeta \leqslant 1$; $\theta = \arccos \zeta$.

latter resembles the response of a first-order system as $t \to \infty$, there is an important difference: for all values of ζ the derivative of the response at time $t=0$ is zero. In particular this means that the impulse response of (3.31) will be continuous at time $t=0$.

Exercise 3.9 Plot the impulse response of system (3.31).

Given its practical importance we are going to analyze the oscillatory response in some detail. In this case the transfer function (3.32) has a pair of conjugate poles of the form $-\zeta\omega_n \pm j\omega_n \sqrt{(1-\zeta^2)}$ whose position in the complex plane is indicated in Fig. 3.9.

From (3.34) we conclude that the real part of the poles is the decay ratio and their imaginary part is the frequency of oscillation $\omega_d = \omega_n \sqrt{(1-\zeta^2)}$. Notice $\omega_d \leqslant \omega_n$ and that $\omega_d = \omega_n$ when $\zeta = 0$ ($\theta = \pi/2$).This is why ω_n, the distance of the poles from the origin, is called the *undamped natural frequency* of the system and ω_d the *damped frequency*. When $\zeta = 1$ ($\theta = 0$) the frequency of oscillation is zero – the response increases monotonically – and the system has a double pole on the real axis. For $\zeta > 1$ there are two distinct real poles.

Of particular interest is the maximum value of $y(t)$ and the time when it occurs, here denoted by t_p. Equating the derivative of (3.34) to zero we get an infinite number of roots of the form $k\pi/(\omega_n\sqrt{(1-\zeta^2)})$, $k=0,1,2,\ldots$; it is not hard to see that the maximum occurs for $k=1$; hence

$$t_p = \pi/(\omega_n\sqrt{(1-\zeta^2)}) \tag{3.37}$$

The *maximum value* of y results from (3.34) by substituting t_p for t:

$$y(t_p) = 1 + \exp(-\zeta\pi/(\sqrt{(1-\zeta^2)}) \tag{3.38}$$

Let us define *overshoot* as

$$M_p = \frac{\text{maximum value of the response} - \text{final value of the response}}{\text{final value of the response}}$$

Then

$$M_p = \exp(-\zeta\pi/(\sqrt{(1-\zeta^2)}) \tag{3.39}$$

Notice that (3.37) and (3.38) enable us to identify the parameters of the transfer function (3.32) from experimental measurements. If the response is non-oscillatory, the identification is more difficult and the reader is referred to Rake (1980).

From (3.34) we conclude that the time required for the oscillatory system response to settle within a given neighbourhood of its final value, the settling time, is determined by $\zeta\omega_n$. If we select an interval of $\pm 1.8\%$ of the final value, it can be shown that $y(t)$ remains in such a neighbourhood for $t \geq 4/\zeta\omega_n$. Therefore we have for the settling time $t_s = 4/\zeta\omega_n$, i.e. the settling time equals four time constants.

Another standard feature of the step response in the *rise time* (T_r), which can be defined as the time required for the response to rise from 10% to 90% of the final value. Although there is no analytical expression relating T_r with the pole locations, the formula

$$T_r \cong \frac{\exp(\theta/\tan\theta)}{\omega_n} \tag{3.40}$$

constitutes a good approximation. $\zeta = \cos\theta$ as in Fig. 3.9.

In summary: because the overshoot is a function of ζ only, we can have systems with different settling times and identical overshoots. With reference to Fig. 3.9 this is the case with systems whose poles lie on the same straight line through the origin; furthermore, the greater the real part of the poles, the shorter the system settling time. The ramp response will be studied later in the book in conection with feedback systems.

Exercise 3.10 Show that, in steady-state, the response of (3.32) to the unit ramp follows the input with a constant deviation equal to $2\,\zeta/\omega_n$.
Hint: Compute $E(s) = U(s) - Y(s)$ and apply the final value theorem of Laplace transform.

3.4 HIGHER-ORDER SYSTEMS

The response of a higher-order system can be obtained as a linear combination of responses of lower-order systems. In fact, if we make a partial fraction expansion of the general transfer function (3.4) we get

$$W(s) = b_0 + \frac{a_{11}}{(s + p_1)} + \ldots + \frac{a_{1m1}}{(s + p_1)^1} + \ldots + \frac{a_{l1}}{(s + p_l)} + \ldots + \frac{a_{lml}}{(s + p_l)^{ml}} \tag{3.41}$$

and the result follows; $-p_1, -p_2, \ldots, -p_l$ are the distinct poles of $W(s)$ with multiplicities $m1, \ldots ml$, respectively. Recall that $m1+m2+\ldots+ml=n$. Each term on the right hand side of (3.41) will contribute to the impulse response, $w(t)$, with a term of the form

$$\mathcal{L}^{-1}[k/(s + p)^m] = k\, t^{(m-1)} \exp(-\,pt)/(m - 1)!$$

If $\mathrm{Re}(-p)<0$ this term goes to zero as $t \to \infty$, and the larger $|\mathrm{Re}(-p)|$, the faster this convergence. In particular this means that w(t) will be *dominated* by the contribution of the poles closer to the origin because these give rise to transients that take longer to die away. The impulse responses of the systems described by the transfer functions on the right hand side of (3.41) are called natural *modes* of the system with transfer function $W(s)$. Figure 3.10 shows how the modes associated with the multiple pole$(-p_k)$ of multiplicity m_k can be generated on an analog computer. Figure 3.11 represents (3.41) with $n=4$ and three distinct poles.

Exercise 3.11 Check the impulse response of the dynamical system diagram in Fig. 3.10.

This type of diagram provides, *by simple inspection*, all the information about the dynamic properties of the system. This is not true of other types of dynamical diagrams, such as Fig. 3.1 for example.

When some of the poles are complex, the above partial fraction expansion yields complex feedback coefficients in the associated mode generators. This is a rather undesirable feature when we wish to simulate the system, e.g. on an analog

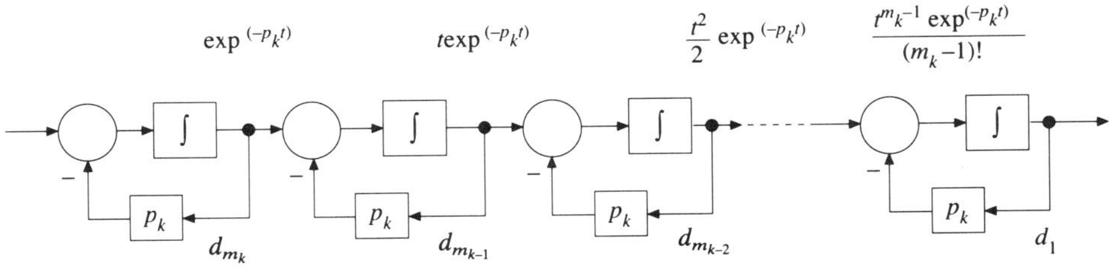

Figure 3.10 Generator of the modes associated with pole $(-p_k)$ of multiplicity m_k.

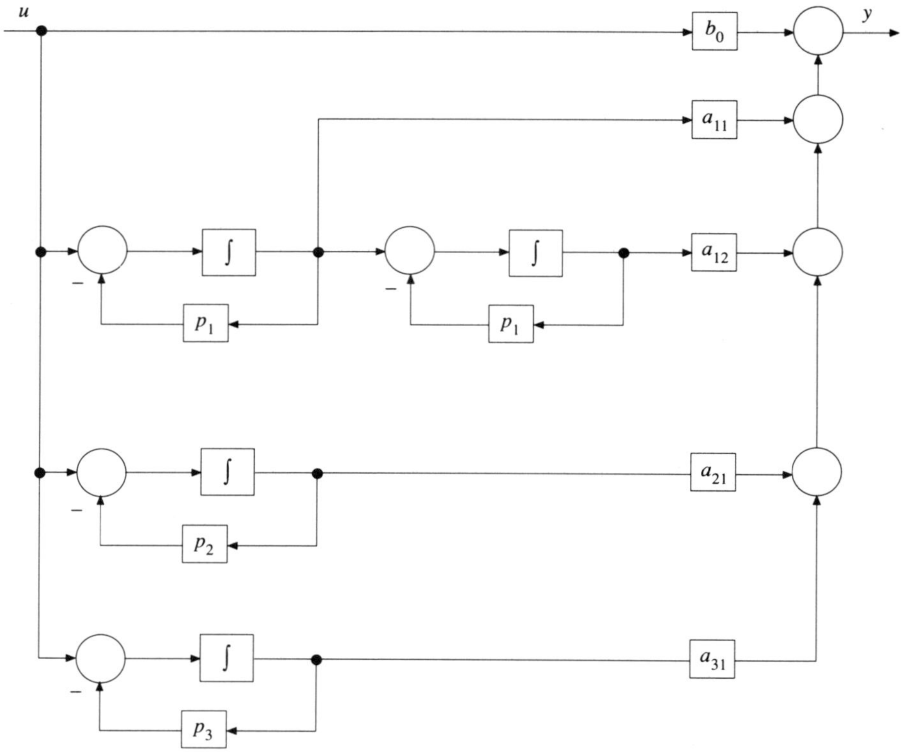

Figure 3.11 Flow diagram of (3.41) with $n=4$ and three distinct poles.

computer. This problem can be easily solved by grouping the mode generators of conjugate poles as shown below. For simplicity, let us assume

$$W(s) = \frac{1}{(s + p)(s + p^*)}, \text{ with } p = a + jb$$

Then we can write

$$W(s) = \frac{j}{2b} \left(\frac{1}{s + p} - \frac{1}{s + p^*} \right)$$

whose impulse response is

$$\begin{aligned}\mathcal{L}^{-1}(W(s)) &= j/(2b)\{\exp[-(a + jb)t] - \exp[-(a - jb)t]\} \\ &= 2\text{Re}\{j/(2b)\exp[-(a + jb)t]\} = (\text{e}^{-at} \sin bt)/b\end{aligned}$$

Figure 3.12 shows a pair of cross coupled integrators with *real* feedback coefficients generating this impulse response, and the *modes* $\text{e}^{-at} \cos bt$ and

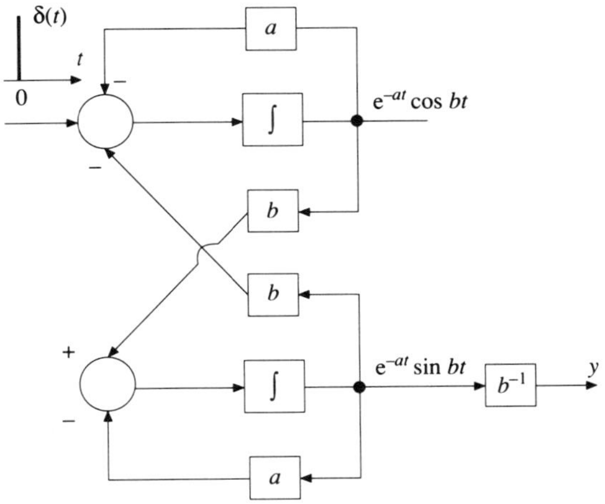

Figure 3.12 Mode generator for a pair of complex-conjugate simple poles.

$e^{-at}\sin bt$. Notice that it is b, the imaginary part of the pole, that is responsible for the coupling between the integrators and therefore for the existence of oscillations.

In Chapter 7 this analysis will be carried out in a systematic form by means of state-space representations which are more general than input–output representations. The state-space encompasses all the modes of the system whether they are present in the output or not.

3.5 THE ROUTH–HURWITZ STABILITY CRITERION

From the partial fraction decomposition just carried out, we conclude that a time signal, whose Laplace transform is a rational function, is amplitude bounded if and only if the poles of the latter either have real parts strictly less than zero, or are *simple* poles on the imaginary axis.

If we then define a system as *stable*, when any bounded input produces a bounded output (b.i.b.o. stability) we conclude that a system with a rational transfer function $W(s)=N(s)/D(s)$ is stable if and only if all the roots of the denominator polynominal $D(s)$ have *negative* real parts. Poles on the imaginary axis lead to unbounded outputs when they are also poles of (the Laplace transform of) the input signal.

The Routh–Hurwitz criterion is an algebraic test that tells us if a given polynomial has roots with positive or non-negative real parts. As shown below, the derivation of a necessary condition for a polynomial to have all the roots with negative real parts is straightforward. Such a polynomial is called a *Hurwitz polynomial*. In fact if we let

$$D(s) = s^n + a_{n-1}s^{n-1} + \ldots + a_1 s + a_0 = (s - r_1)(s - r_2)\ldots(s - r_n) \quad (3.42)$$

we can write

$$D(s) = s^n - (r_1 + r_2 + \ldots + r_n)s^{n-1} + (r_1r_2 + r_1r_3 + \ldots + r_2r_3 + r_2r_4 + \ldots)s^{n-2} - (r_1r_2r_3 + r_1r_2r_4 + \ldots + r_2r_3r_4 + r_2r_3r_5 + \ldots)s^{n-3} + \ldots + (-1)^n r_1r_2r_3\ldots r_n \quad (3.43)$$

The roots of $D(s)$, when complex, occur in conjugate pairs because $a_0,a_1,a_2,\ldots,a_{n-1}$ are real. Then we may conclude, from the right hand side of (3.43), if all the roots of D(s) have negative real parts, all the coefficients of $D(s)$ are positive. For example, the polynomial s^3+3s+1 has at least one root with a non-negative real part because the coefficient of s^2 is zero. Only if a polynomial satisfies the above necessary condition do we apply the Routh-Hurwitz stability test. It can be described in tabular form as follows:

$$\begin{array}{c||llll} n & a_n & a_{n-2} & a_{n-4} & \ldots \\ n-1 & a_{n-1} & a_{n-3} & a_{n-5} & \ldots \\ n-2 & b_{n-1} & b_{n-3} & b_{n-5} & \ldots \\ n-3 & c_{n-1} & c_{n-3} & c_{n-3} & \ldots \\ \ldots & \ldots & & & \\ 0 & h_{n-1} & & & \end{array}$$

where a_n denotes the coefficient of s_n in (3.42) if different from 1 and the $b_i's$, $c_i's,\ldots, h_{n-1}$ are computed recursively by the formulae:

$$b_{n-1} = \frac{a_{n-1}\,a_{n-2} - a_n\,a_{n-3}}{a_{n-1}} = \frac{-1}{a_{n-1}}\begin{vmatrix} a_n & a_{n-2} \\ a_{n-1} & a_{n-3} \end{vmatrix}$$

$$b_{n-3} = \frac{-1}{a_{n-1}}\begin{vmatrix} a_n & a_{n-4} \\ a_{n-1} & a_{n-5} \end{vmatrix}$$

$$c_{n-3} = \frac{-1}{b_{n-1}}\begin{vmatrix} a_{n-1} & a_{n-3} \\ b_{n-1} & b_{n-3} \end{vmatrix}$$

$$\vdots$$

The Routh-Hurwitz criterion states that if the coefficients in the first column of the table are not zero then the number of sign changes is equal to the number of roots of $D(s)$ with positive real parts. However, if some of them are equal to zero then there are roots with non-negative real parts.

Example 3.1 $D(s)=s^3+s^2+2s+20$

$$\begin{array}{c||cc} 3 & 1 & 2 \\ 2 & 1 & 20 \\ 1 & -18 & 0 \\ 0 & 20 & \end{array}$$

There are two sign changes in the first column. The polynomial has two roots with positive real parts.

Example 3.2 $D(s)=s^3+3s^2+s+3$

$$\begin{array}{c||cc} 3 & 1 & 1 \\ 2 & 3 & 3 \\ 1 & 0 & 0 \\ 0 & 3 & \end{array}$$

There is at least one root with a non-negative real part. In fact the roots of this polynomial are (-3) and $\pm j$.

The reader may question at this stage the usefulness of this method in the presence of efficient computational algorithms solving for all the roots of the polynomial, even on the cheapest programmable calculator. In fact this test is quite useful for constructing admissible ranges of variation for parameters that are uncertain or whose value has not yet been decided, as illustrated in the following example.

Assume we are given a system shown with transfer function

$$\frac{Y(s)}{R(s)} = \frac{K}{s^3 + 3s^2 + 2s + K}$$

and we wish to compute the range of values of the amplifier gain K that keeps the system stable. One way is to compute the poles of the transfer function for several values of K. A rather less time-consuming method is to apply the Routh–Hurwitz criterion to the denominator polynomial. From the table

$$\begin{array}{c||cc} 3 & 1 & 2 \\ 2 & 3 & K \\ 1 & \frac{6-K}{3} & 0 \\ 0 & K & \end{array}$$

we conclude that the system is b.i.b.o stable for $0<K<6$. For a proof of the Routh–Hurwitz criterion see Chen (1984).

3.6 THE EFFECT OF ZEROS ON THE STEP RESPONSE

So far we have not explicitly addressed the role of the zeros of a transfer function. From (3.41) we conclude that they have no influence on the stability, which is solely determined by the poles. However, the mode weighting factors a_{ij}, in expansion (3.41), depend on the transfer function zeros. For example if $(-p_k)$ is a simple pole of $W(s)$, then

$$a_{k1} = \lim_{s \to -p_k} (s + p_k)W(s)$$

Now, if $W(s)$ has also a zero at $(-z)$ such that $|z| \cong p_k$, a_{kI} will be very small, and the contribution of mode $\exp(-p_k t)$ to the total response will be negligible; in this case we say the zero has 'almost' cancelled the pole.

A zero may also significantly affect the shape of the transient response, particularly in the presence of fast modes, as the following simple argument shows: let $y_1(t)$ and $y_2(t)$ denote the response of the systems with transfer functions $W_1(s)$ and $W_2(s)$ respectively, to a given input. If

$$W_2(s) = (s + 1)\ W_1(s)$$

then

$$y_2(t) = \frac{d}{dt} y_1(t) + y_1(t) \tag{3.44}$$

If the poles of $W_1(s)$, as seen from the zero at (-1), are slow, the derivate term will have a negligible contribution. As the poles become faster, the zero will significantly change the overshoot and, to a lesser extent, reduce the peak time.

Let us now analyze quantitatively the effects of an additional zero on first- and second-order systems. In this study we shall consider only a zero of magnitude 1. This represents no loss of generality because, by an appropriate time-scaling, we can always place the zero at ± 1. (See exercise 3.12.) Therefore, what matters in our analysis is not the absolute magnitudes of the system zero and poles, but the speed of the poles as seen from the zero.

Exercise 3.12

(a) Given two transfer functions in the time-scale t, viz.

$$H_1(s) = p\frac{1 + \frac{s}{z}}{s + p} \tag{3.45}$$

and

$$H_2(s) = \frac{\omega_n^2 \left(1 + \frac{s}{z}\right)}{s^2 + 2\zeta\omega_n s + \omega_n^2} \tag{3.46}$$

show that time-scaling with $t' = ct$, they become

$$H_1'(s) = p'\left(\frac{1 + \frac{s}{z'}}{s + p'}\right) \tag{3.47}$$

and

$$H_2'(s) = \frac{\omega_n'^2 \left(1 + \frac{s}{z'}\right)}{s^2 + 2\zeta\omega_n' s + \omega_n'^2} \quad (3.48)$$

where

$$p' = \frac{p}{c}, \; z' = \frac{z}{c} \text{ and } \omega_n' = \frac{\omega_n}{c}$$

(b) Show that $H_1(s)$ and $H_2(s)$ can always be reduced to

$$p' \frac{1 + s}{s + p'} \quad (3.49)$$

and

$$\frac{\omega_n' (1 + s)}{s^2 + 2\zeta\omega_n' s + \omega_n'^2} \quad (3.50)$$

by an adequate time-scaling.
(c) Assuming in (3.46):
 (i) $\omega_n=1$ and $z=0.01$
 (ii) $\omega_n=1$ and $z=100$
compute the corresponding forms (3.50). Then convince yourself that the effect of the zero can be judged from 'the speed of the pole(s) as seen from the zero'.
Answers

$$(3.46) = H_2(s) = \frac{Y(s)}{R(s)} \Leftrightarrow \frac{d^2y}{dt^2} + 2\zeta\omega_n \frac{dy}{dt} + \omega_n^2 y = \omega_n^2 \left(\frac{dr}{zdt} + r \right)$$

because $t' = ct$ we get

$$\frac{d^2y}{dt^2} = \frac{d^2y}{dt'^2} c^2 \; ; \frac{dr}{dt} = \frac{dr}{dt'} c$$

Therefore

$$c^2 \frac{d^2y}{dt'^2} + 2c\,\zeta\,\omega_n \frac{dy}{dt'} + \omega_n^2 y = \omega_n^2 \left(\frac{c}{z} \frac{dr}{dt'} + r \right)$$

or equivalently

$$\frac{Y(s)}{R(s)} = \frac{(\omega_n/c)^2\,((c/z)s + 1)}{s^2 + 2\,\zeta\, s\,(\omega_n/c) + (\omega_n/c)^2} = \frac{(\omega_n')^2\,(1 + (s/z'))}{s^2 + 2\,\zeta\,\omega_n' s + (\omega_n')^2}$$

where $z' = z/c$ and $\omega_n' = \omega_n/c$.

Similar arguments apply to (3.45).

The analysis for first-order systems is obvious, and is illustrated in Figs 3.13 and 3.14. In the latter, the zero is on the right-half plane, giving rise to a response, already encountered in Fig. 1.9, known as 'inverse-type response', because the response starts in the opposite direction from the input.

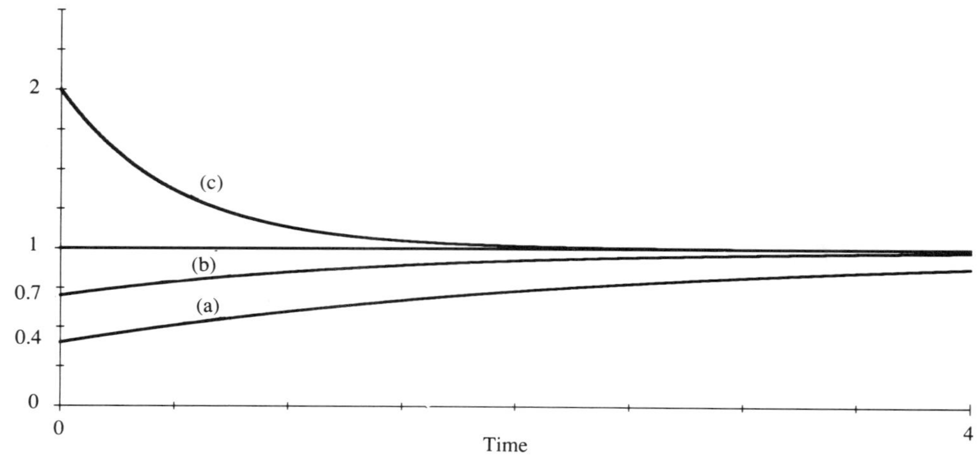

Figure 3.13 Unit-step responses of $\alpha(s+1)/(s+\alpha)$ for: (a) $\alpha=0.4$; (b) $\alpha=0.7$; (c) $\alpha=2$.

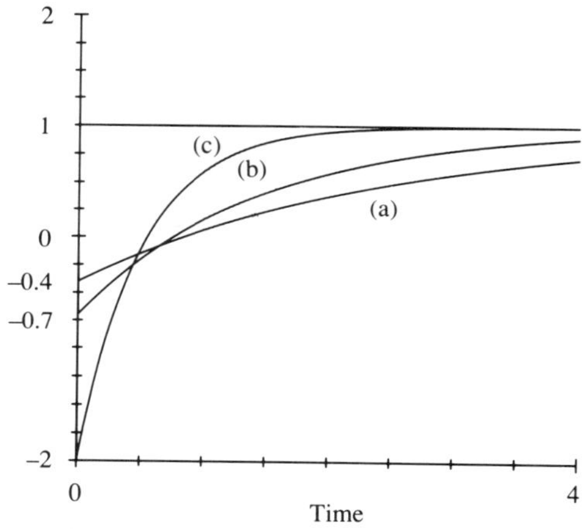

Figure 3.14 Unit-step responses of $\alpha(1-s)/(s+\alpha)$ for: (a) $\alpha=0.4$; (b) $\alpha=0.7$; (c) $\alpha=2$.

The role of an additional zero in a second-order system needs to be well understood for a satisfactory compensator design, particularly by the root-locus method. Our study is based on the transfer function

$$\frac{\omega_n^2\,(1 + s)}{s^2 + 2\zeta\omega_n s + \omega_n^2} \tag{3.51}$$

that has a pair of poles at ($\zeta\omega_n \pm j\,\omega_n \surd(1-\zeta^2)$), with magnitude ω_n, and a steady-state gain of unit. Its step responses were analyzed for the following values

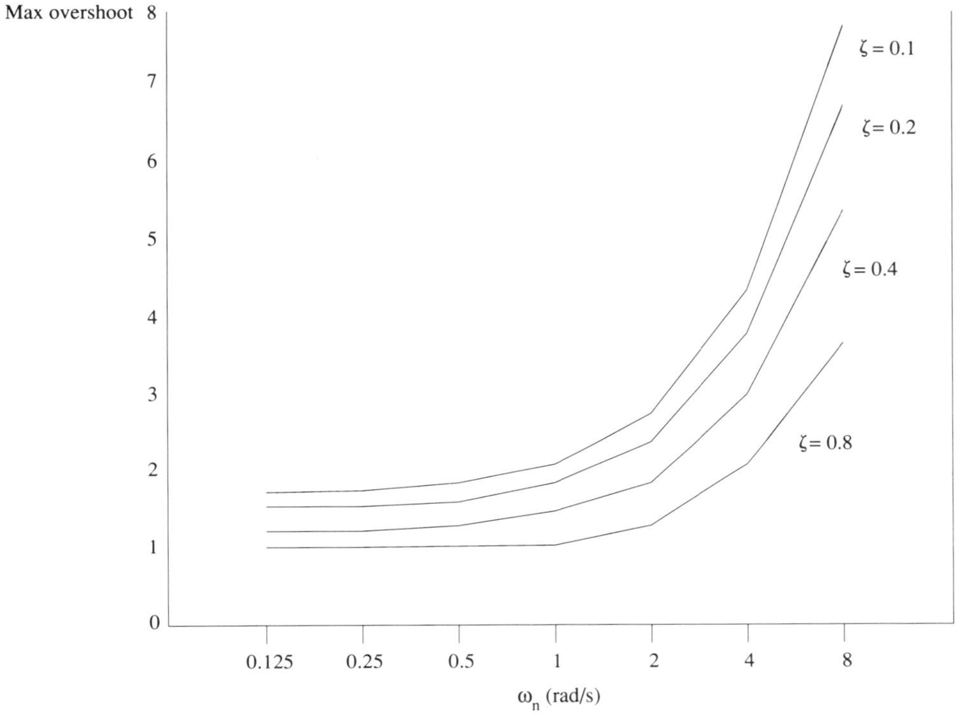

Figure 3.15 Maximum overshoot, as a function of ω_n, of the unit-step response of system (3.51).

$$\omega_n \in \left\{ \frac{1}{8}; \frac{1}{4}; \frac{1}{2}; 1; 2; 4; 8 \right\}$$

and

$$\zeta \in \{0.1; 0.2; 0.4; 0.8\}$$

and are summarized in Figs 3.15 and 3.16.

Figure 3.15 shows the maximum overshoot, in the unit-step response, as a function of ω_n, for several values of ζ. Recall that in a second-order system without a zero, these curves would be horizontal lines. As expected, the overshoot increases with the speed of the system, i.e. ω_n, and with decreasing ζ. While the poles remain slower than the zero, i.e. $\omega_n<1$, the curves are approximately horizontal lines, showing little influence of the zero. As ω_n becomes greater than one, the situation changes, showing a drastic increase in overshoot for $\omega_n>2$; this is natural, in the light of expression (3.44) because now we have a fast system, as seen from the zero.

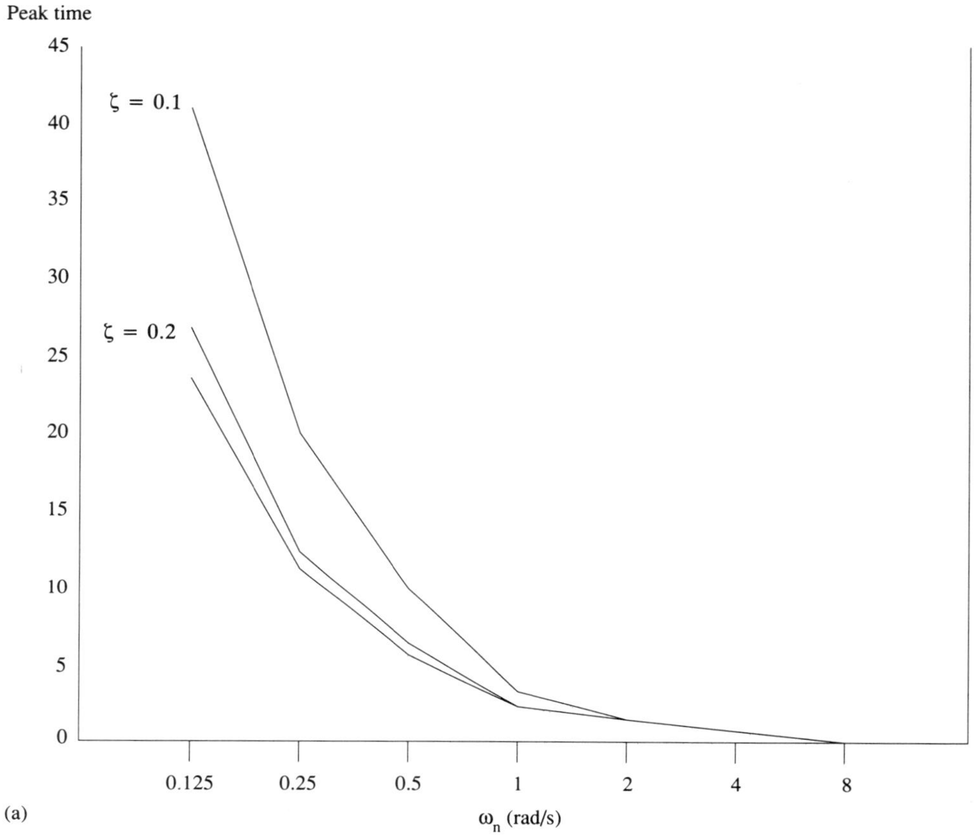

Figure 3.16 Peak times of unit-step responses of: (a) system (3.51) and (b) second-order system without zero.

On the other hand, it was found that the zero hardly alters the peak-time of the pure second-order system, as shown in Figs 3.16(a) and (b), which are almost indistinguishable. The latter is a plot of the peak time t_p, for a pure second-order system, as given by the formula

$$t_p = \frac{\pi}{\omega_n \sqrt{(1 - \zeta^2)}} \tag{3.52}$$

The changes in peak time are only significant (in relative terms) for large values of ω_n. In Fig. 3.17 we compare the step responses of

$$\frac{64(1 + s)}{(s^2 + 1.6s + 64)} \quad \text{and} \quad \frac{64}{(s^2 + 1.6s + 64)}$$

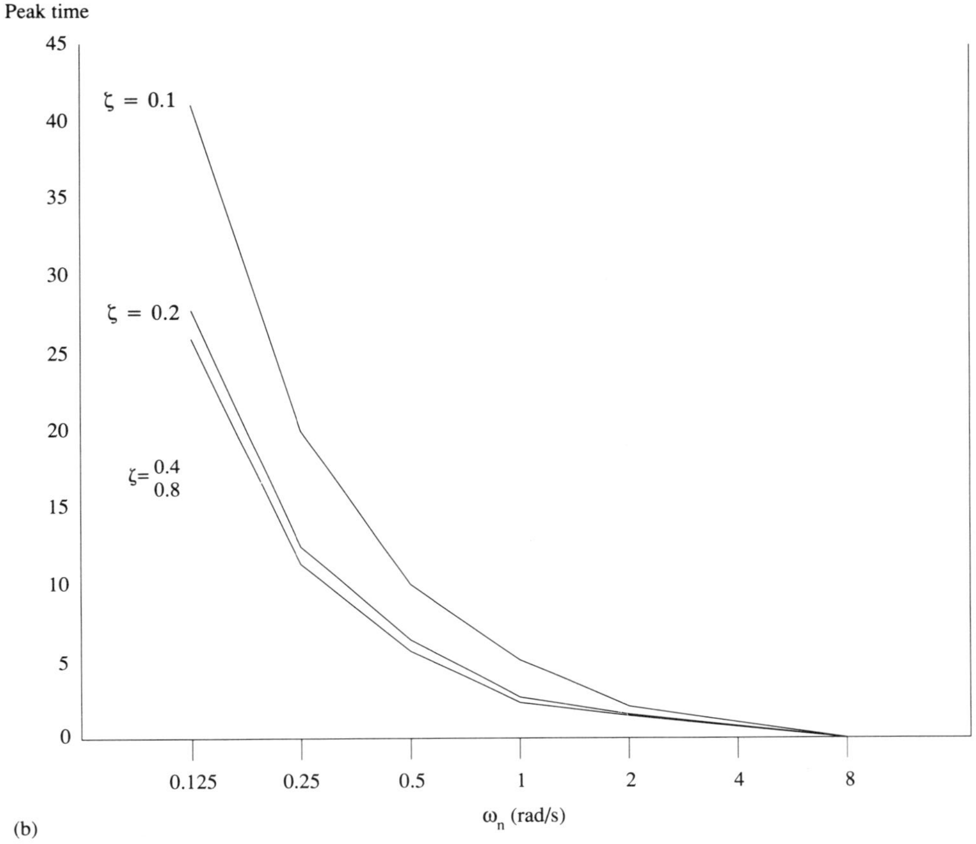

Figure 3.16(b) Note that the curves for ζ=0.4 and ζ=0.8 are almost indistinguishable.

It can be seen that the peak times are 0.2 and 0.4 respectively. Notice that the curves intersect at t=0.4. (Why?)

The study that has just been carried out for left half-plane (LHP) zeros is also applicable to right half-plane (RHP) zeros, with a simple modification: while a LHP zero induces an overshoot, the RHP zero, of identical magnitude, will induce an *undershoot*, i.e. the response will start out in the opposite direction from the input. This is obvious from (3.44) if we recall that it assumes the form

$$W_2(s) = (1 - s)\, W_1(s)$$

in the case of a RHP zero.

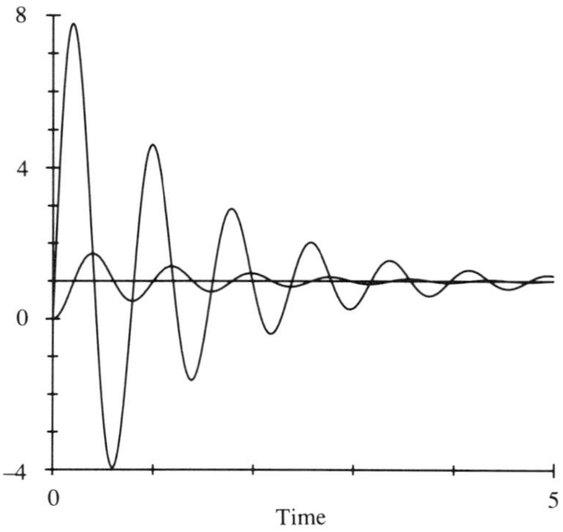

Figure 3.17 Unit-step responses of $64/(s^2+1.6s+64)$ and $64(1+s)/(s^2+1.6s+64)$.

This is illustrated of Fig. 3.18 where we can see that the step response of

$$\frac{64(1 - s)}{(s^2 + 1.6s + 64)}$$

starts out in the opposite direction from the response of

$$\frac{64(1 + s)}{(s^2 + 1.6s + 64)}$$

which we have just seen in Fig. 3.17.

As shown later in the book, this property has important implications on the achievable performance of closed-loop systems, namely those with RHP zeros or RHP poles, in the open-loop transfer function.

3.7 SOLUTION OF THE DYNAMICAL EQUATION $\dot{x}=Ax+Bu$

A key concept in linear system dynamics is the exponential of a matrix. This concept, together with its relevant properties, are discussed below before embarking on the solution of the equation $\dot{x}=Ax+Bu$.

3.7.1 Matrix Exponential and Cayley–Hamilton Theorem

Any square matrix $A\in\mathbb{R}^{n\times n}$ can be regarded as a linear operator from $\mathbb{C}^n$ into $\mathbb{C}^n$.If a vector v has an image colinear with v i.e. $Av=\lambda v$, $\lambda\in\mathbb{C}$ it is called an

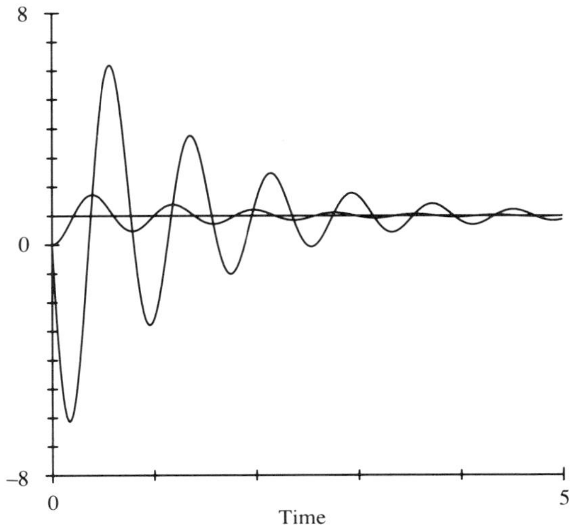

Figure 3.18 Unit-step responses of $64/(s^2+1.6s+64)$ and $64(1-s)/(s^2+1.6s+64)$.

eigenvector of A with eigenvalue λ.Two results follow immediately:

1. Any vector with the same direction as v is also an eigenvector of A.
2. v is also an eigenvector of A^k, k integer, with eigenvalue λ^k.

An equivalent condition to $Av=\lambda v$ is $(\lambda I-A)v=\mathbf{0}$. Therefore, for a non-trivial solution, i.e. $v\neq 0$, we require $\det(\lambda I-A)=\mathbf{0}$. The solutions of the last equation are the eigenvalues of A. Note that $\det(\lambda I-A)$ is a polynomial in λ (cf. exercise 3.3), of degree n, with real coefficients whose roots are therefore the eigenvalues of the matrix A, i.e.

$$\det(\lambda I-A) = \prod_{k=1}^{d} (\lambda - \lambda_k)^{n_k} = p(\lambda) \tag{3.53}$$

where $n_1+n_2+\ldots+n_d=n$.

The *Cayley–Hamilton theorem* states that any square matrix satisfies its own characteristic equation, i.e.

$$p(A) = \prod_{k=1}^{d} (A - \lambda_k I)^{n_k} = A^n + c_{n-1} A^{n-1} + \ldots + c_1 A + c_0 I = \mathbf{0}$$

An important corollary of this theorem is as follows:

Corollary For any integer p, A^p is a linear combination of $I, A, A^2, \ldots, A^{n-1}$.

Exercise 3.13 Prove this corollary.

Notice that there may exist a polynomial l of a lesser degree than p for which $l(\boldsymbol{A})=\boldsymbol{0}$. The polynomial m of the least degree such that $m(\boldsymbol{A})=\boldsymbol{0}$ is known as the *minimal polynomial* of $\boldsymbol{A}$.

3.7.2 Matrix Exponential

By analogy with the scalar power series definition of exponential, we define for $\boldsymbol{A}\in\mathbb{R}^{n\times n}$, $t\in\mathbb{R}$,

$$\exp \boldsymbol{A}t = \boldsymbol{I} + \boldsymbol{A}t + \boldsymbol{A}^2t^2/2! + \ldots = \sum_{k=0}^{\infty} \boldsymbol{A}^k t^k/k! \tag{3.54}$$

The following properties follow from the definition:

1. $\exp(\boldsymbol{0})=\boldsymbol{I}$.
2. $\exp((\boldsymbol{A}_1+\boldsymbol{A}_2)t)=\exp(\boldsymbol{A}_1t)\ \exp(\boldsymbol{A}_2t)$, if and only if $\boldsymbol{A}_1\ \boldsymbol{A}_2=\boldsymbol{A}_2\ \boldsymbol{A}_1$.
3. $\exp(\boldsymbol{A}(t_1+t_2))=\exp(\boldsymbol{A}t_1)\ \exp(\boldsymbol{A}t_2)$.
   In particular for $t_2=-t_1$ it follows that $[\exp(\boldsymbol{A}t_1)]^{-1}=\exp(-\boldsymbol{A}t_1)$.
4. $d/dt\ [\exp(\boldsymbol{A}t)]=\boldsymbol{A}\ \exp(\boldsymbol{A}t)=\exp(\boldsymbol{A}t)\ \boldsymbol{A}$.
5. If $\boldsymbol{v}$ is an eigenvector of $\boldsymbol{A}$ with eigenvalue λ, then $\boldsymbol{v}$ is an eigenvector of $\exp \boldsymbol{A}t$ with eigenvalue $\exp(\lambda t)$.

The proof of these properties is left to the reader as an exercise.

Given a matrix $\boldsymbol{M}(t)$ whose entries $m_{ij}(t)$ are functions of time, the matrix with entries $\mathcal{L}\{m_{ij}(t)\}$ is called the Laplace transform of $\boldsymbol{M}$. We are now in a position to state a further property of $\exp \boldsymbol{A}t$:

6. $\mathcal{L}(\exp(\boldsymbol{A}t))=(s\boldsymbol{I}-\boldsymbol{A})^{-1}$.

We now prove this result. Taking the Laplace transform of both sides of (3.54) we get

$$\mathcal{L}(\exp(\boldsymbol{A}t)) = s^{-1} \sum_{k=0}^{\infty} s^{-k}\boldsymbol{A}^k \tag{3.55}$$

because $\mathcal{L}(t^k/k!)=s^{-(k+1)}$. The well-known scalar equality

$$(1 - r)^{-1} = \sum_{k=0}^{\infty} r^k, \text{ for } |r| < 1$$

also holds when r is a square matrix, provided its eigenvalues have magnitudes less than one (the following section will clarify this requirement, in connection with matrix diagonalization). Then (3.55) becomes:

$$\mathcal{L}\ (\exp \boldsymbol{A}t) = s^{-1}(\boldsymbol{I} - s^{-1}\boldsymbol{A})^{-1} = (s\boldsymbol{I} - \boldsymbol{A})^{-1}$$

QED.

3.7.3 Evaluation of the Exponential of a Matrix

The matrix exponential plays a key role in the study of linear system dynamics. Its numerical computation is in general not easy and this has led to the development of a variety of methods.

In Moler and Van Loan (1978) 19 different methods are presented, together with a discussion of their relative merits and pitfalls. Here we shall pay scant attention to this vast problem by mentioning just the most straightforward ones.

3.7.3.1 Method 1: Series expansion

$$\exp \boldsymbol{A}t = \sum_{k=0}^{\infty} \boldsymbol{A}^k\, t^k/k!$$

This method is a straightforward application of the definition. In general we stop adding new terms when $\| \mathbf{A}^l t^l/l! \|$ *becomes less then some prespecified quantity* $\varepsilon>0$. Here $\|.\|$ denotes some suitable matrix norm, like the largest eigenvalue for example.

3.7.3.2 Method 2: Matrix diagonalization

$$\exp \boldsymbol{A}t = \boldsymbol{M} \exp(\boldsymbol{\Lambda}t)\, \boldsymbol{M}^{-1} \tag{3.56}$$

where $\boldsymbol{\Lambda}$ is a 'block diagonal' matrix, i.e. a matrix with the structure

$$\boldsymbol{\Lambda} = \begin{bmatrix} A_1 & & & \bigcirc \\ & A_2 & & \\ & & \ddots & \\ \bigcirc & & & A_d \end{bmatrix} \tag{3.57}$$

d is the number of distinct eigenvalues of $\boldsymbol{A}$ and $\boldsymbol{A}_1, \boldsymbol{A}_2, \ldots, \boldsymbol{A}_d$ are square matrices. Remember that, for a matrix $\boldsymbol{\Lambda}$ as in (3.57),

$$\exp \boldsymbol{\Lambda} = \begin{bmatrix} \exp A_1 & & & \bigcirc \\ & \exp A_2 & & \\ & & \ddots & \\ \bigcirc & & & \exp A_d \end{bmatrix}$$

When A has distinct eigenvalues, $d=n$, $A_1=\lambda_1$, $A_2=\lambda_2, \ldots, A_n=\lambda_n$ and the columns of matrix M are the eigenvectors of A; hence

$$\exp \boldsymbol{\Lambda} t = \begin{bmatrix} e^{\lambda_1 t} & & & \bigcirc \\ & e^{\lambda_2 t} & & \\ & & \ddots & \\ \bigcirc & & & e^{\lambda_n t} \end{bmatrix}$$

The case of multiple eigenvalues will be dealt with in Chapter 7 (Jordan decomposition theorem). See also exercise 7.6.

3.7.3.3 Method 3: Cayley–Hamilton theorem

$$\exp At = \sum_{i=0}^{n-1} \alpha_i(t)\, A^i \tag{3.58}$$

By applying the above corollary to the definition of exp At the result follows. Because the eigenvalues of A also satisfy the characteristic equation of A, we also have

$$\exp \lambda t = \sum_{i=0}^{n-1} \alpha_i(t)\, \lambda^i,$$

where λ is an eigenvalue of A. If A has distinct eigenvalues then we have enough equations to compute the α_i.

Example 3.3 Let

$$A = \begin{bmatrix} 0 & -2 \\ -1 & -1 \end{bmatrix}$$

whose characteristic equation is $\lambda^2+\lambda-2=0$, giving $\lambda_1=1$ and $\lambda_2=-2$. Since $A \in \mathbb{R}^{2\times 2}$ we have

$$e^{\lambda t} = \alpha_0(t) + \alpha_1(t)\lambda$$

By substituting the computed eigenvalues we get

$$e^{t} = \alpha_0(t) + \alpha_1(t)$$

$$e^{-2t} = \alpha_0(t) - 2\alpha_1(t)$$

or equivalently

$$\alpha_0(t) = 1/3(2e^{t} + e^{-2t})$$

$$\alpha_1(t) = 1/3(e^{t} - e^{-2t})$$

Hence

$$e^{At} = \alpha_0(t)\mathbf{I} + \alpha_1(t)A = \frac{1}{3}\begin{bmatrix} (2e^{t} + e^{-2t}) & -2(e^{t} - e^{-2t}) \\ -(e^{t} - e^{-2t}) & (e^{t} + 2e^{-2t}) \end{bmatrix}$$

Exercise 3.14 Show that in (3.58), and for any square matrix A, $\alpha_0(0)=1$ and $\alpha_1(0)=\alpha_2(0)=\ldots=\alpha_{n-1}(0)=0$.

Exercise 3.15 Show how to evaluate the inverse of a matrix A by the Cayley–Hamilton theorem. *Hint*: multiply both sides of the characteristic equation by A^{-1}.

3.7.3.4 Method 4: Laplace transform

$$e^{At} = \mathcal{L}^{-1}\{(sI - A)^{-1}\} \tag{3.59}$$

Example 3.4 Let A be the same as in the previous example; then

$$(sI - A) = \begin{bmatrix} s & 2 \\ 1 & s+1 \end{bmatrix}; \quad (sI - A)^{-1} = \frac{1}{s^2 + s - 2}\begin{bmatrix} s+1 & -2 \\ -1 & s \end{bmatrix}$$

$$(sI - A)^{-1} = \begin{bmatrix} \frac{1}{3}\left(\frac{2}{s-1} + \frac{1}{s+2}\right) & -\frac{2}{3}\left(\frac{1}{s-1} - \frac{1}{s+2}\right) \\ \frac{1}{3}\left(\frac{-1}{s-1} + \frac{1}{s+2}\right) & \frac{1}{3}\left(\frac{1}{s-1} + \frac{2}{s+2}\right) \end{bmatrix}$$

$$\mathcal{L}^{-1}\{(s\boldsymbol{I} - A)^{-1}\} = 1/3 \begin{bmatrix} 2e^{t}+e^{-2t} & -2(e^{t}-e^{-2t}) \\ -e^{t}+e^{-2t} & e^{t}+2e^{-2t} \end{bmatrix} = e^{At}$$

3.7.4 Solution of the Equation $\dot{\boldsymbol{x}} = \boldsymbol{Ax} + \boldsymbol{B}u$

Given a vector of time functions $\boldsymbol{x}(t)=(x_1(t), x_2(t),\ldots, x_n(t))^T$, we define its integral as

$$\int_0^t \boldsymbol{x}(t')\,dt' = \left(\int_0^t x_1(t')\,dt', \ldots, \int_0^t x_n(t')\,dt'\right)^T$$

When $\boldsymbol{A}=\boldsymbol{0}$ in equation $\dot{\boldsymbol{x}}=\boldsymbol{Ax}+\boldsymbol{B}u$ the solution is simply given by

$$\boldsymbol{x}(t) = \int_0^t \boldsymbol{B}u(t')\,dt' + \boldsymbol{x}(0) \tag{3.60}$$

If we can find a change of variable such that $\dot{\boldsymbol{x}}=\boldsymbol{Ax}+\boldsymbol{B}u$ is transformed into a form $\dot{\boldsymbol{h}}=f(u)$, we shall be in a position to compute its solution. Let us then define a new variable $\boldsymbol{h}(t)$ as

$$\boldsymbol{h}(t) = \exp(-\boldsymbol{A}t)\,\boldsymbol{x}(t) \tag{3.61}$$

Computing the derivative of both sides we get

$$\begin{aligned} \dot{\boldsymbol{h}}(t) &= -\boldsymbol{A}\exp(-\boldsymbol{A}t)\boldsymbol{x}(t) + \exp(-\boldsymbol{A}t)\,\dot{\boldsymbol{x}}(t) \\ &= -\boldsymbol{A}\exp(-\boldsymbol{A}t)\boldsymbol{x}(t) + \exp(-\boldsymbol{A}t)(\boldsymbol{Ax}(t) + \boldsymbol{B}u(t)) \\ &= \exp(-\boldsymbol{A}t)\,\boldsymbol{B}u(t) \end{aligned}$$

From (3.60) we have

$$\boldsymbol{h}(t) = \boldsymbol{h}(0) + \int_0^t \exp(-\boldsymbol{A}t')\,\boldsymbol{B}\,u(t')\,dt'$$

and from (3.61)

$$\boldsymbol{x}(t) = \exp(\boldsymbol{A}t)\,\boldsymbol{x}(0) + \int_0^t \exp(\boldsymbol{A}(t-t'))\,\boldsymbol{B}\,u(t')\,dt' \tag{3.62}$$

The first term on the right is the zero input response and the second term the zero state response.

Exercise 3.16 Show, by direct substitution, that (3.62) is in fact a solution of $\dot{\boldsymbol{x}}=\boldsymbol{Ax}+\boldsymbol{B}u$. Note that

$$d/dt \int_0^t f(t,t')\,dt' = f(t,t) + \int_0^t \partial/\partial t\, f(t,t')\,dt'$$

Observation: The scalar convolution integral (3.20) is a particular case of (3.62) with zero initial condition; in fact the response of the system (3.16), when $D=0$, can be written as

$$y(t) = \int_0^t \boldsymbol{C}\exp(\boldsymbol{A}(t-t'))\boldsymbol{B}\,u(t')\,dt'$$

which, when compared with (3.20), shows that $\boldsymbol{C}\exp(\boldsymbol{A}t)\boldsymbol{B}$ is the system weighting function.

3.8 CONCLUSIONS

It has been shown that a system described by an ordinary linear constant coefficient differential equation, of the type (3.1), is amenable to several types of representation, namely transfer function and state-space. The advantage of the former was simplicity; not only was the solution of the differential equation reduced to a simple algebraic problem, but also the dynamical properties of the system were summarized in a finite set of numbers – the transfer-function poles and zeros.

The state-space representation has revealed that the dynamics of a linear system are generated by the interconnection (linear combinations) of elementary subsystems – the mode generators. Furthermore it was possible to condense, at a given time instant t, all the information about the past behaviour of the system, in a single entity known as the state vector, $\boldsymbol{x}(t)$, and to compute transitions between states in an explicit manner.

REFERENCES

Chen, C.T. (1984) *Linear System Theory and Design*, Holt, Rinehart, Winston.
Godfrey, K.R. (1980) 'Correlation methods', *Automatica*, **16**, 527–34.
Moler, C. and C. Van Loan (1978) 'Nineteen Dubious Ways to Compute the Exponential of a Matrix', *SIAM Review*, **20**(24), October, pp. 801–35.
Rake, H. (1980) 'Step response and frequency response methods', *Automatica*, **16**, 519–26.

4

Time-domain Analysis of Feedback Systems

In the previous chapter we looked at different ways of describing the dynamic behaviour of a plant by means of linear models. In this chapter the plant is inserted in a feedback loop, as shown in Fig. 4.5, and our objective is to study the stability, steady-state and transient behaviour of this more complex system.

4.1 BLOCK DIAGRAM ALGEBRA

The analysis of complex feedback systems can be greatly simplified if we reduce their block diagrams to simpler ones, as shown in Fig. 4.1(a); such a form is known as the 'canonical form' of a feedback system.

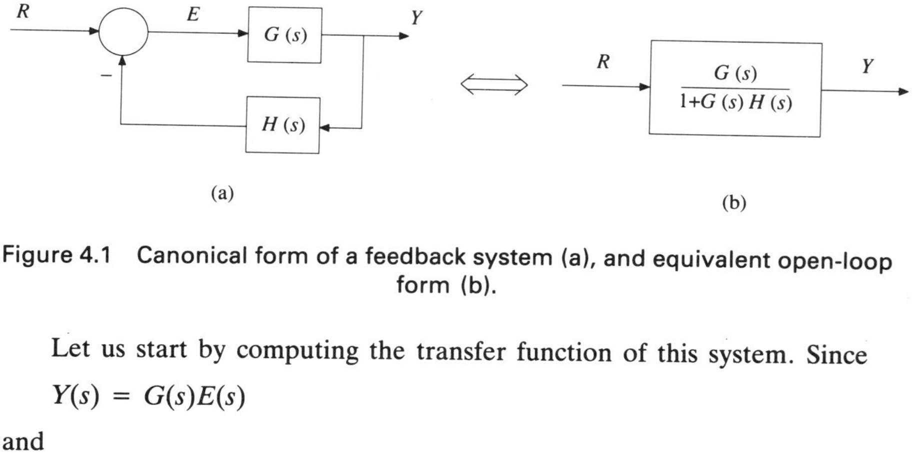

Figure 4.1 Canonical form of a feedback system (a), and equivalent open-loop form (b).

Let us start by computing the transfer function of this system. Since

$$Y(s) = G(s)E(s)$$

and

$$E(s) = R(s) - H(s)Y(s)$$

we obtain

$$Y(s) = G(s)R(s) - G(s)H(s)Y(s)$$

Consequently

$$\frac{Y(s)}{R(s)} = \frac{G(s)}{1 + G(s)H(s)}$$

i.e. the block diagram of Fig. 4.1(a) is equivalent to the 'open-loop' shown in part (b).

From the first of the above equations we also conclude that

$$\frac{E(s)}{R(s)} = \frac{1}{1 + G(s)H(s)}$$

The next example is shown in Fig. 4.2(a). By rearranging summing points, we get the equivalent diagram as in (b). This transformation is justified from the fact that

$$M = R + G_1R - Y$$

The block diagram can be further transformed into (c), from which we immediately have

$$\frac{Y}{R} = \frac{(1 + G_1)G_2}{(1 + G_2)}$$

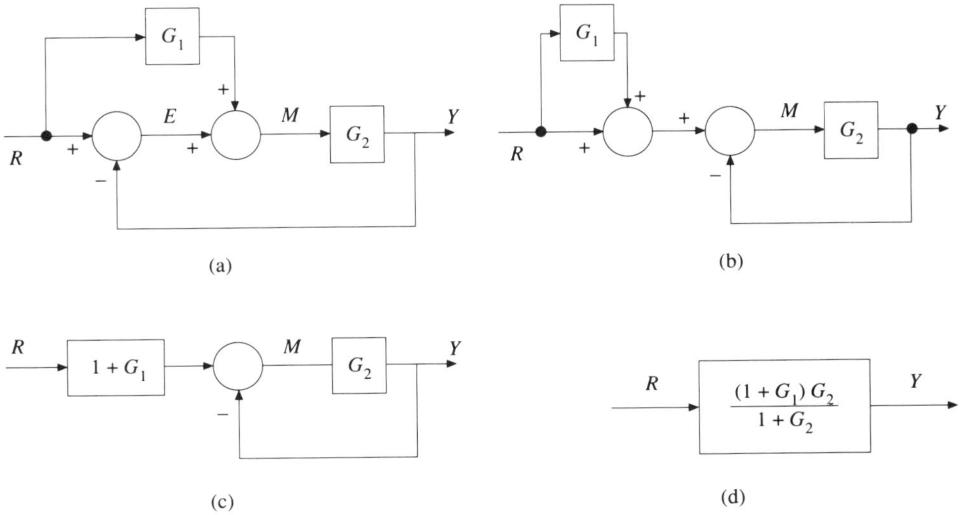

Figure 4.2 Example of block diagram simplification.

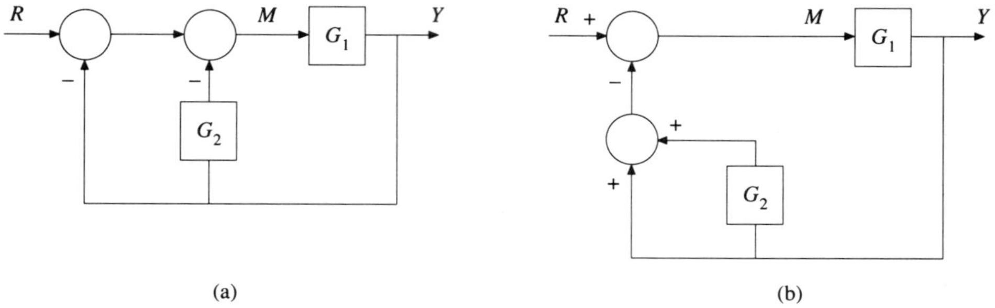

Figure 4.3 Equivalent block diagrams.

A similar situation is depicted in Fig. 4.3(a). Again, by rearranging summing points we arrive at diagram (b). The closed-loop transfer function is therefore

$$\frac{Y}{R} = \frac{G_1}{1 + G_1(1 + G_2)}$$

Let us now attempt a more complex exercise, by reducing the block diagram of Fig. 4.4(a) to open-loop form. In part (b) of the figure, two of the summing

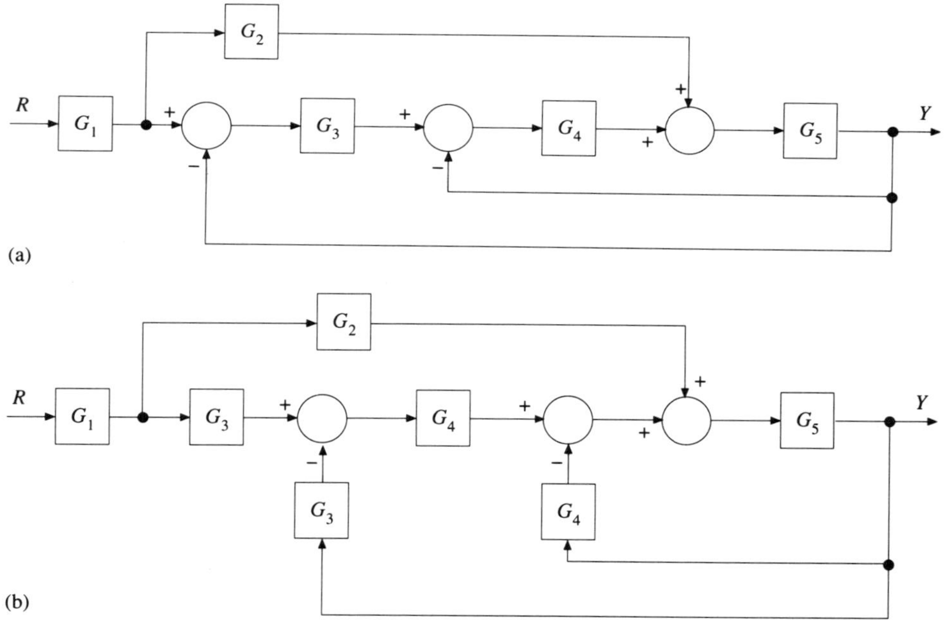

Figure 4.4 Reduction of a block diagram to open-loop form.

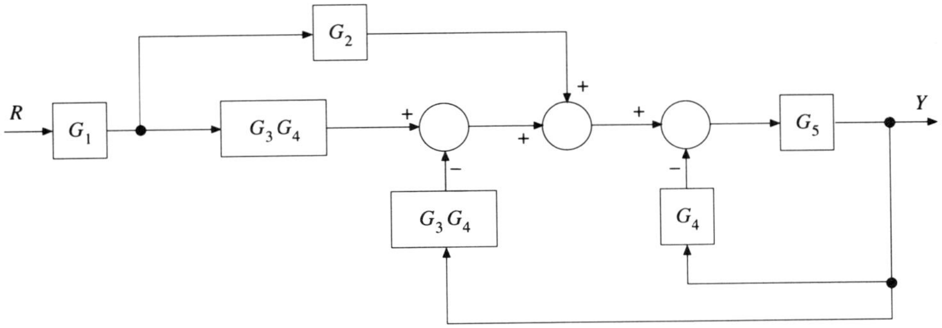

(c)

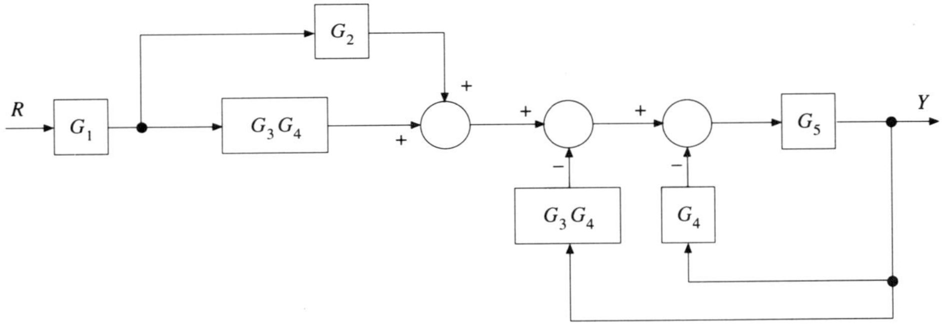

(d)

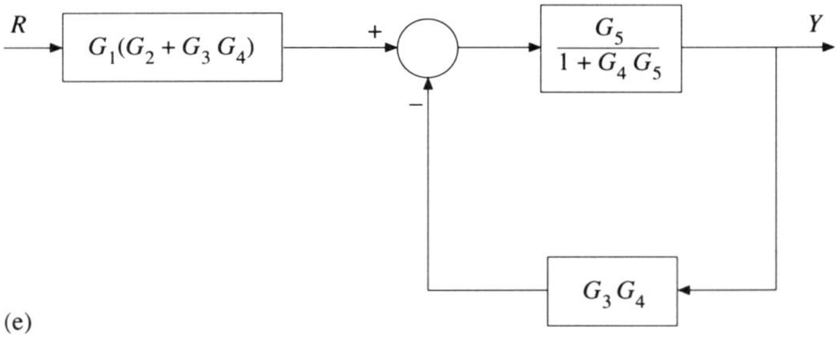

(e)

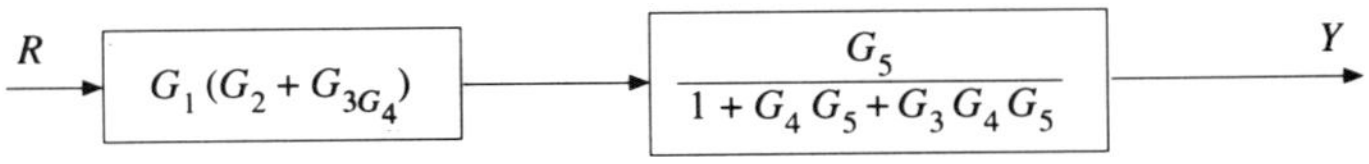

(f)

Figure 4.4 (cont.)

points were moved beyond blocks G_3 and G_4. In (c) the last two summing points were rearranged and the summing point on the left was moved beyond block G_4. In (d) two summing points were rearranged, namely the first two. Then the inner feedback loop is eliminated and two parallel blocks are combined in step (e). Finally in (f), the remaining feedback loop is eliminated, and we end up with two blocks in cascade.

4.2 STEADY-STATE ANALYSIS

In this section we consider the feedback system of Fig. 4.5, with $H(s)=1$, and assume it is stable. From this figure we have

$$C(s) = \frac{K(s)\ W(s)}{1 + K(s)\ W(s)\ H(s)} R(s) + \frac{1}{1 + K(s)\ W(s)\ H(s)} D(s) \tag{4.1}$$

and

$$E(s) = \frac{1}{1 + K(s)\ W(s)\ H(s)} (R(s) - H(s)\ D(s)) \tag{4.2}$$

Exercise 4.1 Compute equality (4.2).

Before going into mathematical detail, let us imagine that Fig. 4.5 represents the temperature control system of a room discussed in Chapter 1. $W(s)$ and $K(s)$ represent the transfer functions of the room and controller respectively, and d represents disturbances, such as variations in the outside temperature. If $K(s)$ is a proportional controller with gain K, $W(s)=1/(s+1)$, and $d=0$, and if at time $t=0$ a step change of 20 °C is made in the reference value, then from (4.1) and (4.2) and

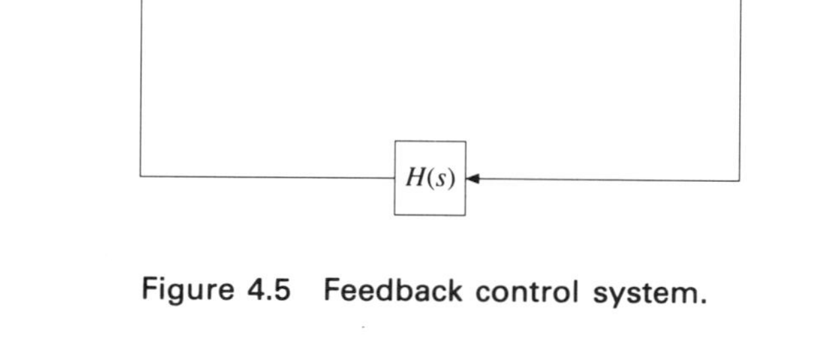

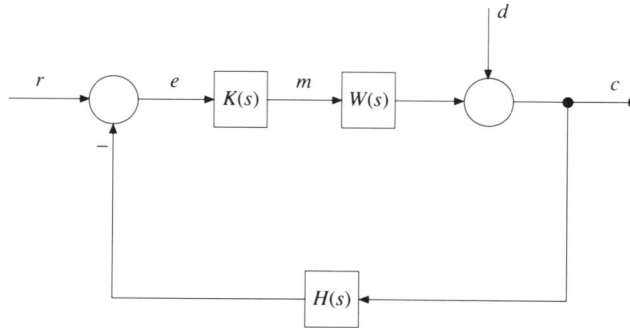

Figure 4.5 Feedback control system.

the final value theorem we have

$$e_{ss} = \lim_{t \to \infty} e(t) = \frac{20}{1 + K} \tag{4.3}$$

and

$$\lim_{t \to \infty} e(t)\ c(t) = \frac{K}{1 + K}\ 20 \tag{4.4}$$

as illustrated in Fig. 4.6. Now assume that at some later time, t_d, the outside temperature rises by 5 °C. The room temperature $c(t)$ will then evolve towards

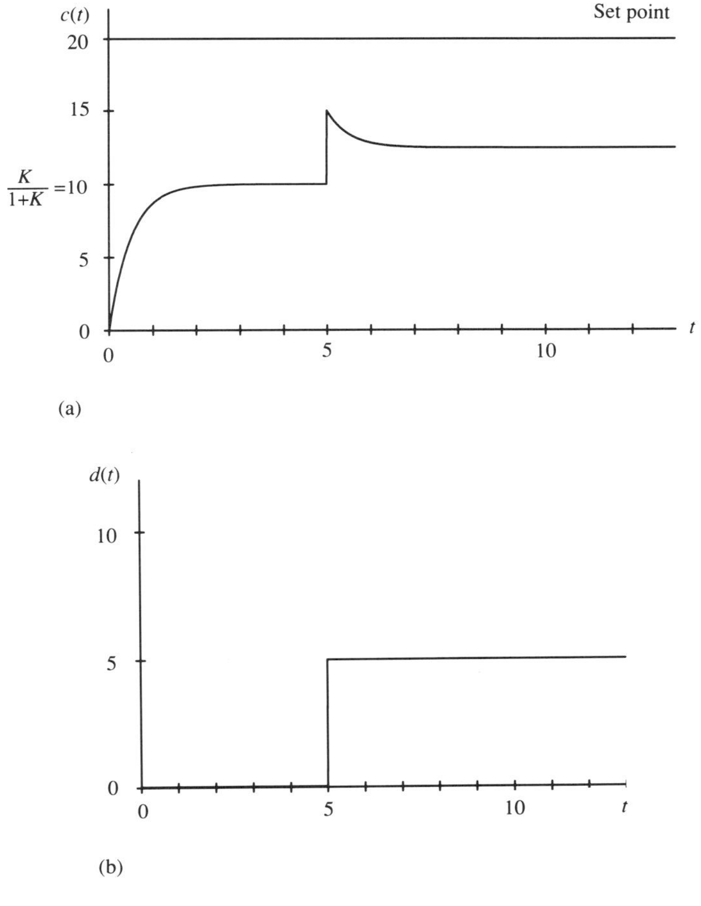

Figure 4.6 Evolution of room temperature (a) following a set point change and a disturbance (b) t_d instants later, assuming $K(s)=1$ and $t_d=5$.

$(20K+5)/(1+K)$and the error will depend on the size of the disturbances. In our case this means, for example, that the stabilizing value of the temperature will be smaller on cooler days than on warmer days, for the same reference value.

Exercise 4.2 Show that, in Fig. 4.6, $c(t)$ evolves according to

$$c(t) = \begin{cases} 10\ (1 - e^{-2t});\ t < t_d \\ 10\ (1 - e^{-2t}) + 2.5\ (1 + e^{-2(t-t_d)});\ t \geqslant t_d \end{cases}$$

From (4.3) above we conclude that the larger the gain K the smaller the error. Because there are practical limitations to the size of K, we must look for alternatives if we wish to eliminate the error. A possible solution is to use instead an integral controller whose transfer function is $K(s)=K/s$ or equivalently $m(t)=K \int_0^t e(t')dt'$. Because the output of this controller only stops when the error becomes zero, a non-zero steady-state error e_{ss} is impossible, as the following argument shows. If e_{ss} is positive then the output of the controller is increasing linearly and so is the room temperature; but this contradicts the assumption of a constant steady-state error, because r is also constant. Therefore e_{ss} must be zero. Figure 4.7 shows the effect of the introduction of an integral controller. The steady-state error is eliminated, whether the disturbance is present or not; however the system takes longer to stabilize. This is true for any stable unit feedback system whose open loop transfer function has at least one pole at the origin. It is now natural to ask what will happen if the reference is not a constant signal; if we are considering the behaviour of an aircraft following radar, for example, then a linearly increasing reference signal is a more accurate simulation of the real situation. In the following we shall analyze the steady-state response to the unit step, ramp and parabola inputs, of stable unit feedback systems as shown in Fig. 4.5; in contrast

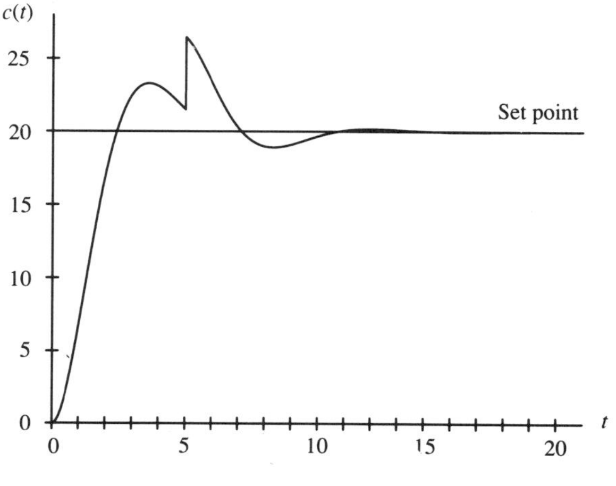

Figure 4.7 Evolution of $c(t)$ in Fig. 4.5 with the same data as Fig. 4.6 except $K(s)=1/s$.

with the transient behaviour, which is determined by the order of the system, the steady-state response only depends upon the number of poles at the origin of the open-loop transfer function, i.e. the system type. For example, in Fig. 4.6 we have a first-order *type zero* system and in Fig. 4.7 a *type one*, second-order system. In the following discussion we assume without any loss of generality that $d=0$ and denote by $G(s)$ the open-loop transfer function $K(s)\ W(s)$ in Fig. 4.5. Then we can write

$$e_{ss} = \lim_{t\to\infty} e(t) = \lim_{s\to 0} \frac{s\ R(s)}{1 + G(s)} \tag{4.5}$$

4.2.1 Unit-step Reference: $R(s)=1/s$

In this case $e_{ss}=1/(1+G(0))$, because $G(s)$ always has the form:

$$\frac{K(1 + b_1 s + b_2 s^2 + \ldots + b_m s^m)}{s^l(1 + a_1 s + a_2 s^2 + \ldots + a_n s^n)}$$

with $l \geqslant 0$ and $l+n \geqslant m$, then

$$\lim_{s\to 0} G(s) = \begin{cases} K;\ \text{if } l = 0 \\ \infty;\ \text{if } l \geqslant 1 \end{cases} \tag{4.6}$$

The above limit is usually known as *static position error coefficient* and denoted by K_p; then

$$e_{ss} = 1/(1 + K_p)$$

Therefore a system of type l, $l \geqslant 1$, has no steady-state error (if stable), for a constant reference input. Cf. Figs 4.6 and 4.7.

4.2.2 Unit-ramp Reference: $R(s)=1/s^2$

$$e_{ss} = \lim_{s\to 0} \frac{s\ (1/s^2)}{1 + G(s)} = \lim_{s\to 0} \frac{1}{sG(s)} \tag{4.7}$$

In this case we define the *static velocity error coefficient* K_v as

$$K_v = \lim_{s\to 0} sG(s) \tag{4.8}$$

Then e_{ss} will be infinite for a type 0 system, finite for a type 1 system and zero for a system of type 2 or higher.

Notice that in steady-state the output of a linear stable feedback system has the same form as the input. In particular if the input is a ramp then e_{ss} will be

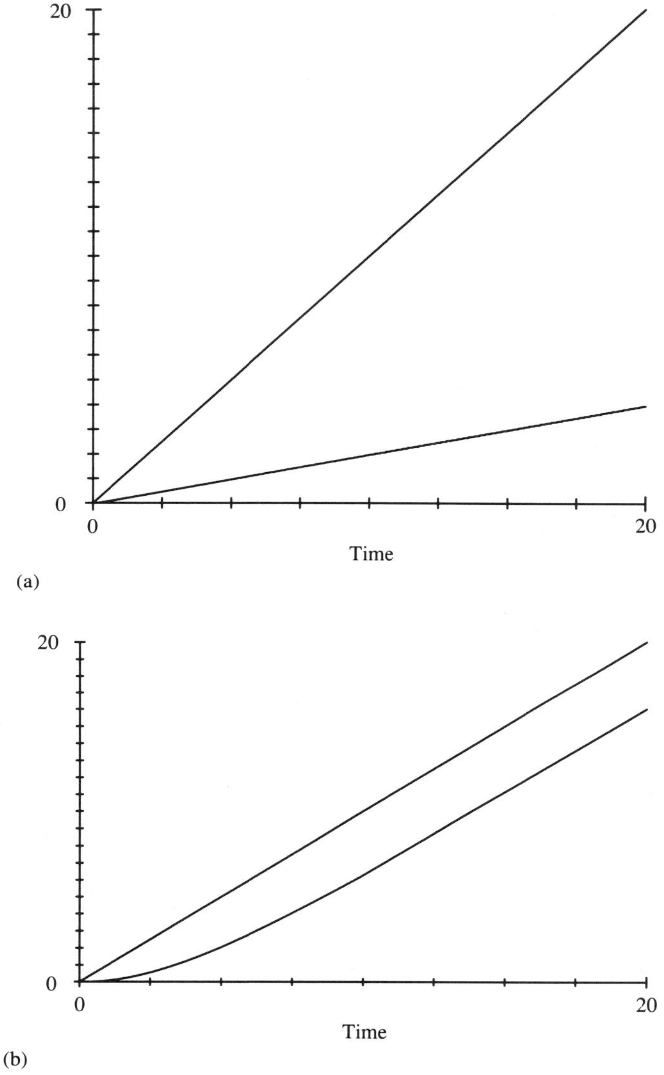

Figure 4.8 Response to a unit ramp of a stable unit feedback system of type: (a) zero; (b) one; (c) two.

infinite if the output converges to a ramp with a slope smaller than the input slope, and it will be finite if it converges to a ramp lagging and parallel to the input as shown in Fig. 4.8. In the case of a parabolic input, $r(t)=at^2$, we define the

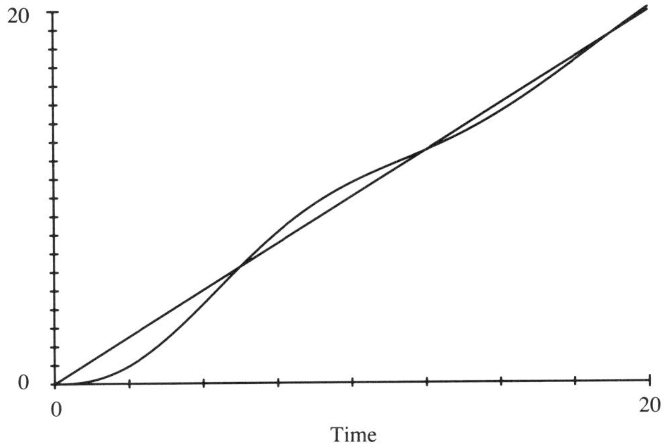

(c)

Figure 4.8 (cont.)

static acceleration error coefficient K_a as

$$K_a = \lim_{s \to 0} s^2 G(s) \tag{4.9}$$

Exercise 4.3 Compute the steady-state errors for a stable unit feedback system of type l, $l=0,1,2,\ldots$ when the reference input is $t^2/2$, i.e the unit parabola (Laplace transform $1/s^3$).

The results of our analysis are summarized in Table 4.1. Adding integrators has a detrimental effect upon the stability, as we have already noticed in Fig. 4.7. This is not hard to accept if we bear in mind that adding an integrator to a controller slows down its response. In general it is not feasible to add more than two integrators and alternative techniques exist to reduce the error in the steady state. The root-locus method discussed in the next section will describe exactly the effect of extra integrators upon the system's stability and transient response.

Table 4.1 Steady-state errors for a stable system

Form of input signal	*Type 0 system*	*Type 1 system*	*Type 2 system*	*Type l system, $l \geq 3$*
Unit step	$\frac{1}{1+K_p}$	0	0	0
Unit ramp	∞	$1/K_v$	0	0
Unit parabola	∞	∞	$1/K_a$	0

4.3 THE ROOT-LOCUS METHOD

In this section we consider the feedback system of Fig. 4.5 with $K(s)$ real and non-negative. Our purpose is to devise a set of rules that enables us to plot the loci of the poles of $C(s)/R(s)$, as K varies from 0 to $+\infty$, without solving any polynomial equation. Then the feedback system will be stable for $K\in[0,\infty]$ if and only if the loci lie strictly inside the left half s-plane. Let us start by analyzing a simple example.

Example 4.1 Assume $H(s)=1$ and $W(s)=1/(s(s+1))$. Then

$$\frac{C(s)}{R(s)} + \frac{K}{(s^2 + s + K)}$$

For $0\leqslant K\leqslant 0.25$ this transfer function has a pair of real poles given by $-1\pm\sqrt{(1-4K)}$ and for $K>0.25$ its poles are complex conjugates of the form $-0.5\pm j\sqrt{((4K-1)/2)}$. This is illustrated in Fig. 4.9, which not only shows that the feedback system is stable for all admissible values of K, but also predicts the closed-loop behaviour. For example, we see that the step response will increase monotonically for $K\in[0,0.25]$ and become oscillatory for $K>0.25$.

Some general features of these loci are already apparent in Fig. 4.9: the loci are symmetrical with respect to the real axis, they start at the open-loop poles and leave the real axis at certain points, breakaway points, where K reaches a relative maximum as a function of s, and s real. This function is defined by $s^2+s+K=0$ and, as illustrated in Fig. 4.10, for s outside $[-1,0]$ there are no positive values of K that satisfy this equation.

Exercise 4.4 For $K\in[0,\infty)$ plot the loci of the poles of the closed-loop system transfer function of Fig. 4.5 with

$$H(s) = (s + 2),\ W(s) = \frac{1}{(s + 1)}$$

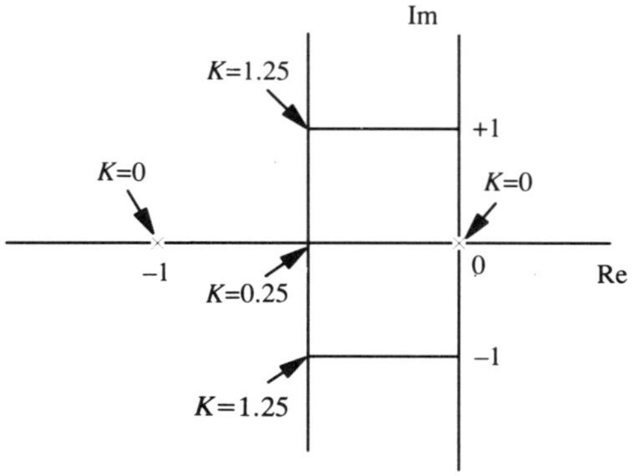

Figure 4.9 Loci of the poles of the closed loop transfer function in example 4.1.

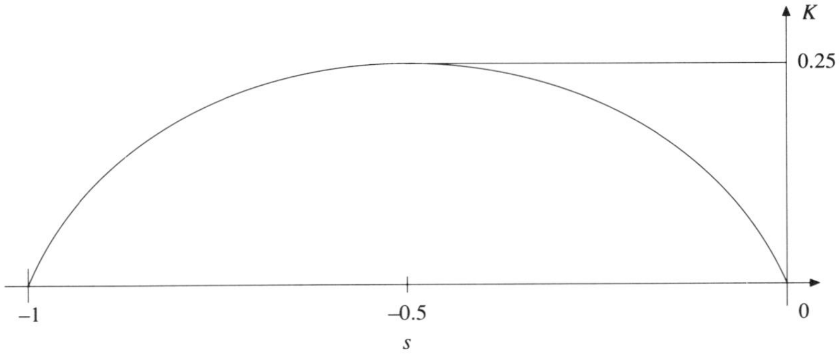

Figure 4.10 Plot of K as a function of s, s real, in example 4.1.

4.3.1 Construction of the Root-loci

Let $N(s)$ and $D(s)$ denote the numerator and denominator polynomials of $W(s) H(s)$, i.e.

$$W(s)\ H(s) = \frac{N(s)}{D(s)} = \frac{\prod_{i=1}^{n} (s - z_i)}{\prod_{i=1}^{d} (s - p_i)} \tag{4.10}$$

where $d \geqslant n$. Then from the closed loop transfer function

$$\frac{C(s)}{R(s)} = \frac{KW(s)}{1 + K\ W(s)H(s)} \tag{4.11}$$

we conclude that the characteristic equation

$$1 + KW(s)H(s) = 0 \tag{4.12}$$

has d roots; in fact this equation is equivalent to $D(s)+K\ N(s)=0$ where the left hand side is a polynomial (the characteristic polynomial) of degree d, because deg. $D \geqslant$deg. N. Furthermore the roots of this equation vary continuously with K. Hence:

> *Rule 1* The number of root-loci is equal to the number of poles of the open-loop transfer function.
>
> *Rule 2* The loci are continuous curves.

These rules are illustrated by the previous example. From (4.12) and (4.10) we can see a point s_0 belonging to the roots-loci is such that

$$\frac{\prod_{i=1}^{n} (s_0 - z_i)}{\prod_{i=1}^{d} (s_0 - p_i)} = \frac{-1}{K} \tag{4.13}$$

This shows that for very small values of K, s_0 must be near a pole p_i; for very large values of K, s_0 is either close to a zero z_i or $|s_0|$ must be very large. If we define the start of the loci at the points at which $K=0$ we have the following rule:

> *Rule 3* The root-loci begin at the poles of the open-loop transfer function and end either at its zeros or at infinity.

The closed-loop poles, being the roots of a polynomial with real coefficients, the characteristic polynomial, occur in conjugate pairs when complex. Therefore

> *Rule 4* The root-loci are symmetrical with respect to the real axis.

From (4.12) we conclude that a point s_0 on the root-loci must also satisfy the following angle condition:

$$K\, W(s_0)\, H(s_0) = (2l + 1)180°,\ l = 0, \pm 1, \pm 2, \ldots \tag{4.14}$$

which is equivalent to

$$\sum_{i=1}^{n} (s_0 - z_i) - \sum_{i=1}^{d} (s_0 - p_i) = (2l + 1)\ 180° \tag{4.15}$$

If s_0 belongs to the real axis then the angular contribution of a pair of complex (conjugate) zeros or poles in (4.15) is zero; real poles or zeros situated to the left of s_0 also contribute with zero arguments. Hence we have

> *Rule 5* A point s_0 on the real axis belongs to the root-loci if and only if the number of open-loop poles and zeros on the real axis, to the right of s_0, is odd.

Example 4.2 Let $H(s)\,W(s)=(s+2)/(s(s+1))$ in Fig. 4.5. With the help of the above rules we are already in a position to plot the trajectories of the poles of $C(s)/R(s)$ as K varies from 0 to ∞. From rule 5 we can see $(-\infty,-2]$ and $[-1,0]$ belong to the loci. Rule 3 states they start at $(-1+j0)$ and $(0+j0)$ and end at $(-2+j0)$ and $(-\infty+j0)$. Because the loci are continuous curves (cf. rule 2) they must leave and re-enter the real axis somewhere in the intervals $(-1,0)$ and $(-\infty,-2)$ respectively. The result is shown in Fig. 4.11. A comparison with Fig. 4.9 reveals that the addition of a zero at -2 to the open-loop transfer function was beneficial, since a gain increase now moves the closed-loop poles away from the right half-plane. Unlike the poles at the origin, the addition of zeros has a beneficial effect upon the system's stability. The next rule tells us how to compute the coordinates of b_0 and b_1 in Fig. 4.11, i.e. break-away and break-in points on the real axis.

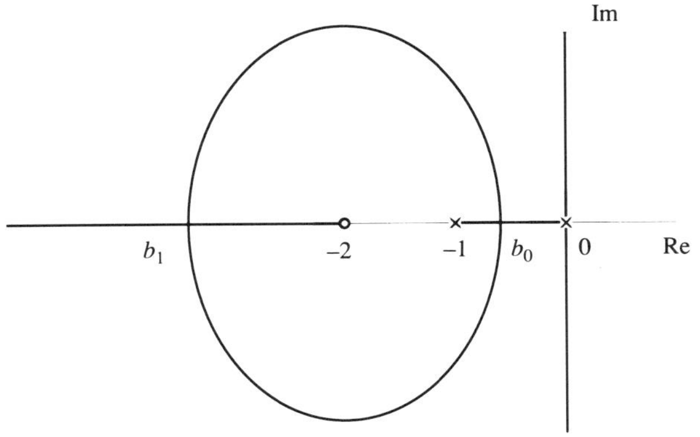

Figure 4.11 Root-loci for example 4.2; $b_0=-0.586$ and $b_1=-3.414$.

Rule 6 The points where the loci leave or re-enter the real axis are points where K, regarded as a function of s and s real, reaches a local maximum or minimum respectively.

This property has already been discussed in example 1 and is illustrated in Fig. 4.10. The coordinates of break-away and break-in points can therefore be obtained by setting $\mathrm{d}K/\mathrm{d}s$ equal to zero. From the characteristic equation in the previous example we have

$$K = \frac{-\,s(s+1)}{(s+2)} \tag{4.16}$$

Then $\mathrm{d}K/\mathrm{d}s=0$ gives $s\in\{-3.414, -0.586\}$, i.e. $b_0=-0.586$ and $b_1=-3.414$ in Fig. 4.11. From (4.16) we can see the values of K at the break-away and break-in points are 0.172 and 5.828 respectively. This means that for $0\leqslant K\leqslant 0.172$ and $K>5.828$ the poles of the closed-loop transfer function are real.

The next rule is useful when the open-loop transfer function has complex poles or zeros, because it gives us the direction in which the locus leaves the complex pole (angle of departure) or enters the complex zero (angle of arrival). Imagine an open-loop transfer function with poles and zero as shown in Fig. 4.12. Consider also a small ball centred at p_2 of radius r much smaller than the minimum of $\{|p_2-z|, |p_2-p_3|, |p_2-p_1|\}$. Then for a point s inside this ball we can replace, with negligible error, $(s-z)$, $(s-p_3)$ and $(s-p_1)$ by (p_2-z), (p_2-p_3) and (p_2-p_1) respectively. In particular, if s belongs to the locus starting at p_2, from (4.15) we can write

$$\angle(s-p_2) \cong (2l+1)180° + \angle(p_2-z) - \angle(p_2-p_1) - \angle(p_2-p_3) \tag{4.17}$$

which constitutes a fair approximation of the angle of the tangent to the locus at p_2. This result is stated in the following rule:

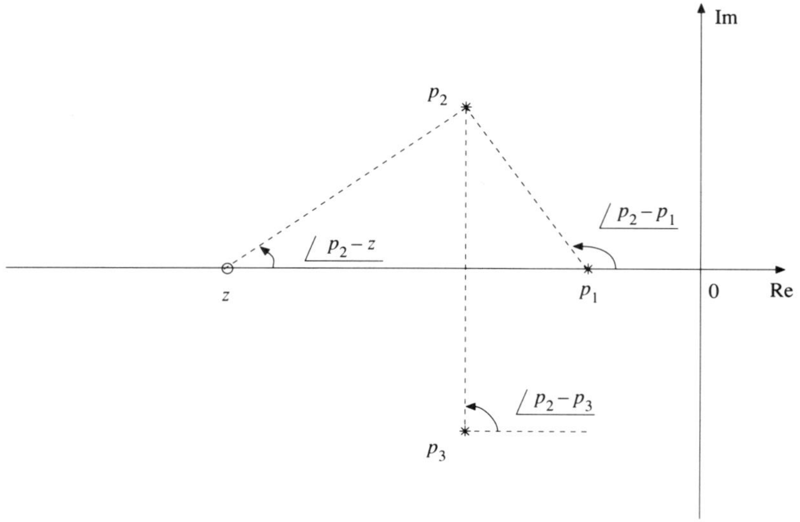

Figure 4.12 Angle of departure at a complex pole.

Rule 7 If the real and imaginary axes have identical scales, then the angle of the tangent to the root locus at an open-loop pole or zero can be obtained by adding up, according to the angle condition (4.15), the angles from all the other poles and zeros to the pole or zero in question and adding 180° to this sum.

We stress the fact that the above rule was obtained on the assumption that the real and imaginary axis are drawn on the same scale. Before considering the next rule, let us look at the following example.

Example 4.3 Consider $H(s)=1$ and $W(s)=1/(s-p)^m$ in Fig. 4.5. We wish to plot the root-loci for $m=1,2,3,4$. The rules studied so far are not very helpful in this case and we must go back to the definitions. Let $(s-p)=r \exp(j\theta)$; then

$$\underline{/K\, H(s)\; W(s)} = -\, m\theta$$

and

$$|KH(s)\; W(s)| = \frac{K}{r^m}$$

For a point $s_0=r_0 \exp(j\theta_o)$ on the roots-loci the angle condition (4.14) implies

$$-\, m\theta_o = (2l + 1)180°$$

i.e. the root-loci are made of straight lines intersecting at p with angles to the real axis given by $(2l+1)\,180°/m$, $l=0,\pm1,\pm2,\ldots$ These are represented in Fig. 4.13 for m up to 4.

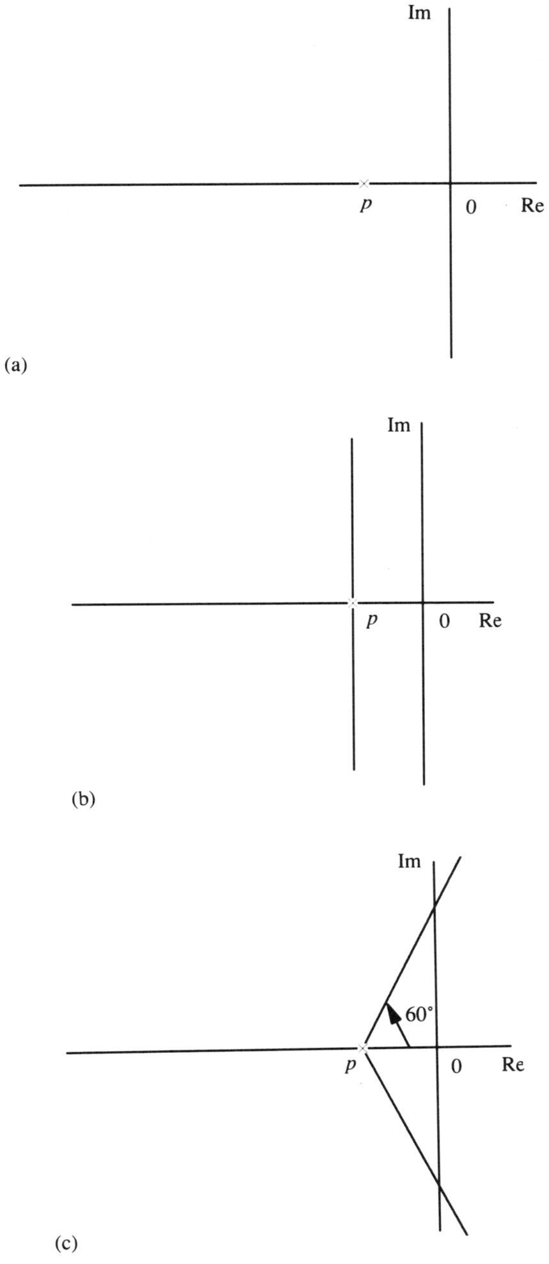

Figure 4.13 Root-loci for example 4.3 with (a) $m=1$, (b) $m=2$, (c) $m=3$, (d) $m=4$.

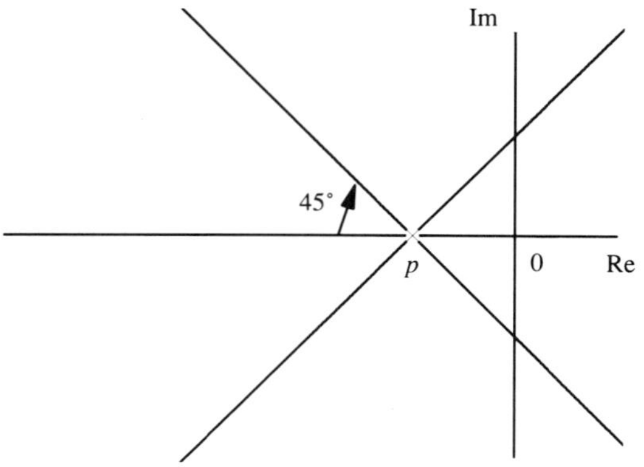

(d)

Figure 4.13 (cont.)

When the open-loop transfer function has more poles than zeros there will be $(d-n)$ loci ending at infinity. From (4.10) we have for $|s|$ very large

$$W(s)\ H(s)\ \cong \frac{1}{s^{d-n}}$$

which shows that for large values of the roots the loci can be replaced by those of example 4.3 with $p=0$ and $m=d-n$. A better approximation can be obtained as follows. From (4.10) we can write

$$KW(s)\ H(s) = K\frac{s^n - \left(\sum_{i=1}^{n} z_i\right)s^{n-1} + \ldots + \prod_{i=1}^{n}(-\ z_i)}{s^d - \left(\sum_{i=1}^{d} p_i\right)s^{d-1} + \ldots + \prod_{i=1}^{d}(-\ p_i)}$$

Dividing both terms by the numerator we get

$$K\ W(s)\ H(s) = \frac{K}{s^{d-n} + \left(\sum_{i=1}^{d} p_i - \sum_{i=1}^{n} z_i\right)s^{d-n-1} + \ldots}$$

But

$$(s - p)^{d-n} = s^{d-n} - (d - n)ps^{d-n-1} + \ldots + (-\ p)^{d-n}$$

This shows that the $(d-n)$ branches of the loci ending at infinity can be replaced, for large values of $|s|$, by the loci in example 4.3 with $m=d-n$ and

$$p = \frac{\left(\sum_{i=1}^{d} p_i - \sum_{i=1}^{n} z_i\right)}{d - n}$$

This leads to

Rule 8 For large values of the roots, $(d-n)$ loci are asymptotic to straight lines with angles to the real axis given by $(2l+1)\ 180°/\ (d-n)$ and intersecting the real axis at the point

$$\sigma = \frac{\left(\sum_{i=1}^{d} p_i - \sum_{i=1}^{n} z_i\right)}{d - n} \tag{4.18}$$

When the loci cross the imaginary axis the closed loop system changes from stable to unstable (or vice versa); the Routh–Hurwitz criterion is particularly suited to computing the value of K at these points as shown below.

The characteristic equation in example 4.3, for $d=3$, is $s^3-3ps^2+3p^2s-p^3+K=0$, leading to the Routh array:

$$\begin{vmatrix} 1 & 3p^2 \\ -3p & -p^3+K \\ (+8p^3+K)/3p & \\ -p^3+K & \end{vmatrix}$$

The stability conditions are $p<0$, $K>p^3$ and $8p^3+(K/3p)>0$. If $p<0$, the third inequality becomes $K<-8p^3$. If $p=-1$, for example, the closed-loop system is stable for $-1<K<8$; the points of intersection with the imaginary axis occur for $K=8$ which implies $r^3=8$, i.e. the loci intersect the imaginary axis at $0\pm j\sqrt[3]{8}\sin 60°=0\pm j\sqrt{3}$. Hence

Rule 9 When loci cross the imaginary axis, the crossover points and the gain at these points can be determined by the Routh–Hurwitz criterion.

Exercise 4.5 Consider $H(s)=1$ and $W(s)=1/s(s+1)(s+2)$

(a) Compute the angles and the intersection of the asymptotes with the real axis.
(b) Plot the root-loci.
(c) Compute the crossover points and the gain at these points.
(d) Compute the break-away point.

Answers: (a) $\pm 60°$ and $180°$; $\sigma=-1$, (b) $0\pm j\sqrt{2}$ and $K=6$, (c) -0.423.

The root-loci for the above exercise are shown in Fig. 4.14.

A set of more elaborate rules could be devised, particularly for break-away and break-in points outside the real axis. However, their practical usefulness is questionable; when the loci become very involved it is advisable to use a computer to do the job.

4.4 ANALYSIS OF TIME-DELAY FEEDBACK SYSTEMS

A plant whose output y is a time-shifted version of its input u, i.e. $y(t)=u(t-T)$ is called a *pure time delay* (or transport delay) where T is the delay time. The typical response of such a plant is shown in Fig. 4.15; obviously, the transfer function of this plant must be e^{-Ts}. The reader has already been introduced to some of the problems arising when plants with transport delays are inserted in a feedback loop (see Figs 1.6–1.8): such a feedback system has a tendency to oscillate and can

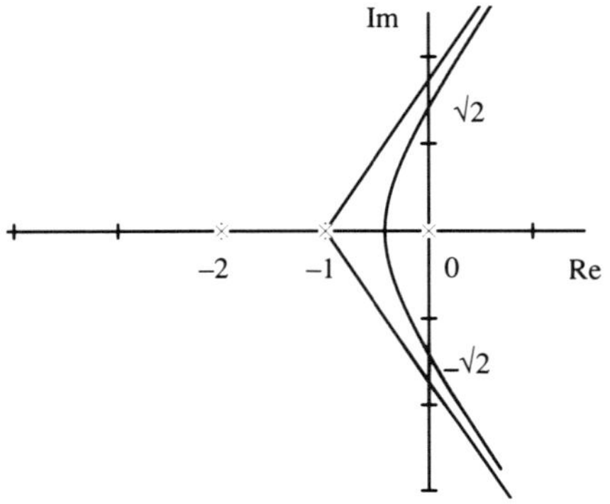

Figure 4.14 Roots-loci for exercise 4.5.

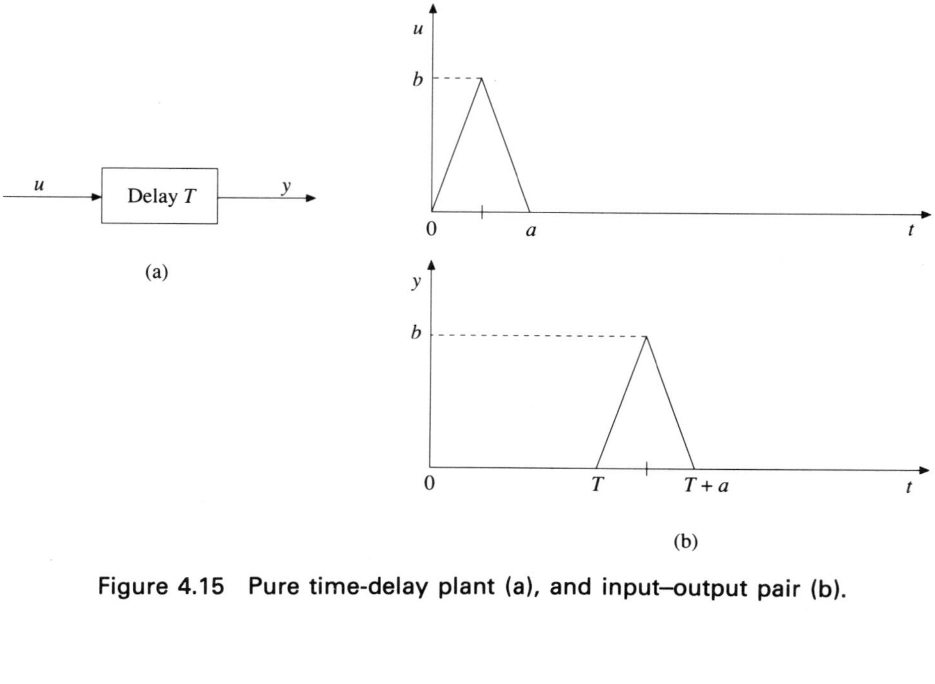

Figure 4.15 Pure time-delay plant (a), and input–output pair (b).

easily become unstable. In this section we will analyze the transient behaviour of a few simple examples.

We start by analyzing the unit impulse response of the system in Fig. 4.16. If $r(t)=\delta(t)$, i.e. the Dirac δ function applied at time $t=0$, then we have at time $t=T$ a Dirac impulse with area K in c, and a Dirac impulse with area $-K^2$ in m; at time $t=2T$ we have a Dirac impulse with area $-K^2$ in c and so on. The result is shown in Fig. 4.17 for several values of the parameter K and can be expressed analytically by

$$y(t) = \sum_{l=1}^{\infty} (-K)^l \, \delta(t - lT) \tag{4.19}$$

An important conclusion is that a loop gain of 1 sustains an oscillation whose period is twice the dead time T, or equivalently, with frequency π/T.

Exercise 4.6
(a) Show that with a loop gain of 0.5 we get a 'quarter-amplitude damping'.
(b) Show that a sustained sinusoidal oscillation $m(t)=A \sin \pi/T\, t$ is also possible in the system of Fig. 4.16 for $K=1$, $r=0$ and A any positive constant.
Hint: notice that at this frequency a 180° phase shift takes place in the time-delay element.

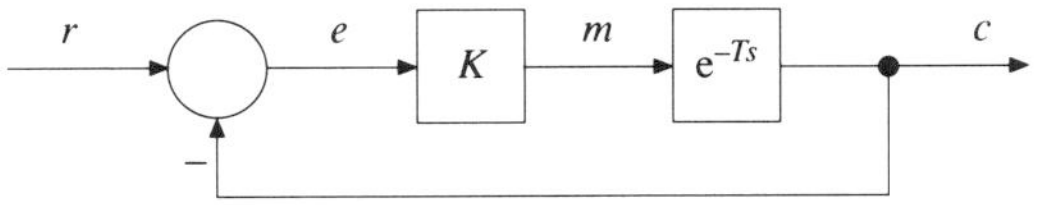

Figure 4.16 Pure time-delay plant under feedback control.

Formula (4.19) also expresses the application of the superposition principle to our (linear) system. From Fig. 4.17 we can easily obtain the step response; in fact the unit step being the integral of the unit impulse and the system being linear, the step response will be the integral of the impulse response. Figure 4.18 shows the unit-step response for $K<1$ and $K=1$, with the former converging to $0.8/(1+0.8)=0.444$.

Exercise 4.7 Show that $c(t)$ in Fig. 4.18, for $lT \leq t < (l+1)T$ is given by

$$\frac{K(1 - (-K)^l)}{(1 + K)}$$

Hint: recall that $s_n=r+r^2+...+r^n$ is given by $S_n=r(1-r^n)/(1-r)$.

From this simple analysis, we can conclude that our feedback system is unstable for $K>1$. This conclusion can also be reached from the closed-loop transfer function:

$$\frac{C(s)}{R(s)} = \frac{K\,e^{-Ts}}{1 + K\,e^{-Ts}} \tag{4.20}$$

as shown in the following.

Exercise 4.8

(a) Show (4.20) is the Laplace transform of (4.19).

(b) From (4.20), and assuming the system is stable, show that

$$\lim_{t\to\infty} c(t) = \frac{K}{1 + K}$$

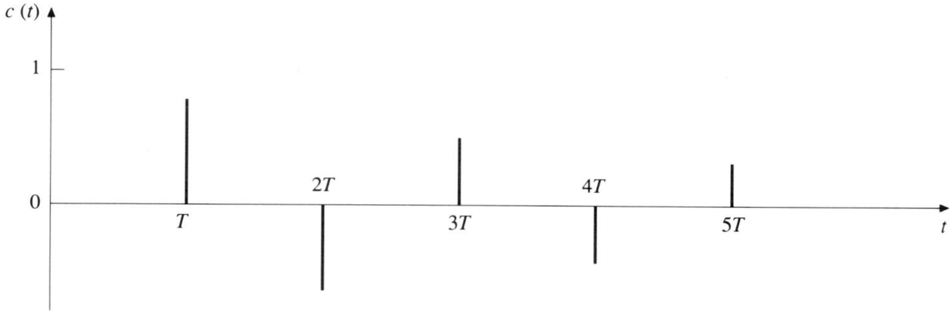

(a)

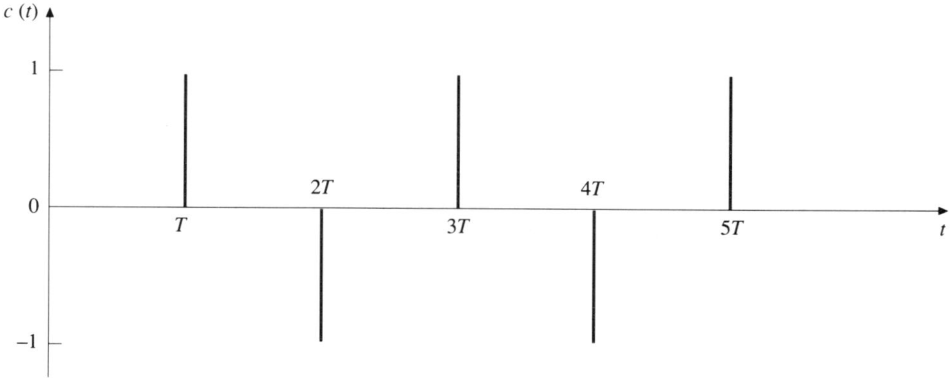

(b)

Figure 4.17 Unit impulse responses of the system in Fig. 4.16 with (a) K=0.8; (b) K=1; (c) K=1.2.

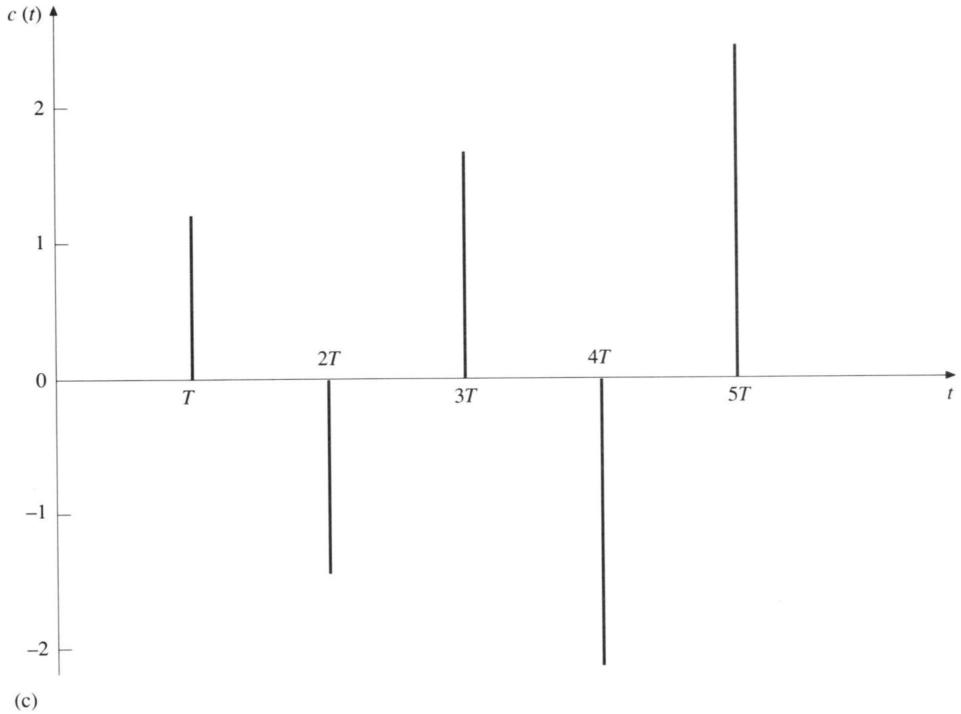

Figure 4.17 (cont.)

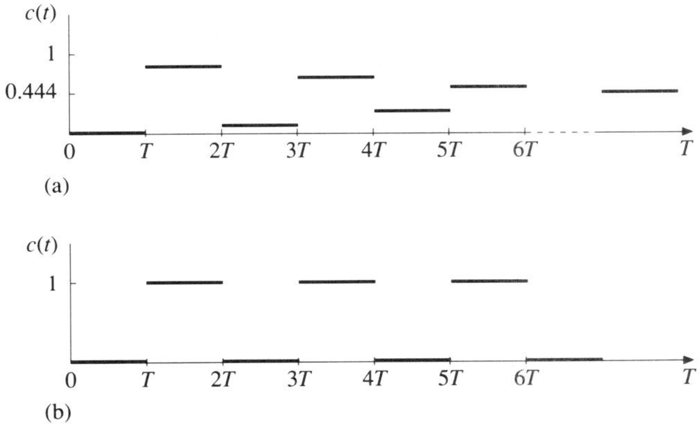

Figure 4.18 Unit step responses of the system of Fig. 4.16 with (a) $K=0.8$; (b) $K=1$.

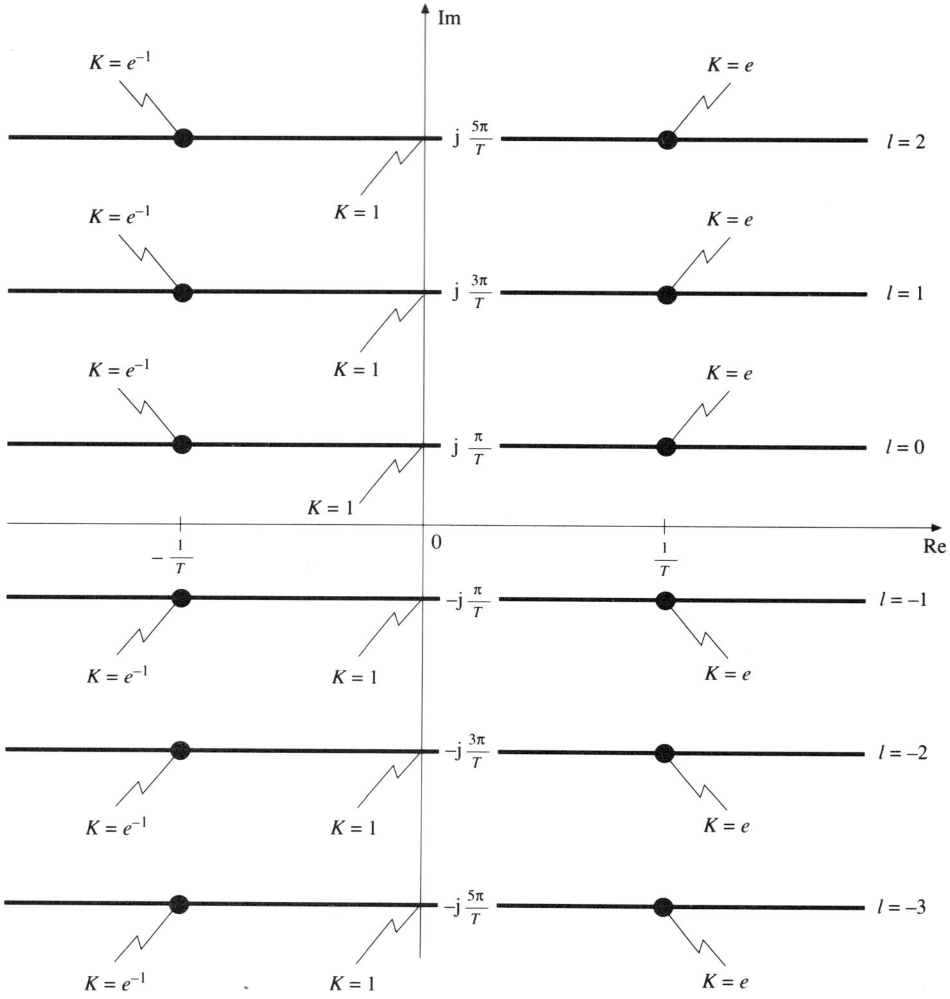

Figure 4.19 Root-loci of the system shown in Fig. 4.16.

when $r(t)$ is the unit step.

(c) Compare this limit with

$$\lim_{l \to \infty} \frac{K(1 - (-K)^l)}{1 + K}$$

in exercise 4.7.

In contrast with those encountered so far, transfer function (4.20) is not a rational function. It has an infinite number of poles of the form

$$s = \frac{1}{T}[\ln K + j(2l + 1)\pi],\ l \text{ integer} \tag{4.21}$$

The loci, graduated in terms of K, are shown in Fig. 4.19; for $K<1$ all the poles are in the left half s-plane, i.e. the system is stable. For $K=1$ the poles are on the imaginary axis. The corresponding step response oscillates with angular frequency π/T which is precisely the modulus of the first pair of poles. Notice that all the branches start at the open-loop pole $(-\infty+j0)$ as expected.

After this example, the reader may have realized that, as a consequence of the superposition principle, a linear time-delay feedback system can be solved, for each time segment $(lT,(l+1)T)$, as a sequence of open-loop problems with appropriate inputs. With this result in mind, we now analyze the transient behaviour of the plant with transfer function $\exp(-Ts)/(s+p)$, $p\in\mathbb{R}$, which is of great interest in process control studies. Step responses of such a system have already been shown in Fig. 1.8 with $p=1$ and the plant under proportional control; the salient feature was a less damped response for increasing gains and eventually instability. If $r(t)$ is the unit impulse then we have from Fig. 4.20:

$$C(s) = \frac{K\,e^{-Ts}}{(s+p)} - \frac{K^2\,e^{-2Ts}}{(s+p)^2} + \frac{K^3\,e^{-3Ts}}{(s+p)^3} + \dots - \left(\frac{-K}{(s+p)}\right)^l e^{-lTs} + \dots \tag{4.22}$$

Inverting, we get

$$\begin{aligned} c(t) = {} & K\,e^{-p(t-T)}\,u(t-T) - \frac{1}{1!}K^2\,(t-2T)\,e^{-p(t-2T)}\,u(t-2T) \\ & + \frac{1}{2!}K^3\,(t-3T)^2\,e^{-p(t-3T)}\,u(t-3T) \\ & - \dots - \frac{1}{(l-1)!}(-K)^l(t-lT)^{l-1}e^{-p(t-lT)}\,u(t-lT) + \dots \end{aligned} \tag{4.23}$$

where $u(t)$ denotes the unit step:

$$u(t) = \begin{cases} 0, & t<0 \\ 1, & t\geq 0 \end{cases}$$

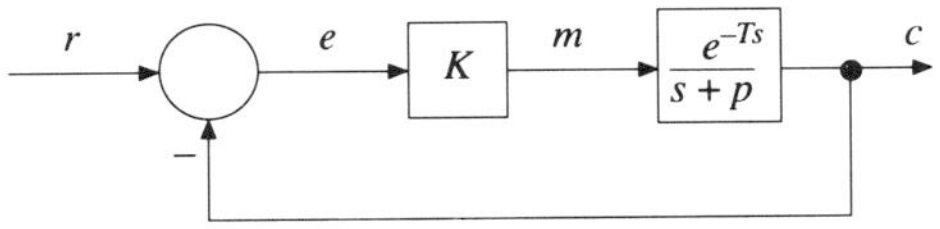

Figure 4.20 A first-order process with pure-transport delay under feedback control.

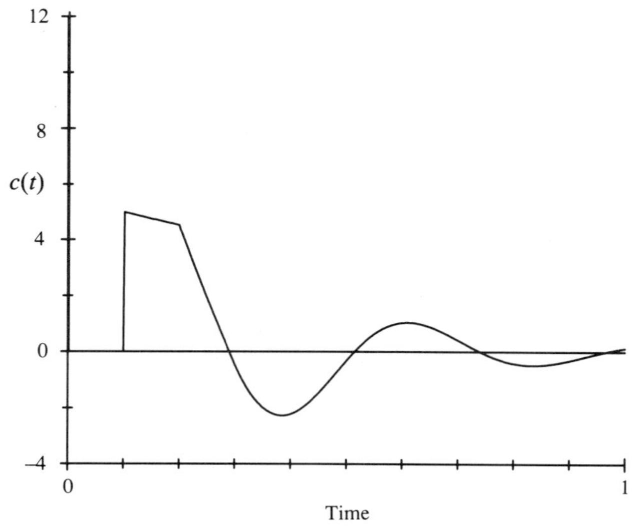

Figure 4.21 Unit impulse response of the system shown in Fig. 4.20 with $K=10$, $T=0.1$ and $p=1$.

A plot of $c(t)$ for $p=1$, $K=10$ and $T=0.1$ is represented in Fig. 4.21; the integral of this curve is shown in Fig. 1.8(a) for $r=1$. The Laplace transform of $c(t)$, given by (4.22), can also be obtained from the closed-loop transfer function

$$\frac{C(s)}{R(s)} = \frac{(K\mathrm{e}^{-Ts})/(s+p)}{1+(K\mathrm{e}^{-Ts})/(s+p)} \tag{4.24}$$

Recalling the identity $(1+a)^{-1} = 1-a+a^2-a^3+a^4+\ldots$, for $|a|<1$, and the fact that $R(s)=1$, we get from (4.24):

$$C(s) = \frac{K\,\mathrm{e}^{-Ts}}{s+p}\left(1 - \frac{K\,\mathrm{e}^{-Ts}}{s+p} + K^2\frac{\mathrm{e}^{-2Ts}}{(s+p)^2} - K^3\frac{\mathrm{e}^{-3Ts}}{(s+p)^3} + \ldots\right)$$

which is precisely (4.22).

This system has also an infinite number of poles, whose calculation is now discussed. The poles of (4.24) are complex numbers $s=x+\mathrm{j}y$ such that $K(\mathrm{e}^{-Ts})/(s+p)=-1$; hence *equating moduli*

$$|s+p| = K\,\mathrm{e}^{-Tx} \tag{4.25}$$

and *equating arguments*:

$$(s+p) = (2l+1)\pi - Ty, \quad l \text{ integer} \tag{4.26}$$

For each value of l there will be a set of loci. In particular for $l=0$ we can see all points of the real axis to the left of $(-p)$ belong to the loci because (4.26) is

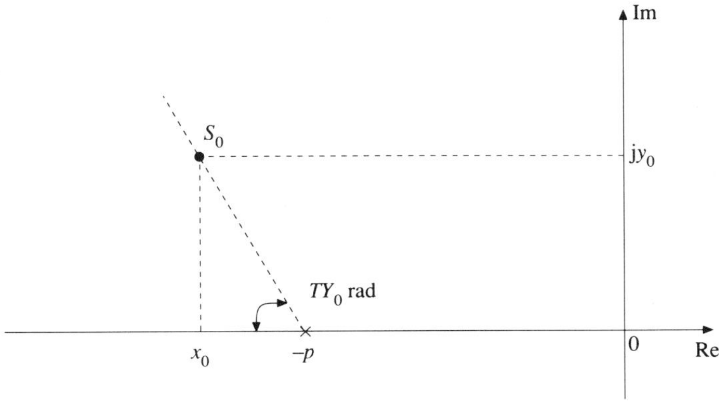

Figure 4.22 Construction of the loci for system in Fig. 4.20.

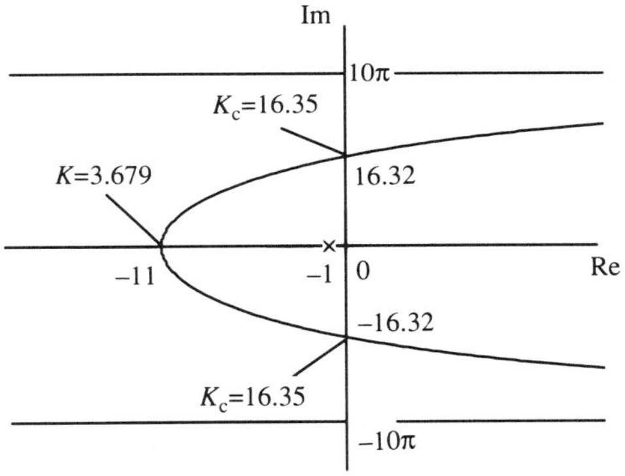

Figure 4.23 Root-loci of the system shown in Fig. 4.20, for $l=0$ in (4.26), $p=1$ and $T=0.1$.

satisfied; but there must also be branches outside the real axis, because $y=\pm\pi/T$ is a solution when $x\to\infty$. Such loci can easily be determined from (4.26): for example, assuming $l=0$ and $y_0\in[-\pi/T,\ \pi/T]$ then x_0 must lie on the line containing $(-p)$ and at an angle $(\pi-y_0T)$ rad with the real axis, as shown in Fig. 4.22; using this rule we get the loci shown in Fig. 4.23; the point of intersection with the imaginary axis, jy_c, was calculated from the equation $\tan(\pi-y_0T)=y_c/p$; the gain at this point, K_c, is computed from (4.25) with $x=0$.

From the characteristic equation, we get $K=-(s+p)\,e^{Ts}$; the break-away point is the solution of the equation $dK/ds=0$, i.e. $s=-11$. For $K>16.35$ the system is therefore unstable; however, to make sure it is stable for $K<16.35$ we must look at the other branches of the root-loci. In fact the system is stable for $K<16.35$ as the following argument shows: for a given non-zero integer l_0 we have

$(\infty+j(2l_0+1)(\pi/T))$ and $(-\infty+j2l_0(\pi/T))$ are solutions of (4.26); consequently the loci will evolve as shown in Fig. 4.24, for $l_0 \in \{-2,-1,0,1,2\}$, the gain at the point jy_{l_0} (intersection with the imaginary axis) is given by $|jy_{l_0}+p|$ and therefore increases with l_0. Then the system is stable for $K<16.35$. As expected, the addition of a time constant makes the system slower; in fact, while in Fig. 4.16 the system was capable of a sustained oscillation of $\pi/T=31.42$ rad/s, now this frequency has been lowered to 16.32 rad/s as shown in Fig. 4.23.

Exercise 4.9 Plot the root-locus of the system shown in Fig. 4.20 with $T=0$, i.e. no time delay. What is the effect of a gain increase upon the system transient response when $T=0$? And when $T>0$?

Exercise 4.10 Assume $p=0$ in Fig. 4.20, i.e. a plant formed by a pure delay and a capacity. Then

(a) Redraw the loci of Fig. 4.23.

(b) Assuming

$$m(t) = A \sin\frac{\pi t}{2T}$$

compute $\int m(t)dt$.

(c) Show that a sustained oscillation with $m(t)$ as in (b) is possible provided $K=\pi/(2T)$. What is the loop gain in this case?

Answers: (a) Similar to Fig. 4.23, starting at $-\infty+j0$ and $0+j0$ and intersecting the imaginary axis at $\pm j(\pi/2T)$; at this point $K=\pi/(2T)$; (b) $A(2T/\pi)\sin(\pi(2T)^t-\pi/2)$; (c) one.

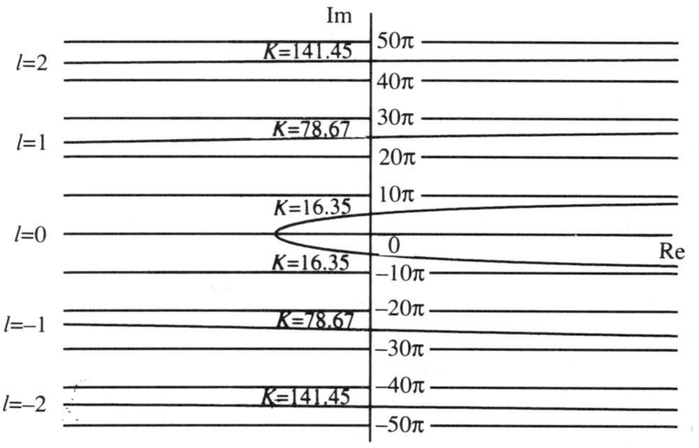

Figure 4.24 Root-loci of the system in Fig. 4.20 for $p=1$ and $T=0.1$.

4.5 FEEDBACK CONTROL MODES

The archetypal configuration of a feedback control system was introduced in Chapter 1 (cf. Fig. 1.4), and throughout the text the reader has been introduced to some of the most common control modes. These are:

1. on–off (or two-position)
2. proportional
3. integral
4. derivative

The first is a non-linear action and a typical application has already been analyzed in Chapter 1 (Figs 1.2 and 1.3); because the output of an on–off controller has only two possible values, it can only produce extreme values for the manipulated variable, i.e. too high or too low; as a result, cycling occurs when such a controller is inserted in a feedback loop. Because of its simplicity, an on–off controller is the cheapest approach to feedback control; its applicability is only limited by the desired accuracy, i.e. the maximum deviation allowable for the controlled variable. (When studying non-linear systems we shall see that any loop whose gain varies inversely with amplitude – the on–off controller is a typical example – is prone to oscillations.)

Proportional, integral and derivative control modes are linear control actions which are implemented in the majority of commercial controllers according to the following law:

$$m(t) = K(e + \frac{1}{T_i} \int_0^t e(t')\mathrm{d}t' - T_d \frac{\mathrm{d}c}{\mathrm{d}t}) \tag{4.27}$$

which is known as three mode or PID controller. The user can modify the dynamic properties of this controller by acting on the (adjustable) parameters K, T_i and T_d.

Exercise 4.11 Given a PID controller characterized by (4.28), in series with a second-order plant with transfer function

$$G_p(s) = \frac{K_p}{(s + p_1)(s + p_2)}$$

show how we can make the closed-loop system look like a first order one. Is this beneficial from a practical viewpoint?
Hint: draw the root-locus when the controller zeros coincide with the plant poles.
Answer: $(p_1+p_2)=1/T_d$; $p_1 p_2=1/(T_i T_d)$. In this case p_1 and p_2 remain poles of the load transfer function C/D (cf. Fig. 4.37). Therefore cancellation may cause problems if p_1 and p_2 are slow poles.

4.5.1 Proportional Control

If $T_d=0$ and $T_i=\infty$ we have a *proportional* controller; its effect upon the steady-state and transient behaviour of a feedback system has already been discussed in sections 4.2 and 4.3. In essence, we have shown that this control mode is unable to eliminate the steady-state error, or *offset*, when controlling type 0 (or self-regulating) plants; it was also shown that the magnitude of the offset was a function of the load (cf. Fig. 4.6) and that the offset could be reduced by a gain increase; however, this gives rise in general to an increase in settling time and eventually instability, as revealed by root-loci analysis. In practice there are limitations to the size of controller input and output, say $|m(t)|\leqslant M/2$ and $|e(t)|\leqslant E/2$, for all values of t. Such a restriction immediately implies that the (linear) models used so far are only valid when e and m are inside these ranges. Another consequence is, for example, that the larger the controller gain K, the narrower the range of allowable variation for e as shown in Fig. 4.25. Therefore a more useful measure of proportional action is given in terms of *proportional band* (PB),which is defined as the percent change of error which will cause a 100% change in controller output; therefore we have $K=100\%/\text{PB}$ where K is the (dimensionless) gain in (4.27); typically $K\in[0.2, 50]$ or equivalently PB $\in[2\%, 500\%]$.

The reader may already have noticed that most of the time it would not be feasible to have negative controller output values; for example this would require a negative supply of heat, i.e. heat removal, in temperature control problems, for example; in fact the variables in our models represent deviations from steady-state

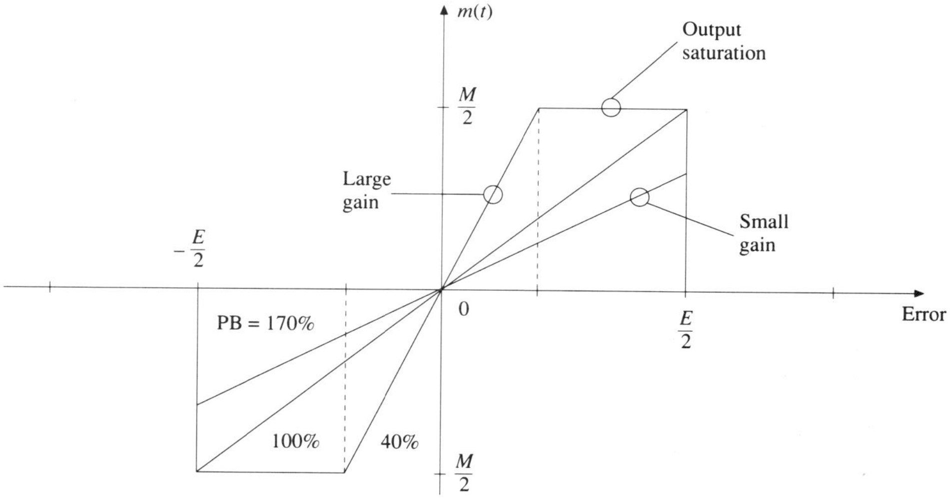

Figure 4.25 Relationship between gain and proportional band of a proportional controller.

operation, in order to simplify the analysis. Regardless of the value of the gain the real output P of the proportional controller is normally biased at a value P_0 to produce 50% of the output range when the error is zero, i.e. $P=100/\text{PB}\ e + P_0$.

4.5.2 Proportional and Integral Control

If only $T_d=0$ in (4.27), we have a PI controller; the integral part produces a controller transfer function with a pole at the origin (type one system) thereby eliminating steady-state errors to constant reference inputs, whatever the plant under control, provided the feedback system is stable. A comparison of Figs 4.6 and 4.7 illustrates the main features of the integral mode: the steady-state error no longer exists, despite the (possible) presence of sustained disturbances, but the system takes longer to reach the steady-state, i.e. the integral action increases the settling time. The combination of proportional and integral modes is advantageous because it reduces the destabilizing effects of the integral mode while retaining the ability to eliminate the offset; if in the system discussed in Figs 4.6 and 4.7, for example, we use instead a PI controller with transfer function $(1+1/s)$, we get the result shown in Fig. 4.26.

The parameter T_i in (4.27) is known as the *integral time* and quantifies the rate at which the output of the PI controller is driven when its input is a step: it is the time required for the contribution of the integral part to equal *or repeat* the contribution of the proportional mode, as shown in Fig. 4.27. Most instrument manufacturers express T_i in units of min/repeats or repeats/min; typically

$$0.02 \text{ repeats/min} \leqslant T_i \leqslant 50 \text{ repeats/min}$$

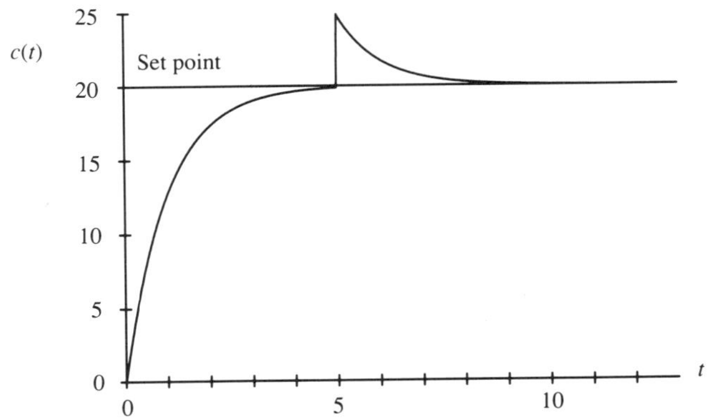

Figure 4.26 Evolution of $c(t)$ in Fig. 4.5 with the same data as Fig. 4.6 except $K(s)=(1+1/s)$.

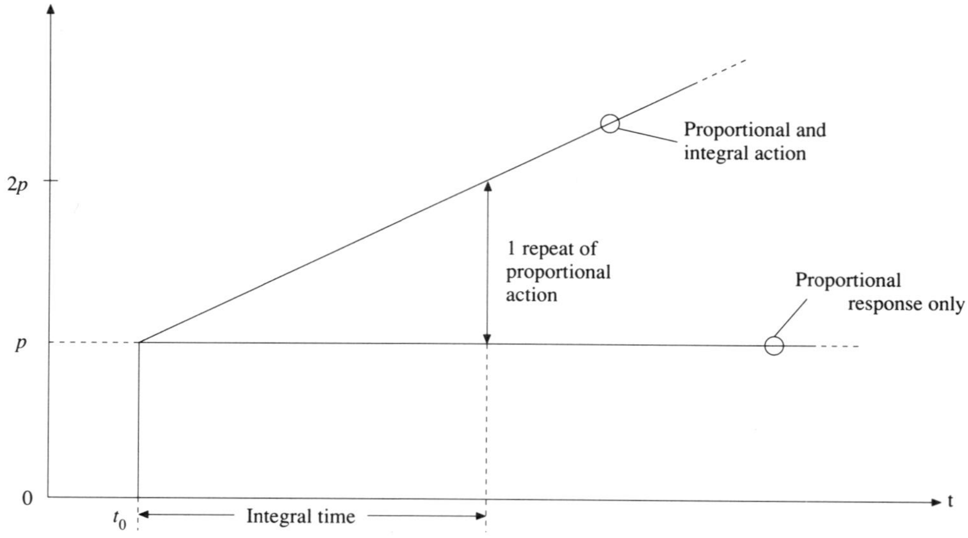

Figure 4.27 Illustration of integral (or reset) time.

4.5.3 Integral Windup

Actuators have a limited range of linear operation and mostly exhibit an input–output characteristic as depicted in Fig. 4.28 (saturation type non-linearity); control valves are typical examples, where the limits correspond to a fully open or a fully closed valve. Therefore, when a controller produces large output values, the use of linear models may lead to the wrong conclusions as a simple analysis of Fig. 4.28 reveals. This is particularly important in batch processes during start-up, where a large set point change normally takes place, or in processes with large load changes (disturbances). In both cases, we have a large initial error and eventually actuator saturation. Consequently the controlled variable will take longer to reach the desired set point; the integral mode, having more time to

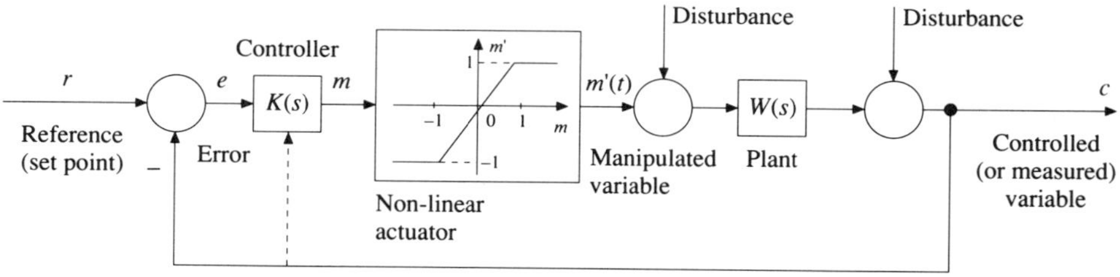

Figure 4.28 Control system with a non-linear (saturation type) actuator.

integrate the error, increases monotonically to a (possibly) very large value (integral windup) until it begins to decrease just before the error changes sign; because of its large initial value it may take a long time for the controller output to change sign, and before this occurs the controlled variable will be increasing. The inevitable result is a large and sustained overshoot. This situation is illustrated by curves (b) of Figs 4.29 and 4.30 which describe the evolution of the system in Fig. 4.28 following a step change, with amplitude 10, in the reference signal. However, if the controller has a device that stops integration once saturation takes place, we get a much better result as shown by curve (c); such a device is known as a *batch unit* because the windup problem is primarily associated with batch processes.

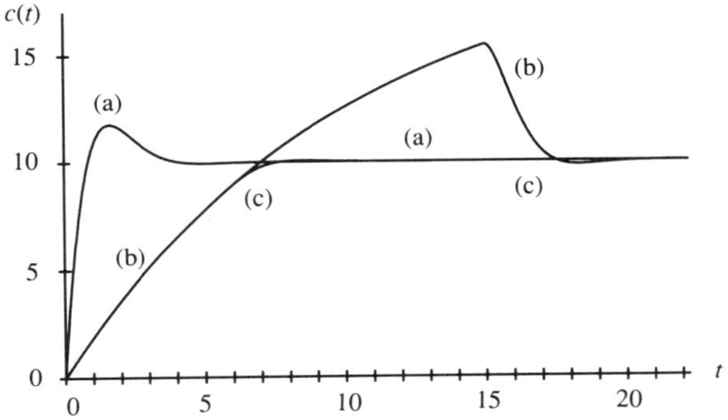

Figure 4.29 Response of system shown in Figure 4.28 to a step input of amplitude 10, with $K(s)=(1+1/s)$; $W(s)=2/(s+0.1)$ and: (a) linear actuator with unit gain; (b) actuator with saturation as shown; (c) controller stopping integration once saturation takes place (batch switch).

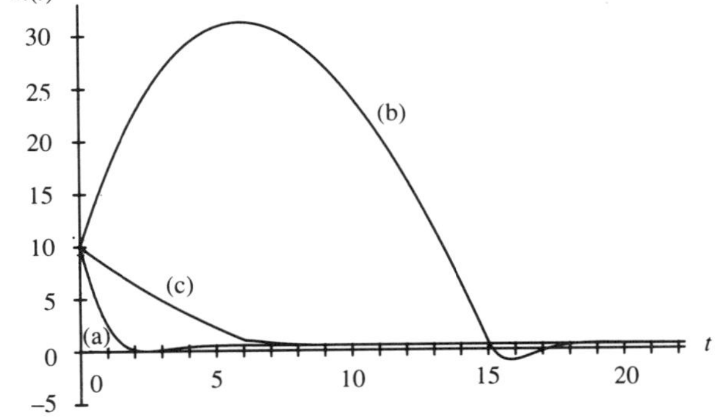

Figure 4.30 Evolution of $m(t)$ for the example shown in Fig. 4.29.

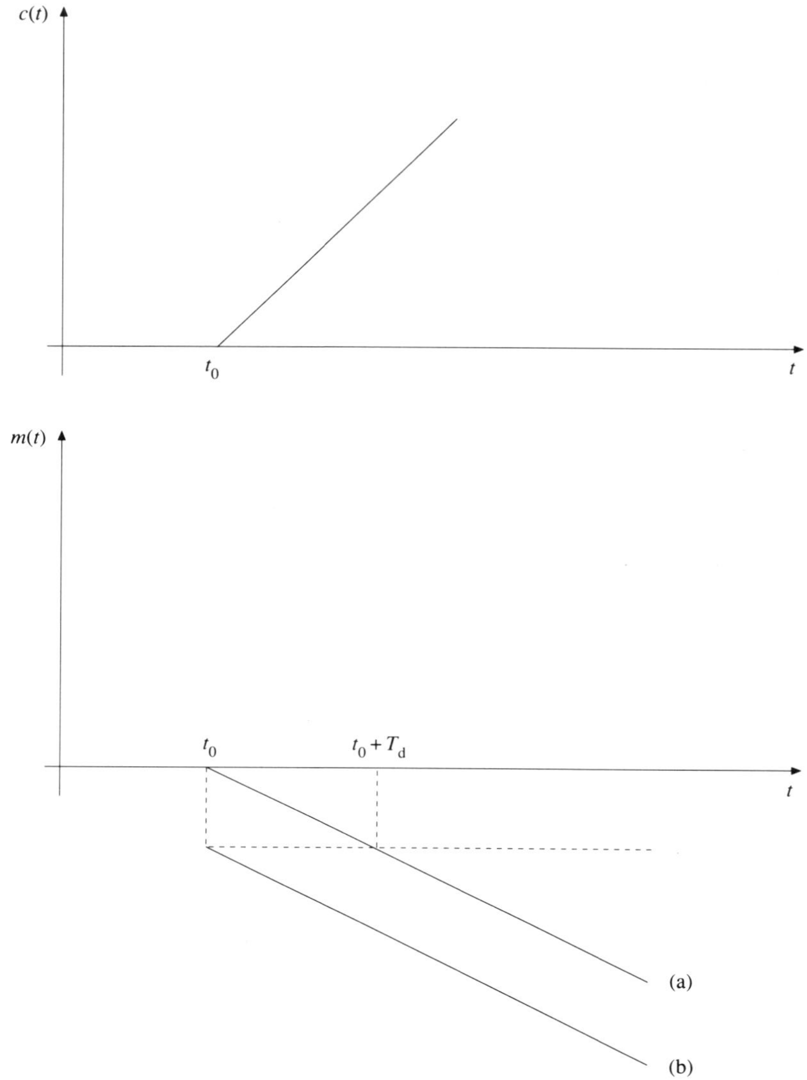

Figure 4.31 Open-loop response of a controller with: (a) proportional mode only; (b) proportional and derivative mode.

Nowadays process controllers have built-in simple logic algorithms, in addition to the traditional PID continuous algorithms, which can be programmed to cope with the automatic start of batch processes and other situations. Curve (a) shows a far better performance, which is the one we would have without saturation.

Practical ways to avoid controller windup can be found in Section 8.3.

4.5.4 Derivative Action

When T_i and T_d are non-zero, (4.27) becomes a PID controller, also known as a three mode controller. The inclusion of a term of the form $T_d\, dc/dt$ overcomes the limitation of proportional and integral actions which require either a significant error or time interval to produce a sizeable response; by responding to the rate at which the controlled variable is changing the controller becomes more alert even in the presence of small errors, and hopefully will reduce overshoot and settling time. The parameter T_d is a measure of this alertness, as a simple analysis of the open-loop response of the controller immediately reveals; in fact, assuming $r=0$ and c increasing linearly we get the response (a) in Fig. 4.31, with proportional mode only; adding derivative action we get curve (b). Notice that the latter is curve (a) *advanced* by T_d units of time with the restriction of being zero for $t \leqslant t_0$.

In many textbooks one finds derivative action applied to the error signal instead of the controlled variable, i.e. a controller characterized by the equation

$$m(t) = K\left(e + \frac{1}{T_i}\int_0^t e(t')\mathrm{d}t' + T_d\frac{\mathrm{d}e}{\mathrm{d}t}\right) \tag{4.28}$$

In this case a low pass filter (a device with step response of the type shown in Fig. 2.3) must be used with the derivative mode to avoid 'bumping' the process every time a step change in set point takes place. This is why the majority of present-day controllers apply derivative action to the controlled (or measured) variable: one achieves basically the same response to disturbances as with (4.28) but sharp set point variations are smoothed out (filtered) by the process before reaching the derivative mode. Figure 4.32 illustrates the result of what has been said above when implemented on the system of Fig. 4.28, assuming a linear actuator, and the two types of controllers. Unlike the integral action, the derivative mode is not used alone; it is always combined with, at least, proportional action.

The benefits of derivative action can also be assessed from a root-locus analysis; in general the addition of this mode 'pushes' the root-loci to the left, given a proper choice of parameters. For example, the loci of the poles of the transfer function $C(s)/D(s)$ in Fig. 4.33, with $T_d=0$, are as shown in Fig. 4.9. However, with $K=2T_d$ they become those of Fig. 4.11.

Exercise 4.12 Illustrate the last mentioned statement.

Another important conclusion from the root-loci analysis is that plants with pure delays or with right half-plane zeros (cf. Figs 1.8 and 1.9) will not benefit

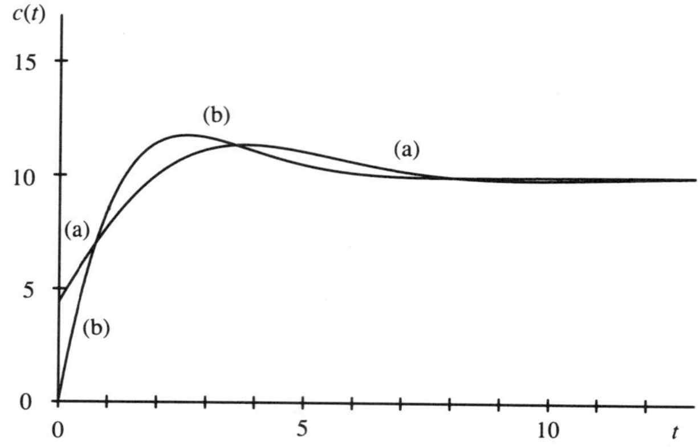

Figure 4.32 Response of the system shown in Fig. 4.28 to a step change, of magnitude 10, in reference, assuming a linear actuator $W(s)=2/(s+0.1)$, $K_c=T_i=T_d=1$ and: (a) controller (4.28); (b) controller (4.27).

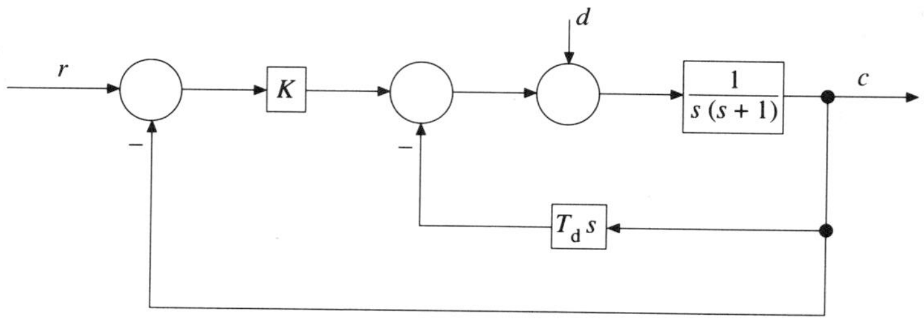

Figure 4.33 Illustration of the effect of derivative action.

much from derivative action. In fact, whether or not derivative action is present, there will always be loci crossing into the right half-plane; this is not suprising because, as already noted in Chapter 1, such plants require *cautious* control. An obvious practical limitation of derivative action is its sensitivity to noise; loosely speaking, noise is an undesirable and sustained signal with low amplitude and very irregular evolution and consequently an even more irregular derivative (when it exists); therefore adequate filtering must be provided when applying it to noisy measurements. Summing up, we can say derivative action is not beneficial in any of the following situations:

1. (Predominantly) first order processes.
2. Processes with large pure transport delays.
3. Inverse-response-type plants.

4.5.5 Controller Tuning

The question now is the selection of appropriate values for K, T_i and T_d in (4.27) to obtain a satisfactory performance when controlling a given process. To accomplish such a task the first thing we need is a mathematical model of the plant. The accurate identification of a plant is complex and time-consuming, and the situation is even worse if the plant is time-varying. However, the problem is greatly simplified if we restrict ourselves to a given class of low-order models. Experience has shown that a process model consisting either of a time constant τ and a transport delay T, i.e. a transfer function

$$\frac{K_p e^{-Ts}}{\tau s + 1} \tag{4.29A}$$

or a pure integrator in series with a time delay

$$\frac{K'_p e^{-Ts}}{s} \tag{4.29B}$$

is adequate in the majority of industrial process control problems. A tuning technique will then consist of the following:

1. A test to estimate the model parameters.
2. A set of formulae that relate the controller settings K, T_i and T_d to the model parameters, in order to provide a response with desired characteristics.

Tuning techniques are generally classified as open-loop or closed-loop methods. The former explicitly identifies the plant parameters while the latter requires the value of the gain on the limit of stability, the ultimate gain K_u, and the ultimate or natural period of oscillation P_u.

Open-loop Methods

The parameters in (4.29A) and (4.29B) are estimated from the plant response $c(t)$ to a step change in input, of magnitude M, as shown in Fig. 4.34. In model (4.29A), with the exception of K_p which is given by $K_p = M'/M$, the values of T and τ will depend on the goodness of fit criteria selected. In our case we will use a simple graphical criterion, based on the tangent of maximum slope as illustrated in Fig. 4.34(a): the point of its intersection with the time axis yields the time delay T. Because the response of

$$\frac{K_p}{\tau s + 1}$$

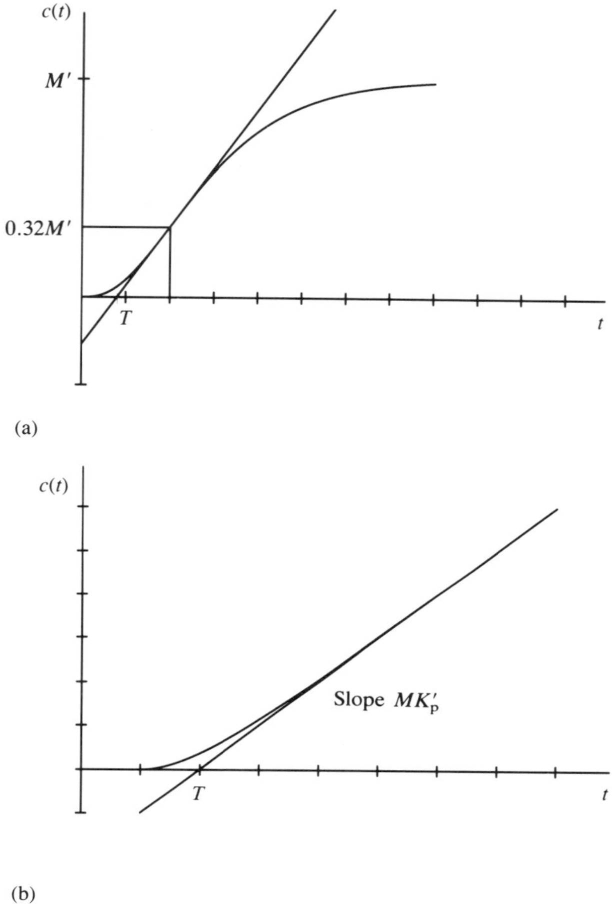

Figure 4.34 Data for open-loop tuning methods estimated from the plant response to a step input, of magnitude M applied at time $t=0$: (a) assuming model (4.29A); (b) assuming model (4.29B).

to a step input applied at time $t=0$ reaches 63.2% of its final value at time $t=\tau$ (cf. Fig. 3.6), the time constant in (4.29A) will be given by $\tau=t'-T$ where $c(t')=0.632M'$. The resulting fit is shown in Fig. 4.35, where $c(t)$ is reproduced for comparison. The ratio MK_p/τ is the maximum slope of the approximating model. For a unit-step input such a slope will be K_p/τ.

Again model (4.29B) is easily identified by the steady-state value of the slope of the response, to a step input of size M, which is MK_p', as indicated in Fig. 4.34(b). For a unit-step input this slope will be K_p' and is called the *reaction rate*, R_r.

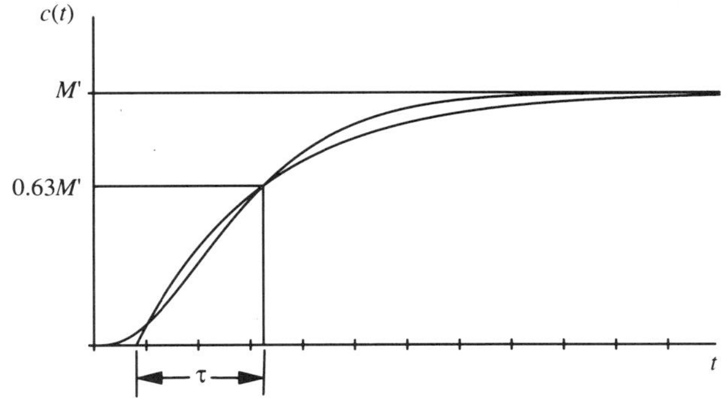

Figure 4.35 Approximation of $c(t)$ by model (4.29).

Closed-loop Methods

With the feedback loop closed (controller in 'automatic') and the integral and derivative modes adjusted to minimum action, i.e. T_d and $(T_i)^{-1}$ as small as possible, the gain is gradually increased until the controlled (or measured) variable cycles with constant amplitude; this gain is the ultimate gain, K_u, and the period of oscillation is the natural or ultimate period, P_u. This situation corresponds to a pair of conjugate poles on the imaginary axis, whose magnitude is the frequency of oscillation, as shown in Fig.4.23. Although this method is easy to apply, it has practical limitations: many processes will not tolerate oscillations for very long; furthermore such a test is very time-consuming if applied to slow processes. It is also very important to make sure that no element in the control loop saturates during cycling, which would render the results meaningless.

The formulae for the controller settings will depend upon the desired response following a step change in load or reference; typical performance measures are: integrated absolute error (IAE), integrated square error (ISE), integrated time multiplied by absolute error (ITAE), etc. Here, the tuning relations are derived in order to minimize the following error functionals:

$$\int_0^\infty |e(t)| \, dt$$

$$\int_0^\infty e(t)^2 \, dt$$

$$\int_0^\infty t \, |e(t)| \, dt$$

respectively.

Unfortunately there are no 'global optimum settings': these depend not only on the criteria used but also on the type of disturbance under consideration. A good illustration can be found in Miller *et al.* (1967) where six controller tuning

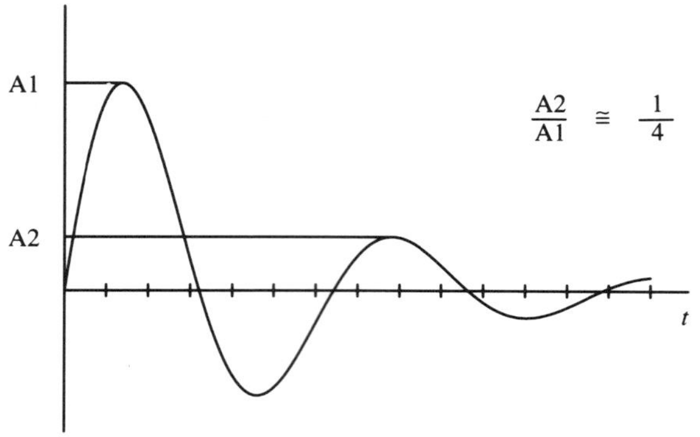

Figure 4.36 Quarter amplitude damping.

techniques are compared. The most popular formulae for controller settings were developed by Ziegler and Nichols (1942), Shinskey (1979), and Cohen and Coon (1953). Their objective was to determine settings that would produce a quarter decay ratio; while the first two assume the process model (4.29B) (a non-self-regulating process), Cohen and Coon assume model (4.29A) (a self-regulating process). Although their formulae yield different controller settings, in the majority of practical applications, we regard a controller as being optimally tuned when the response of the feedback system to a step change in load has a *quarter-amplitude damping* as shown in Fig. 4.36.

Table 4.2 Formulae for controller tuning

Controller	*Setting*	*Ziegler–Nichols (closed-loop)*	*Shinskey*	*Ziegler–Nichols (open-loop)*	*Cohen–Coon*
P	K	$0.5\ K_u$	$0.5\ K_u$	$\frac{1}{T\ R_r}$	$\frac{\tau}{TK_p}\left(1 + 0.33\frac{T}{\tau}\right)$
PI	K	$0.45\ K_u$	$0.5\ K_u$	$\frac{0.9}{T\ R_r}$	$\frac{\tau}{TK_p}\left(0.9 + 0.082\frac{T}{\tau}\right)$
	T_i	$0.833\ P_u$	$0.43\ P_u$	$3.33\ T$	$T\left(\frac{3.33 + 0.3T/\tau}{1 + 2.2T/\tau}\right)$
PID	K	$0.6\ K_u$	$0.5\ K_u$	$\frac{1.2}{T\ R_r}$	$\frac{\tau}{TK_p}\left(1.35 + 0.27\frac{T}{\tau}\right)$
	T_i	$0.5\ P_u$	$0.34\ P_u$	$2\ T$	$T\left(\frac{2.5 + 0.5T/\tau}{1 + 0.6T/\tau}\right)$
	T_d	$0.125\ P_u$	$0.08\ P_u$	$0.5\ T$	$T\left(\frac{0.37}{1 + 0.2T/\tau}\right)$

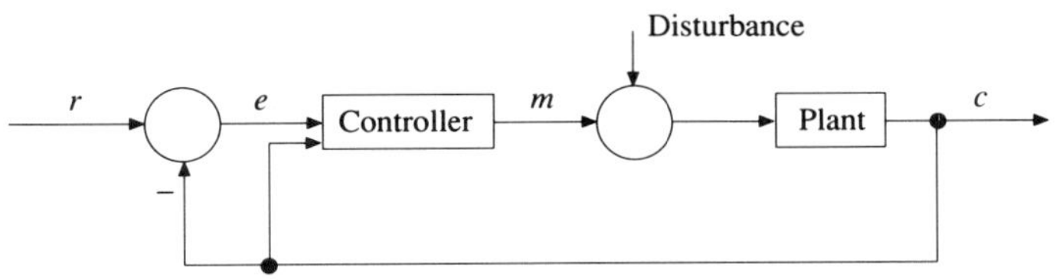

Figure 4.37 Feedback control configuration for comparison of the controller tuning methods in Table 4.2.

Controller settings for each of these methods are summarized in Table 4.2. In general these formulae will not produce exactly the desired type of response; they should be regarded instead as good initial guesses for an iterative approach. Practical hints for such a procedure can be found in Shinskey (1979). A glance at this table reveals that the introduction of integral action is accompanied by a gain decrease, as expected, to compensate for its destabilizing effect. However, the relative contribution of the proportional and integral modes is reinforced when derivative action is present. As mentioned earlier, derivative action has an adverse effect on the transient behaviour of a predominantly dead time process, i.e. a process where the ratio T/τ is large. Neither method, with the exception of that of Cohen and Coon, evaluates this ratio; therefore it is not surprising to discover in the examples that follow that Cohen and Coon's method has an overall better performance. Three examples are now presented in order to illustrate the relative merits and pitfalls of the formulae in Table 4.2. They represent the evolution of $c(t)$, following a unit-step load disturbance in the feedback control configuration of Fig. 4.37, for three different plants.

Example 4.4 This example attempts to show the influence of a first-order approximation of a higher-order plant. The plant selected has transfer function

$$G_p = \frac{1}{(s+1)^3}$$

whose unit step response is depicted in Fig. 4.34: the maximum slope is 0.271 at point $t=2$; the tangent at this point intersects the time axis at $t=0.806$. Then we have $K_p=1$, $T=0.806$, $\tau=2.45$, for approximating model (4.29A), and $T=0.806$, $K_p=R_r=0.271$ for approximating model (4.29B).

A unit feedback loop around G_p, in series with a proportional controller, produces a closed-loop transfer function with a pair of conjugate poles on the imaginary axis for $K_c=8$ leading to an ultimate period

$$P_u = \frac{2\pi}{\sqrt{3}}$$

The resulting controller settings can now be computed and are shown in Table 4.3. Figure 4.38 shows the response, to a unit-step disturbance, of the plant under

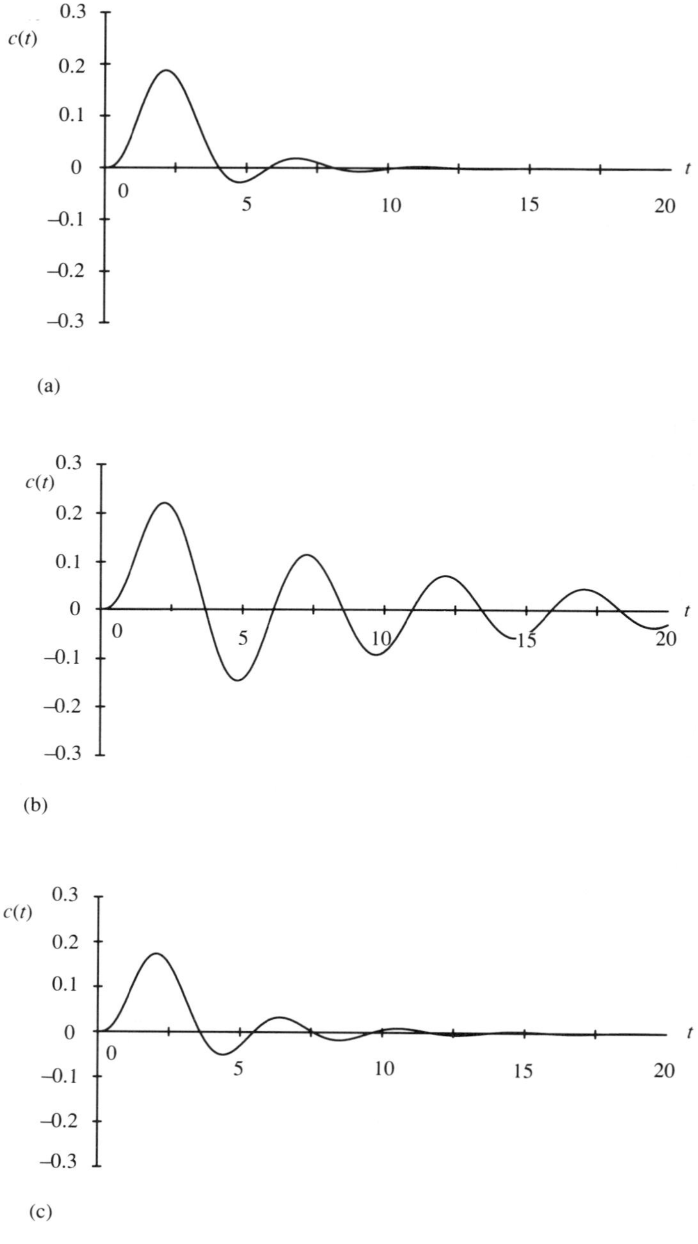

Figure 4.38 Simulations for example 4.4: (a) Ziegler and Nichols (closed-loop); (b) Shinskey; (c) Ziegler and Nichols (open-loop); (d) Cohen and Coon.

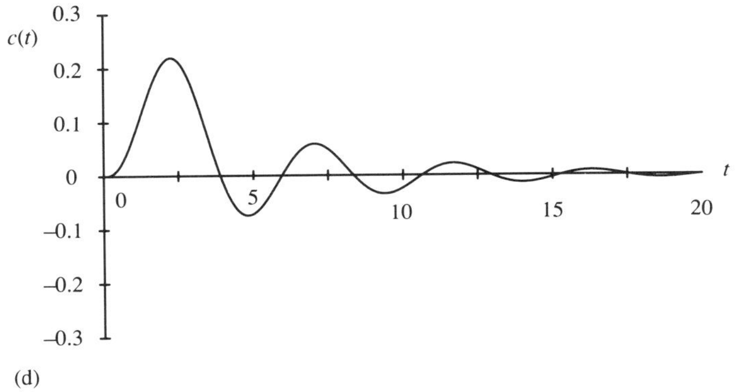

Figure 4.38 (cont.)

PID control for each of these controller settings. The Ziegler–Nichols (closed-loop) setting produced the best result, given the higher contribution of the derivative mode and the absence of time delay.

Exercise 4.13

(a) Show that the Ziegler–Nichols (closed-loop) choice produces a double zero at $-4/P_u$ for the controller (4.28).

(b) In example 4.4 compute the zeros of the PID controller (4.28) resulting from the Ziegler–Nichols (closed-loop) and Shinskey's settings.

(c) Check that although the Ziegler–Nichols settings have 'almost' cancelled two of the plant poles, the controller zeros still lie to the left side of the plant poles. What are the values of T_i and T_d that would cause an exact cancellation? (Recall exercise 4.11.) Later in the book we will analyze the relevance of this result.

(d) Plot and compare the roots-loci, parametrized as a function of controller gain, for both of the above settings.

Answers: (b) Ziegler–Nichols: -1.103 and -1.103; Shinskey: -1.31 and -2.13. (c) $T_i=2$ and $T_d=0.5$. (d) Ziegler–Nichols (closed-loop) settings produce a stable system for all values of the controller gain K ($K \geqslant 0$), while the system becomes unstable with Shinskey's settings for $K \geqslant 2.5$. Note that in this case the loci asymptotes, parallel to the imaginary axis, intersect the real axis at the point $+0.22$.

With the next two examples we can compare the efficiency of these techniques as a function of the ratio T/τ, for a given ultimate period. In both examples the plant transfer function has the form

$$G_p = \frac{e^{-Ts}}{\tau s + 1}$$

Table 4.3 Controller settings for example 4.4

Controller	Setting	Ziegler–Nichols (closed-loop)	Shinskey	Ziegler–Nichols (open-loop)	Cohen–Coon
P	K	4	4	4.59	3.37
PI	K	3.6	4	4.13	2.82
	T_i	3.02	1.56	2.68	1.60
PID	K	4.8	4	5.50	4.38
	T_i	1.8	1.23	1.61	1.79
	T_d	0.453	0.290	0.403	0.280

Example 4.5 The plant selected for this example has transfer function

$$G_p(s) = \frac{e^{-0.880s}}{0.183\,s + 1}$$

with a relatively large T/τ ratio ($T/\tau=4.81$). Under proportional control we have an ultimate gain $K_u=1.14$ with a period of oscillation $P_u=2.09$. The reaction rate for approximate model (4.29B) is $R_r=(0.183)^{-1}$. This leads to the controller settings shown in Table 4.4. Simulations of the response to a unit-step disturbance are plotted in Fig. 4.39. The poor performance of the Ziegler–Nichols (open-loop) method suggests that its values for T_i and T_d are too high for a dominant dead time process.

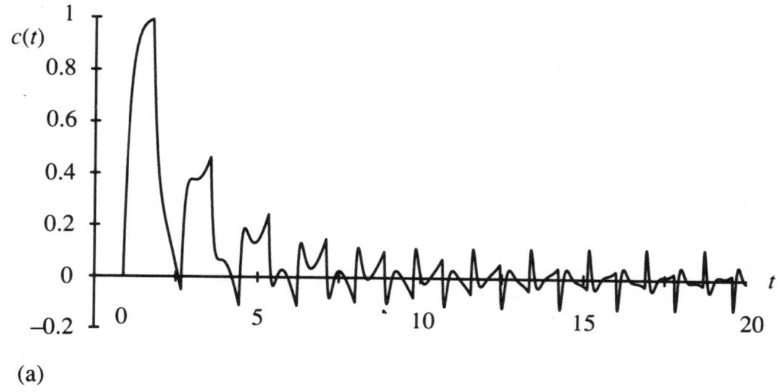

Figure 4.39 Simulations for example 4.5: (a) Ziegler and Nichols (closed-loop) (b) Shinskey; (c) Ziegler and Nichols (open-loop); (d) Cohen and Coon.

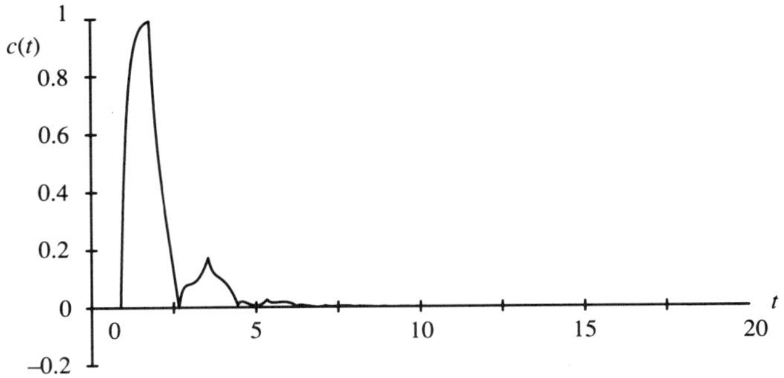

(b)

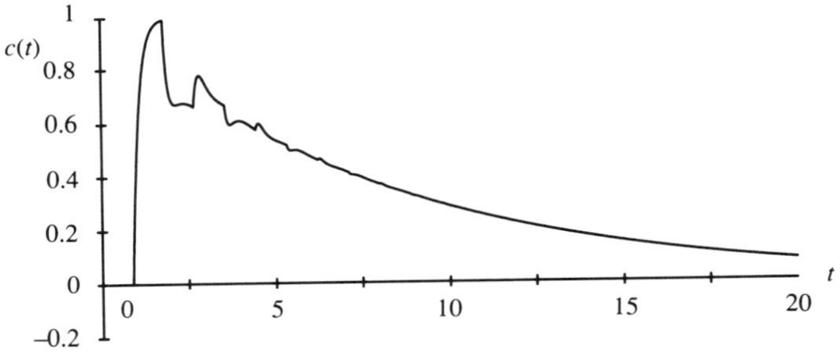

(c)

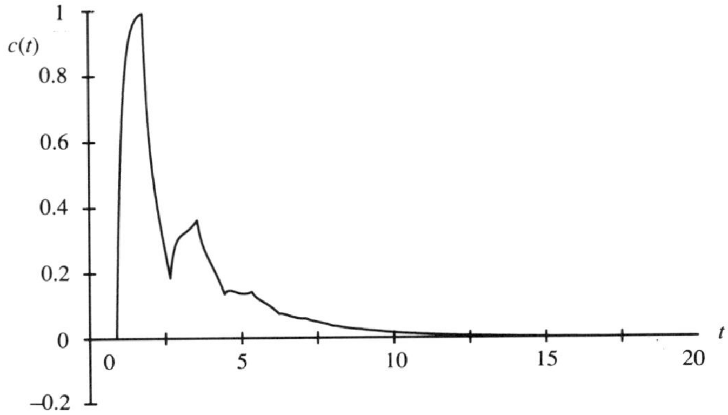

(d)

Figure 4.39 (cont.)

Table 4.4 Controller settings for example 4.5

Controller	*Setting*	*Ziegler–Nichols (closed-loop)*	*Shinskey*	*Ziegler–Nichols (open-loop)*	*Cohen–Coon*
PID	K	0.688	0.571	0.250	0.551
	T_i	1.05	0.712	1.76	1.11
	T_d	0.262	0.168	0.440	0.166

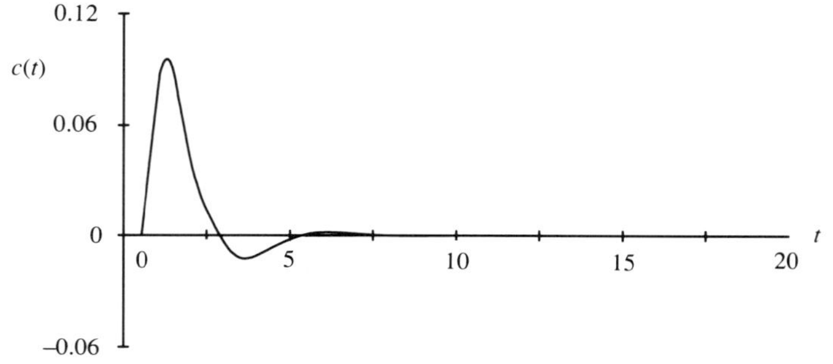

(a)

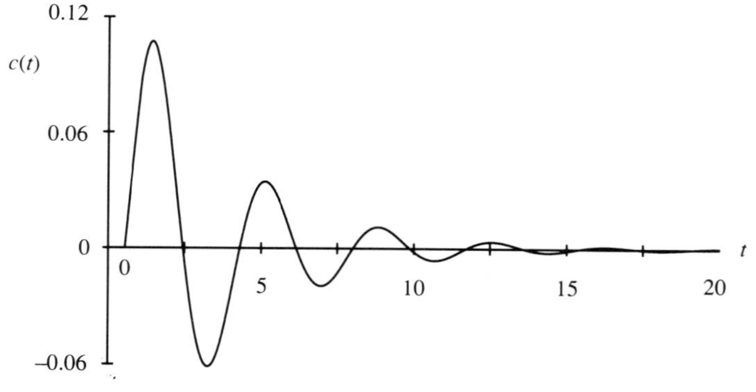

(b)

Figure 4.40 Simulations for example 4.6: (a) Ziegler and Nichols (closed-loop); (b) Shinskey; (c) Ziegler and Nichols (open-loop); (d) Cohen and Coon.

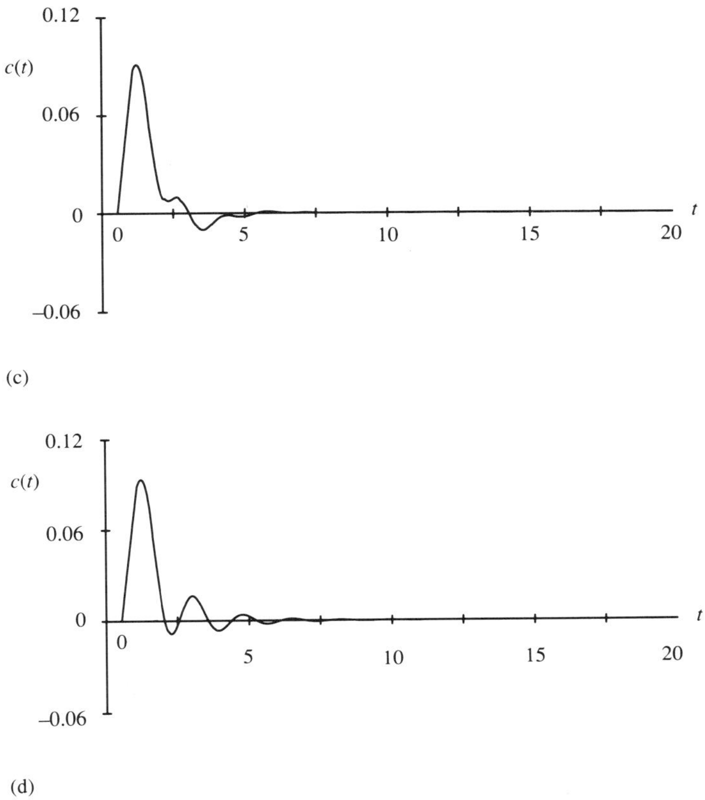

Figure 4.40 (cont.)

Example 4.6 The plant with transfer function

$$G_p(s) = \frac{e^{-0.542s}}{6s + 1}$$

has a T/τ ratio 53 times smaller ($T/\tau=0.0903$), an ultimate gain $K_u=18.03$ and the same ultimate period $P_u=2.09$. The settings for a PID controller are shown in Table 4.5 and the response following a unit-step disturbance, for each setting, is plotted in Fig. 4.40. By contrast, the Ziegler–Nichols settings exhibit this time the best results, in particular the open-loop method, although Cohen and Coon's method again performs very satisfactorily. As expected, the process with the highest ratio of dead time to lag was the most difficult to control.

Exercise 4.14 Explain why the overshoots are higher in example 4.5 than in example 4.6.

Table 4.5 Controller settings for example 4.6

Controller	*Setting*	*Ziegler–Nichols (closed-loop)*	*Shinskey*	*Ziegler–Nichols (open-loop)*	*Cohen–Coon*
	K	10.9	9.01	13.3	15.3
PID	T_i	1.05	0.712	1.08	1.31
	T_d	0.262	0.168	0.271	0.197

Exercise 4.14 Explain why the overshoots are higher in example 4.5 than in example 4.6.

4.6 CASCADE AND FEEDFORWARD CONTROL

In the feedback control scheme discussed so far the manipulated variable is set as a function of the deviation (error) of the measured variable only. Although such a scheme can cope with *all* types of disturbances, corrective action can only start after their effects show up at the process output (controlled variable). This leads to poor performance when disturbances are large and frequent in relation to process dead time or dominant time constant. However, most of the time, it is possible to measure the major disturbances either directly or through their effects at intermediate points in the process; using this extra information in a convenient way to set the manipulated variable, we can make the controlled variable less sensitive to these disturbances particularly when they are responsible for most of the control effort. Then the feedback controller will be left with the much lighter task of trimming the inadequacies of the selected procedure and (minor) unmeasurable disturbances. That is the purpose of the cascade and feedforward control strategies which are discussed below.

4.6.1 Cascade Control

Assume we have a process with a measurable intermediate variable; then the process can be broken apart as shown in Fig. 4.41, where c_2 is such a variable. An obvious way to speed up the recovery from a disturbance in d_2 is to use information from c_2 to start corrective action. A possible set-up is shown in Fig. 4.42: by means of the secondary controller corrective action starts before the disturbance reaches the controlled variable c_1, and hopefully will make c_1 less sensitive to d_2. (Because the output of one controller adjusts the set point of the other, the two controllers are said to be in *cascade*.)

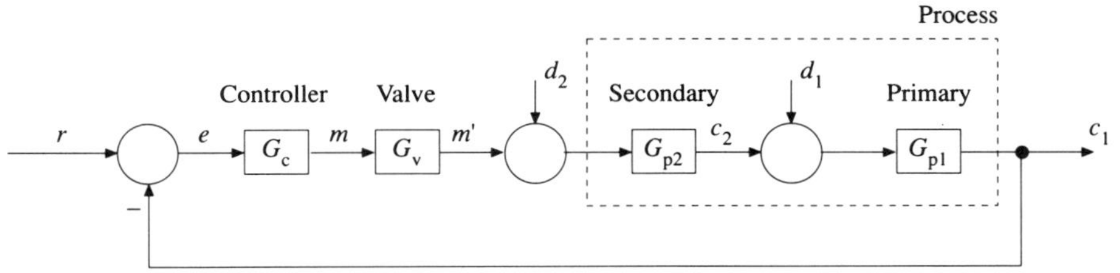

Figure 4.41 Single loop control of a process.

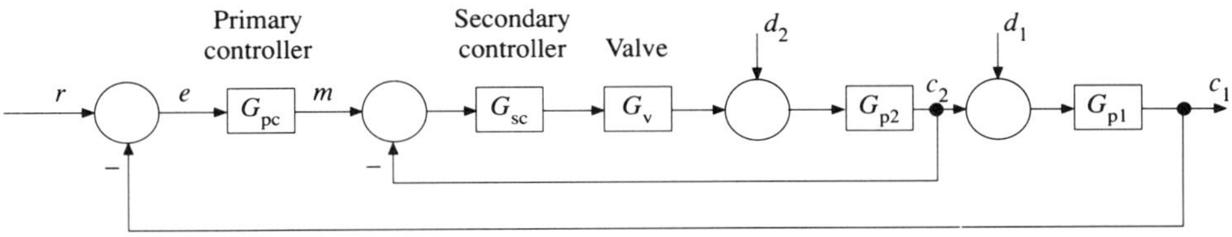

Figure 4.42 Classic cascade control structure for feedback controllers that are nested within one another.

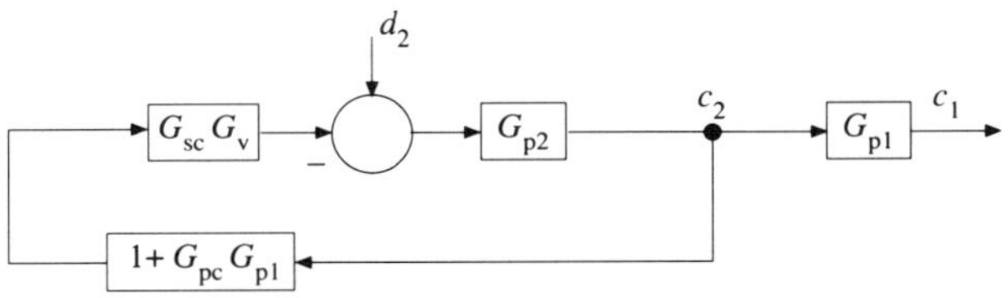

Figure 4.43 Diagram for exercise 4.15.

This strategy is particularly effective when G_{p2} is a predominantly time constant process because the influence of d_2 on c_2 (and c_1) is greatly attenuated by a simple gain increase of controller G_{sc}. In fact if we assume $G_{sc}=K_s$ (proportional controller) the transient performance of the inner loop improves as the gain increases, because its pole is pushed into the left half of s-plane; at the same time the sensitivity of c_2 to d_2 decreases because K_s appears in the denominator of $C_2(s)/D_2(s)$, while $C_2(s)/M(s)$ converges to a unit gain.

Exercise 4.15 Assume, in Fig. 4.42, that $G_v=1$, $G_{sc}=K_s>0$, $G_{p2}=K_2/(\tau_2 s+1)$, $G_{p1}=K_1/(\tau_1 s+1)$ and $G_{pc}=K_p(1+1/(T_i s))$. Then:

(a) Compute the transfer function $C_2(s)/M(s)$ and plot the locus of its pole as a function of K_s.

(b) Compute $C_2(s)/D_2(s)$, assuming $r=0$. *Hint*: use the equivalent diagram shown in Fig. 4.43.
(c) Assuming $r=0$ and d_2 a unit step, applied at time zero, show that C_2 converges to zero if $T_i \neq \infty$ and to $1/(1+K_s K_p K_1 K_2)$ when $T_i=\infty$.
(d) Show that C_2 converges to $T_i/(K_s K_p K_1 K_2)$ when d_2 is a unit ramp, applied at time $t=0$, and $T_i \neq \infty$.

Answers:

(a) $$\frac{C_2}{M} = \frac{G_{p2} G_{sc} G_v}{1 + G_{p2} G_{sc} G_v}$$

(b) $$\frac{C_2}{D_2} = \frac{G_{p2}}{1 + G_{sc} G_{p2} G_v(1 + G_{pc} G_{p1})}$$

A process that is particularly tailored for cascade control is the jacket-cooled stirred reactor shown in Fig. 4.44. There, an exothermic reaction takes place but it is necessary to keep its temperature constant; this is achieved by removing the heat by means of a controlled flow of water circulating through the jacket. When a disturbance occurs, e.g. a sudden increase in the temperature of the cooling water (represented by d_2 in Fig. 4.41), it has to propagate through three time constants in series, viz. the heat capacitances of the jacket contents, the wall (not considered in Fig. 4.41) and the reactor contents, before being detected by the temperature bulb inside the reactor. As a result of the delayed corrective action, a large deviation in reactor temperature will take place. From what has been said above, it is now obvious how to improve this state of affairs: it can be done by measuring the jacket temperature c_2 in order to provide a feedback signal to a second controller that controls the valve as shown in Fig. 4.42. By doing so, the effects of d_2 are restricted to the immediate vicinity of its point of entry, or equivalently, the reactor temperature c_1, is now *less sensitive* to d_2. This is apparent in the example below.

Example 4.7 In this example the advantages of cascade control are illustrated by comparing the outputs of Figs 4.41 and 4.42 when d_2 is a unit step disturbance, assuming

$$G_{p1}(s) = \frac{1}{20s + 1}, \quad G_{p2}(s) = \frac{1}{10s + 1}, \quad G_v = 1$$

and

$$G_c(s) = G_{pc}(s) = \left(1 + \frac{1}{15s}\right)$$

The parameters in these transfer functions were computed assuming the time expressed in seconds. Figure 4.45(a) shows such a response in the absence of cascade control; (b) and (c) show the same response after the addition of a second (proportional) controller with gain 6 and 10 respectively. Cascade control has improved the system settling time and the 'sensivity' of the controlled variable to the disturbance d_2 has been reduced by a factor of 8 (0.4/0.05) in (b); a further increase of the gain of G_{sc}, reduces the original sensivity 13 times as shown in (c).

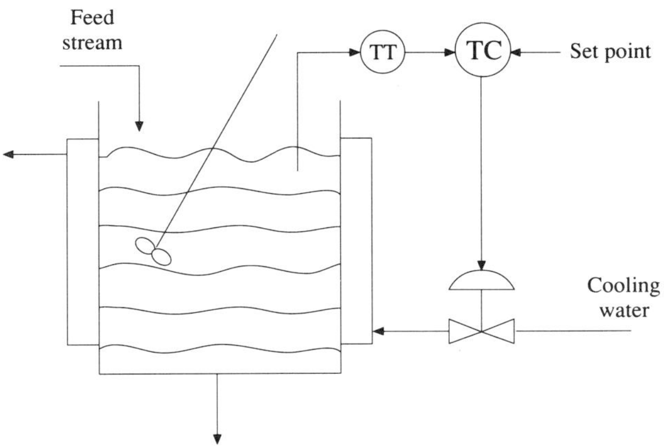

Figure 4.44 Jacketed reactor.

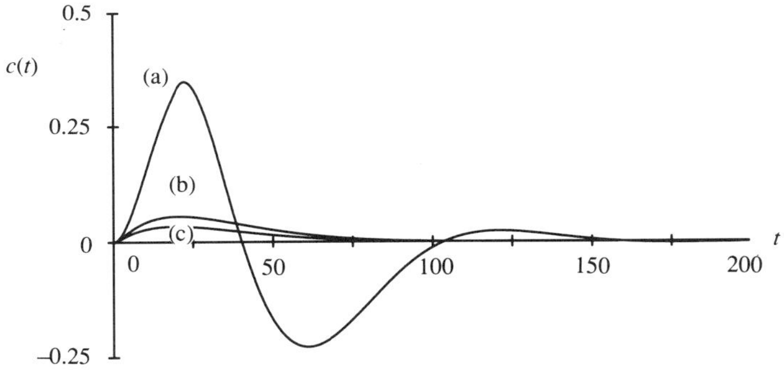

Figure 4.45 Responses to a unit step disturbance, d_2, of systems of: (a) Fig. 4.41; (b) Fig. 4.42 with $G_{sc}=6$; (c) Fig. 4.42 with $G_{sc}=10$.

In practice, precautions have to be taken with the set-up of Fig. 4.42 to protect the primary controller from integral saturation (windup) when the secondary controller, G_{sc}, is placed in manual, i.e. its output is held at a constant value. In this case the control system operates in an open-loop configuration and a sustained error at the primary controller input is unavoidable. If the integral mode of G_{pc} is not disconnected it will eventually saturate, with the undesirable consequences we already know. Fortunately most current commercially available controllers, allowing cascade configuration, are already programmed to prevent integral saturation in the primary controller.

4.6.2 Feedforward Control

The purpose of the cascade control strategy was to desensitize the controlled variable to a disturbance by speeding up the corrective action around its point of entry. Because load disturbances can often be measured before they enter the process, we can conceive another control strategy that 'prepares' the system for the impact of the disturbances, by changing the manipulated variable directly as a function of load-variable changes. Such a strategy is called *feedforward,* and it is summarized in Fig. 4.46 for a single disturbance. The corrective action generated by the feedforward controller $G_f(s)$, in response to a load disturbance d, will cancel its effect upon the controlled variable if

$$G_f(s) = \frac{-G_L(s)}{G_v(s)G_p(s)} \tag{4.30}$$

The difficulty in applying such a strategy is that it requires the knowledge of G_v, G_p and G_c which are seldom known *a priori*; when the process is stabilized, we can, for example, determine $G_L(s)$, in the absence of any other disturbances, by holding m at the required value (feedback loop open) and disturbing the process in d (Fig. 4.46).

Exercise 4.16 Compute $G_f(s)$ and discuss the feasibility of its implementation in the following cases:

(a) $G_L(s) = \dfrac{\exp(-T_L s)}{s+a}$, $G_p(s) = \dfrac{\exp(-T_p s)}{s+a}$; $T_p > T_L$

(b) $G_p(s) = \dfrac{(s-b)}{(s+a)}$, $b > 0$

Answer: neither case is feasible because (a) leads to an anticipatory G_f and (b) to an unstable G_f.

The additional advantages of imposing cascade and/or feedforward control upon the feedback strategy are well illustrated in the classical 'heat exchanger control' as represented in Fig. 4.47. The objective of such a system is to heat the cold fluid (water), W, at a desired temperarure (set point) by manipulating the steam flow rate W_s. (It can be shown that the (mass) flow rate of steam required to heat a mass flow rate of water, W, at temperature T_1, to a temperature T_2 is proportional to $W(T_2-T_1)$.) The process is subjected to a variety of disturbances, namely *supply* disturbances, such as fluctuations in steam flow rate, and *load* disturbances, such as changes in water flow rate and inlet temperature. Because the outlet temperature responds more slowly to changes in steam flow rate, disturbances in this variable can be appropriately dealt with by cascade control.

Exercise 4.17 Show how cascade control can be added to the configuration of Fig. 4.47 to cope with changes in steam flow rate.

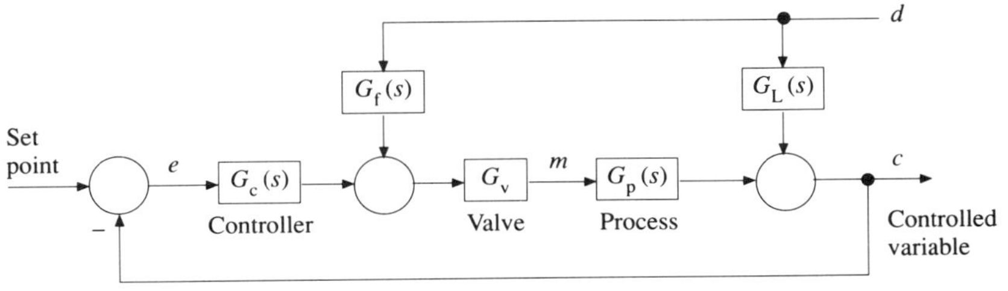

Figure 4.46 Feedforward control strategy.

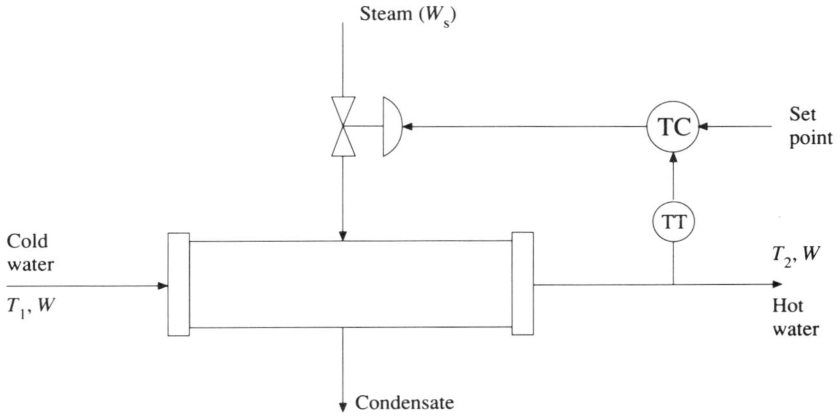

Figure 4.47 Heat exchanger process.

A practical control configuration, where both cascade and feedforward control are included for improved efficiency over the single feedback loop, is represented in Fig. 4.48. The block diagram of Fig. 4.49 represents a linearization of this system around a normal operating point: $W=\bar{\omega}$, $T_1=\bar{\tau}_1$ and $T_2=T_r$; under such conditions the error signals in the flow rate and temperature controllers are zero, and controllers are *biased* to produce an output equal to their set points, i.e. $W_c=K\bar{\omega}(T_r-\bar{\tau}_1)$, and $T_c=T_r$; also $W_f=\bar{\omega}$.

An important advantage of the set-up of Fig. 4.48 is the ability to *adapt* the feedback loop to variations in process gain. In our process, dead time, time constant and steady-state gain all vary with input flow W. A similar situation was already encountered in Chapter 2 where we found that the time constant of the liquid level system was a function of pressure head (cf. equation (2.11)). The steady-state gain of the heat exchanger, i.e. the steady-state increment in T_2 for a given change in heat input, varies inversely with the flow W, which is natural because a given increment of heat will produce a smaller temperature rise in a

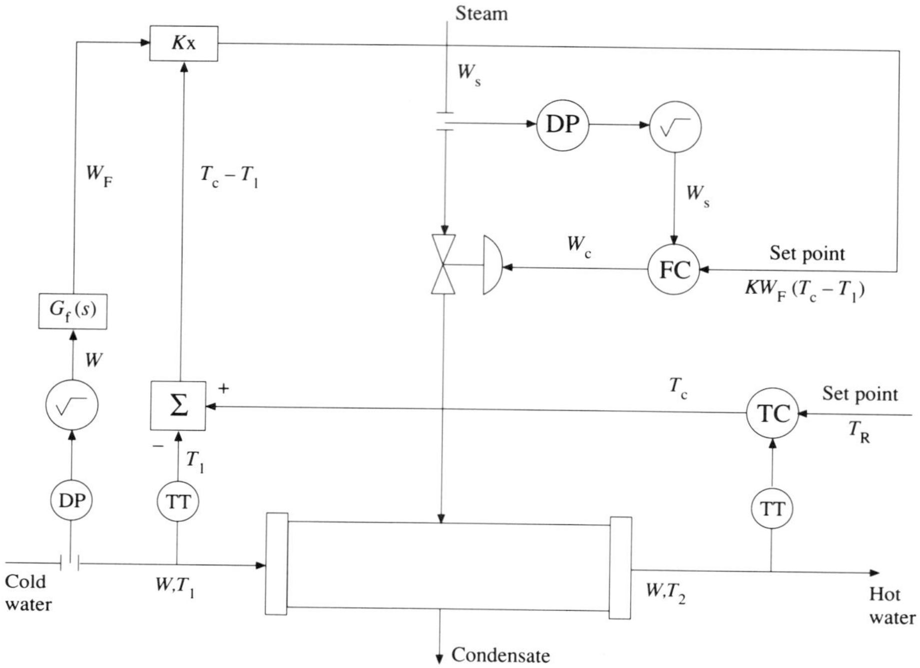

Figure 4.48 Complete control system for the heat exchanger. $G_f(s)$ denotes the transfer function of the feedforward controller.

larger flow. This implies, for example, that in Fig. 4.47 (a process under feedback control only) variations ΔW of input flow rate around a value $\bar{\omega}_1$ will cause shorter transients Δt_2 in T_2 than variations around $\bar{\omega}_2$ if $\bar{\omega}_2 < \bar{\omega}_1$. In the set-up of Fig. 4.48 the resulting adjustments in set point for the flow rate controller (FC), given a similar disturbance, will be:

$$\Delta(\text{set point}) = K\Delta W_F(T_c - T_1) + K\,\bar{\omega}\Delta T_c.$$

The second term on the right shows that the contribution of the feedback controller, ΔT_c, is in fact multiplied by the product flow rate, the net result being as if the feedback loop (see Fig. 4.49) had constant gain. The system of Fig. 4.48 also provides a *static* feedforward compensation for changes ΔT_1 in inlet temperature T_1, which is sufficient if T_1 changes slowly as it usually does. The feedback temperature controller (TC) will look after other (minor) disturbances and steady-state errors that would occur if only feedforward was used. This is shown in Fig. 4.50, which represents T_2 following a step change in cold water flow rate with the system under feedforward control only (Tc=constant): modelling errors and sustained minor disturbances like heat losses, calibration errors, parameter changes, etc., are responsible for the illustrated offset.

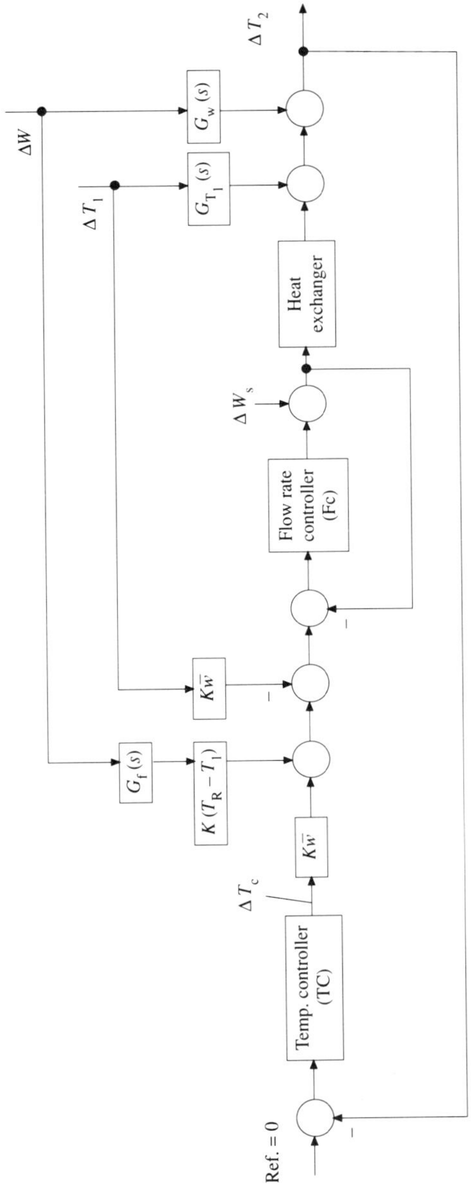

Figure 4.49 Block diagram of a linearization of system of Fig. 4.48 around a normal operating point.

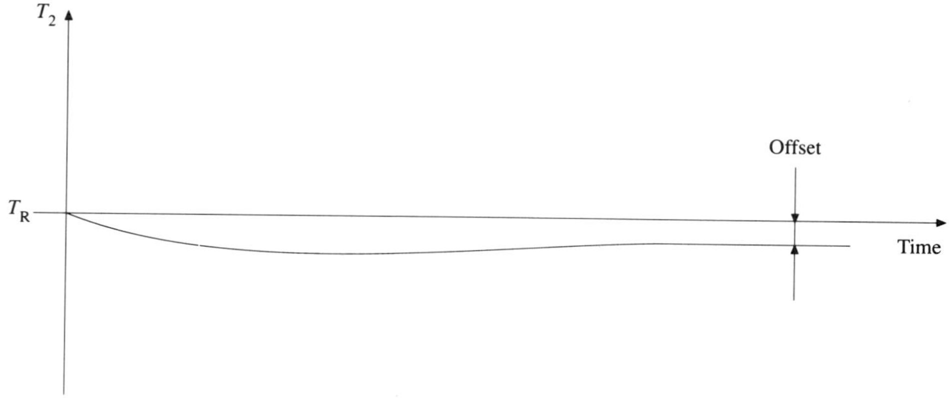

Figure 4.50 Typical profile of outlet temperature T_2 in Fig. 4.48, following a step disturbance in cold water flow, in the absence of feedback control.

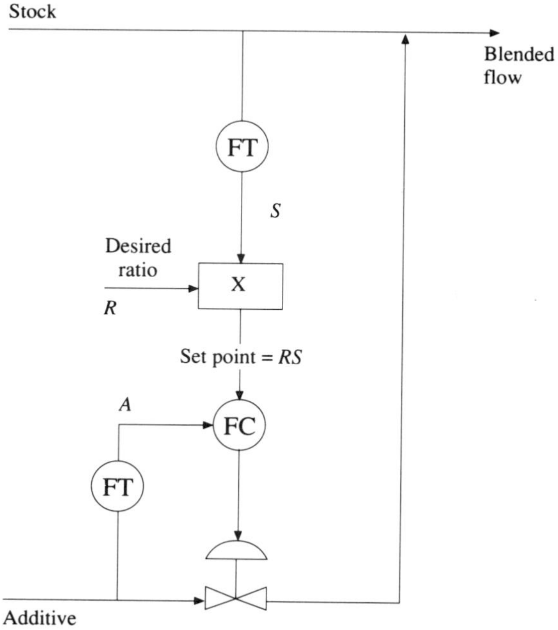

Figure 4.51 Ratio control of a blending process.

Another version of (static) feedforward control which is commonly encountered is *ratio control*, where one variable is controlled in ratio with another 'wild' variable. For example, in the blending of solid or liquid ingredients, on a continuous basis, it is often required to maintain a ratio among them to keep the blend composition constant. Figure 4.51 illustrates such an application where a flow of additive A is to be added to a varying flow S of stock, at a fixed ratio $R=A/S$. Notice that the set point of the additive flow controller is generated by the multiplier.

4.7 CONCLUSIONS

An important practical problem is to make the output of a plant follow a desired trajectory. In the absence of uncontrollable disturbances this could be achieved as follows: let $u(t)$ and $H(s)$ denote the plant input and transfer function respectively, and $r(t)$ the desired output; then

1. $u(t) = \mathcal{L}^{-1}\{H^{-1}(s)\, R(s)\}$ is a possible solution.
2. Inserting, in series with the plant, a compensator with transfer function $H^{-1}(s)$, to get an overall *unit* transfer function.

Both methods, besides being unable to cope with uncontrollable disturbances, have a serious drawback: they require a precise knowledge of the plant transfer function. Feedback allows not only the realization of virtually a unit transfer function, with a limited knowledge about the plant dynamics, but also the ability to cope with uncontrollable disturbances. When the plant contains a pure transport delay T, the control problem that arises could only be overcome by means of an anticipatory block. Because such a device is not possible in practice, the best we can expect from the above methods is a unit transfer function in series with e^{-Ts}. Later in the book we shall see algorithms for minimizing the effects of transport delays in feedback systems, such as the Smith predictor algorithm.

REFERENCES

Cohen, G.H. and G.A. Coon (1953) 'Theoretical consideration of retarded control', *Trans. Am. Soc. Mech. Engrs.*, July, 827.

Miller, J.A., A.M. Lopez, C.L. Smith, P.W. Hurrill (1967) 'Comparison of controller tuning techniques', *Control Engineering*, **14**(12), 72–5.

Shinskey, F.G. (1979) *Process Control Systems*, 2nd ed., McGraw-Hill.

Ziegler, I.G. and N.B. Nichols (1942) 'Optimum Settings for Automatic Controllers', *Trans. ASME*, **64**, November, 759–68.

5

Frequency-domain Analysis

The purpose of this chapter is twofold. First we analyze the steady-state response of stable linear systems to sinusoidal inputs. Then we learn how to use such information to infer the properties of a closed-loop system (stability, transient response, etc.) from its open-loop characteristics.

5.1 OPEN-LOOP ANALYSIS

In Chapter 3, the reader's attention was drawn to the fact the steady-state response of a stable linear system to a sinusoidal input is also a sinusoid and, furthermore, that two sinusoidal inputs of equal amplitudes, but different frequencies, will most likely produce outputs of different amplitudes (cf. exercise 3.5). This result has important implications, namely the fact that such systems have 'memory', i.e. the response, at time t, is not only a function of the input at a single time instant, but also a function of its behaviour on some non-zero time interval; otherwise two sinusoids of identical amplitude would always produce outputs of the same amplitude.

The realization of such a simple property had a tremendous impact on the development of linear systems theory. In essence, we can say that any linear dynamical system is wholly determined by knowledge of its steady-state responses to sinusoidal inputs $\sin \omega t$, $\omega \in [0,\infty)$. If we denote such time responses by

$$A(\omega) \sin(\omega t + \varphi(\omega)) \tag{5.1}$$

then we can show that

$$A(\omega) \exp[j(\varphi(\omega))] = G(j\omega) \tag{5.2}$$

where $G(s)$ denotes the system transfer function. In other words, what we have just said implies that the transfer function, $G(s)$, is *entirely* determined by the values it assumes along the imaginary axis.

For the mathematically minded reader, we can add that this is a consequence of Cauchys integral formula, which shows the value of an analytical function in a region is determined throughout the region by its values on the boundary. Furthermore $G(j\omega)$ has a nice physical interpretation; from (5.2) we conclude that the magnitude of $G(j\omega)$ is the 'amplification' for the input sinusoid, $\sin \omega t$, and arg $(G(j\omega))$ is the phase shift in steady-state, between the input and output sinusoids. The result is shown below.

Let $G(s)$ denote a strictly proper, rational, transfer function of a stable linear system, with input $u(t) = \sin \omega t$, $t \geqslant 0$, and output $y(t)$. From Chapter 3, namely from the partial fraction expansion (3.41), we know that the output will be the sum of the system natural modes, which converge to zero (we are assuming the system is stable) plus the input modes.

Since

$$\mathcal{L}(\sin \omega t) = \frac{\omega}{s^2 + \omega^2}$$

we can write

$$Y(s) = \left(\frac{a}{s - j\omega} + \frac{\bar{a}}{s + j\omega}\right) + \left(\frac{b_1}{(s - p_1)} + \ldots + \frac{b_n}{(s - p_n)}\right) \quad (5.3)$$

where $p_1, \ldots, p_n$ are the poles of $G(s)$. Assuming the system has only simple poles we have in the time domain

$$Y(t) = (a \exp(j\omega t) + \bar{a} \exp(-j\omega t) + (b_1 \exp(p_1 t) + \ldots + b_n \exp(p_n t) \quad (5.4)$$

If the system is stable, eventually we are left with the term

$$(a \exp(j\omega t) + \bar{a} \exp(-j\omega t) \quad (5.5)$$

Let us then compute a and $\bar{a}$. Multiplying both sides of (5.3) by $(s+j\omega)$ and computing the limit when $s \to -j\omega$ we arrive at

$$a = \lim_{s \to -j\omega} G(s) \frac{\omega(s + j\omega)}{s^2 + \omega^2} = \lim_{s \to -j\omega} G(s) \frac{\omega}{(s - j\omega)} = \frac{1}{2} j\, G(-j\omega) \quad (5.6)$$

Similarly

$$\bar{a} = \frac{-1}{2} j\, G(j\omega) \quad (5.7)$$

If we express G in polar form, namely

$$\begin{aligned} G(j\omega) &= |G(j\omega)| \exp(j\varphi) \\ G(-j\omega) &= |G(j\omega)| \exp(-j\varphi) \end{aligned} \quad (5.8)$$

then (5.5) becomes

$$\begin{aligned} a \exp(-j\omega t) + \bar{a} \exp(j\omega t) &= \frac{1}{2} j\, |G(j\omega)| (\exp(-j(\varphi + \omega t)) - \exp(+j(\varphi + \omega t))) \\ &= |G(j\omega)| \sin(\varphi + \omega t) \end{aligned}$$

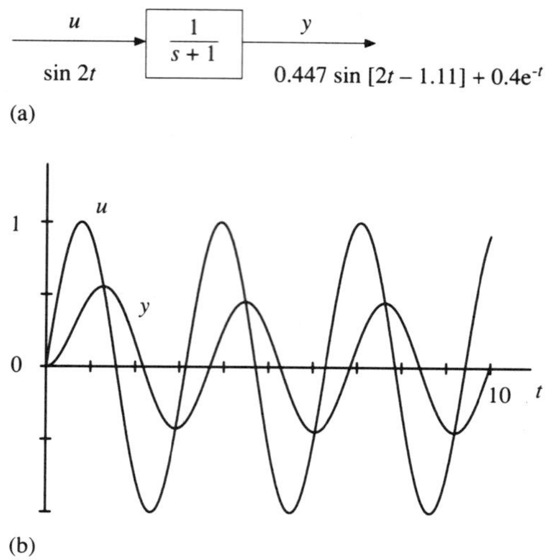

Figure 5.1 Response, y, of system with transfer function $1/(s+1)$ to the input $u=\sin 2t$, $t \geq 0$.

Summing up, we have shown that a stable linear system, with a strictly proper rational transfer function $G(s)$, and input $\sin \omega t$, $t \geq 0$, exhibits an output $y(t)$ that converges to

$$|G(j\omega)| \sin(\omega t + \arg G(j\omega)) \tag{5.9}$$

Figure 5.1 depicts the response y, of the plant with transfer function $1/(s+1)$, to the input, $u(t)=\sin 2t$: we can see that after the initial transient has died down, the output becomes a sinusoid of the same period, has a smaller amplitude and is out of phase with the input.

The ratio of the output to input amplitudes is the *gain* at that frequency, and the corresponding phase shift between the input and the output is the phase angle. The geometric construction of Fig. 5.2 shows how to compute these values as a function of frequency: the inverse of the magnitude of the line segment $\overline{AB}$ is the gain, and $(-\varphi)$ the phase angle, at frequency ω.

Exercise 5.1 Compute the response $y(t)$ of the transfer function $(1/s)$ to the input $u(t)=\sin \omega t$, $t \geq 0$, under steady-state conditions. Recall $\mathcal{L}(\sin \omega t)=\omega/(s^2+\omega^2)$. The response is not purely sinusoidal. Why?
Answer: The response is

$$\frac{1}{\omega}\left(1 - \sin\left(\omega t - \frac{\pi}{2}\right)\right)$$

Since the system is unstable, the contribution of the system mode does not vanish, as time goes to infinity. In fact, $\mathcal{L}^{-1}(1/s)=1$, $t>0$.

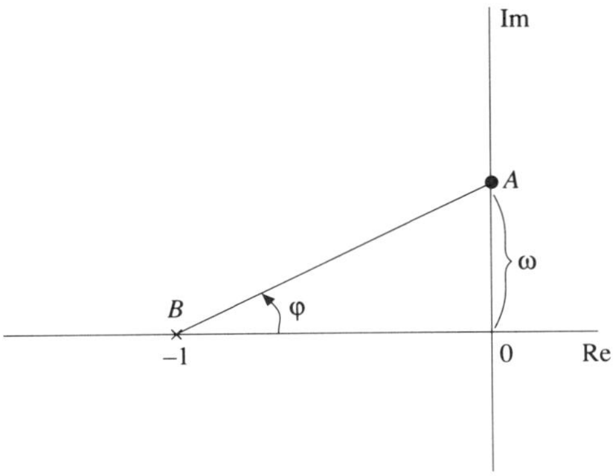

Figure 5.2 Geometric construction to compute gain and phase angle, as a function of the input frequency ω, for the transfer function $1/(s+1)$.

Exercise 5.2 For the system with transfer function $1/(s^2+0.02s+1)$
(a) Compute the system poles.
(b) Draw the geometric construction that enables one to compute the gain and phase as a function of frequency (cf. Fig. 5.2).
(c) Compute the gain at the frequencies $\omega \in \{0.1; 1; 1.01; 10\}$
Note: the poles are very close to the imaginary axis.
Answer:

ω	Gain
0.1	0.99
1	51 (approximately)
10	0.01
1.01	greater than 4.5×10^4

The previous exercise illustrates an important phenomenon known as 'resonance': when a system has a pair of very lightly damped poles (i.e. very near the imaginary axis), it exhibits a very large amplitude of oscillation when excited by a frequency close to the poles' magnitude. Such large oscillations normally impose an enormous stress upon the system and may even cause its destruction.

5.1.1 Graphical Representation of Frequency Response

Since $G(j\omega)$ is a complex function of a real variable, it can be represented either in the complex plane, with ω as a parameter (polar plot), or with separate representations for the magnitude and for the phase as a function of ω. In general terms we can say that the polar plot is more convenient for analysis, of stability for example, while separate graphs for the magnitude and phase are preferable for controller design.

5.1.2 Polar Plots

As mentioned above, the polar plot for a transfer function $G(s)$ is a graphical representation, in the complex plane, of $G(\mathrm{j}\omega)$, $\omega \in (0,\infty)$.

Assume, for example,

$$G(\mathrm{j}\omega) = \frac{p}{p + \mathrm{j}\omega} \tag{5.10}$$

which can also be written as

$$G(\mathrm{j}\omega) = \frac{p(p - \mathrm{j}\omega)}{(p + \mathrm{j}\omega)(p - \mathrm{j}\omega)} = \frac{p^2}{p^2 + \omega^2} - \mathrm{j}\frac{\omega p}{p^2 + \omega^2} = x + jy$$

As ω varies from 0 to infinity, $G(\mathrm{j}\omega)$ describes a semicircle centred at $(-1/2+\mathrm{j}0)$, with radius 1/2, as in Fig. 5.3, no matter what the value of p is. In fact, if we compute the magnitude of

$$(x + \mathrm{j}y) - \left(-\frac{1}{2} + \mathrm{j}0\right)$$

we find that it equals 1/2, as shown below.

Since

$$(x + \mathrm{j}y) - \left(\frac{1}{2} + \mathrm{j}0\right) = \left(x - \frac{1}{2}\right) + \mathrm{j}y = \frac{p^2 - \omega^2}{2(p^2 + \omega^2)} - \mathrm{j}\frac{\omega p}{p^2 + \omega^2}$$

the magnitude squared of $(G(\mathrm{j}\omega)-(1/2+\mathrm{j}0))$ is given by

$$\left(\frac{p^2 - \omega^2}{2(p^2 + \omega^2)}\right)^2 + \left(\frac{\omega p}{p^2 + \omega^2}\right)^2 = \frac{p^4 - 2p^2\omega^2 + \omega^4}{4(p^2 + \omega^2)^2} + \frac{4\omega^2 p^2}{4(p^2 + \omega^2)} = \frac{1}{4}$$

and the result follows.

In Fig. 5.3 we can see the position of $G(\mathrm{j})$ for $p=1$ and $p=10$ revealing that, for a given frequency in the range $(0,\infty)$, the gain increases with increasing $|p|$. This is not unexpected since the larger $|p|$, the faster the system. It also reveals that the gain decreases monotonically with increasing frequency.

The situation for second-order systems with transfer function

$$G(\mathrm{j}\omega) = \frac{\omega_n^{\ 2}}{(\mathrm{j}\omega)^2 + 2\zeta\omega_n(\mathrm{j}\omega) + \omega_n^{\ 2}} \tag{5.11}$$

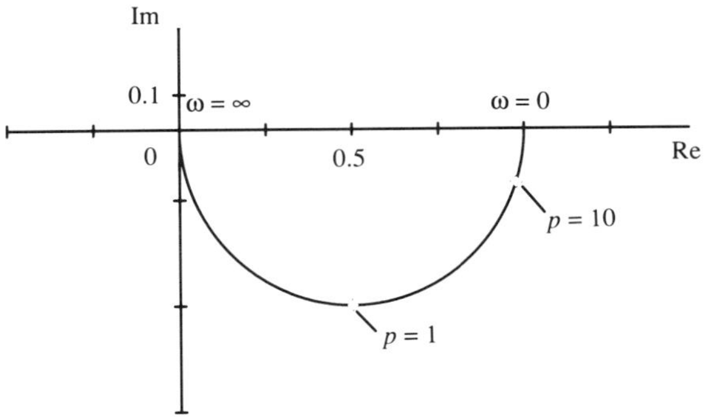

Figure 5.3 Polar plot of the transfer function $p/(s+p)$. The indicated values correspond to ω=1rad/s.

is illustrated in Fig. 5.4, for $\omega_n=1$, and $\zeta \in \{2, 1, 0.7, 0.4, 0.1, 0.05\}$. In contrast to the previous case, there are situations where the gain does not decrease monotonically. In fact it can be shown that, for $\zeta<0{,}707$, $|G(j\omega)|$ has a maximum at $\omega=\omega_n\sqrt{(1-2\zeta^2)}$; this is the resonance effect mentioned earlier.

Although the interpretation we gave above for the meaning of $G(j\omega)$ assumed $G(s)$ a rational transfer function, it can be shown the result is also applicable when $G(s)$ contains terms of the form e^{-sT}. For the pure transport delay $T, G(j\omega)=e^{-j\omega T}$, the polar plot is a circle of unit radius no matter what the value of T is; because its phase decreases linearly with ω, this term can change drastically the shape of other polar plots.

Figure 5.5 shows the polar plots for $\exp(-j\omega T)/(j\omega+1)$ for $T=1$ and $T=10$. The reader is invited to compare them with the plots in Fig. 5.3.

5.1.3 Bode Plots

This representation uses two separate graphs for the representation of $G\,(j\omega)$: One for the magnitude versus frequency, and another for the phase versus frequency. The use of logarithmic scales makes these plots an invaluable tool in the analysis and design of control systems.

Let us start by rewriting the transfer function

$$G(s) = \frac{K\,(s + z_1)\,.\,.\,.\,(s + z_m)}{s^r\,(s + p_1)\,.\,.\,.\,(s + p_n)}\,e^{-Ts} \tag{5.12}$$

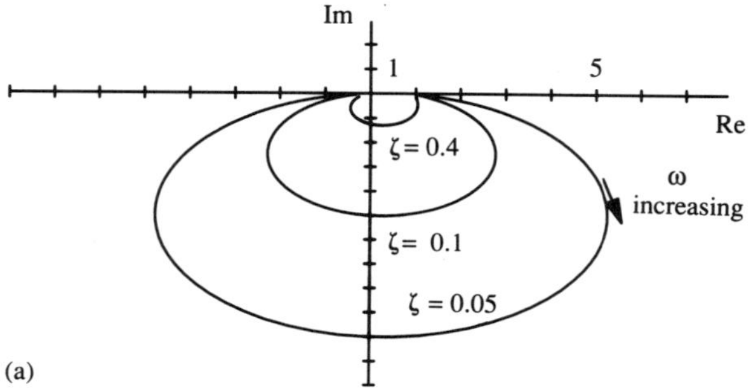

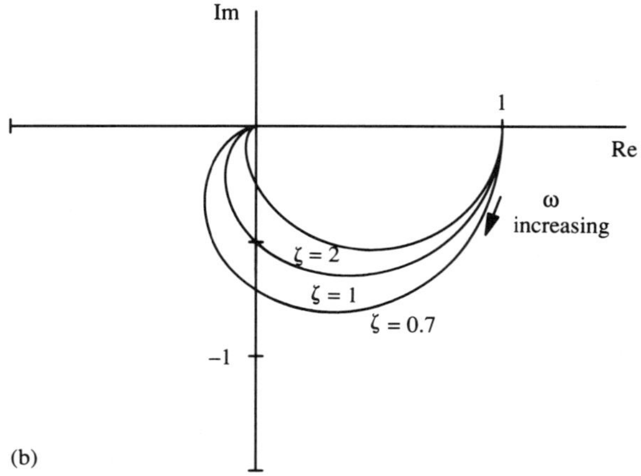

Figure 5.4 Polar plots for $1/(s^2+2\zeta s+1)$ and: (a) $\zeta \in \{0.05, 0.1, 0.4\}$; (b) $\zeta \in \{0.7, 1, 2\}$.

in the form

$$G(s) = K_B \frac{(1 + s/z_1)(1 + s/z_2) \ldots (1 + s/z_m)\, e^{-Ts}}{s^r (1 + s/p_1)(1 + s/p_2) \ldots (1 + s/p_n)} \tag{5.13}$$

assuming $(n+r) \geqslant m$, $|p_i| > 0$, $|z_i| > 0$.

Exercise 5.3 Given

$$G(s) = \frac{K(s + z_1) \ldots (s + z_m)}{s^r (s + p_1) \ldots (s + p_n)} e^{-Ts}$$

re-express it in the form (5.13) and compute K_B.
Answer:

$$K_B = \frac{K(z_1 z_2 \ldots z_m)}{p_1 p_2 \ldots p_n}$$

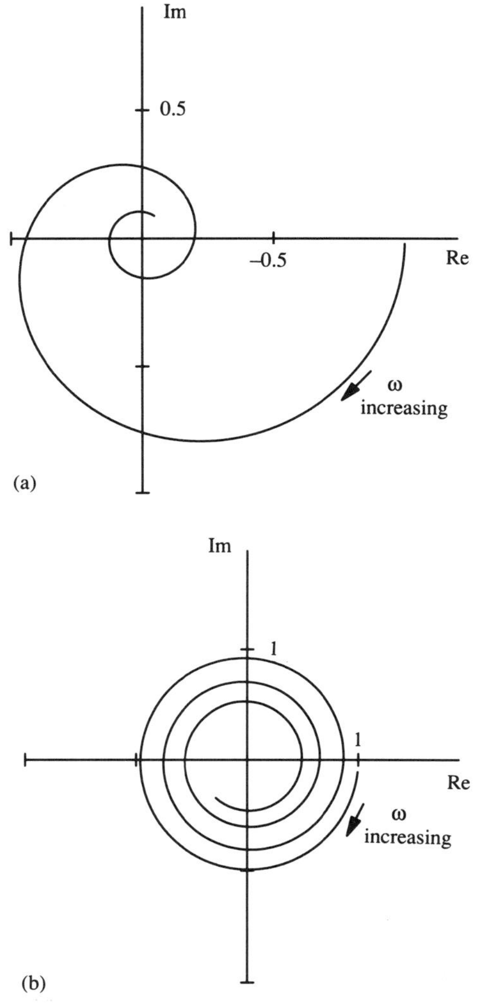

Figure 5.5 Polar plots for exp($-j\omega T$)/($j\omega+1$): (a) $T=1$ and $\omega \in [0.01; 10]$; (b) $T=10$ and $\omega \in [0.01; 2]$.

Then $\log \mid G\,(j\omega) \mid$ is the sum of the following terms:

$$\log \mid K_B \mid \tag{5.14}$$

$$\log \mid 1 + j\omega/z_i \mid, i = 1,2, \ldots, m \tag{5.15}$$

$$-\log \mid 1 + j\omega/p_i \mid, i = 1,2, \ldots, n \tag{5.16}$$

$$-r \log \mid j\omega \mid = - r \log \omega \tag{5.17}$$

When transfer function (5.13) has a pair of complex-conjugate poles, say $p_1=p$ and $p_2=\bar{p}$, it is convenient to group them in the form

$$\frac{1}{(1 + j\omega/p)\,(1 + j\omega/\bar{p})} = \frac{1}{1 + 2\zeta\,(j\omega/\omega_n) + (j\omega/\omega_n)^2} \tag{5.18}$$

where

$$p, \bar{p} = \zeta\omega_n \pm \sqrt{(1 - \zeta^2)}$$

As for arg $(G(j\omega))$ we have

$$\arg\,(G(j\omega)) = \sum_{i=1}^{m} \arg\left(1 + \frac{j\omega}{z_i}\right) - \sum_{i=1}^{m} \arg\left(1 + \frac{j\omega}{p_i}\right) - r\frac{\pi}{2} - \omega T \tag{5.19}$$

The magnitude $|G(j\omega)|$ will be plotted in *decibel* (dB) units. The value of $|G(j\omega)|$ in decibels is given by $20 \log_{10} |G(j\omega)|$. The magnitude of $G\,(j\omega)$, in decibels, versus ω on a log scale is called the *Bode magnitude plot.* Arg($G\,(j\omega)$) versus ω on a log scale is the *Bode phase-angle plot.*

Consequently, the Bode plots for any transfer function of the form (5.13) can be easily obtained by addition of the Bode plots of the following factors:

1. K_B.
2. $(1+j\omega/z)^{\pm 1}$.
3. $(j\omega)^{\pm r}$.
4. $\dfrac{1}{1 + 2\zeta\,(j\omega/\omega_n) + (j\omega/\omega_n)^2}$.
5. $e^{-j\omega T}$.

The Bode plots for each of these factors is now analyzed in detail. The constant K_B has a constant magnitude, and a phase angle of 0° if positive or −180° if negative. Therefore its Bode plots are simply horizontal lines at these values. This immediately reveals one of the advantages of the Bode diagrams. The multiplication by a constant K_g simply translates the magnitude plot along the vertical axis, by $20 \log_{10} |K_g|$ dB.

The magnitude Bode plot for the factor $(j\omega)^{\pm r}$ is a straight line with slope $\pm 20r$ dB/decade. In fact

$$20 \log_{10} |\,j10\omega\,|^{\pm r} \text{ dB} = \pm\, r\,(20 \log_{10} \omega \text{ dB} + 20 \text{ dB})$$

showing that for each decade of frequency increase, the magnitude changes by $\pm 20r$ dB. As for the phase, it is obvious that it is constant and equal to $\pm r$ 90°. Figure 5.6 shows the magnitude and phase Bode plots for $(j\omega)^{-r}$, for $r = 1, 2, 3$.

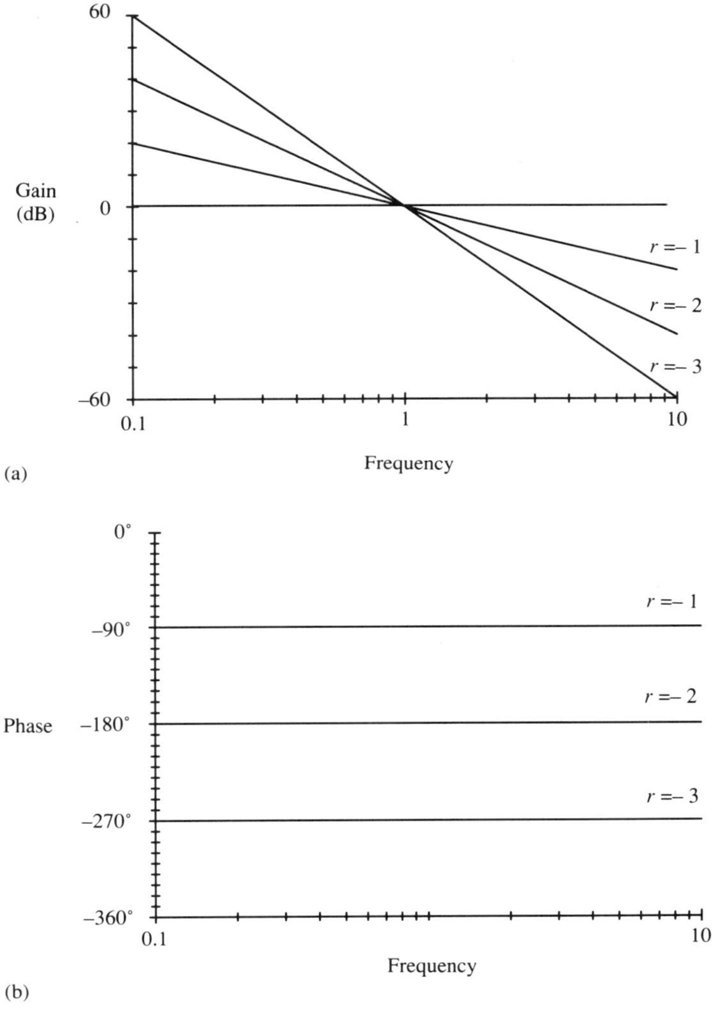

Figure 5.6 Bode plots for $(j\omega)^{-r}$, r=1, 2, 3.

Exercise 5.4 Show that the magnitude Bode plot for $(j\omega/p)^{-r}$ is a straight line with slope $-20r$ dB/decade, intersecting the 0 dB line at $\omega = p$.

Often the slope of the magnitude Bode plot is expressed in dB/octave. If we recall that a frequency ω_2 is one *octave* above ω_1 if $\omega_2 = 2\omega_1$, then we conclude that

$$\pm\, s \text{ dB/decade} = \pm\, (0.3)^{-1} s \text{ dB/octave}$$

since

$$\log_{10}(2\omega_1) - \log_{10}\omega_1 = \log_{10}2 = 0.3$$

In particular

$$\pm\ 20 \text{ dB/decade} = \pm\ 6 \text{ dB/octave}$$

Exercise 5.5 Construct a graph for the conversion of dB/decade into dB/octave, and viceversa.
Answer: in Cartesian coordinates, with abscissa in dB/decade and ordinate in dB/octave, such a graph is a straight line through the origin with slope 0.3.

Let us now analyze the factor $1/(1+j\omega/p)$. Noting that

$$(1 + j\omega/p)^{-1} \cong 1, \ \omega << 1$$

$$(1 + j\omega/p)^{-1} \cong (j\omega/p)^{-1}, \ \omega >> 1$$

we conclude the amplitude Bode diagram for this factor can be approximated, for very low and very high frequencies, by two straight lines, namely the 0 dB line, and the −20 dB/decade slope line intersecting the 0 dB line at $\omega=p$. These are known as the low and high frequency *asymptotes*, respectively. The Bode plots for $(1+j\omega/p)^{-1}$ are given in Fig. 5.7.

The frequency $\omega=p$ is known as the 'corner frequency'. At $\omega=p$ we have

$$|\ (1 + j\omega/p)^{-1}\ | = -\ 3 \text{ dB}$$

$$\arg\ (1 + j\omega/p)^{-1} = -\ 45°$$

The Bode plot for $(1+j\omega/p)$ can now be easily derived. Since

$$20 \log(x^{-1}) = -\ 20 \log x$$

i.e. when expressed in decibels the reciprocal of a number is the negative of the number, the Bode magnitude plot for $(1+j\omega/p)$ is the reflection, about the 0 dB line, of the magnitude plot for $(1+j\omega/p)^{-1}$.

As for the phase, we have that

$$\arg[(1 + j\omega/p)^{-1}] = -\ \arg(1 + j\omega/p)$$

Therefore, the phase plot is also the reflection, about the 0° line, of the phase plot for $(1+j\omega/p)^{-1}$.

Figure 5.8 depicts the Bode plots for $(1+j\omega/p)^r$, $r=-2, -1, 1, 2$. Of interest to note is the symmetry of these plots, relative to the 0 dB and 0° lines, and the 'quantum leap' of ±20 dB/decade between the asymptotes of the magnitude plots.

Exercise 5.6

(a) Compute the errors at $\omega=p$, of the asymptotic approximations of $(1+j\omega/p)^{-r}$, for $r=-2,-1, 1, 2$.

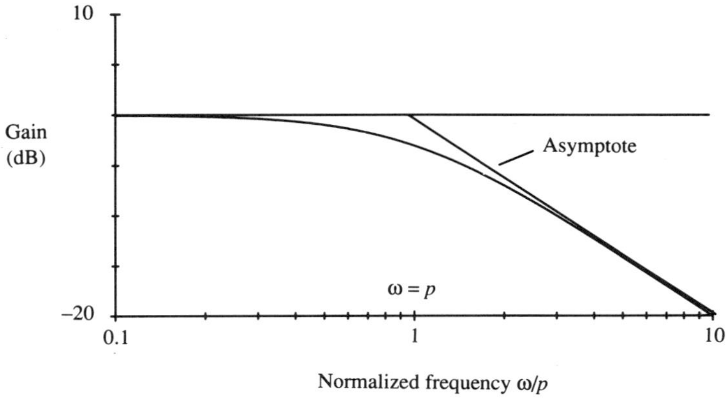

(a)

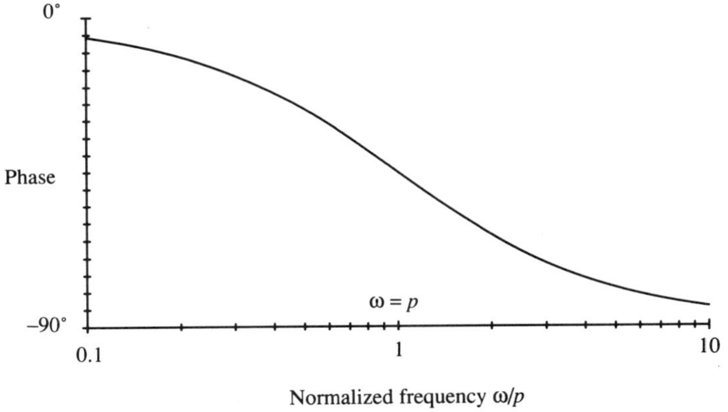

(b)

Figure 5.7 Bode plots for $(1+j\omega/p)^{-1}$.

(b) Compute the argument of $(1+j\omega/p)^{-r}$, for r as in (a), at $\omega=p$.
Answers: (a) −6 dB, −3 dB, +3 dB and +6 dB respectively; (b) −90°, −45°, 45°, 90° respectively.

The corner frequencies are very important, because they allow us to identify the system poles (or zeros) from its frequency response. Figure 5.9 depicts the magnitude Bode plot for

$$G(s) = \frac{10}{(s+1)(s+10)} \tag{5.20}$$

and the corresponding asymptotic approximations. These are the result of adding the asymptotes for $1/(s+1)$ and $10/(s+10)$.

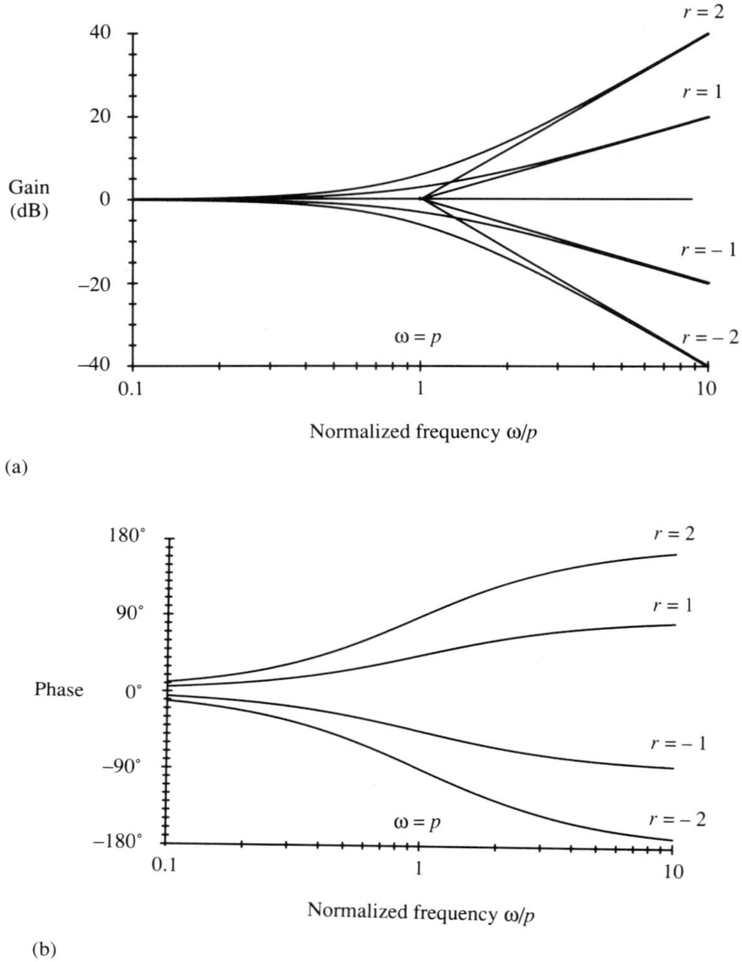

Figure 5.8 Bode plots for $(1+j\omega/p)^r$, for $r=-2$, -1, 1, 2: (a) magnitude; (b) phase.

Let us look now at a second-order system with complex poles, namely

$$\frac{1}{1 + 2\zeta\,(j\omega/\omega_n) + (j\omega/\omega_n)^2} \tag{5.21}$$

where $0<\zeta\leqslant 1$. As $\omega\to 0$, the magnitude approaches 0 dB and the phase goes to 0°. As $\omega\to\infty$ the phase converges to $-180°$ and the magnitude approaches a straight line with -40 dB/dec slope, intersecting the 0 dB line at $\omega=\omega_n$. Figure 5.10 shows the magnitude and phase Bode plots for (5.21) and $\zeta \in \{0.1,0.2,0.35,0.5,0.7\}$.

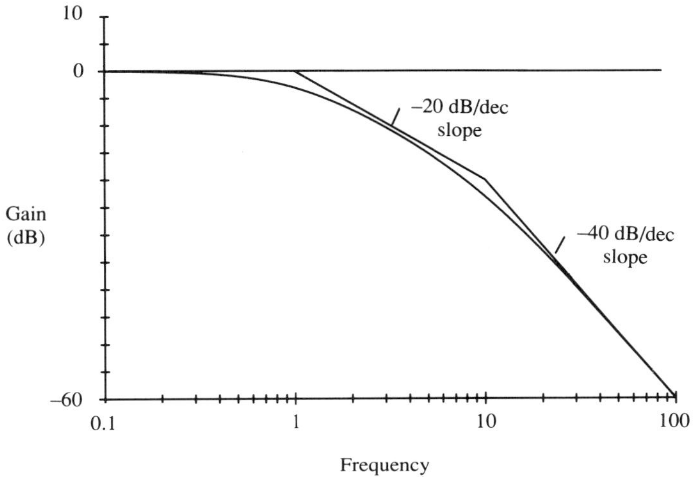

Figure 5.9 Magnitude Bode plot for 10/(s+1) (s+10)).

As mentioned earlier, the magnitude Bode plot ceases to be monotonically decreasing for $\zeta<0.707=\sqrt{2}/2$, exhibiting a maximum at

$$\omega = \omega_r = \omega_n \sqrt{(1 - 2\zeta^2)} \tag{5.22}$$

Recall that (5.21) exhibits an impulse response with a damped transient term of frequency

$$\omega_n \sqrt{(1 - \zeta^2)} \tag{5.23}$$

At $\omega=\omega_r$ the (maximum) magnitude value, the *resonant peak* M_r, is given by

$$M_r = \frac{1}{2\zeta \sqrt{(1 - \zeta^2)}} \tag{5.24}$$

Naturally $M_r=1$ for $\zeta>\sqrt{2}/2$.

As $\zeta\rightarrow 0$, M_r tends to infinity. This is in agreement with the geometric interpretation of frequency response given earlier. As the poles approach the imaginary axis, the peak magnitude increases, becoming infinite when the poles are on the imaginary axis. The geometric interpretation also shows that for low values of ζ, the resonant peak occurs in the vicinity of the magnitude of the poles. As for the phase curve, we see from Fig. 5.10 that the $-180°$ step, located at $\omega=\omega_n$ is a good approximation when ζ is small. We invite the reader to compare this analysis with the time-domain study in Chapter 3, namely formulae (3.37)–(3.39).

To complete this analysis, the phase Bode plots for the pure transport delay factor $e^{-j\omega T}$ are shown in Fig. 5.11 for $T=0.5$ and $T=1$. The magnitude plot is obviously a straight line at 0 dB.

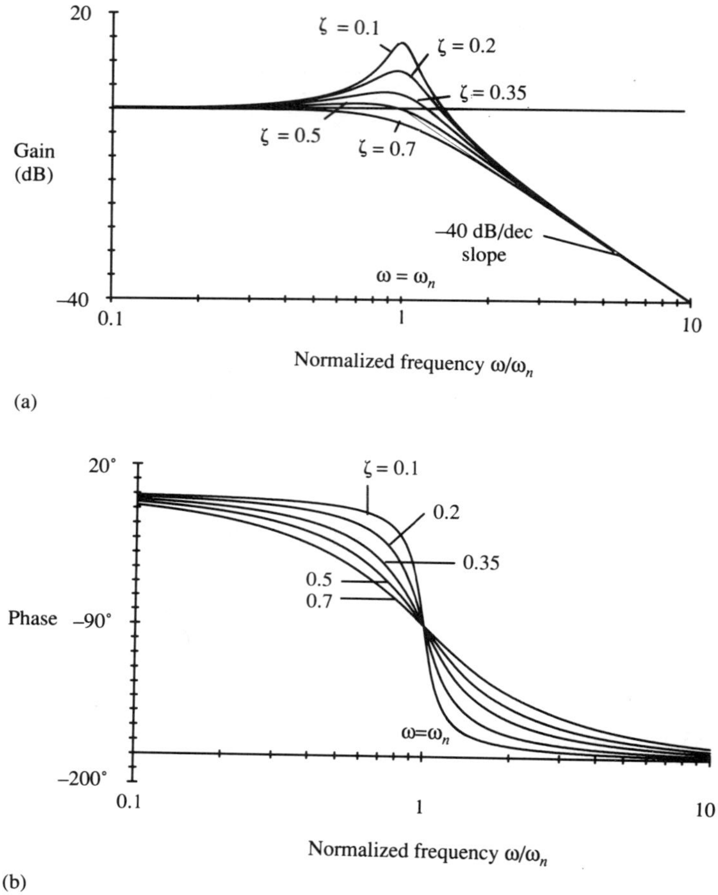

Figure 5.10 Bode plots for $1/(1+2\zeta\,(j\omega/\omega_n)+(j\omega/\omega_n)^2)$.

Despite its apparent simplicity, the pure transport delay factor may induce interesting characteristics when combined with rational factors. This is the case for the transfer function $(1-e^{-Ts})/s$, which plays an important role in the analysis of sampled-data systems, namely in connection with the problem of continuous-time reconstruction of (digital) computer signals, as seen later in the book.

Exercise 5.7 Compute the unit-step response of $(1-e^{-Ts})/s$.
Answer: the response is a continuous piecewise-linear function, consisting of a line segment through the origin, with unit slope, between 0 and T, and a horizontal line afterwards, with ordinate T.

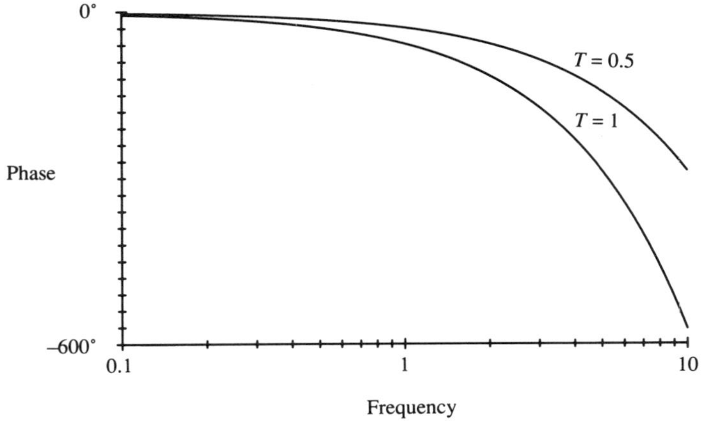

Figure 5.11 Phase Bode plots for exp $(-j\omega T)$.

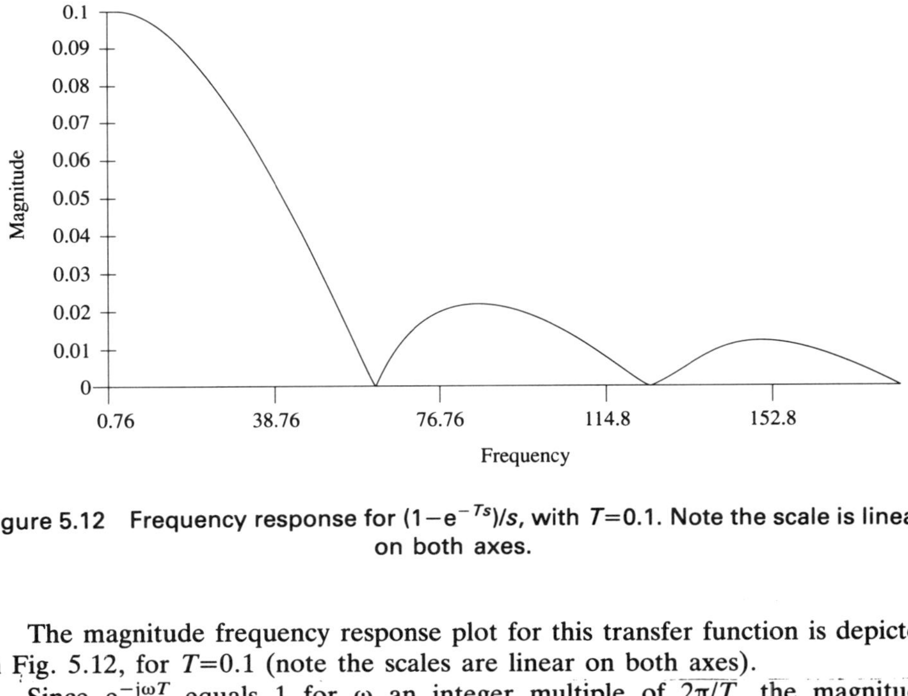

Figure 5.12 Frequency response for $(1-e^{-Ts})/s$, with T=0.1. Note the scale is linear on both axes.

The magnitude frequency response plot for this transfer function is depicted in Fig. 5.12, for T=0.1 (note the scales are linear on both axes).

Since $e^{-j\omega T}$ equals 1 for ω an integer multiple of $2\pi/T$, the magnitude frequency response vanishes at these frequencies (on a Bode magnitude plot the ordinates at such frequencies would be minus infinity).

For $\omega=k\pi/T$, $k\in\{1, 3, 5, 7, \ldots\}$, $e^{-j\omega T}$ equals -1, and the magnitude of

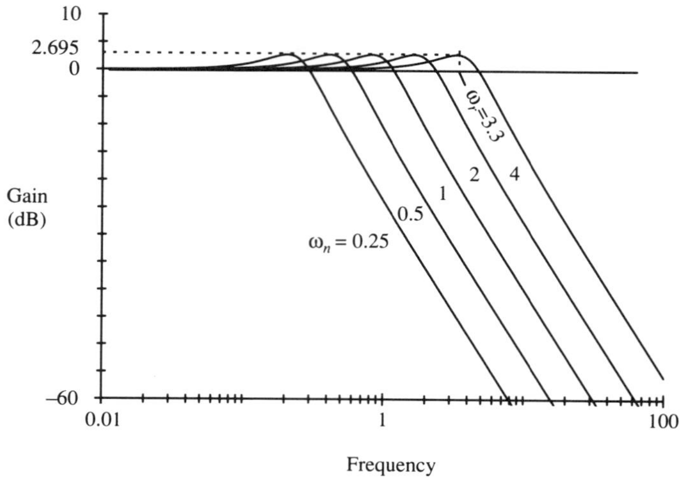

Figure 5.13 Bode magnitude plots for $\omega_n^2/(s^2+2\zeta\,\omega_n\,s+\omega_n^2)$, with ζ=0.4 and $\omega_n \in \{0.25, 0.5, 1,2,4\}$.

$(1-e^{-j\omega T})$ reaches the maximum. Since $1/\omega$ is a monotonic decreasing function, these maxima become the local maxima of magnitude $(1-e^{-j\omega T})/j\omega$, as shown in Fig. 5.12.

5.1.4 The Role of Zeros on the Frequency Response

The role of zeros on transient response can also be analyzed in the frequency domain. The analysis carried out in Chapter 3 has shown that the effect of a zero can be judged from 'the speed of the poles as seen from the zero'. Figure 3.15 illustrates this property for the system

$$\frac{\omega_n^2(1 + s)}{s^2 + 2\zeta\,\omega_n\,s + \omega_n^2} \tag{5.25}$$

Let us repeat this analysis in the frequency domain by plotting the Bode magnitude curves for (5.25) and for

$$\frac{\omega_n^2}{s^2 + 2\zeta\,\omega_n\,s + \omega_n^2} \tag{5.26}$$

These are shown in Figs 5.13 and 5.14: comparing them clearly reveals that the role of the zero is to increase the resonant peak; furthermore, as the speed of poles increases (i.e. ω_n) the more pronounced the effect of the zero at (-1).

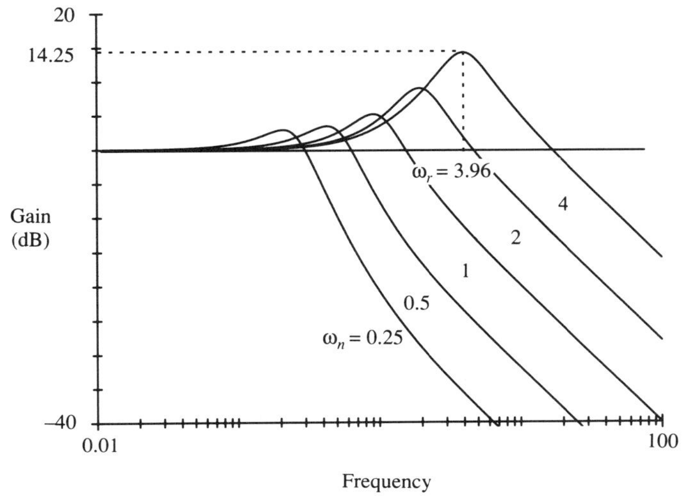

Figure 5.14 Bode magnitude plots for $\omega_n^2\,(s+1)/(s^2+2\zeta\,\omega_n\,s+\omega_n^2)$, with $\zeta=0.4$ and $\omega_n \in \{0.25, 0.5, 1,2,4\}$.

When the zero is 'very fast, as seen from the poles', it has a negligible effect, since its contribution appears only when the magnitude of (5.26) is already very small. This happens, for example, for $\omega_n=0.25$ in Fig. 5.14. In any case, the zero increases the slope of the high frequency asymptote by +20 dB/dec.

Note that Bode plots also help to quantify the idea of 'the speed of a zero z as seen from a pole p'. As we have seen, such speed is judged in terms of their distances on the logarithmic scale, i.e.

$$\log|p| - \log|z| = \log\left|\frac{p}{z}\right|$$

and not by $|p-z|$. In a sense, the logarithmic scale measures their *relative* effects. We offer another example to convince the more sceptical reader.

Example 5.1 The frequency response of

$$\frac{16\,(s+1)}{(s+3.2 \pm \mathrm{j}\,2.4)} \tag{5.27}$$

has a maximum of 14.24 dB at ω_r=3.9 rad/s. Also

$$|p-z| = |\bar{p}-z| = |-3.2+\mathrm{j}\,2.4+1| = 3.3$$

and

$$|p/z| = 4$$

On the other hand

$$\frac{40\,(s + 10)}{s^2 + 1.6\,s + 400} \tag{5.28}$$

exhibits a smaller resonance, namely 9 dB at ω_r=19.5 rad/s, although it has a greater pole-zero distance:

$$|\,p - z\,| = |\,\bar{p} - z\,| = |\,-8 + j\,18.33 + 10\,| = 18.44$$

However, the pole-zero ratio is twice as large in the first case.

Exercise 5.8

(a) Sketch the magnitude Bode plots for

(i) $\dfrac{0.4\,(s + 1)}{s + 0.4}$

(ii) $\dfrac{0.7\,(s + 1)}{s + 0.7}$

(iii) $\dfrac{2\,(s + 1)}{(s + 2)}$

(b) What conclusions can be drawn from the previous diagrams?

Answers: (b) When the pole becomes greater than the zero, the gain starts to *increase* monotonically as $\omega \to \infty$. Consequently the response to a unit step at t=0, for example, will assume larger values at $t=0^+$ than at $t=\infty$. Cf. Fig. 3.13.

A comparison of the Bode plots for the transfer functions (5.25) and (5.26), for $\omega >> 1$, reveals that the addition of the zero raised the slope of the high frequency asymptote by +20 dB/decade and produced a 'phase lift' of +90°. This is a consequence of the fact that the frequency response of any rational transfer function

$$\frac{b_0 s^m + b_1 s^{m-1} + \ldots + b_{m-1} s + b_m}{s^n + a_1 s^{n-1} + \ldots + a_{n-1} s + a_n}$$

converges to

$$\frac{b_0}{(j\omega)^{n-m}}$$

as ω gets larger and larger. In other words, we can say that, in the high frequency region, any proper rational transfer function behaves as a pure gain in series with $(n-m)$ integrators. Cf. the Bode plots of Fig. 5.6.

Recalling our simulation diagram of Fig. 2.1, what we have just said means that at high frequencies what counts is the 'fastest path' between input and output. For example, when $m=n$ the fastest path is the direct path through the pure gain block b_0. If $(n-m)=1$ there is no such path between u and y, because b_0

has now x_n (the output of the first integrator) as its input. The 'fastest path' is therefore through the integrator with output x_n. And so on, for larger values of $(n-m)$.

5.1.5 Transfer Function Identification from Bode Plots

Assume, for simplicity, that transfer function (5.12) is strictly proper and rational, i.e. $T=0$ and $(n+r)>m$.

It was mentioned earlier that $G(s)$, s complex, is wholly determined by $G(j\omega)$, $\omega \in [0,\infty)$, on the basis of Cauchys integral formula. We have therefore two equivalent representations for a transfer function:

1. By means of Bode plots.
2. By means of a quotient of two polynomials.

The former is an infinite dimensional representation, while the latter constitutes a finite dimensional one; in fact we can always establish a one-to-one correspondence between $^{(n+r+m)}$ and the polynomial coefficients of $G(s)$.

Usually, Bode plots are obtained from experimental tests, while the rational form of $G(s)$ is the result of Laplace transforming a differential equation model of the system obtained by physical considerations. Both representations have their merits and pitfalls when it comes to controller design, although time-domain methods are more amenable to routine design procedures. Therefore it is of practical interest to compute the poles and zeros of a transfer function defined by its Bode plot. Unfortunately this is not easy to achieve in practice, except in rare situations where the poles and zeros are well apart, say at least by one decade. In these cases, their calculation is straightforward on the basis of the corner frequencies (cf. exercise 5.6). Figure 5.15 illustrates an entirely opposite situation (Lopes dos Santos and Martins de Carvalho, 1990). It gives the Bode plots for the transfer functions:

$$G_1(s) = \frac{(s + 3)\,(s + 4)}{(s + 1)\,(s + 2)\,(s + 20)} \tag{5.29}$$

and

$$G_2(s) = \frac{(s + 5.554)}{(s + 0.874)\,(s + 21.341)} \tag{5.30}$$

Despite the differences between the two transfer functions, namely in the degrees of the numerators and denominators, the plots are almost indistinguishable!

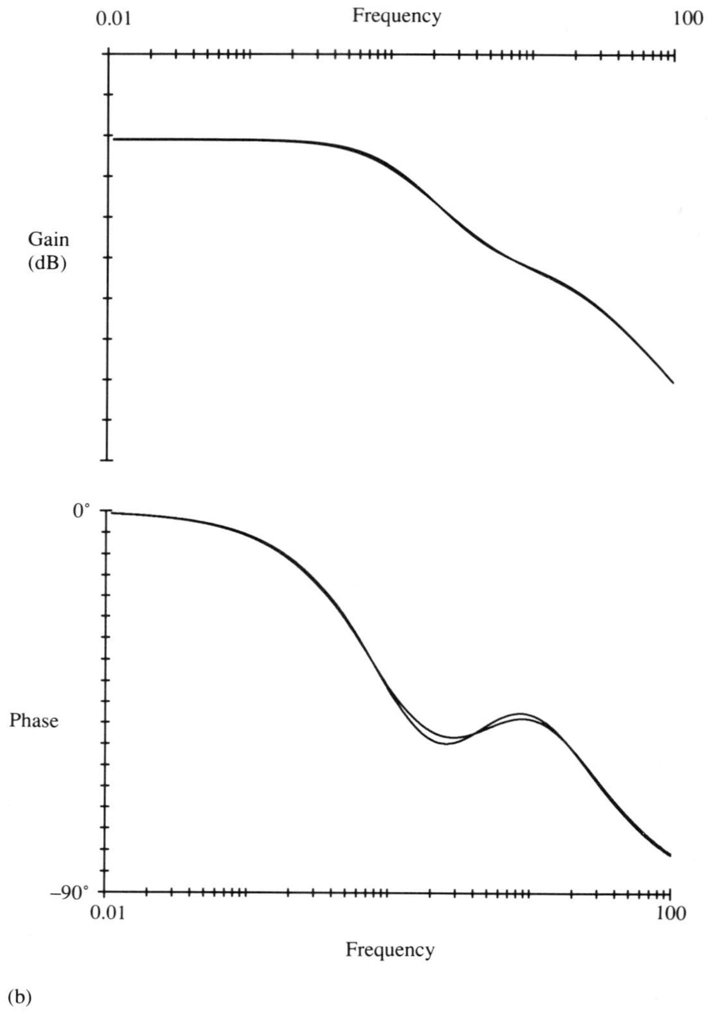

Figure 5.15 Bode plots for the transfer functions $(s+3)\,(s+4)/((s+1)\,(s+2)\,(s+20))$ and $(s+5.554)/((s+0.874)\,(s+21.341))$: (a) magnitude; (b) phase.

5.2 CLOSED-LOOP ANALYSIS

We begin this section by addressing the following problem: given the frequency response for a transfer function

$$G(s)\ H(s) = \frac{N(s)}{D(s)}$$

where polynomials $N(s)$ and $D(s)$ are unknown, is there a way to analyze the stability of the negative feedback system shown in Fig. 5.16? (Obviously if $N(s)$ and $D(s)$ were known, the problem could be solved by computing the zeros of $1+G(s)H(s)$.) The answer is yes and is given by Nyquist stability criterion. The underlying result is the so-called *principle of the argument* which can be stated as follows. Let F be a function of a complex variable which is analytic in a region D of the complex plane, except for a finite number of poles, and Γ a clockwise oriented closed contour in D containing no poles and zeros of F (a contour is a continuous chain of a finite number of smooth arcs). Then the number of clockwise encirclements N of the origin, of the image of Γ by F, is $(Z-R)$, where Z and R denote the number of zeros and poles of F (multiplicity accounted for) inside Γ.

Let us assume, for example,

$$F(s) = \frac{s - a}{s - b}$$

and Γ a clockwise oriented closed contour as shown in Fig. 5.17. Then for a point s' on Γ we have

$$\arg F(s') = \arg \left(\frac{s' - a}{s' - b} \right) = \alpha$$

Let us take a closer look at part (a) of this figure, where Γ encloses only one pole, i.e. $Z=0$ and $R=1$. When s' is at P_1 we have $\alpha=-\pi$. As s' travels along Γ in the indicated direction we find α equals $-\pi/2$, 0, $\pi/2$, π when s' is at P_2, P_3, P_4 and P_1 respectively. Consequently, $N=-1$. Since $Z=0$ and $R=1$ we have confirmed in this example the above-mentioned principle. A similar analysis applies to cases (b) and (c).

The usefulness of this result is now apparent with reference to the feedback system depicted in Fig. 5.16. If Γ is a contour as depicted in Fig. 5.18, with r

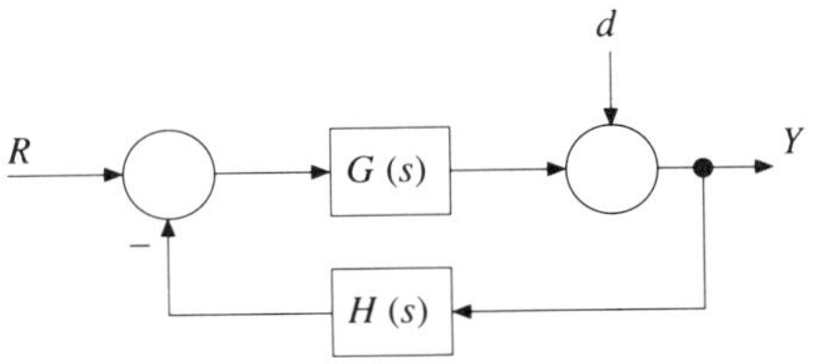

Figure 5.16 Negative feedback configuration.

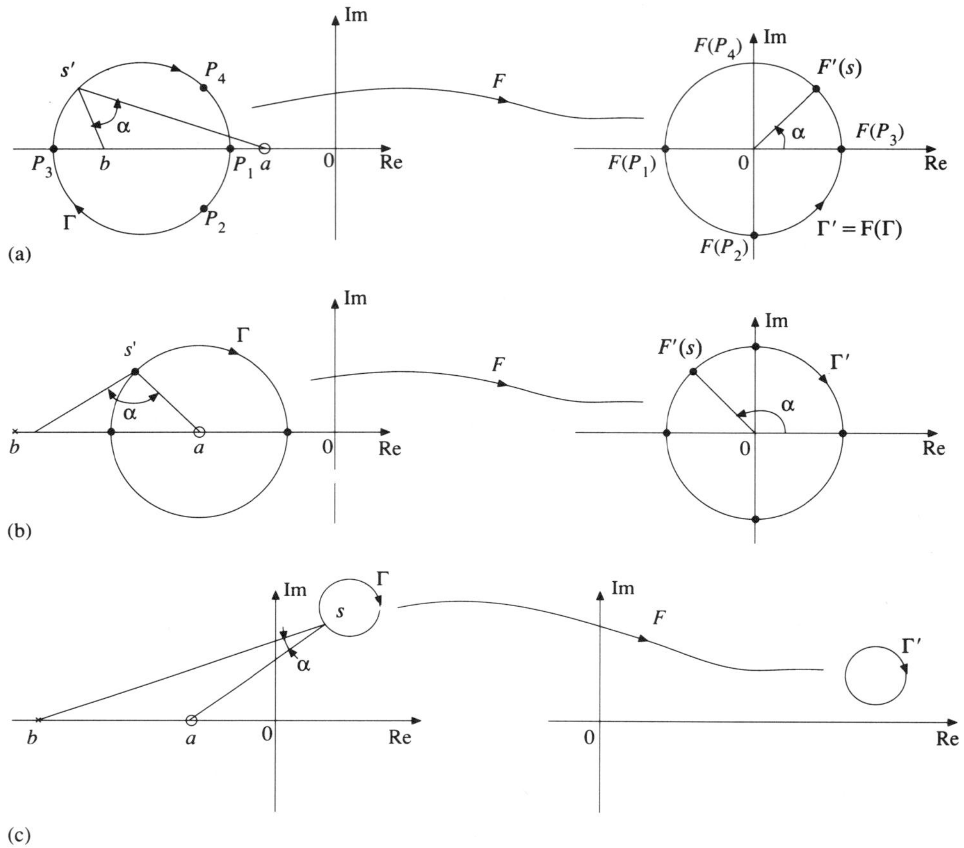

Figure 5.17 Illustration of the principle of the argument with $F(s)=(s-a)/(s-b)$.

sufficiently large to encompass all the right half-plane (RHP) poles and zeros of $F(s)=1+G(s)H(s)$, then the number of zeros of $(1+G(s)H(s))$ in the right half-plane (unstable closed-loop poles) equals the number of poles of $(1+G(s)H(s))$ in the RHP (which is zero if open-loop stable) plus the number of clockwise encirclements of the origin of the image of Γ by $(1+G(s)H(s))$.

Since we do not know, in general, any upper bound for the magnitude of the poles and zeros of $G(s)H(s)$, the semicircle of radius r must encompass all the RHP, i.e. we must compute the limiting image of this contour when $r \to \infty$.

If $G(s)H(s)$ is strictly proper, then the image of the semicircle with infinite radius will be zero. Otherwise it will be a finite constant (cf. exercise 3.2).

Consequently the image of the contour in Fig. 5.18, when $r=\infty$, can be computed without the knowledge of the numerator and denominator polynomials of $G(s)H(s)$: it suffices to know $G(s)H(s)$ along the imaginary axis, i.e. $s=j\omega$, $\omega \in$

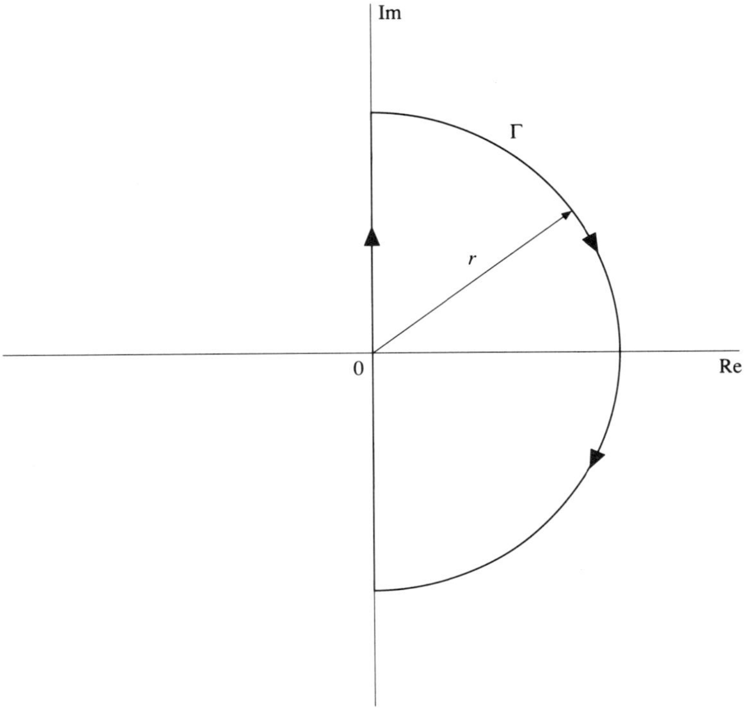

Figure 5.18 A practical way to define a contour encircling all the right half-plane zeros and poles of a given transfer function.

$(-\infty,\infty)$. But again we have a further simplification. Since the numerator and denominator polynomials in $G(s)H(s)$ have real coefficients, $G(j\omega)$ is the conjugate of $G(-j\omega)$; therefore we only need to compute the image of the positive imaginary axis. For practical reasons, we plot $G(j\omega)H(j\omega)$ instead of $(1+G(j\omega)H(j\omega))$. Since the former is obtained from the latter by subtracting $(-1+j0)$, the number of encirclements of the origin by $(1+G(j\omega)H(j\omega))$ is the same as the number of encirclements of $(-1+j0)$ by $G(j\omega)H(j\omega)$.

We are now in a position to state the *Nyquist stability criterion*: for the system of Fig. 5.16 let

N = number of clockwise encirclements of the $(-1+j0)$ point by $G(j\omega)H(j\omega)$, $\omega \in (-\infty+\infty)$;
R = number of poles of $G(s)H(s)$ in the right half-plane;
Z = number of zeros of $(1+G(s)H(s))$ in the right half-plane (unstable closed-loop poles).

Then

$$Z = N + R$$

Consequently, if $G(s)H(s)$ is a stable transfer function ($R=0$) the closed-loop system is also stable ($Z=0$) if and only if the locus of $G(j\omega)H(j\omega)$, $\omega \in (-\infty+\infty)$, does not encircle the $(-1+j0)$ point. When $G(s)H(s)$ has R right half-plane poles the $G(j\omega)H(j\omega)$ locus must encircle the $(-1+j0)$ point R times anti-clockwise to ensure stability of the closed-loop system.

Although the proof is more involved, it can be shown that the Nyquist stability criterion also holds when the open-loop transfer function has the form

$$G(s)H(s) = \frac{N(s)}{D(s)}e^{-Ts}$$

There is still another point that deserves attention, which concerns the case of $G(s)H(s)$ having poles or zeros on the imaginary axis. In this situation, the contour Γ, depicted in Fig. 5.18, violates our assumptions. A practical alternative is to modify Γ by means of a detour into the RHP, consisting of a semi-circle centred at such points, with a very small positive radius ε. Then, by making ε go to zero, all the closed-loop RHP poles, if any, will eventually be enclosed. The following examples will clarify these issues.

Example 5.2 In Fig. 5.16, let

$$G(s) = \frac{1}{s\,(s+1)}, \quad H(s) = 1$$

The open-loop transfer function has a pole on the imaginary axis and consequently the contour Γ, of Fig. 5.18, must be modified. A possible alternative is shown in Fig. 5.19: we take a detour around the origin, by means of a semicircle of radius ε, $0<\varepsilon<<1$, and leave the remaining part of Γ unchanged.

In order to compute the image of Γ by $G(s)$, we set

$$s = \varepsilon e^{j\theta}, \; s \in \text{arc } ABC$$

$$s = j\omega, \; s \in \text{imaginary axis}$$

$$s = re^{j\alpha}, \; s \in \text{arc } DEF$$

The result is shown in Fig. 5.20(a) assuming a small positive ε and a large r: when s travels along the semicircle ABC, θ changes from $-\pi/2$ to $\pi/2$ and r remains constant; also $G(s) \cong (1/\varepsilon)e^{j\theta}$; then arg $G(s)$ will change from $\pi/2$ to $-\pi/2$, approximately.

When s belongs to arc DEF, we can write

$$G(s) \cong \frac{1}{s^2} = \frac{1}{r^2}e^{-j2\alpha}$$

because r is very large. Since the argument of s will then decrease by π rad, the argument of $G(s)$ will increase by approximately 2π.

In the limiting case, i.e. $r=\infty$ and $\varepsilon=0$, we have the situation illustrated in Fig. 5.20(b), showing that $N=0$. Because R is also zero (no open-loop right half-plane poles), we have $Z=0$. The closed-loop system is therefore stable. This comes

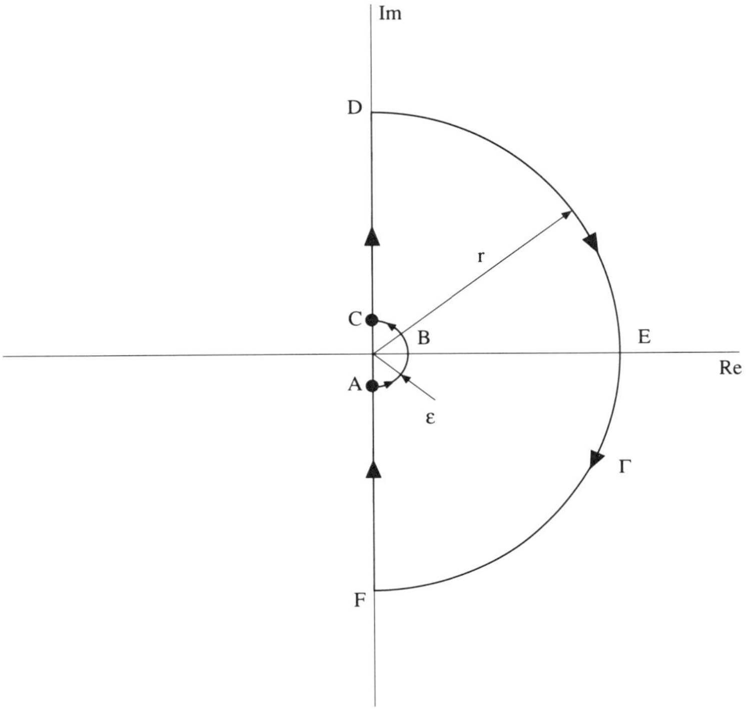

Figure 5.19 A closed contour to avoid the pole at the origin.

as no surprise since we already know, from the root-locus analysis, that the system of Fig. 5.16 is stable when $G(s)H(s)=K/[s\ (s+1)]$, for all non-negative K.

However, as K increases, the curve on Fig. 5.20(b) will pass closer to the $(-1+j0)$ point. Since a large K induces a highly oscillatory step response, we are led to the conclusion that a lower bound for the distance of the $G(j\omega)\ H(j\omega)$ curve to this point must be imposed, in order to ensure a satisfactory closed-loop performance. We will resume this point later in the chapter.

Example 5.3 In this example we take $H(s)=1$ and

$$G(s) = \frac{K}{s\ (s+1)^2}$$

To analyze the stability of the feedback system we must use the s-plane contour of Fig. 5.19 since we have again a pole at the origin. Its image by $G(s)$, for a small positive ε and a large r, is given in Fig. 5.21(a). Note that for $\omega=1$, $G(j\omega)=-K/2$. Also, as ω approaches 0, the real part of $G(j\omega)$ converges to $-2K$.

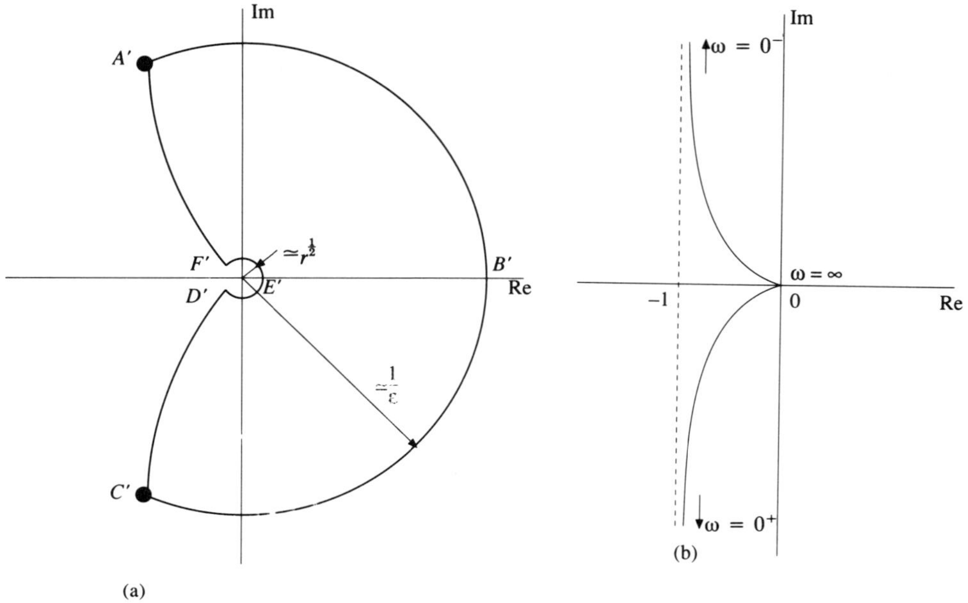

Figure 5.20 Image of the contour of Fig. 5.19 by the transfer function $1/(s(s+1))$: (a) $0 < \varepsilon << 1$; $1 << r < \infty$; (b) $\varepsilon = 0$; $r = \infty$.

Exercise 5.9 Compute the real part of $G(j\omega)$ in example 5.3, and its limit when $\omega \to 0$.

Answer:

$$G(j\omega) = \frac{K}{j\omega\,(1-\omega^2) - 2\omega^2} = \frac{K(-2\omega^2 - j\omega(1-\omega^2))}{4\omega^4 + \omega^2\,(1-\omega^2)^2}$$

In fact

$$\lim_{\omega\to 0} \mathrm{Re}(G(j\omega)) = \lim_{\omega\to 0} \frac{-2K\omega^2}{4\omega^4 + \omega^2\,(1-\omega^2)^2} = -2K$$

For $\omega \cong 0$ the denominator is dominated by ω^2. The result follows.

The limiting situation is depicted in Fig. 5.21(b). If $K<2$, there are no encirclements of the $(-1+j0)$ point; since $P=0$ we have $Z=0$. The closed-loop system is therefore stable for $K<2$. For $K=2$ the closed-loop system has a pair of poles on the imaginary axis, namely $\pm j$. This is confirmed by the root-locus shown in Fig. 5.22. Also of interest is the fact that the ordinate of the intersection of the loci with the imaginary axis is the value of ω for which $G(j\omega)\,H(j\omega)$ intersects with the negative real axis. This has an interesting interpretation: from the root-locus we conclude thathe closed-loop conjugate pole pair on the imaginary

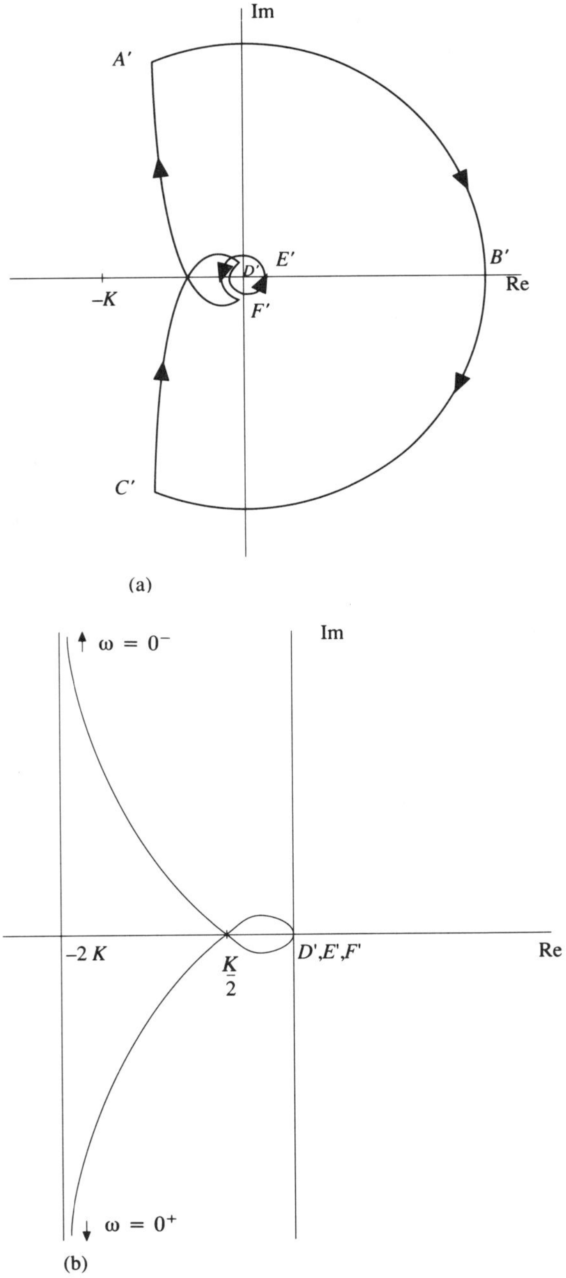

Figure 5.21 Image of the contour of Fig. 5.19 by the transfer function $G(s)=K/(s(s+1)^2)$: (a) $0<\varepsilon<<1$; $1<<r<\infty$; (b) $\varepsilon=0$; $r=\infty$.

axis will make the system oscillate sinusoidally with a frequency equal to the pole magnitude, in the absence of input. From the Nyquist analysis of Fig. 5.21, we can see that the closed-loop sinusoidal gain becomes infinite, at frequency $\omega=1$ when $K=2$; consequently the system is capable of exhibiting a steady oscillation of finite amplitude and frequency $\omega=1$, when the input is zero.

Example 5.4 Let us now consider a system with a pure transport delay, namely the system of Fig. 4.16. We already know the closed-loop system is unstable, for $K \geq 1$, and it has an infinite number of poles, as shown by the root-loci in Fig. 4.19. It would be interesting to reach these conclusions by means of the Nyquist criterion. With this in mind, we plot the images of the contours shown in Fig. 5.23(a) and (b), by the function $K \exp(-Ts)$: it can be seen that, as the radius of the semicircle (ABC) goes to infinity, the image of point B, i.e. B', goes to zero; the image of the segment of Γ on the imaginary axis remains on the circle of radius K, and exhibits a new complete turn around the origin every time the radius of the semicircle (ABC) is incremented by π/T. Consequently there is no limit for the image of the semicircle ABC.

When $K<1$, point $(-1+j0)$ is never encircled, and the closed-loop system is stable, as confirmed by the loci of Fig. 4.19. When $K>1$, we find in Fig. 5.23(a) that point $(-1+j0)$ is not encircled. This is in agreement with Fig. 4.19 since the s-plane contour Γ does not yet enclose any closed-loop pole. However the contour Γ, in part (b), already encloses a pair of closed-loop poles, which is confirmed by the two encirclements of $(-1+j0)$. As the radius of the semicircle (ABC) goes to infinity, the image of Γ will encircle the $(-1+j0)$ an infinite number of times. The closed-loop system has therefore an infinite number of right half-plane poles. This is again confirmed by the loci in Fig. 4.19.

Quite naturally, for $K=1$ the closed-loop poles will lie on the imaginary axis, since they change continuously with K. In this case the image of the imaginary axis contains the $(-1+j0)$ point.

Example 5.5 In the same line of thought, let us now analyze a slightly more complex example, namely the system of Fig. 4.20. The root-loci depicted in

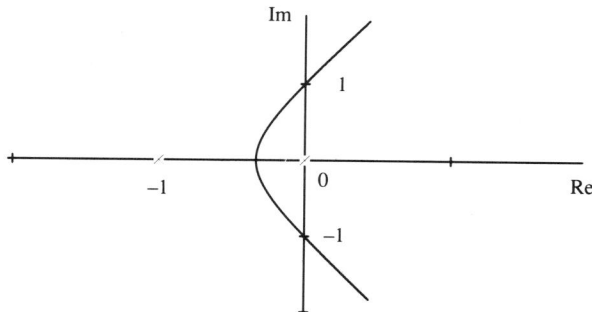

Figure 5.22 Root-locus for $G(s)H(s)=K/(s(s+1)^2)$.

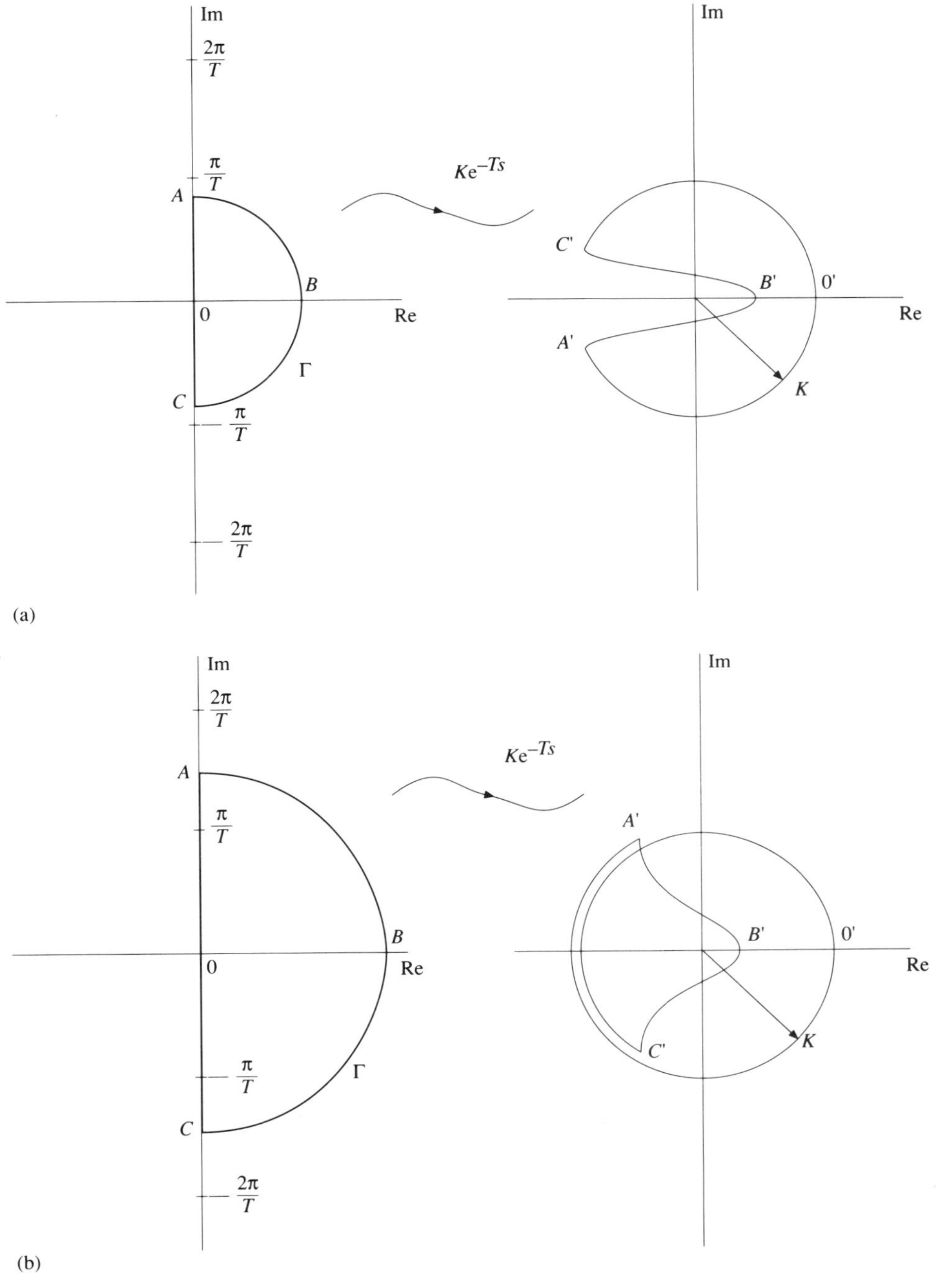

Figure 5.23 Mapping an s-plane contour by Ke^{-Ts}, $K>0$.

Fig. 4.24 show that the closed-loop system is unstable for $K \geq K_c$. In contrast to the previous example, here the number of closed-loop right half-plane poles (if any) is always finite: there is an increasing sequence of positive values for the gain K,

$$(K_n)_{n \in \mathbb{N}} \rightarrow \infty$$

such that the closed-loop system is stable for $K < K_1 = K_c$, and it has $2(n-1)$ right half-plane poles when $K_{n-1} \leq K < K_n$ for $n \geq 2$.

The polar plots of Fig. 5.5 confirm this result. Also bearing in mind the image of the negative imaginary axis (not represented), we conclude that the $(-1+j0)$ point is always encircled an even (and finite) number of times.

From now on, we shall consider only the image of the positive imaginary axis in stability studies. As mentioned earlier, the images of $j\omega$ and $-j\omega$ by any transfer function are complex conjugates; therefore no clarity is lost by doing so.

5.2.1 Relative Stability

In practice we are not only interested in knowing if the system is stable but also how far it is from instability. Because the distance of the $G(j\omega)\ H(j\omega)$ locus to the $(-1+j0)$ point is the magnitude of the denominator of the closed-loop (sinusoidal) transfer function Y/R, i.e.

$$|\ 1 + G(j\omega)H(j\omega)\ |$$

we can say that the closer the $G(j\omega)H(j\omega)$ locus comes to encircling the $(-1+j0)$ point, the more oscillatory the closed-loop system response will be. Such closeness is usually 'measured' in terms of *phase margin* and *gain margin*.

The gain margin (GM) indicates how much the gain must be raised to drive the system to the verge of instability, i.e. to make the loop gain equal to unit when the phase is equal to $-180°$.

The phase margin (PM) is the angle

$$180° + G(j\omega_c)\ H(j\omega_c)$$

where ω_c, the crossover frequency, is the value of ω at the intersection of the $G(j\omega)\ H(j\omega)$ locus with the origin-centred unit circle. Figures 5.24 and 5.25 show how these margins can be determined from the polar and Bode plots. The latter are particularly tailored to determine these measures since the decibel scale converts gain multiplications into additions, in particular 1/GM dB$=-$GM dB. Consequently a stable closed-loop system will have, in decibels, a positive gain margin.

Stability also requires a positive phase margin; with reference to Fig. 5.24, if we multiply the $G(j\omega)H(j\omega)$ locus by a positive constant K, such that the $(-1+j0)$ is encircled, the closed-loop system will be unstable and the angle

$$180° + KG(j\omega_c)H(j\omega_c)$$

will then be negative.

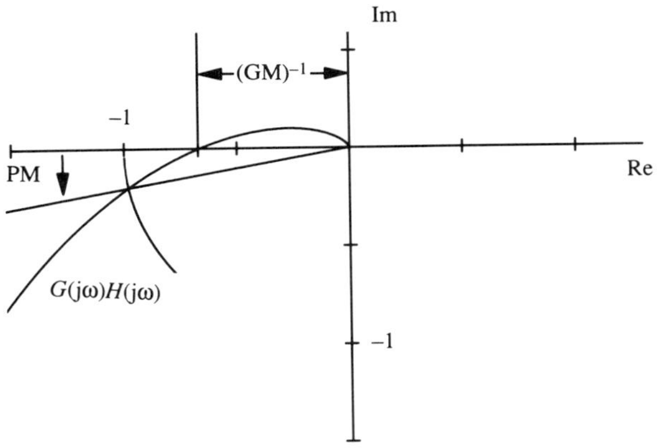

Figure 5.24 Gain and phase margins from the polar plot.

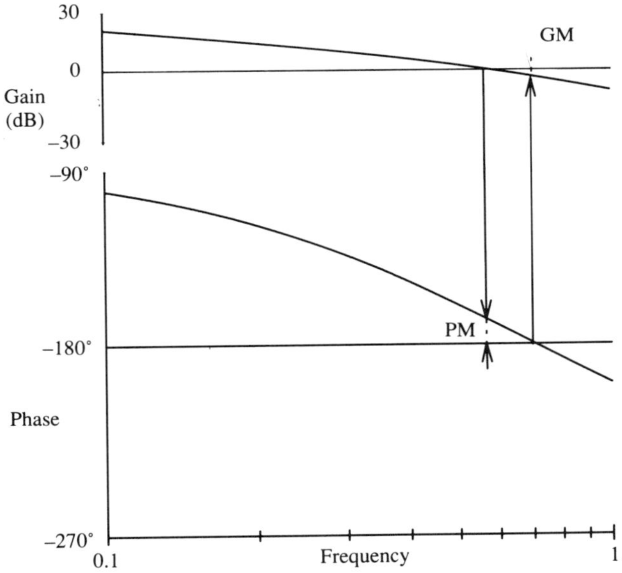

Figure 5.25 Gain and phase margins from the Bode plots.

When the magnitude and the phase of $G(j\omega)\ H(j\omega)$ decrease monotonically with increasing ω, as depicted in Fig. 5.24, the notions of gain margin and phase margin are well defined, and can be used as measures of stability. This is the situation in the great majority of the cases. However there are exceptions. Two

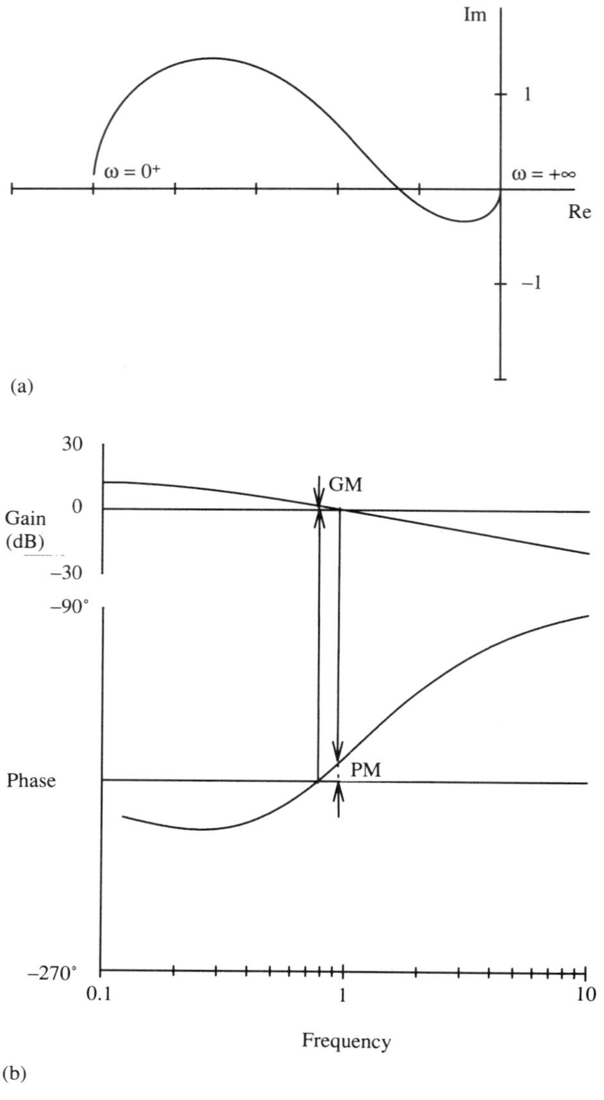

Figure 5.26 Polar (a) and Bode (b) plots for the open-loop transfer function $G(s)H(s)=(s+1)/((s+0.2)(s-1))$.

such cases will be addressed in the following, namely the case of more than one crossing of the unit circle and the case of anti-clockwise encirclements of the $(-1+j0)$ point. A case of anti-clockwise encirclements of the $(-1+j0)$ point is shown in Fig. 5.26(a): although only the image of the positive imaginary axis is shown, it is easy to see that the image of the entire jω axis encircles the $(-1+j0)$ point once in the anti-clockwise direction. Therefore $N=-1$. Since we have one right half-plane pole, $P=1$. Consequently $Z=0$, i.e. the closed-loop system is stable. However, the gain margin (in decibels) is negative as shown in Fig. 5.26 (b)!

The transfer function

$$\frac{s+1}{(s+0.2)(s-1)}$$

is known as a *non-minimum* phase transfer function because the net phase change, for the entire frequency range, is greater for the same magnitude, than if the pole were in the left half-plane. Quite naturally, all transfer function with all the poles and zeros in the left half-plane are minimum phase transfer functions. In fact, the Bode magnitude curves for

$$\frac{s+1}{(s+0.2)(s-1)}$$

and

$$\frac{1}{(s+0.2)}$$

are identical; however, the former undergoes a larger net change in phase as ω goes from 0 to ∞.

Exercise 5.10 In Fig. 5.27 we have the complete polar plot for an open-loop stable transfer function. What are the gain-margin and the stability properties of the closed-loop system, assuming:
(a) $P_1=(-1+j0)$;
(b) $P_2=(-1+j0)$.
Answer: the gain margin concept is meaningless in (a). In (a) the closed-loop system is stable and in (b) it is unstable.

When there is more than one crossing of the unit circle (and the open-loop transfer function has all its poles and zeros in the left half-plane), a meaningful definition of phase margin can be achieved in connection with the largest pure time-delay the system can withstand without losing stability. Consider, for example,

$$G(s)\,H(s) = \frac{(s^2+0.7s+1600)\,1000}{s\,(s+100)\,(s+160)}$$

Its Bode and polar plots are depicted in Figs 5.28 and 5.29. Since there are three values of frequency, namely 32.33 rad/s, 50.47 rad/s and 979.4 rad/s for which $|\,G(j\omega)\,H(j\omega)\,| = 1$, we have the candidate phase margin values 63°, 223.58°, 105.07°, respectively. At these frequencies, the values of T such that ωT equals

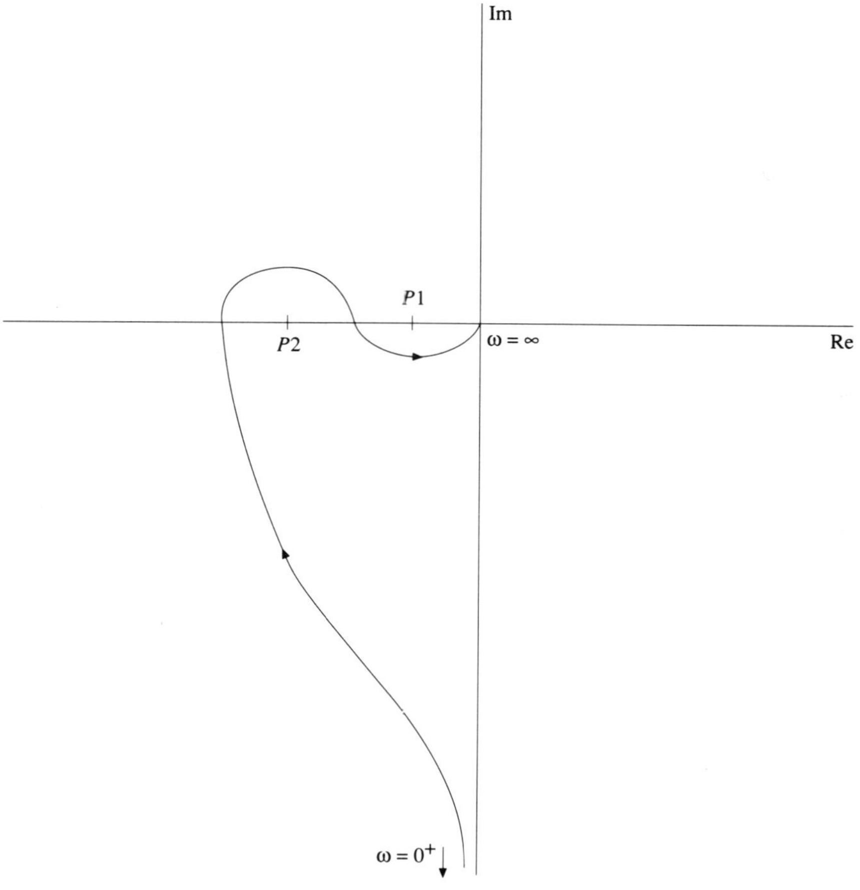

Figure 5.27 Illustration of ambiguities in defining the gain margin.

the candidate phase margins are, respectively, 0.034 s, 0.077 s and 0.001 873 s.

Therefore, the pure transport delay the system can withstand without losing stability must be less than 0.001 873 s. In Fig. 5.30 we can see the polar plot of

$$G(j\omega)\ H(j\omega)\ e^{-j\omega(0.001\ 873)}$$

with $G\ H$ as in Fig. 5.29: the closed-loop system is on the verge of stability, as expected. Consequently the phase margin is 105.07 degrees.

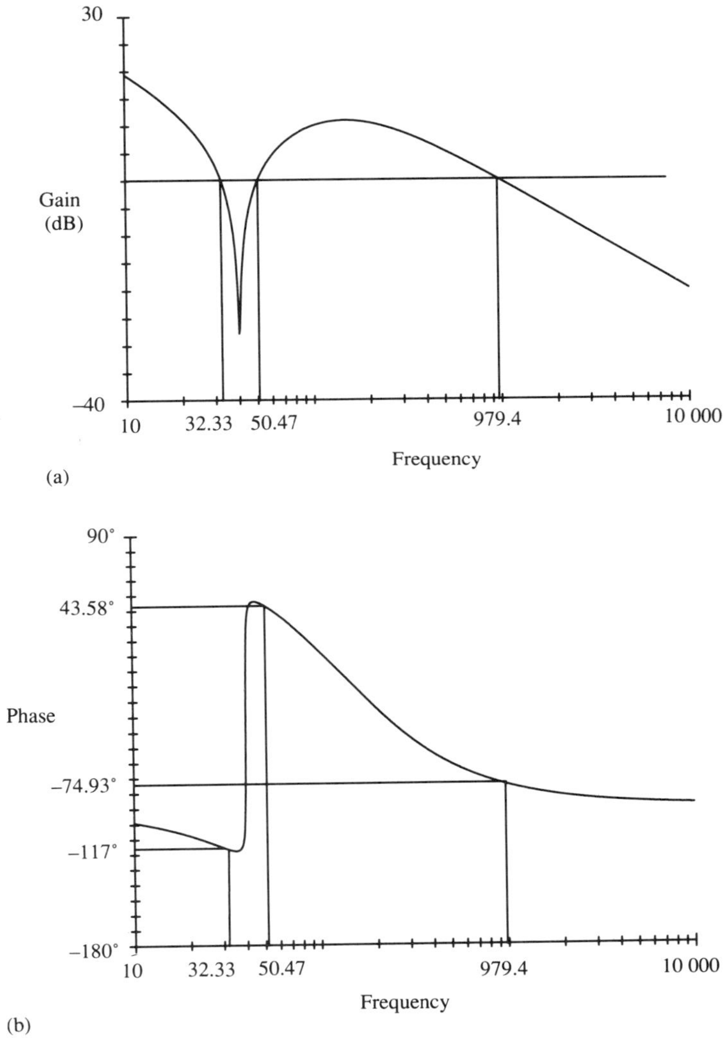

Figure 5.28 Bode plots of the transfer function $(1000\ (s^2+0.7s+1600))/(s\ (s+100)\ (s+160))$.

5.2.2 Connections Between Open-loop and Closed-loop Frequency Response Characteristics in a Feedback System

Given a feedback controlled system of the type shown in Fig. 5.16, the desired performance is usually expressed in terms of closed-loop specifications, namely

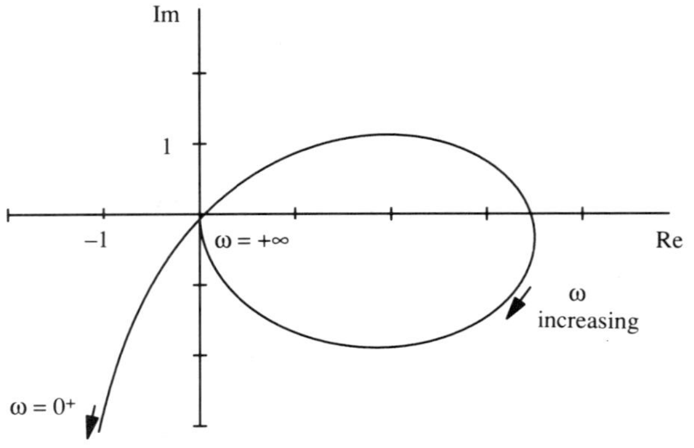

Figure 5.29 Polar plot of the transfer function $(1000\ (s^2+0.7s+1600))/(s\ (s+100)\ (s+160))$.

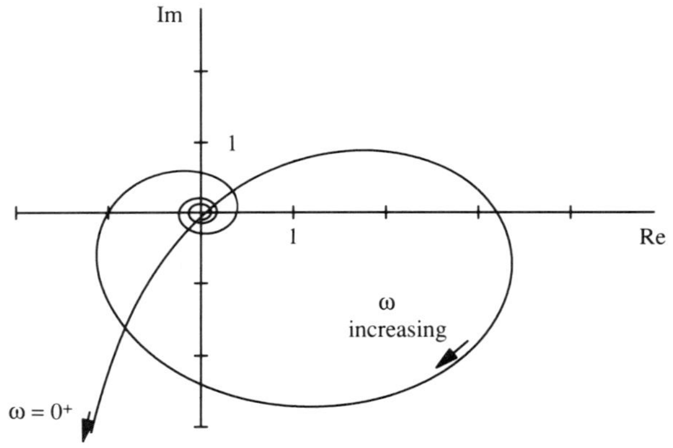

Figure 5.30 Polar plot of $(1000\ (s^2+0.7s+1600)\exp\ (-0.001\ 873\ s))/(s\ (s+100)\ (s+160))$.

for the transfer function

$$\frac{Y}{R} = \frac{GH}{1 + GH}$$

If $G=G_cG_p$, where G_c and G_p denote the compensator (controller) and plant transfer function, we can see that it is easier to design G_c to satisfy given open-loop characteristics because the Bode plots of G_cG_pH depend linearly on G_c, while Y/R does not. Therefore it is of interest to know how to relate closed-loop

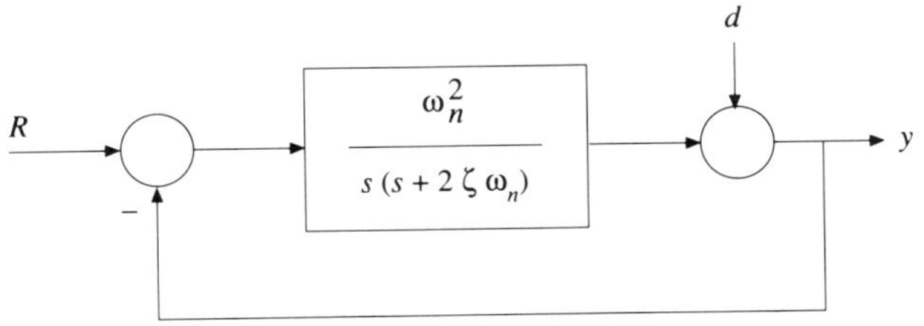

Figure 5.31 Unit feedback system with closed-loop transfer function $Y/R=\omega_n^2/(s^2+2\zeta\,\omega_n\,s+\omega_n^2)$.

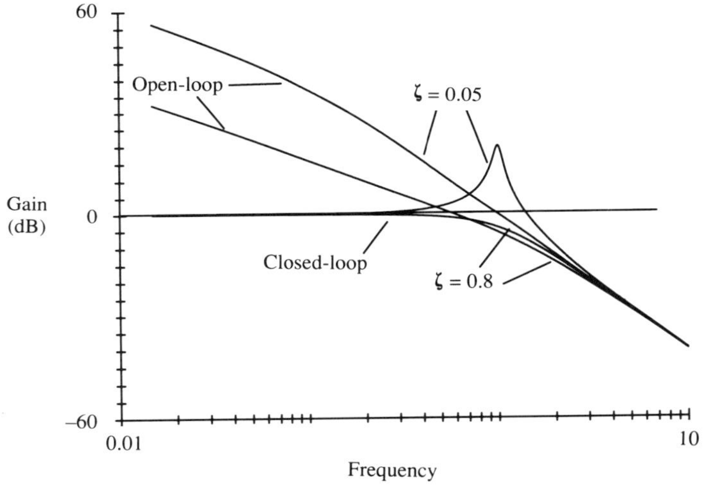

Figure 5.32 Bode magnitude plots for the system of Fig. 5.31, for $\zeta=0.05$ and $\zeta=0.8$, assuming $\omega_n=1$.

specifications with open-loop characteristics. This topic will be treated in some detail in the next chapter.

Here we shall pay special attention to a particular case, namely the system of Fig. 5.31 whose closed-loop transfer function is

$$\frac{Y}{R} + \frac{\omega_n^{\,2}}{s^2 + 2\zeta\,\omega_n\,s + \omega_n^{\,2}}$$

Despite its simplicity, this model constitutes a good working approximation in many situations.

Roughly speaking, we can say that the properties of the closed-loop system are determined by the open-loop characteristics in the vicinity of the crossover frequency ω_c (recall $|\,G\,H\,(j\omega_c)\,| = 1$).

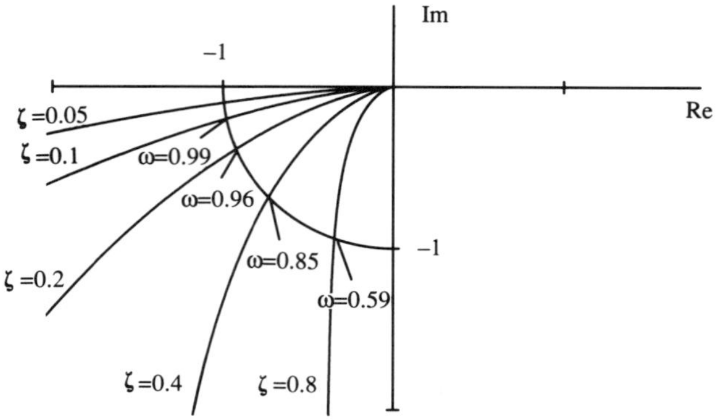

Figure 5.33 Polar plots for $G(s)H(s)=1/(s(s+2\zeta))$ and $\zeta \in \{0.05, 0.1, 0.2, 0.4, 0.8\}$.

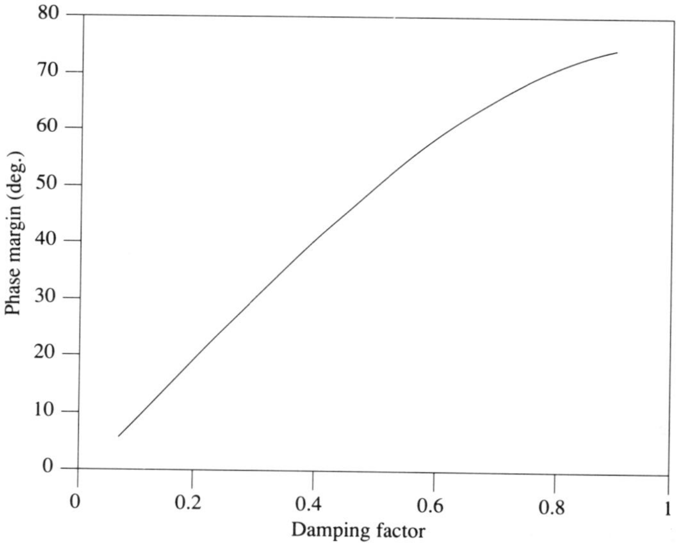

Figure 5.34 Phase margin for the second-order system of Fig. 5.31 as a function of ζ.

In Fig. 5.32 we have the open- and closed-loop magnitude logarithmic plots for the above system in two rather distinct situations. Despite this fact, the closed-loop curves are indistinct in the low and high frequency ranges, i.e. for $\omega << \omega_c$ and $\omega >> \omega_c$ respectively. The salient features occur around ω_c.

Let us then look in more detail at the evolution of the open-loop transfer functions around ω_c. Figure 5.33 depicts such behaviour for several values of the damping factor (assuming $\omega_n=1$). Quite clearly the gain margin is infinite. The phase margin increases with the damping factor and is almost a linear function in the range $0\leqslant\zeta\leqslant0.5$, as shown in Fig. 5.34.

The usefulness of ω_c comes from the fact that it gives us an estimate of the closed-loop bandwidth ω_B. Recall that ω_B is defined as the lowest frequency such that the closed-loop gain is 3 dB below its steady-state value, i.e.

$$\frac{Y}{R}(j0) = \sqrt{2}\,\frac{Y}{R}(j\omega_B)$$

For example, in Fig. 5.32 we have $\omega_B\cong0.9$ rad/s for $\zeta=0.8$, and $\omega_B\cong1.6$ rad/s for $\zeta=0.05$. The system bandwidth gives us an indication of its speed of response. Although there are no exact expressions relating the system rise time t_r and ω_B, a frequently used 'rule of thumb' (Middleton and Goodwin, 1990) is

$$t_r\ \omega_B\cong2.3$$

Therefore the larger the bandwidth, the faster the speed of elimination of disturbances. For example, in practice we find that $\omega_c\leqslant\omega_B\leqslant2\ \omega_c$.

Figure 5.35 depicts the variation of ω_c and ω_B, in our second-order example, as a function of the damping factor ζ: as expected, ω_B is always greater than ω_c and both decrease with increasing damping.

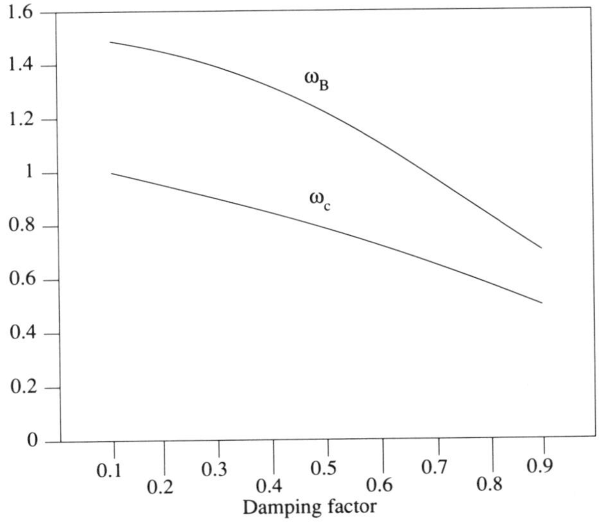

Figure 5.35 Crossover frequency ω_c, and bandwidth ω_B, for the system of Fig. 5.31, assuming $\omega_n=1$.

Exercise 5.11 Compute ω_B and ω_c when
(a) $G\,H = 1/s$
(b) $G\,H = 1/s^2$
Answer: (a) $\omega_c=\omega_B=1$ rad/s; (b) $\omega_c=1$ rad/s; $\omega_B=1.55$ rad/s

Exercise 5.12 When $\omega_B \neq 1$, how can we use the information contained in Fig. 5.35?
Answer: since

$$\frac{\omega_n^2}{(j\omega)^2 + 2\zeta\omega_n(j\omega) + \omega_n^2} = \frac{1}{(j\,\omega/\omega_n)^2 + 2\zeta(j\omega/\omega_n) + 1}$$

all we have to do is to multiply the curves by ω_n. For example, if $\omega_n=2$, then for $\zeta=0.7$ we have $\omega_c=2\times0.65=1.3$ rad/s and $\omega_B=2\times1.009=2.018$ rad/s.

5.2.3 Non-minimum Phase Transfer Functions

Given a transfer function

$$\frac{s - z}{(s - p_1)\,(s - p_2)}$$

we have seen above that the Bode magnitude plot will not change, if the zero z or pole p_i, $i=1, 2$, is replaced by $-z$ or $-p_i$. However, things were different with respect to phase. If, say, the transfer function pole, $p_i=a+jb$, has negative real part, then the phase associated with it is

$$\phi_l = \tan^{-1}\left(\frac{\omega - b}{|a|}\right) \in \left[-\frac{\pi}{2}, +\frac{\pi}{2}\right].$$

If p_i has positive real part, then this phase becomes

$$\phi_l = \tan^{-1}\left(\frac{\omega - b}{-|a|}\right) \in \left[-\frac{\pi}{2}, +\frac{3\pi}{2}\right].$$

Therefore the pole on the left half-plane is the one with *minimum phase*.

Naturally, a transfer function with a non-minimum phase pole is unstable. However, non-minimum phase zeros can occur in stable systems; for example, the transfer function

$$\frac{1 - s}{(s + 1)\,(s + 0.2)}$$

has a zero on the right half-plane. Its step response displays a rather peculiar behaviour: it starts out in the opposite direction from the input, and then turns positive as time increases, as depicted in Fig. 5.36. This is explained by the fact that this transfer function exhibits a negative gain at 0.83 rad/s, and a positive

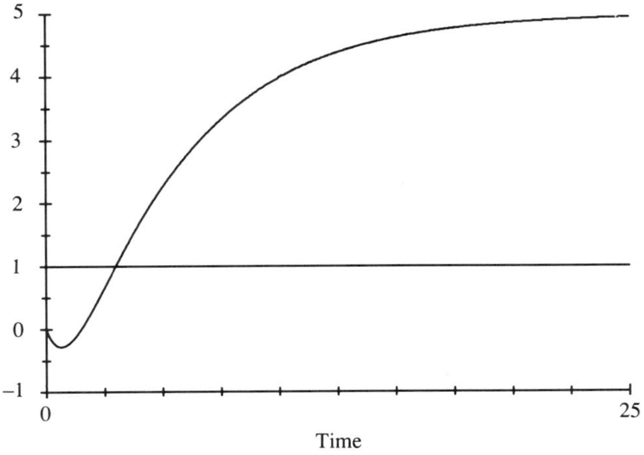

Figure 5.36 Unit-step response of transfer function $1-s/((s+1)(s+0.2))$.

gain at very low frequencies. When used with a (negative) feedback configuration, this plant may cause instability, unless continuous design is exercised to avoid *positive* feedback at high frequencies. Any transfer function of the form

$$\frac{N(s)}{D(s)} e^{-Ts}$$

where $N(s)$ and $D(s)$ are polynomials in s, is also known as a non-minimum phase transfer function. The reason lies in the fact that the pure delay factor, e^{-Ts}, can be approximated with any desired accuracy by a rational, stable, transfer function, with all of its zeros in the right half-plane (cf. Padé approximants in section 6.5).

5.2.4 Steady-state Analysis

In Chapter 4, we learned that steady-state behaviour only depends upon the number of poles at the origin of the open-loop transfer function, the system type, and three static error coefficients were defined, namely K_p, K_v and K_a. Such values can also be determined from the magnitude Bode plot as shown below.

5.2.4.1 Static position error coefficient: K_p

Since

$$K_p = \lim_{s \to 0} |\, G(s)\, H(s) \,| = \lim_{\omega \to 0} |\, G(j\omega)\, H(j\omega) \,|$$

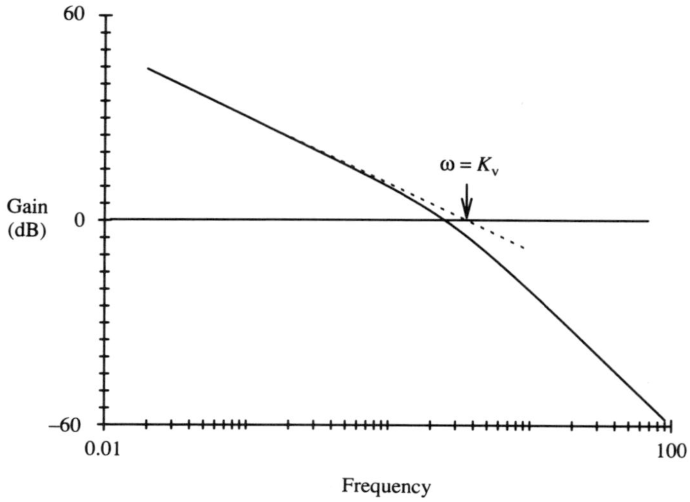

Figure 5.37 Graphical calculation of the static velocity error coefficient for $GH = 10/(s(s+3))$.

we conclude that K_p is determined by the low frequency range of the magnitude Bode plot.

5.2.4.2 Static velocity error coefficient: K_v

By definition

$$K_v = \lim_{\omega \to 0} \omega \mid G(j\omega)\, H(j\omega) \mid$$

i.e. when $\omega \to 0$ the magnitude curve approaches the line K_v/ω. But the magnitude Bode plot of K_v/ω is a straight line with a −20 dB/dec slope that intersects the 0 dB line at $\omega = K_v$. In Fig. 5.37 we show the graphical calculation of K_v for

$$G(s)\, H(s) = \frac{10}{s\,(s+3)}$$

which is obviously equal to 10/3.

5.2.4.3 Static acceleration error coefficient: K_a

When

$$\lim_{\omega \to 0} \omega^2 \mid G(j\omega)\, H(j\omega) \mid = K_a$$

is a non-zero constant, we can show by similar arguments that K_a is the intersection with the 0 dB line of the low frequency asymptote, which in this case has a slope of −40 dB/dec.

Table 5.1 Frequency domain data of example 4.4, under PID control, with settings as in Table 4.3

	Ziegler–Nichols (closed-loop)	*Shinskey*	*Ziegler–Nichols (open-loop)*	*Cohen–Coon*
Gain margin	∞	∞	∞	∞
Phase margin (degrees)	30	10	23	19
Crossover frequency (rad/s)	1.4	1.2	1.4	1.3

Example 5.6 Let us go back to Chapter 4, and analyze, on the basis of frequency-domain concepts just learned, the results of the PID controller tuning, namely example 4.4.

Recall that the plant to be controlled was $1/(s+1)^3$, and the best result, for PID control, was achieved by the Ziegler–Nichols (closed-loop) method. The resulting phase-margin and crossover frequency, for each of the PID settings (cf. Table 4.3), are shown in Table 5.1.

These are in agreement with the time-domain results, showing the largest phase margin for the Ziegler–Nichols (closed-loop) method. In this case all the tuning methods yielded approximately the same crossover frequency, and naturally an infinite gain margin, since the pole excess of the loop transfer function is now equal to two. With proportional control only, a controller gain $K=5$ is required in order to achieve 1.4 rad/s crossover frequency; however the phase margin is only 17° as depicted in Fig. 5.38. With PID control we have been able to achieve the same crossover frequency, a better phase margin and an infinite gain margin.

Under PID control we also achieved a larger bandwidth, as shown in Fig. 5.39, namely $\omega_B=2.18$ rad/s. As expected $\omega_B>\omega_c$.

The effect of integral action is also apparent in Fig. 5.39: since the plant has no poles at the origin, under P control the closed-loop gain does not converge to one as frequency decreases. Therefore it will exhibit a steady-state error for step reference and step disturbance inputs.

In Fig. 5.38 the Nyquist curve for P control comes closer to the $(-1+j0)$ point than for PID control. This results in a higher resonant peak on the closed-loop polar plot as in Fig.5.39. Consequently the response, under P control, for step reference inputs will exhibit larger overshoots.

Also, under P control, the system is on the verge of instability for $K=8$ (ultimate gain) as seen in example 4.4. In this situation the polar plot intersects the negative real axis at the $(-1+j0)$ point, which corresponds to the image of $j\omega$ by $8/(j\omega+1)^3$, for $\omega=\sqrt{3}=1.73$.

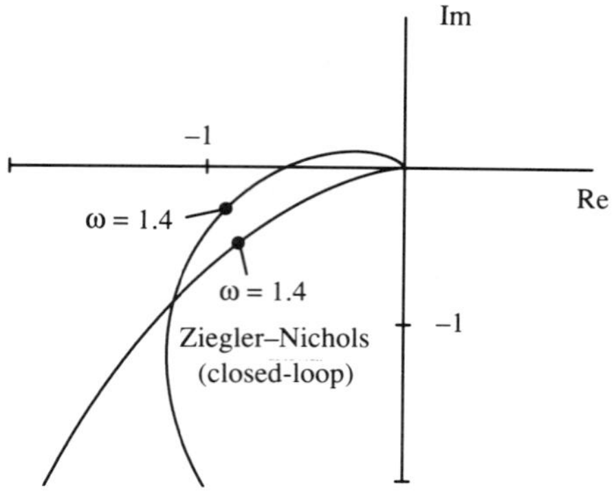

Figure 5.38 Nyquist plots for the system of example 4.4 (plant transfer function $(s+1)^{-3}$) under P control with $K=5$, and PID control with Ziegler–Nichols (closed-loop) settings as in Table 4.3.

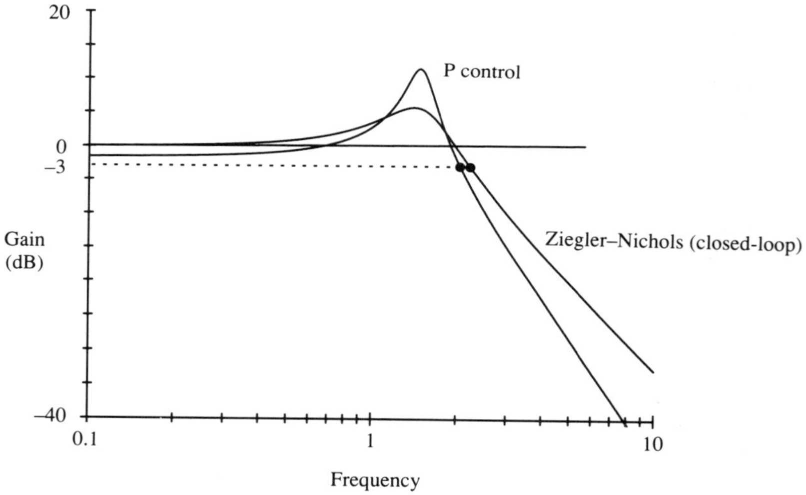

Figure 5.39 Closed-loop Bode magnitude plots for the system of Fig. 5.38. The blobs indicate the −3 dB points (bandwidth).

REFERENCES

Lopes dos Santos, P. and J.L. Martins de Carvalho (1990) 'Automatic transfer function synthesis from a Bode plot', 29th IEEE Conference on Decision and Control, Hawaii, December.

Middleton, R.H. and G.C. Goodwin (1990) *Digital Control and Estimation: A Unified Approach*, Prentice Hall.

6

Classical Controller Synthesis

6.1 INTRODUCTION

In this chapter a variety of controller design methods for continuous-time systems will be addressed. The reader was already introduced in Chapter 4 to a particular type of controller, namely the PID controller, and related empirical tuning techniques. These have proved to be very successful among chemical engineers, e.g. in the control of heat and flow processes; their authors had the ingenuity to translate in a simple set of formulae the needs of these particular users. Unfortunately, the success of these methods is no longer guaranteed once we leave that 'typical' class of processes for which they were tailored.

In order to achieve broader applicability, we must provide information about the process at hand by means of a mathematical model. Such a model may assume the form of a non-parametric representation (frequency response plots), or a parametric form (transfer-function or state-space representation). Consequently, the design method will be a function of the available plant model.

A well-designed controller must ensure the following:

1. A *robust* closed-loop system, i.e. it must be capable of ensuring the desired performance in the presence of modelling errors.
2. Good *regulation* (or stabilization), i.e. low sensitivity to (load and plant) disturbances.
3. Good *tracking*, i.e. the ability to follow the reference input as closely as possible.

The root-locus technique (section 6.3) and state-space design (Chapter 8) primarily address the regulation problem since the specifications are in terms of the systems (dominant) poles.

In the tracking problem both poles and zeros are of interest. Consequently it is better dealt with by means of the algebraic method (section 8.2). Besides the requirement of greater model accuracy, the difficulty with the tracking problem is zero allocation, especially when the plant has unstable or lightly damped zeros.

The frequency-domain design, by means of Bode plots, being a non-parametric representation, does not require the determination of the system order. Therefore it comes as no surprise that frequency-domain designs are, in general, more robust than designs by parametric methods.

The regulation and tracking problem are easily handled in the frequency domain. The desired performance is amenable to a nice interpretation in terms of the shape of the open-loop gain curve; the disadvantage, if any, resides in the absence of explicit formulae relating time-domain with frequency-domain specifications, except in lower-order systems.

The methods described in this section were initially devised for finite dimensional systems. Consequently special precautions must be taken in the presence of non-negligible pure transport delays, because they give rise to a closed-loop transfer function with an infinite number of poles. Such systems are the topic of section 6.5.

6.2 COMPENSATORS

In Chapter 4 we found that the industrial PID controller was a 'universal' device for improving the performance of a feedback control system; in fact its parameters can be adjusted, at any time, simply by tuning a knob or keying in a few instructions.

However, when the process under control is linear and time-invariant we can use instead fixed parameter devices, which are far more economical, and still achieve a satisfactory performance. Such devices are the well-known lag, lead and lag–lead compensators to be discussed in the following. We shall also analyze the notch compensator, which is particularly oriented to high performance servo-mechanisms. Methods to compute their parameters in order to meet the control system's desired performance, will be presented in sections 6.3 and 6.4.

The question of the physical realization of these devices, with either passive or active components, is also addressed.

6.2.1 Lag Compensator

Integral action was introduced as a means of eliminating steady-state errors in a feedback system, by increasing the system type. In Chapter 4 we found that the PI controller was characterized by the equation

$$m(t) = K\left(e(t) + \frac{1}{T_\mathrm{i}} \int_0^t e(t)\mathrm{d}t\right) \tag{6.1}$$

or equivalently by the transfer function

$$\frac{M(s)}{E(s)} = K\frac{(s + 1/T_\mathrm{i})}{s} \tag{6.2}$$

The presence of the pure integrator gives the controller the ability to exhibit, in the steady-state, a constant non-zero output after its input (error signal) has vanished. In frequency-domain terms this means it has an infinite gain at zero frequency; such a property is also apparent from its polar plot depicted in Fig. 6.1(a). If we are prepared to accept only a finite increase in the steady-state gain, then we can do without the pure integrator and all we need is a stable transfer function of the form

$$\frac{s + z}{s + p} \tag{6.3}$$

with $|z|>|p|>0$. The polar plot of this transfer function is represented in Fig. 6.1 (b), and it confirms what we would expect by applying the initial and final value theorems of Laplace transforms: a steady-state gain $|z|/|p|$ times greater than the gain at high frequencies.

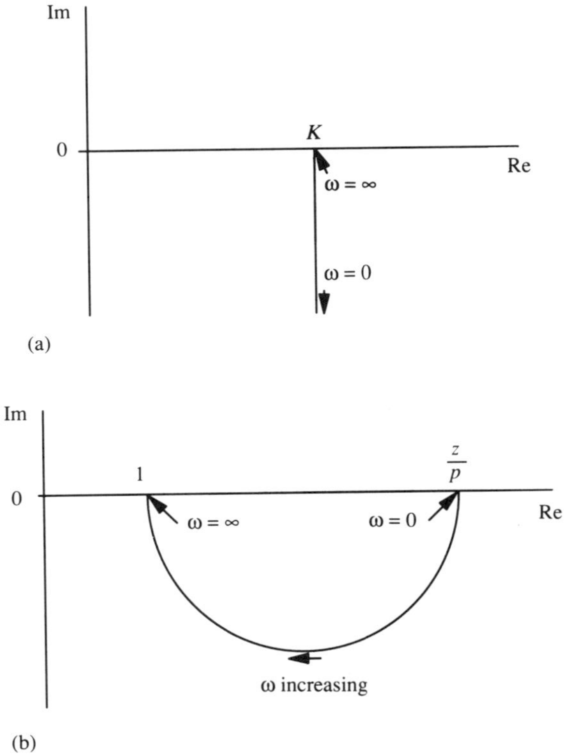

Figure 6.1 Polar plots of: (a) PI controller; (b) lag compensator (6.3).

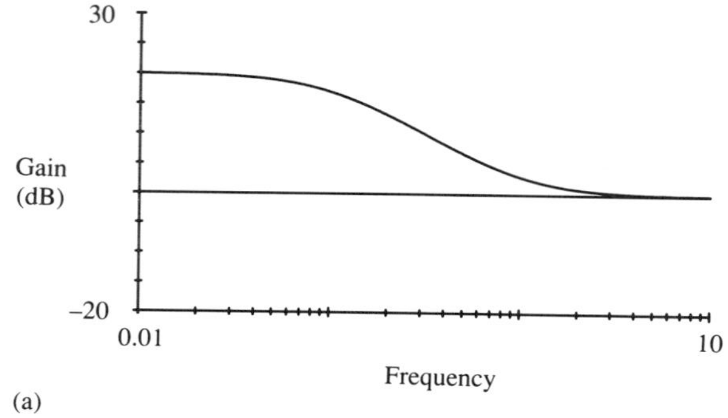

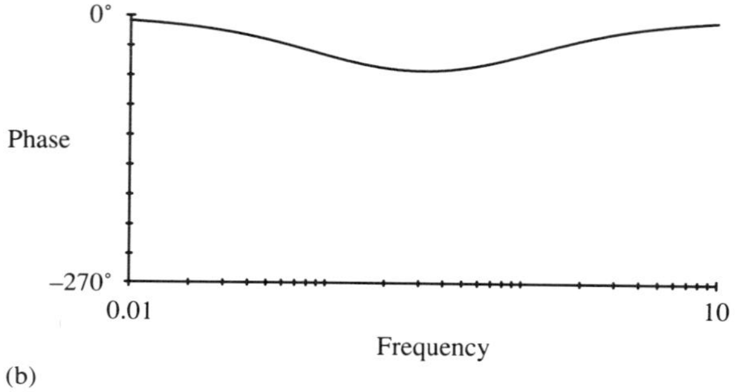

Figure 6.2 Bode plots for $(s+1)/(s+0.1)$.

Figure 6.2 shows the phase and the magnitude Bode plots for the lag compensator

$$\frac{s+1}{s+0.1} \tag{6.4}$$

Needless to say, lag compensation must be performed when a satisfactory reduction in steady-state error, by a simple gain increase, is not possible. Recall that a gain increase may not be possible if it has a detrimental effect on the system's relative stability. This suggests that a well-designed lag compensator must have little 'authority', immediately after a reference change or disturbance takes place, gradually increasing it as the system settles down. In this fashion the lag compensator will not be responsible for the transient period, i.e. for the relative stability. This fact is illustrated by the step response shown in Fig. 6.3 for the compensator

$$\frac{s+1}{s+0.1}$$

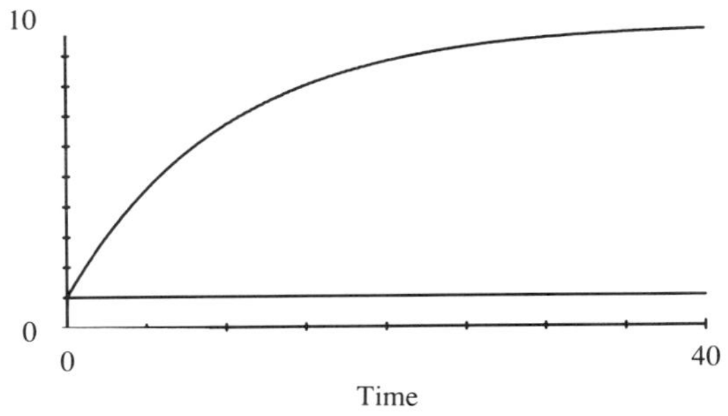

Figure 6.3 Unit-step response of $(s+1)/(s+0.1)$.

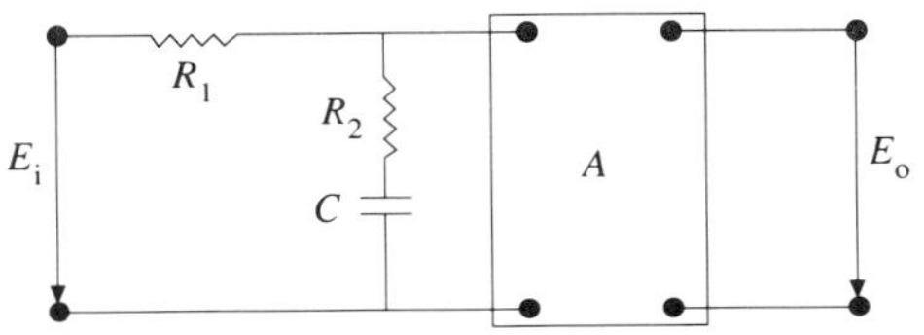

Figure 6.4 Lag compensator realization.

A lag compensator can be realized by the simple network shown in Fig. 6.4, whose transfer function is

$$\frac{E_0(s)}{E_i(s)} = \frac{s + 1/\tau}{s + 1/(\alpha\ \tau)}\ \alpha^{-1}A \tag{6.5}$$

where

$$\alpha = (R_1 + R_2) \,/\, R_2 \tag{6.6}$$

$$\tau = R_2\, C \tag{6.7}$$

and A is the gain of an ideal amplifier.

Exercise 6.1

(a) Show that the minimum phase of the lag compensator (6.3) is a function of $|\ z/p\ |$ only.

(b) On the basis of formulae (6.6) and (6.7) discuss the hardware limitations on the desirable properties of a lag compensator, namely a large $|\ z/p\ |$ and small $|\ z\ |$.

Answers: (a) the result is obvious from Fig. 6.1(b); (b) these are conflicting requirements with respect to R_2. However, R_2 cannot be very small because that would require large capacitors.

6.2.2 Lead Compensator

The purpose of a lead compensator is to improve the transient performance of the system when that is not possible with simple gain adjustments. In this sense it plays a similar role to the PD controller whose transfer function is

$$\frac{M(s)}{E(s)} = K\,(1 + T_{\mathrm{d}}\,s) \tag{6.8}$$

with the polar plot as shown in Fig. 6.5. The presence of the derivative block gives the controller the ability to exhibit large output values, even in the presence of small but fast-changing errors. In this way it can force the system to settle very rapidly, therefore reducing the transient period, i.e. improving stability.

In practice derivative action has some limitations; besides the difficulties associated with its implementation, it cannot be used in form (6.8) if high frequency sensor noise is present. The lead compensator provides a practical alternative to the pure derivative action, at a much lower cost, with the transfer function

$$G(s) = \frac{s + z}{s + p} \tag{6.9}$$

where $|z|<|p|$. The corresponding polar plot is depicted in Fig. 6.6: it shows a bounded monotonic gain increase with frequency. Figure 6.7 shows the angle contribution φ as a function of the pole-zero location. In Fig. 6.8 we can see the Bode plots for the lead compensator

$$\frac{s + 1}{s + 10} \tag{6.10}$$

In contrast with the lag compensator, the lead compensator displays a positive phase at all frequencies, which is beneficial.

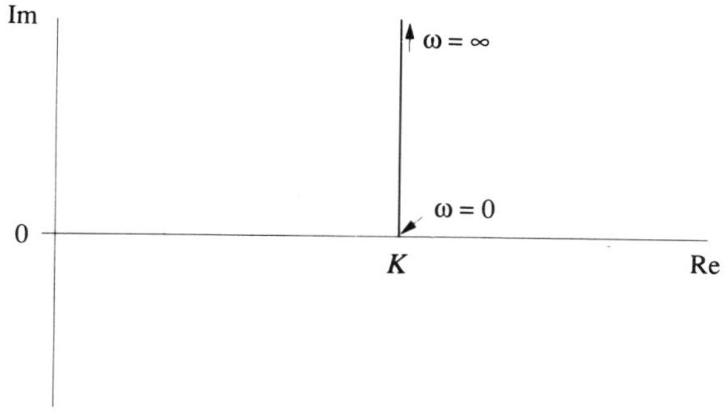

Figure 6.5 Polar plot of the PD controller (6.8).

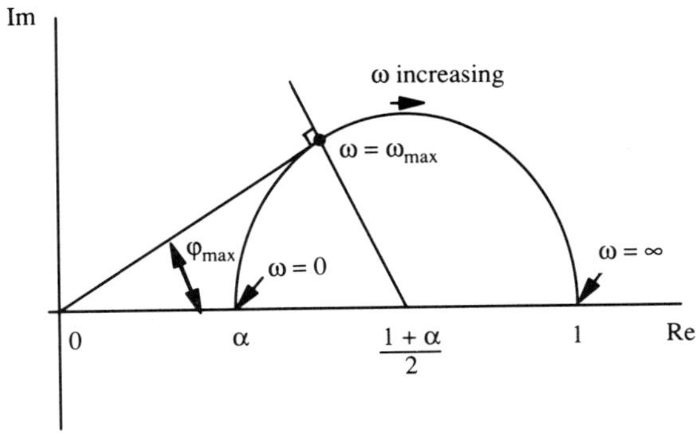

Figure 6.6 Polar plot of the lead compensator (6.9). α denotes $|z|/|p|$.

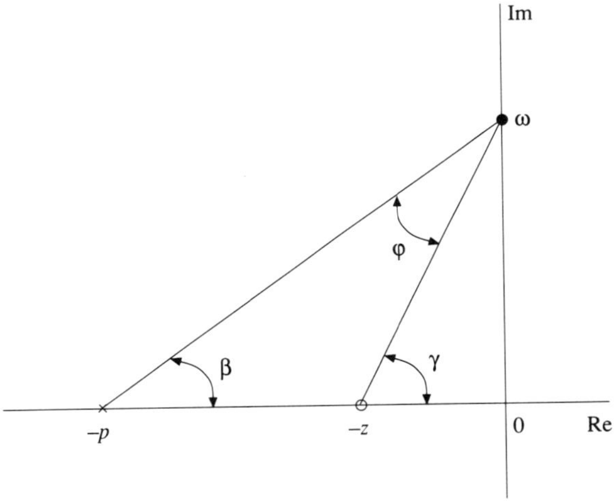

Figure 6.7 Phase, φ, of lead compensator (6.9) at frequency ω. Notice that $\varphi=\gamma-\beta$.

The maximum phase contribution φ_{max} which solely depends on

$$\alpha = \frac{|z|}{|p|}$$

as shown in Fig. 6.6, and the frequency it occurs ω_{max}, are the relevant parameters for this compensator.

Figure 6.9 depicts φ_{max} as a function of α for compensator (6.9).

Exercise 6.2 Plot the time response of lead compensator (6.10) and compare it with the one depicted in Fig. 6.3. What are the most relevant differences?

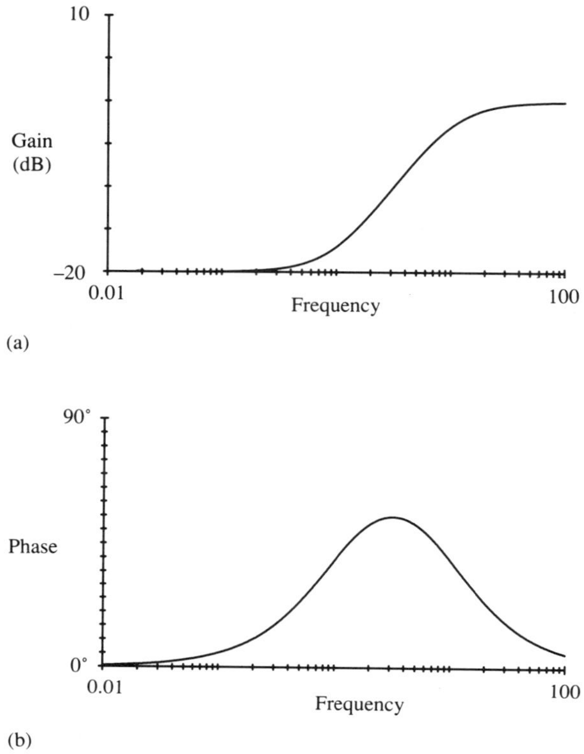

Figure 6.8 Bode plots for $(s+1)/(s+10)$.

Answer: In contrast with the lag compensator, the lead compensator has more 'authority' immediately after a change takes place, gradually reducing it as time passes.

Exercise 6.3

(a) Compute the breakpoints of the asymptotes of the Bode plot in Fig. 6.8.

(b) Check that the maximum phase occurs midway between the breakpoints on the log scale.

(c) Show that the maximum phase contribution φ_{max} of the lead compensator (6.9) occurs at $\omega = z/\sqrt{\alpha}$ and obeys the relation

$$\sin \varphi_{max} = \frac{1 - \alpha}{1 + \alpha}$$

where $\alpha = z/p$.

Answers:

(a) 1 and 10.

(b) $1/2\,(\log_{10} 1 + \log_{10} 10) = 0.5 = \log_{10} \omega_{max} \Rightarrow \omega_{max} = 3.16$.

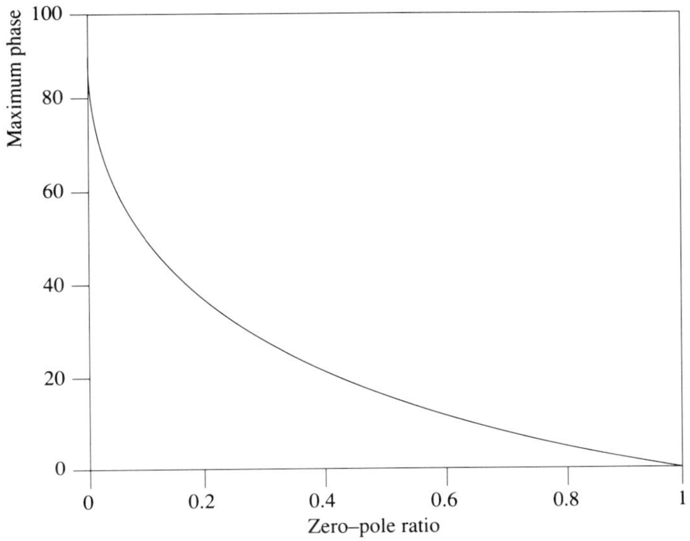

Figure 6.9 Plot of maximum angle contribution φ_{max} of lead compensator (6.9) as a function of $\alpha = z/p$.

(c) Computing sin (φ_{max}) based on Fig. 6.6, we get

$$\sin(\varphi_{max}) = \frac{(1-\alpha)/2}{(1+\alpha)/2} = \frac{1-\alpha}{1+\alpha}$$

However, this proof is not complete because we have not shown that the curve is a circle. An alternative proof is given below.

From Fig. 6.7 we have

$$\angle G(j\omega) = \gamma - \beta$$

But $\gamma = \arctan \omega/z$ and $\beta = \arctan \omega\alpha/z$. Bearing in mind that

$$\frac{d}{dx}\arctan x = \frac{1}{1+x^2}$$

we compute the derivative of $\angle G(j\omega)$ with respect to ω

$$\frac{d}{d\omega}\angle G(j\omega) = \frac{1}{1+(\omega/z)^2}\frac{1}{z} - \frac{1}{1+(\omega\alpha/z)^2}\frac{\alpha}{z}$$

Now

$$\frac{d}{d\omega}\angle G(j\omega) = 0 \Leftrightarrow \omega = \frac{z}{\sqrt{\alpha}}$$

The maximum value of φ is then

$$\angle G(j\omega)\Big|_{\omega=z/\sqrt{\alpha}} = \arctan\frac{1}{\sqrt{\alpha}} - \arctan\sqrt{\alpha} = \varphi_{\max}$$

Because

$$\sin z = \frac{\tan z}{\sqrt{(1+\tan^2 z)}}, \quad \cos z \frac{1}{\sqrt{(1+\tan^2 z)}}$$

we have

$$\sin\gamma_1 = \cos\beta_1 = \frac{1}{\sqrt{(\alpha+1)}}$$

$$\sin\beta_1 = \cos\gamma_1 = \frac{\sqrt{\alpha}}{\sqrt{(\alpha+1)}}$$

where

$$\gamma_1 = \arctan\left(\frac{1}{\sqrt{\alpha}}\right) \text{and } \beta_1 = \arctan(\sqrt{\alpha})$$

Now

$$\sin(\gamma_1 - \beta_1) = \sin\gamma_1\cos\beta_1 - \cos\gamma_1\sin\beta_1$$

Then we can write

$$\sin(\varphi_{\max}) = \sin(\gamma_1 - \beta_1) = \frac{1}{\alpha+1} - \frac{\alpha}{\alpha+1} = \frac{1-\alpha}{1+\alpha}$$

QED.

A lead compensator can be realized by the network depicted in Fig. 6.10. The resulting transfer function is

$$\frac{E_0(s)}{E_i(s)} = A\,\frac{s + 1/\tau_1}{s + 1/(\alpha\,\tau_1)} \tag{6.11}$$

where $\tau_1 = R_1C$ and $\alpha = R_2/(R_2+R_1)$.

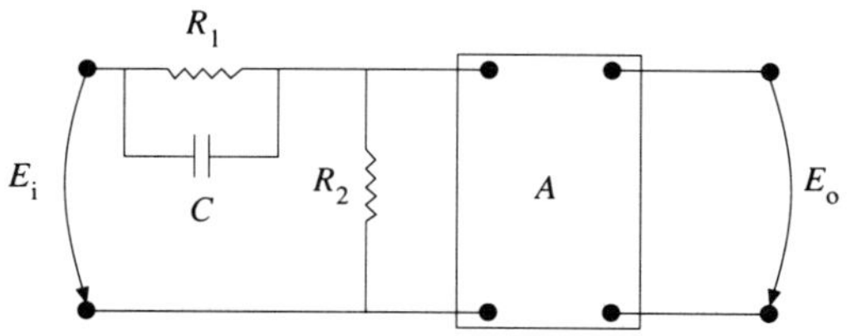

Figure 6.10 Lead compensator realization.

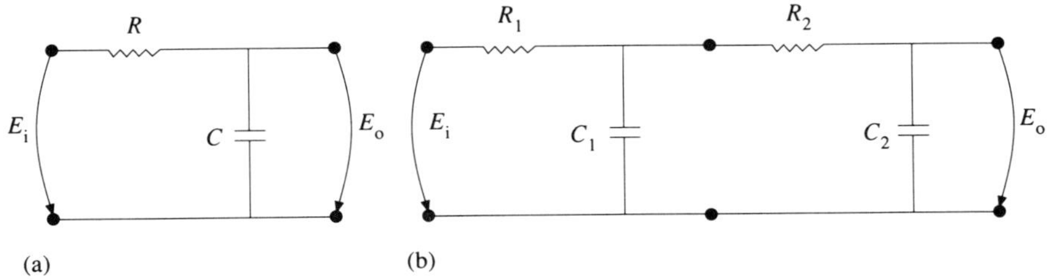

Figure 6.11 (a) A simple lag network; (b) two simple lag networks connected in series.

'A' denotes the gain of an (ideal) isolating amplifier. Its role is not only to compensate for the attenuation of the network, but also to free the compensator from loading effects that could alter its transfer function (see exercise below). Again we have practical limitations: α cannot be very small, and in slow systems we may have problems with the choice of resistor and capacitors.

Exercise 6.4 Figure 6.11(a) shows an alternative, though less satisfactory, implementation of a lag compensator:
(a) Compute its transfer function.
(b) Compute the transfer function of the series connection shown in Fig. 6.11(b).
(c) Compare the result in (b) with the product of the individual transfer functions obtained in (a).

Answers:

(a) $$\frac{E_0(s)}{E_i(s)} = \frac{1}{RCs + 1}$$

(b) $$\frac{E_0(s)}{E_i(s)} = \frac{1}{R_1R_2C_1C_2s^2 + (R_1C_1 + R_1C_2 + R_2C_2)s + 1}$$

The reader may already have noticed that for a compensator with a desired pole-zero pair, the choice of resistors and capacitor is not unique. This extra freedom can be used advantageously to satisfy other requirements, such as impedance matching over the frequency range of interest.

6.2.3 Lag–Lead Compensator

When improvements in both transient and steady-state performance are required, one could use a cascade of a lead and a lag network connected via an isolating amplifier. However, if we are prepared to accept a lead and a lag compensator with reciprocal values of α, then a more compact solution is the single lag-lead network shown in Fig. 6.12. The transfer function of this network is

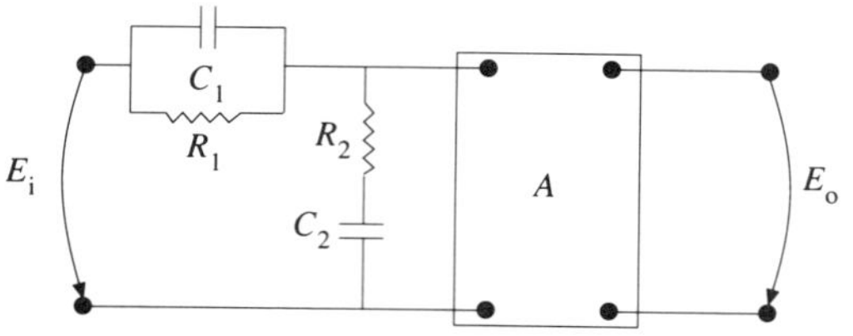

Figure 6.12 Lag–lead network.

$$\frac{E_0(s)}{E_i(s)} = A\frac{\left(s + \dfrac{1}{R_1C_1}\right)\left(s + \dfrac{1}{R_2C_2}\right)}{s^2 + \left(\dfrac{1}{R_2C_2} + \dfrac{1}{R_2C_1} + \dfrac{1}{R_1C_1}\right)s + \dfrac{1}{R_1C_1R_2C_2}} \tag{6.12}$$

or equivalently

$$\frac{E_0(s)}{E_i(s)} = A\frac{\left(s + \dfrac{1}{\tau_1'}\right)\left(s + \dfrac{1}{\tau_2'}\right)}{\left(s + \dfrac{\alpha'}{\tau_1'}\right)\left(s + \dfrac{1}{\alpha'\tau_2'}\right)} \tag{6.13}$$

where

$$\tau_1' = R_1C_1,\ \tau_2' = R_2C_2 \tag{6.14}$$

$$\frac{\alpha'}{\tau_1'} + \frac{1}{\tau_2'\alpha'} = \frac{1}{\tau_1'} + \frac{1}{\tau_2'} + \frac{1}{R_2C_1} \tag{6.15}$$

Expression (6.13) is more illuminating because it immediately reveals that the compensator is made of a lead and a lag section, namely

$$\frac{s + \dfrac{1}{\tau_1'}}{s + \dfrac{\alpha'}{\tau_1'}} \quad \text{and} \quad \frac{s + \dfrac{1}{\tau_2'}}{s + \dfrac{1}{\alpha'\tau_2'}}$$

respectively, if α' is set greater than 1. The polar plot of (6.13) is shown in Fig. 6.13.

The phase angle is zero at the frequency

$$\omega_1 = \frac{1}{\sqrt{(\tau_1'\,\tau_2')}} \tag{6.16}$$

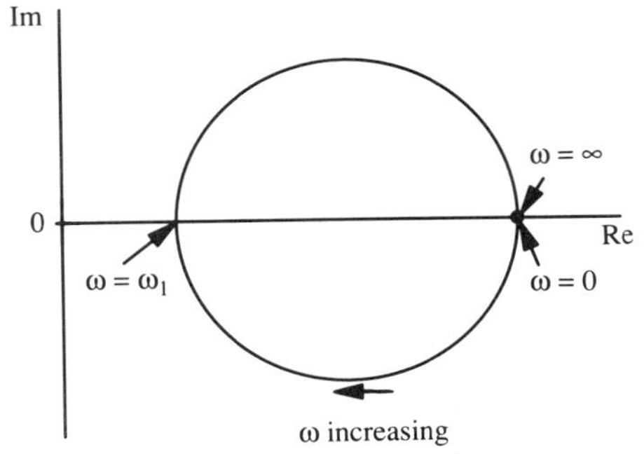

Figure 6.13 Polar plot of the lag–lead network (6.13), with $A=1$.

For $0<\omega<\omega_1$ the network behaves as a lag compensator, while for $\omega_1<\omega<\infty$ it acts as a lead network.

Exercise 6.5 Show that the product of the poles and the product of the zeros in (6.12) are necessarily equal. *Hint*: recall that the constant term in a polynomial is the product of its roots.

In general, the transfer function of the lag–lead compensator will have the form

$$G(s) = K\frac{s/g + 1}{\beta_1 s/g + 1} \cdot \frac{s/d + 1}{s/(\beta_2 d) + 1} \tag{6.17}$$

with β_1 and β_2 greater than 1; typically $\beta_1>\beta_2$. The lag–lead compensator is particularly useful when we are working in the frequency domain, given the flexibility it offers in reshaping the system's frequency response.

The lead portion of the compensator adds phase and increases the phase margin, thereby improving system stability; the lag portion provides attenuation above crossover frequency, thereby allowing an increase of gain at the low frequency range to improve the steady-state response.

A typical Bode diagram of a lag–lead compensator is shown in Fig. 6.14.

6.2.4 Notch Compensator

Very often the desired bandwidth of high performance control equipment may encompass resonant frequencies of the mechanical structure or housing to which the actuators are attached. Robot manipulator arms, computer disk drives, missiles, etc., are typical examples exhibiting resonant modes that can be excited by the action of the actuators, thus giving rise to stability problems. As shown later one way to overcome this problem is by means of a 'notch filter' that can strongly attenuate frequencies in a given narrow band.

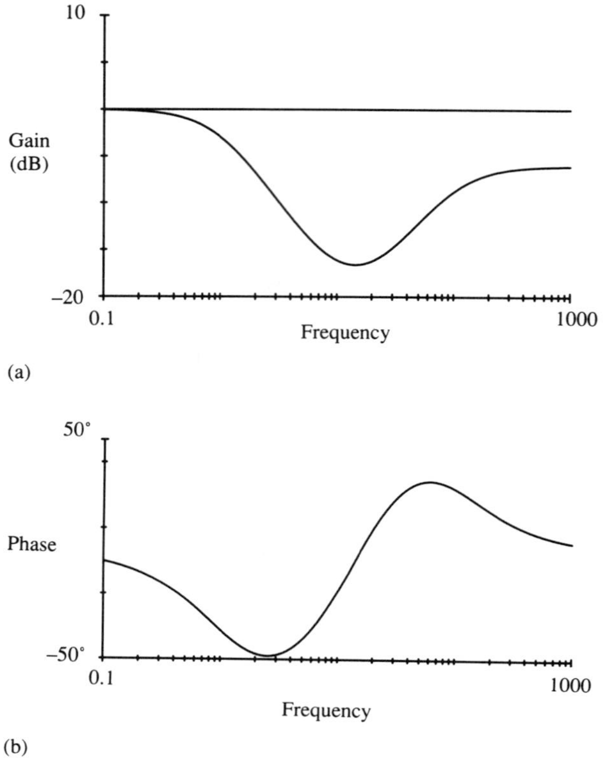

Figure 6.14 Bode diagrams of the lag–lead compensator

$$\frac{(0.1s+1)\,(0.05s+1)}{(s+1)\,(0.01s+1)} = \frac{1}{2}\frac{(s+10)(s+20)}{(s+1)(s+100)}.$$

Figure 6.15 depicts the pole-zero pattern of such a filter with transfer function

$$G(s) = \frac{s^2 + 2s + 100}{s^2 + 20s + 400} \tag{6.18}$$

Figure 6.16 shows the corresponding Bode plots, clearly exhibiting a notch centred around $\omega=10$.

Familiar realizations of notch filters consisting of only resistors and capacitors are shown in Fig. 6.17. However, these implementations are restricted to transfer functions where the pole and zero pairs share a common modulus.

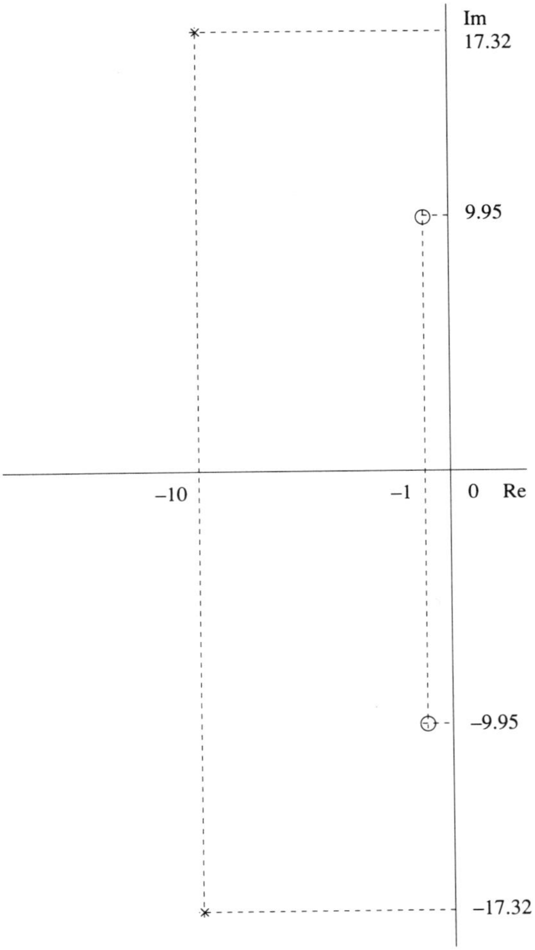

Figure 6.15 Pole-zero diagram of the notch filter (6.18).

Exercise 6.6 Compute and plot the unit-step response of the notch filter with transfer function

$$\frac{s^2 + 2s + 100}{s^2 + 20s + 400}.$$

Answer: $0.25+0.965 \exp(-10t) \sin(\sqrt{300}t+2.25)$. This curve is plotted in Fig. 6.18.

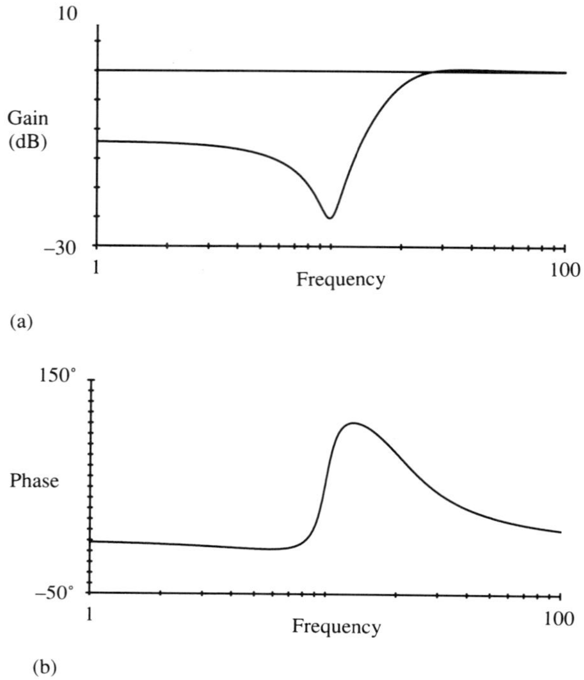

Figure 6.16 Bode plots of the notch filter (6.18).

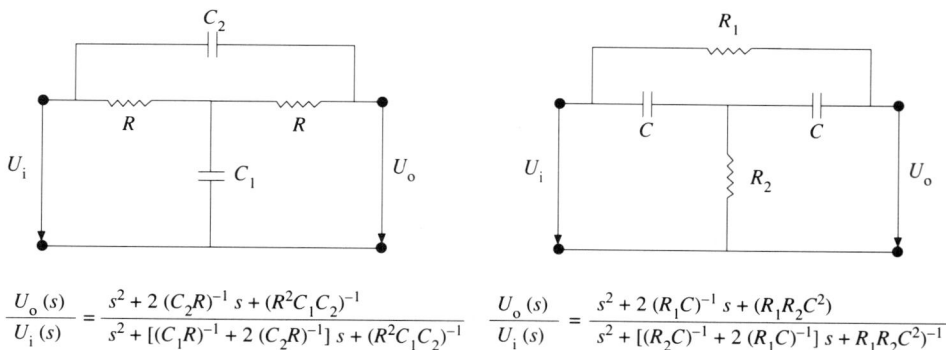

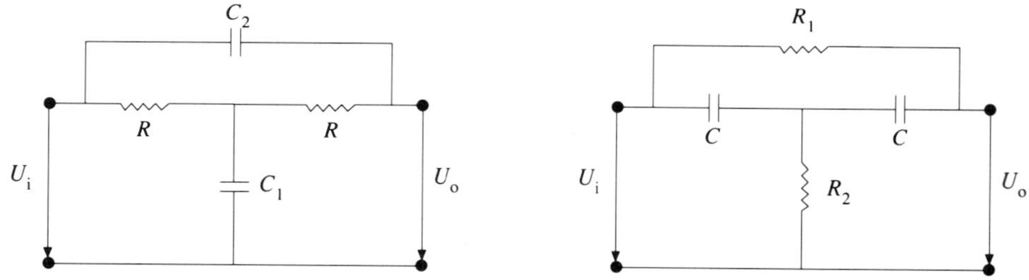

$$\frac{U_o(s)}{U_i(s)} = \frac{s^2 + 2\,(C_2R)^{-1}\,s + (R^2C_1C_2)^{-1}}{s^2 + [(C_1R)^{-1} + 2\,(C_2R)^{-1}]\,s + (R^2C_1C_2)^{-1}} \qquad \frac{U_o(s)}{U_i(s)} = \frac{s^2 + 2\,(R_1C)^{-1}\,s + (R_1R_2C^2)}{s^2 + [(R_2C)^{-1} + 2\,(R_1C)^{-1}]\,s + R_1R_2C^2)^{-1}}$$

Figure 6.17 Realizations of notch filters.

6.2.5 Compensator Realization Using Active Components

The compensation networks presented so far can also be constructed using active elements. Figure 6.19(a) depicts a possible realization of a *lag compensator* using an operational amplifier (cf. Chapter 2).

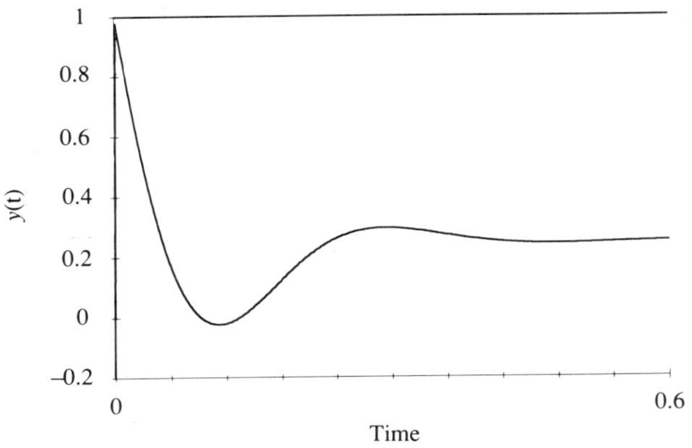

Figure 6.18 Unit-step response of the notch filter (6.18).

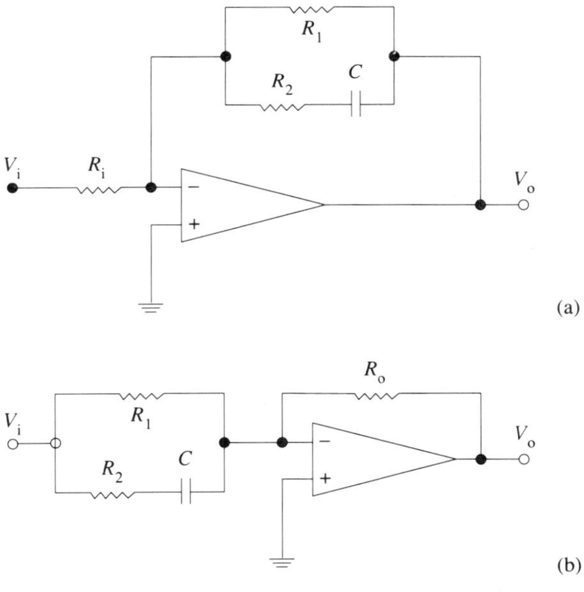

Figure 6.19 Compensator realization using an operational amplifier: (a) lag compensator; (b) lead compensator.

Exercise 6.7 With reference to the network shown in Fig. 6.19(a):

(a) Show that the dynamic behaviour (i.e. the pole and the zero) of the network is necessarily independent of R_i.

(b) Assuming V_i is a step function, what is the steady-state gain? Why is it independent of R_2 and C?

(c) Compute the transfer function $V_0(s)/V_i(s)$.

As shown in the previous exercise this network has a transfer function $(-V_0/V_i)$ of

$$\frac{R_1}{R_i}\;\frac{R_2\left(s+\dfrac{1}{R_2C}\right)}{(R_2+R_1)\left(s+\dfrac{1}{(R_2+R_1)C}\right)} \tag{6.19}$$

clearly showing that the pole is closer to the imaginary axis than the zero.

Similarly we can construct a *lead network* having a transfer function $(-V_0/V_i)$ of

$$\frac{R_0}{R_i}\;\frac{(R_2+R_1)\left(s+\dfrac{1}{C(R_2+R_1)}\right)}{R_2\left(s+\dfrac{1}{CR_2}\right)} \tag{6.20}$$

as shown in Fig. 6.19(b).

As expected we have a DC gain of R_0/R_1 and the dynamic behaviour is independent of R_0. The zero–pole ratio is now $R_2/(R_1+R_2)$.

Higher-order transfer functions require a systematic procedure for their realization and that can be achieved with the help of the corresponding flow diagram. Given the nth order transfer function

$$\frac{Y(s)}{U(s)}=\frac{b_0\,s^n+b_1\,s^{n-1}+\ldots+b_{n-1}\,s+b_n}{s_n+a_1\,s^{n-1}+\ldots+a_{n-1}\,s+a_n} \tag{3.4}$$

the following equivalent equation can be written:

$$Y=b_0\,U+\frac{1}{s}(b_1\,U-a_1\,Y)+\ldots+\frac{1}{s^{n-1}}(b_{n-1}\,U-a_{n-1}\,Y)+\frac{1}{s^n}(b_n\,U-a_n\,Y) \tag{6.21}$$

from which we can draw the flow diagram shown in Fig. 6.20.

Rewriting (6.21) in the form

$$Y=b_0\,U-\frac{1}{s}\left[(b_1\,U-\alpha_1Y)-\frac{-1}{s}\left[(b_2\,U-\alpha_2Y)-\frac{-1}{s}\left[\ldots\right.\right.\right.$$
$$\left.\left.\left.-\frac{-1}{s}\left[(b_{n-1}\,U-\alpha_{n-1}Y)-\frac{-1}{s}(b_n\,U-a_nY)\right]\ldots\right]\right]\right] \tag{6.22}$$

we conclude that the transfer function can be realized with the help of the operational amplifier configurations of Figs 2.15 and 2.16, namely with n integrators and $2n$ subtractors.

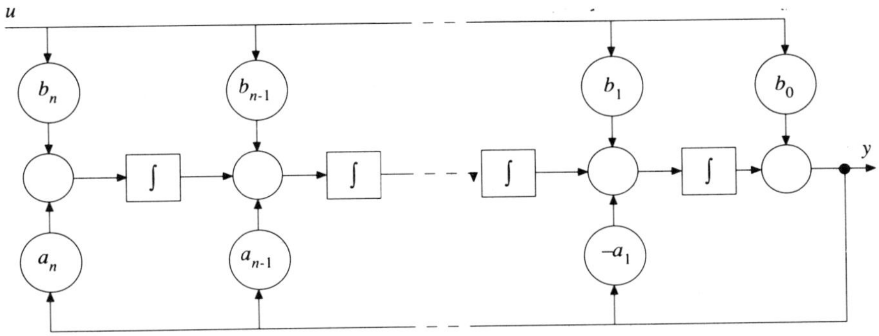

Figure 6.20 Flow diagram of the nth order differential equation
$y^{(n)} + a_1y^{(n-1)} + \ldots + a_{n-1}y^{(1)} + a_ny = b_0u^{(n)} + b_1u^{(n-1)} + \ldots + b_{n-1}u^{(1)} + b_nu.$

We shall now illustrate this procedure by producing a realization for the notch filter

$$\frac{Y}{U} = \frac{s^2 + 2s + 100}{s^2 + 20s + 400} \tag{6.18}$$

We start with the equivalent relation

$$s^2Y + 20s\,Y + 400\,Y = s^2U + 20sU + 400U$$

and divide it by s^2

$$Y + 20\frac{Y}{s} + 400\frac{Y}{s^2} = U + 2\frac{U}{s} + 100\frac{U}{s^2}$$

Solving with respect to the first term on the left

$$Y = U + \left(\frac{2U}{s} - \frac{20Y}{s}\right) + \left(\frac{100U}{s^2} - \frac{400Y}{s^2}\right)$$

or

$$Y = U + \frac{2}{s}\left[(U - 10Y) + \frac{50}{s}(U - 4Y)\right]$$

and finally

$$Y = U - \left(-\frac{2}{s}\right)\left[(U - 10Y) - \left(-\frac{50}{s}\right)(U - 4Y)\right] \tag{6.23}$$

This equation shows that the notch filter (6.18) can be implemented with the help of two integrators and four subtractors. The result is depicted in Fig. 6.21, where $(R_1C_1)^{-1}=50$ and $(R_2C_2)^{-1}=4$.

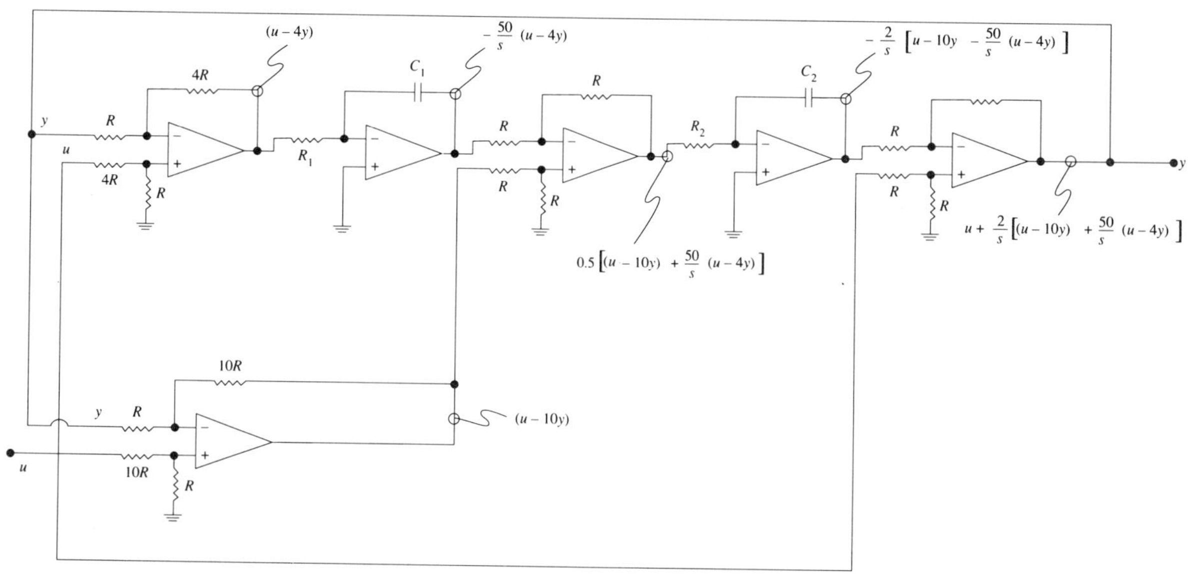

Figure 6.21 Realization of the transfer function $(s^2+2s+100)/(s^2+20s+400)$ with op-amps: $(R_1C_1)^{-1}=50$ and $(R_2C_2)^{-1}=4$.

6.3 ROOT-LOCUS COMPENSATION

The root-locus method is particularly tailored for controller design when the specifications are expressed in the time domain. In this section we shall only consider the simple case where the compensated system has a pair of complex-conjugate dominant poles and the specifications are given for such a pole pair. Recall that, for a second-order underdamped system, simple explicit formulae exist relating the location of the poles and the system's time response, namely peak time and overshoot. The case of dominant real poles will not be considered; however, in the majority of applications an oscillatory response is desirable to overcome the effect of non-linearities. When a monotonic response is required – a robot manipulator is a typical example (Machado and Martins de Carvalho, 1989) – the method described in this section is not recommended. In subsequent sections we shall develop more elaborate methods of compensator design of wider applicability.

If we are going to use simple compensators of the type described in the previous section, having one pole and one zero only, i.e. two degrees of freedom, to shift the system poles, we cannot expect more than the ability to locate two closed-loop poles. The position of the remaining ones will be outside our control, and it may well happen with higher-order plants that the allocated poles by the design exercise are not really dominant.

The disadvantage of the root-locus method is that the information content available for the human designer gradually decreases as the number of branches increases; in such cases the choices are either to perform the design based on reduced-order model (Lopes dos Santos and Martins de Carvalho, 1990) or to work in the frequency domain.

6.3.1 Lead Compensation

Once the dominant closed-loop poles have been specified and it is found that they do not belong to the root-loci of the original system we have to design a lead compensator. In terms of root-loci the effect of the lead compensator is to translate the asymptotes into the left half-plane, along the real axis, by the amount

$$\frac{|\text{compensator pole} - \text{compensator zero}|}{\text{pole excess}} \qquad (6.24)$$

as implied by formula (4.18). As a result the branches of the loci will be 'bent over' to the left hand side of the s-plane as shown in Fig. 6.22; then a simple gain adjustment will be sufficient to get better damped poles.

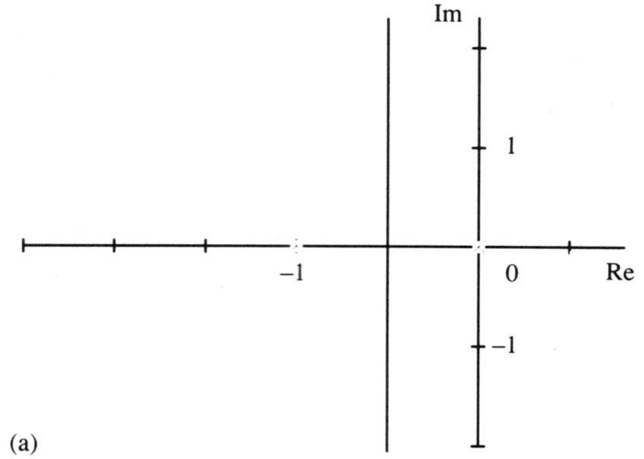

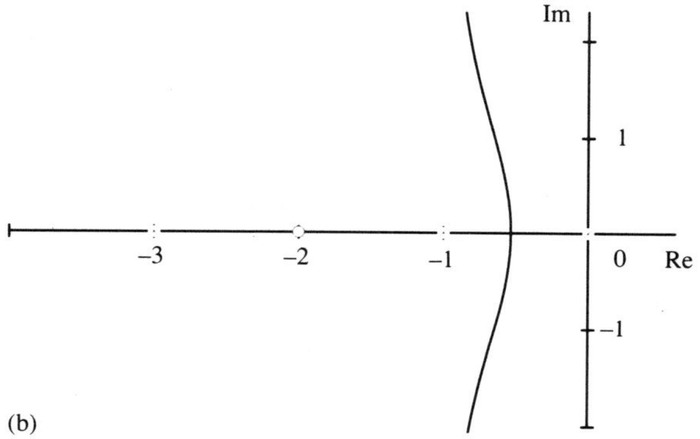

Figure 6.22 Illustrating the effect of lead compensation. Roots-loci of: (a) plant; (b) plant in series with the compensator $(s+2)/(s+3)$.

Let

$$p_d = \sigma_d + j\omega_d$$

and

$$\bar{p}_d = \sigma_d - j\omega_d$$

denote the conjugate pair of desired (dominant) closed-loop poles. To design a compensator we must bear in mind that it must provide at p_d or $\bar{p}_d$ an angle contribution φ_c, such that the open-loop transfer function has an argument which

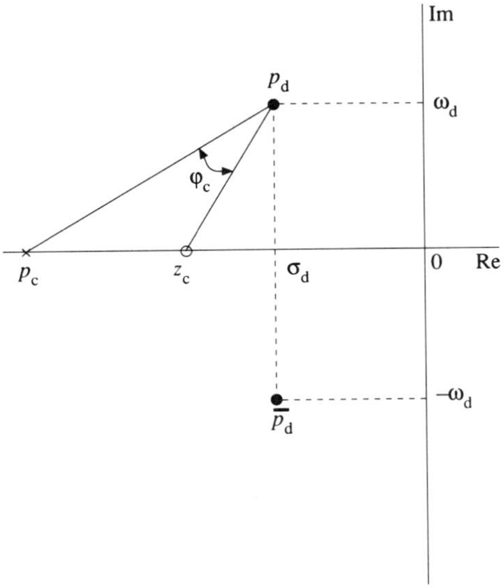

Figure 6.23 Angle contribution, φ_c, of the lead compensator $(s+z)/(s+p)$ at point $s=p_d$.

is an odd multiple of 180°, i.e.

$$\varphi_c = \angle G_c(p_d) = (2l + 1)\,180° - \angle G_p(p_d) \tag{6.25}$$

for some integer l.

Figure 6.23 shows how to relate φ_c with the compensator parameters. Although there are an infinite number of possible choices, practical reasons dictate the range of possible values for z_c and p_c.

Let us illustrate what has just been said by means of an example. Assume that in Fig. 6.25

$$G_p = \frac{500}{s(s+1)}$$

and it is desired to have the closed-loop system with the dominant poles p_d, $\bar{p}_d$ at $-32.5 \pm j\,38$ ($\zeta=0.65$ and $\omega_n=50$).

The first step in the design process is the calculation of the phase contribution φ_c, of the compensator at p_d. Because

$$\angle G_p(s)\Big|_{s=p_d} = -260°$$

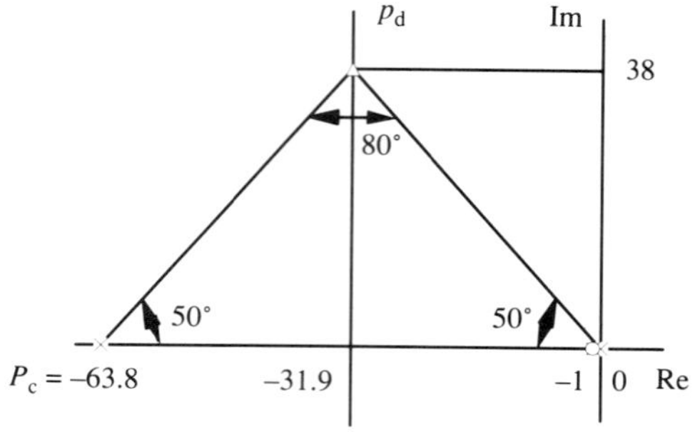

Figure 6.24 Root-locus for $G_c(s)\ G_p(s)=K\ 500/(s(s+63.8))$. For $K=4.93$ we have the closed-loop poles at p_d and $\bar{p}_d$.

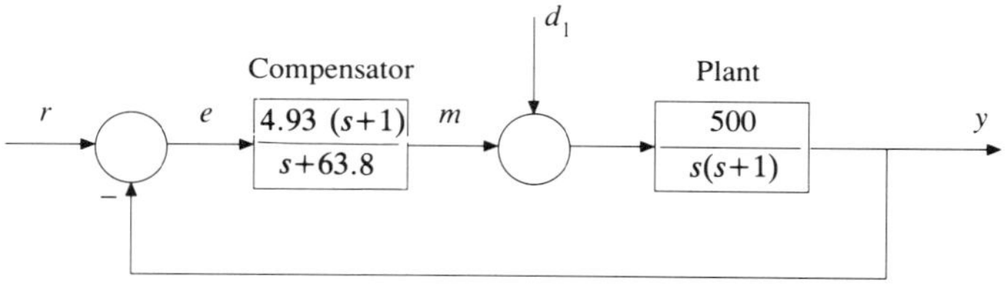

Figure 6.25 Compensation by pole-zero cancellation in the presence of load disturbances.

φ_c must satisfy

$$-260° + \varphi_c = -180° \tag{6.26}$$

i.e.

$$\varphi_c = 80°$$

The second step is the selection of the zero z_c and of the pole p_c of the compensator. From the point of view of computations, a simplification is achieved if we cancel the plant pole at -1 with the compensator zero, i.e. if we set $z_c=-1$. Then p_c can be easily computed as shown in Fig. 6.25, yielding $p_c=-63.8$. The desired closed-loop poles occur for $K=4.93$.

Apparently we have performed a satisfactory design. However, a closer look at the problem reveals that our choice is unsatisfactory. In fact, from Fig. 6.24 we get

$$\frac{Y}{R} = \frac{2465}{(s + 32.5 + j38)(s + 32.5 - j38)} \tag{6.27}$$

$$\frac{Y}{D_1} = \frac{500(s + 63.8)}{(s + 32.5 + j38)(s + 32.5 - j38)(s + 1)} \tag{6.28}$$

This reveals that the 'cancelled' mode e^{-t} will be present in the response to a load disturbance. Therefore, such a technique must not be used in practice when a *large reduction* in settling time is desired.

The following exercise will provide greater insight into these issues.

Exercise 6.8 Figure 6.26 depicts the standard configuration used in series compensation; N_c, D_c, N_p and D_p are polynomials, and K a positive constant.

(a) Compute the transfer functions

$$\frac{Y}{R}, \frac{E}{R}, \frac{M}{R}; \quad \frac{Y}{D_1}, \frac{E}{D_1}, \frac{M}{D_1}; \quad \frac{Y}{D_3}, \frac{M}{D_3}$$

(b) Show that when N_c and D_p have zeros in common they appear as poles of Y/D_1 and E/D_1 (this is one of the reasons why compensation by pole-zero cancellation is not always recommended).

(c) Show that when the closed-loop system is 'much faster' than the plant, the manipulated variable m may assume very large values. (Quite naturally, it takes more actuator authority to achieve faster command response.)

(d) Assuming a stable closed-loop system, show that the response y to a unit-step reference always *undershoots* (i.e. it starts out in the opposite direction from the input) if the plant has a right half-plane zero.

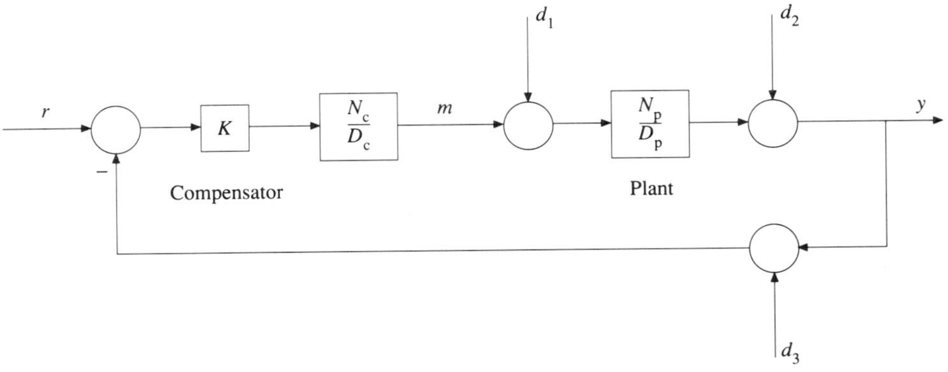

Figure 6.26 Configuration for exercise 6.8.

(e) Show that the output of the closed-loop system (assumed stable) *overshoots* to a step reference input, provided the plant has a right half-plane pole.

Answers:

(a) All the transfer functions have denominator $D=D_cD_p+KN_cN_p$

$$\frac{Y}{R}=\frac{K\,N_c\,N_p}{D};\ \frac{E}{R}=\frac{D_c\,D_p}{D};\ \frac{M}{R}=\frac{K\,N_c\,D_p}{D}$$

$$\frac{Y}{D_1}=\frac{N_p\,D_c}{D};\ \frac{E}{D_1}=-\frac{Y}{D_1};\ \frac{M}{D_1}=-\frac{Y}{R}$$

$$\frac{Y}{D_3}=\frac{M}{D_1};\ \frac{M}{D_3}=-\frac{M}{R}$$

(b) In this case the common zeros of N_c and D_p are also zeros of D, which are not cancelled by the numerators (assuming that the polynomial pairs (N_c,D_c) and (N_p,D_p) are coprime).

(c) Recall our discussion in Chapter 3 about the effect of an additional zero on the transient response and the fact that in this case the (slow) plant poles are zeros of M/R.

(d) In practice an RHP zero cannot be cancelled. Therefore such a zero is also a zero of Y/R . Then apply the results of Chapter 3 about the role of zeros on the transient response.

(e) Cancellation of unstable poles does not work in practice. Consequently such poles become zeros of E/R. Since $E=R-Y$, and $R=1$, an undershoot in E induces an overshoot in Y.

With the insight gained with the previous problem we are now in a better position to carry on with our design exercise. Given that the compensator zero is also a zero of the closed-loop transfer function Y/R, the accepted design heuristics is to choose $|z_c|\simeq|p_d|$. In this way the overshoot of the compensated system will be essentially the same when compared to that of the desired second-order system (recall the discussion in Chapter 3 about the effects of an additional zero). Greater values of $|z_c|$, although beneficial as far as the overshoot is concerned, would reduce the maximum angle contribution of the compensator (cf. Fig. 6.23).

Let us now resume the design exercise. From what has been said we start by setting $-z_c=|p_d|=50$. However, this choice is not possible because we need an angle contribution of 80° and the maximum we could achieve was 65.27° (with the pole at infinity) as shown in Fig. 6.27. At the expenses of a small increase in overshoot we set $-z_c=|p_d|/2=25$. Then the compensator will contribute with 80° at

$$s = p_d \quad \text{if} \quad p_c = -130.6$$

Also

$$G_c = K\frac{s+25}{s+130.6} \tag{6.29}$$

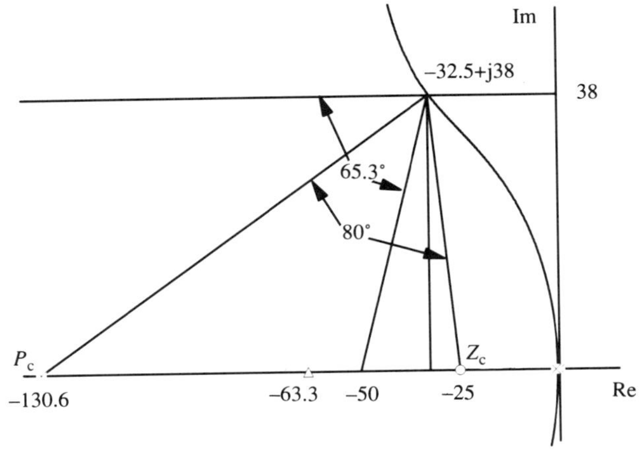

Figure 6.27 Data for the design example of series compensation.

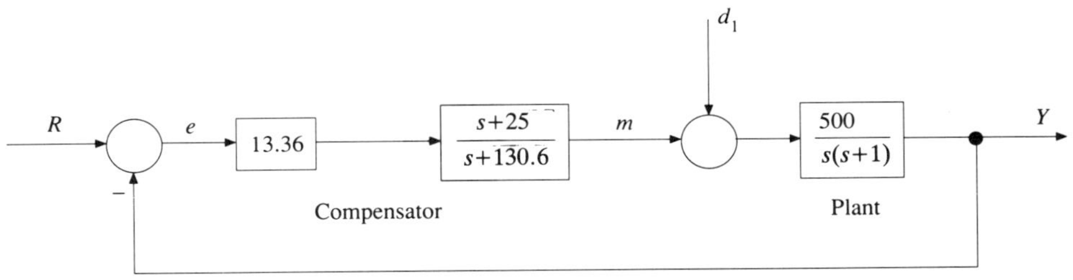

Figure 6.28 Final result of the series compensation example.

For $K=13.36$ we have the desired closed-loop poles and the closed-loop transfer function

$$\frac{Y}{R} = \frac{6680(s + 25)}{(s + 32.17 \pm j38.05)(s + 67.255)} \tag{6.30}$$

Figure 6.28 shows the resulting system. Figure 6.29(a) is a plot of $y(t)$ and $m(t)$ following a unit-step reference change. For comparison, the unit-step response of the desired closed-loop transfer function

$$\frac{2500}{(s + 32.5 \pm j38)} \tag{6.31}$$

and $y(t)$ are plotted in Fig. 6.29(b); we conclude that our compensated system still has an overshoot considerably higher than desired. A possible attempt to improve

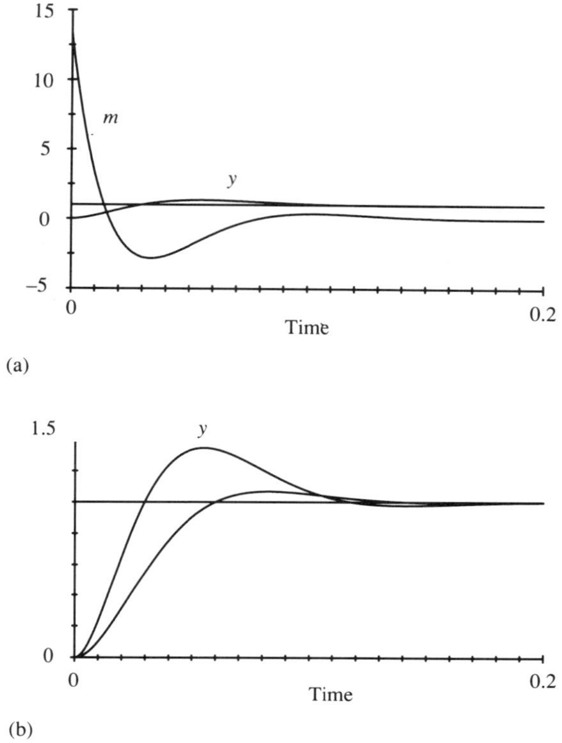

Figure 6.29 Time plots for a unit step change in reference in system of Figure 6.28: (a) $y(t)$ and $m(t)$; (b) $y(t)$ and desired response.

the situation is to place the compensator in the feedback path. The following exercise will clarify the advantages of this alternative.

Exercise 6.9 Consider the compensation scheme shown in Fig 6.30 where the compensator G_f is placed in the feedback path. With the same assumptions as in exercise 6.8, and defining the error as $(R-Y)$:

(a) Compute the transfer functions

$$\frac{Y}{R}, \frac{M}{R}; \quad \frac{M}{D_1}, \frac{Y}{D_1}; \quad \frac{Y}{D_3}, \frac{M}{D_3}$$

and the error when R is the unit ramp.

(b) Show that the common factors of N_f and D_p (if any) are poles of Y/R and Y/D_1. (This fact shows that we cannot avoid cancellation problems.)

(c) Compare Y/R and M/R with their counterparts in exercise 6.8. What appears to be the most important difference? Why?

Answers:

(a) All the Transfer functions have the denominator

$$D = D_f\, D_p + K\, N_f\, N_p$$

$$\frac{Y}{R} = \frac{K\, N_p\, D_f}{D}; \quad \frac{M}{R} = \frac{K\, D_f\, D_p}{D}$$

$$\frac{Y}{D_1} = \frac{N_p\, D_f}{D}; \quad \frac{M}{D_1} = \frac{-\,K\, N_p\, N_f}{D}$$

$$\frac{Y}{D_3} = \frac{M}{D_1}; \quad \frac{M}{D_3} = \frac{-\,K\, N_f\, D_p}{D}$$

$$\text{error} = \frac{1}{s^2}\left(1 - \frac{Y}{R}\right) = \frac{-\,1}{s^2}\left(\frac{KN_p\, N_f + D_f\, D_p - K\, D_f\, N_p}{D_f\, D_p + K\, N_f\, N_p}\right)$$

(c) N_c has been replaced by D_f. This means that in the parallel (feedback) scheme the poles of the compensator become zeros of these transfer functions. This is beneficial because the overshoot will be reduced in case of lead compensation, i.e. when $|p_c| > |z_c|$. On the other hand this fact will be detrimental if G_f is a lag compensator.

In contrast with series compensation, the parallel scheme of Fig. 6.30 has the advantage that the poles of the compensator are now zeros of the transfer functions Y/R and M/R. As a result we can afford $|z_c|$ smaller than $|p_d|$, and consequently achieve a larger angle contribution, without the overshoot penalty.

To illustrate the value of this approach, let us repeat the previous design, with the (same) compensator placed now in the feedback path as shown in Fig. 6.31. The performance of the system is shown in Fig. 6.32, where the actual and desired responses are plotted, showing a drastic improvement in comparison with the series compensation (Fig. 6.29(b)).

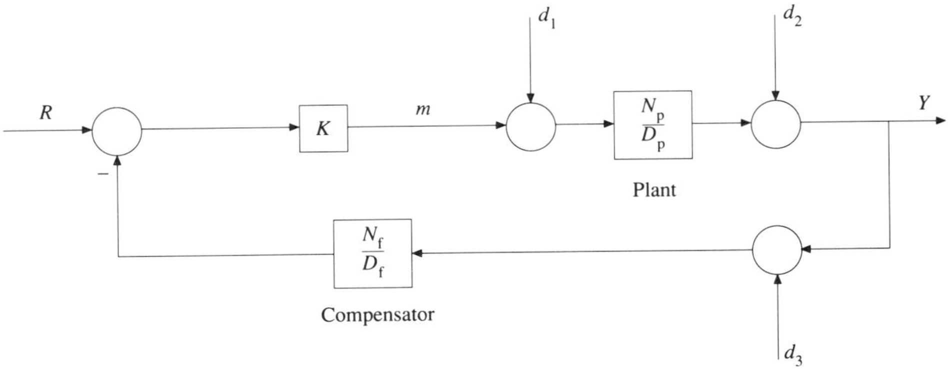

Figure 6.30 Parallel compensation scheme.

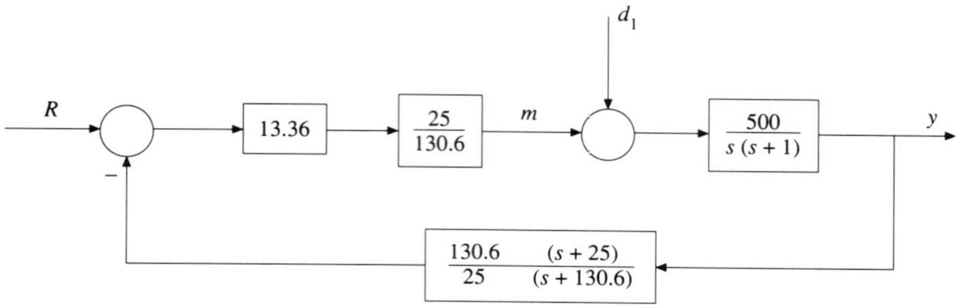

Figure 6.31 Parallel compensation design example: notice the feedback path retains the steady-state gain of unit.

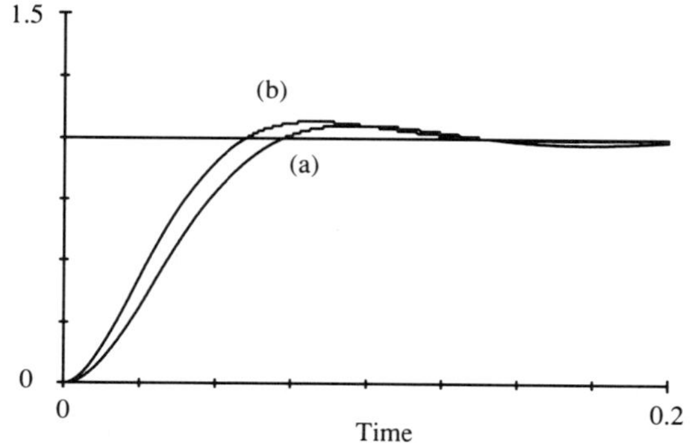

Figure 6.32 Actual (a) and desired (b) responses of system depicted in Fig. 6.31 to a unit-step input.

Exercise 6.10

(a) Compute Y/R and M/R for the system in Fig. 6.31.

(b) Compare the poles and zeros of these transfer functions with their counterparts in the system of Fig. 6.28.

(c) For the system with parallel compensation compute $\lim_{t \to 0^+} m(t)$, when $r(t)$ is the unit step.

Compare the result with that of Fig. 6.29(b).

Answers:

(a) $$\frac{Y}{R} = \frac{1278.71\,(s + 130.6)}{(s^2 + 64.328s + 2482.02)\,(s + 67.273)}$$

$$\frac{M}{R} = \frac{2.557\,s\,(s + 1)\,(s + 130.6)}{(s + 32.164 \pm j38.046)\,(s + 67.273)}$$

(c) $\lim_{t \to 0^+} m(t) = \lim_{s \to \infty} \frac{M(s)}{R(s)} = 2.557$

As shown in the previous exercise and in Fig. 6.33, the benefits of parallel compensation are also extensive to the manipulated variable: for the same reference change, the peak value of $m(t)$ was reduced by a factor of more than five in the parallel scheme.

Conclusion: whenever possible the lead compensator must be placed in the feedback path. Notice, however, that the zero of the compensator remains a zero of the transfer function Y/D_3 in both configurations. Consequently, the closed-loop system will be very sensitive to measurement noise if the closed-loop poles have been 'pushed' too far into the left half-plane. This is related to the trade off, to be discussed in section 6.4, between closed-loop bandwidth and the region where noise power is significant.

So far we have not directly addressed the question of steady-state performance. However, when imposing system (6.31) as the desired one we have implicitly specified the steady-state error coefficients, namely

$$K_p = \infty$$

and

$$K_v = \frac{\omega_n}{2\zeta} = \frac{50}{1.3} = 38.46 \Rightarrow \frac{1}{K_v} = 0.026$$

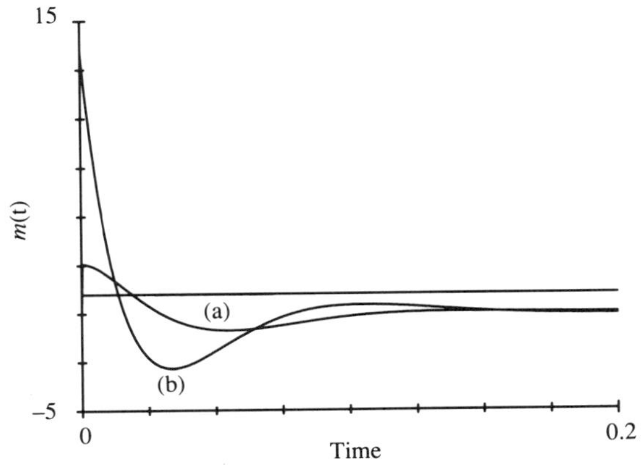

Figure 6.33 Control effort, $m(t)$ following a unit step change in reference, in (a) Fig. 6.31 (parallel compensation); (b) Fig. 6.28 (series compensation).

Exercise 6.11 Compute the steady-state error to the unit-step and unit-ramp inputs in:
(a) System (6.31).
(b) System of Fig. 6.28 (series compensation).
(c) System of Fig. 6.31 (parallel compensation).
Answers: The steady-state errors for the unit-step and unit-ramp inputs are, respectively:
(a) 0 and 2.6×10^{-2}

(b) 0 and $\dfrac{1}{K_v} = \left(\lim_{s \to 0} \dfrac{13.36 \times 500\,(s + 25)}{(s + 130.6)\,(s + 1)} \right)^{-1} = 7.8 \times 10^{-4}$

(c) 0 and 3.3×10^{-2}.

As shown in the previous exercise the steady-state behaviour was also greatly improved in the series compensation, namely the error was reduced by the factor $(2.6/7.8)\times 10^2 = 33.33$. However, in the parallel scheme the error is still 0.033/0.026=1.27 times greater than desired, i.e. the value of K_v is still low. The solution is then to introduce either integral action or a lag compensator. The former was considered in Chapter 4.

6.3.2 Lag Compensation

Because the transfer function of the lag compensator is

$$\frac{s + z_c}{s + p_c} \tag{6.32}$$

with $|z_c| > |p_c|$, the relevant steady-state error constants will be increased by the factor $|z_c|/p_c|$ (provided the compensator is placed in series with the plant). The design of the lag compensator consists then in finding z_c and p_c to satisfy a given ratio. Because for a fixed $|z_c|/|p_c|$, the angle contribution at p_d decreases with $|z_c|$, we must select $|z_c|$ as small as possible, so that the addition of the lag compensator has little influence on the transient performance, i.e. on the roots-loci in the vicinity of the dominant poles. The next exercise shows that the effect of this compensator is slightly to 'bend-over' the initial roots-loci to the right hand side in the s-plane.

Exercise 6.12 Show that the addition of the lag compensator (6.32) shifts the initial roots-loci asymptotes to the right, by the amount

$$\frac{p_c - z_c}{\text{pole excess}}$$

Hint: recall expression (6.24).

In practice we normally choose $|z_c| \simeq \omega_n/10$ where ω_n is the magnitude of the dominant pole pair. As shown in the next section, this is similar to the accepted design heuristics in the frequency domain, to choose the zero of the lag compensator one decade below the crossover frequency of the original system.

Smaller values of $|z_c|$ are not recommended by practical reasons, either in continuous- or discrete-time. As shown in section 6.2, the smaller $|z_c|$, the larger capacitor required for its physical realization. If the compensator is to be discretized for computer control, this hardware limitation no longer exists; however, higher values of $|z_c|$ allow a better numerical accuracy in the discretized model. As shown in Chapter 7, the (zero-order hold) discretization of the transfer function $(s+a)/(s+b)$ is

$$1 + \left(\frac{a-b}{b} \; \frac{1-\exp(-bh)}{z-\exp(-bh)} \right)$$

where h is the sampling period. As b approaches zero, the accuracy of the term $(1-\exp(-bh))$ decreases. Therefore $|b|$ should be as large as possible.

The location of the lag compensator in series with the plant is less detrimental for the transient response. In exercises 6.8 and 6.9 we found that the difference between the series and the parallel configurations resided in the numerators of the transfer functions Y/R and M/R. Bearing in mind that the closed-loop system will have a pole near the compensator zero, the series configuration appears more attractive for lag compensation, because Y/R and M/R will then share the zero with the compensator, therefore providing a more efficient cancellation of that (undesirable) pole.

With reference to the steady-state behaviour, the only sensible location for the lag compensator is in series with the plant. If we were to place it in the feedback path, its effect would be to lower the value of y, therefore increasing the error $e=r-y$. The reader is invited to confirm this result in exercise 6.12.A.

Exercise 6.3.5.A

(a) Assume the structure of Fig. 6.30 (feedback compensation) with $K=1$, $N_p/D_p=500/(s(s+1))$ and compute the steady-state error for a unit ramp reference input, when:

(i) $\dfrac{N_f}{D_f} = \dfrac{s+5}{s+2}$;

(ii) $\dfrac{N_f}{D_f} = \dfrac{2s+5}{5s+2}$

(b) Assume now the lag compensator in series with the plant as in Fig. 6.26, with $K=1$. Compute the steady-state error, for a unit-ramp reference input, with compensator (i) above.

Answers:

(a) (i) $e_{ss}=\infty$;

(ii) $e_{ss}=149/500$.

(b) $e_{ss} = \dfrac{2}{5} \times \dfrac{1}{500}$

6.4 FREQUENCY-DOMAIN COMPENSATION

Frequency-domain design techniques are probably the most popular ones among control engineers in industry. We believe this is mainly due, on the one hand, to the relatively simple hardware, and information processing algorithms, required for frequency response measurements, and on the other to the nature of the data representation employed; in fact, as the data are presented in the form of continuous curves (a non-parametric form), the usefulness of the method is not impaired by the system order. Recall that this is one of the disadvantages of the roots-loci method, where the information content for the human designer is gradually lost as the number of branches increases. But above all the use of logarithmic scales has made frequency response plots even more attractive: the low frequency region (the active range) is expanded and absolute errors on the Bode plot become relative errors on a linear scale (Lopes dos Santos and Martins de Carvalho, 1990). As such, the frequency domain is particularly tailored to handle modelling inaccuracies, either structured or unstructured, and that favours its use in robust design (Polak *et al.*, 1984).

Quite naturally, being a non-parametric and therefore an infinite dimensional representation, we cannot expect explicit formulae relating the frequency response with time response characteristic parameters and vice versa, even assuming a known system order. Consequently the mastering of frequency-domain methods requires considerably more experience than its parametric counterparts (state-space and root-locus).

Except otherwise stated, we shall assume in this section a feedback configuration as shown in Fig. 6.34, and an open-loop transfer function with no transport delays or right half-plane poles.

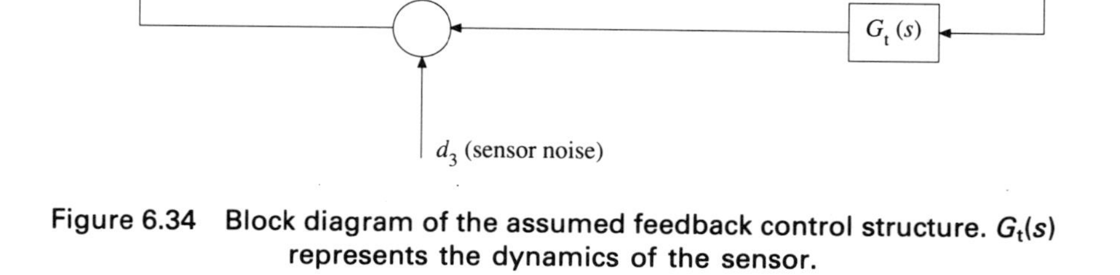

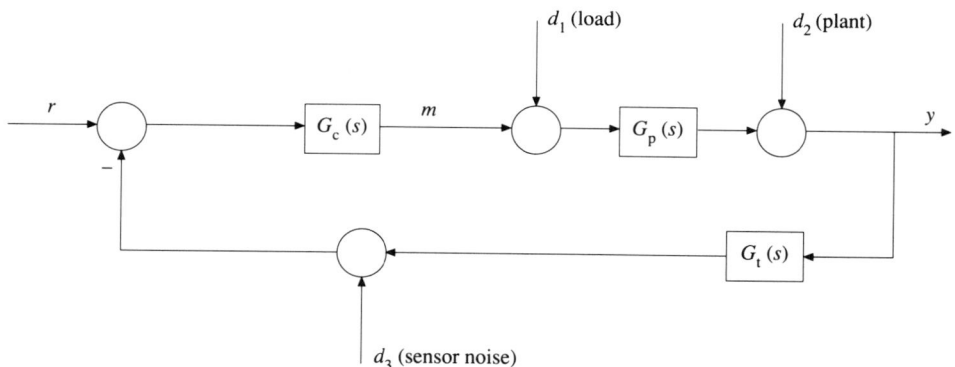

Figure 6.34 Block diagram of the assumed feedback control structure. $G_t(s)$ represents the dynamics of the sensor.

6.4.1 The Feedback Design Problem

The frequency-domain analysis of second-order systems, carried out in the previous chapter, has revealed some advantages of feedback, namely the ability to increase the bandwith and to desensitize the closed-loop system to parameter variations, yielding a transfer function of unit in the range where the open-loop gain is sufficiently high. With the help of Fig. 6.34 let us look again at the problem in some detail. Neglecting sensor dynamics ($G_t(s)=1$), a simple calculation yields:

$$y = G_c G_p(1 + G_c G_p)^{-1} (r - d_3) + (1 + G_c G_p)^{-1} d_2 + G_p(1 + G_c G_p)^{-1} d_1 \tag{6.33}$$

Defining $e=r-y$ we get

$$e = \frac{1}{1 + G_c G_p}(r - d_2) + \frac{C_c G_p}{1 + G_c G_p} d_3 - \frac{G_p}{1 + G_c G_p} d_1 \tag{6.34}$$

Equations (6.33) and (6.34) immediately suggest that feedback design is not a trivial exercise because certain performance tradeoffs and design limitations must be satisfied. One of them concerns reference (r) and disturbance (d_2) error reduction versus sensor noise (d_3) error reduction – the conflict between these two goals is evident in equation (6.34): if we make $|G_c G_p|$ large over a broad frequency range we reduce the errors due to r and d_2. But then $y \approx r - d_3$, i.e. sensor noise is also passed through.

The control effort may also reach unacceptable levels if a very large loop gain is achieved at the expense of the compensator alone, i.e. when $|G_c G_p| >> 1$ and $|G_p|$ small; because

$$m = \frac{G_c}{1 + G_c G_p}(r - d_2 - d_3) - \frac{G_c G_p}{1 + G_c G_p} d_1 \tag{6.35}$$

it becomes in this case

$$m \approx G_p^{-1} (r - d_2 - d_3)$$

Such an extra controller effort in response to commands and disturbances, in order to compensate for the plant attenuation, is not unexpected; however the problem of sensor noise also being amplified is not so intuitive. Therefore the loop gain cannot be made arbitrarily high over arbitrarily wide frequency ranges. In order to satisfy the above tradeoffs we may think of shaping the loop gain as shown in Fig. 6.35: high loop gain in the active frequency range, for good performance, large attenuation in the region where noise power is significant and a steep attenuation rate between these two in order to keep the bandwidth as broad as possible.

Again a major limitation arises: in the case of proper, rational and stable transfer functions, steep attenuation rates are only possible at the expense of

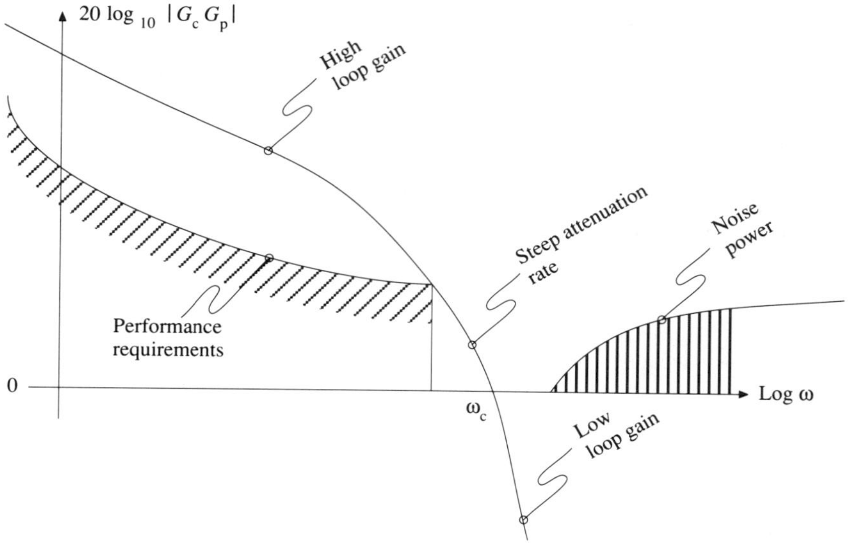

Figure 6.35 Undesirable shape for open-loop gain.

small phase margins. For example, $1/s^2$ has twice the attenuation rate of $1/s$ (viz. −40 dB/dec and −20 dB/dec) but

$$G_cG_p = \frac{1}{s^2}$$

has a phase margin of zero (closed-loop unstable) while

$$G_cG_p = \frac{1}{s}$$

has a phase margin of 90°. Bode has shown (Bode, 1945) that the phase of a rational, stable, proper transfer function $G(s)$, with no right half-plane zeros, and such that $G(0)>0$, is uniquely determined by its magnitude, and he derived various expressions for such a function. One of them gives the phase at frequency ω_0, in terms of a weighted average attenuation rate as follows:

$$\angle G(j\omega_0) = \frac{1}{\pi}\int_{-\infty}^{+\infty} \frac{d\ln|G(j\omega)|}{d\bar{\omega}}\left(\ln\coth\frac{|\bar{\omega}|}{2}\right)d\bar{\omega} \tag{6.36}$$

where

$$\bar{\omega} = \ln(\omega/\omega_0) \tag{6.37}$$

Since

$$\frac{d}{d\bar{\omega}}(\ln|G(j\omega)|) \tag{6.38}$$

is the slope of the magnitude curve in the Bode diagram and ln coth $|\bar{\omega}|/2$ has the form plotted in Fig. 6.36, we conclude that the larger the rate of decrease of $|G(j\bar{\omega})|$, the smaller $G(j\omega_0)$. Furthermore, Fig. 6.36 shows that the phase of $G(j\omega_0)$ is mainly determined by the slope of its log magnitude within a neighbourhood of radius of one decade; in fact for $\omega=10\omega_0$ we already have

$$\ln \coth \frac{|\bar{\omega}|}{2} = 0.2$$

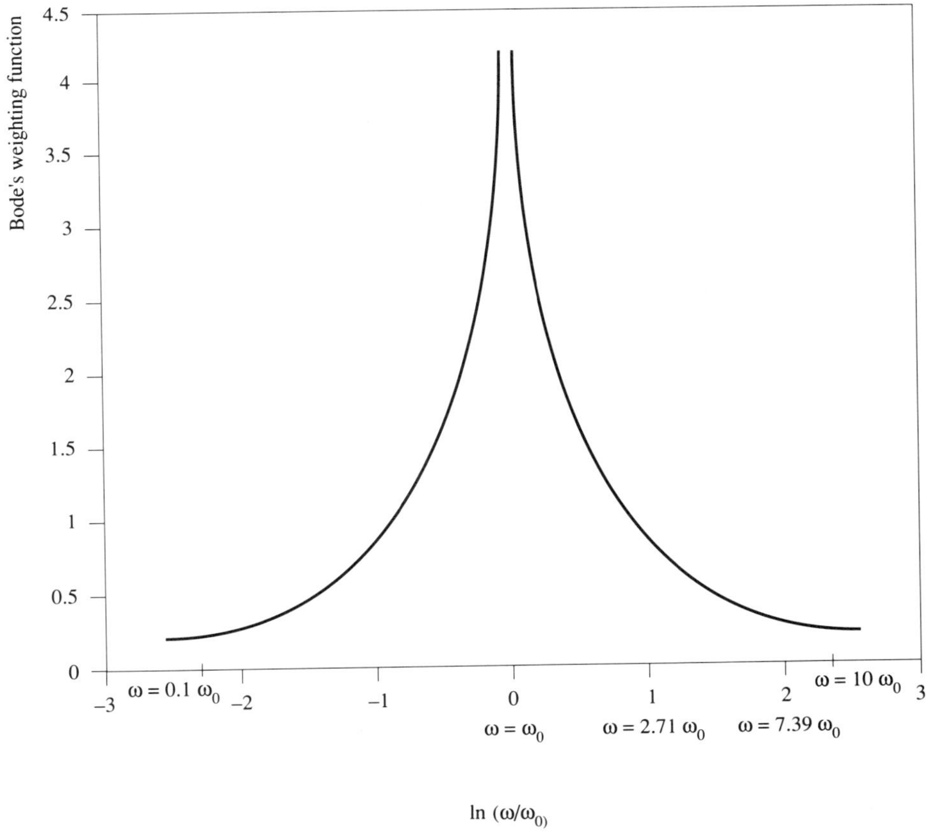

Figure 6.36 Plot of ln coth $|\bar{\omega}|/2$.

while for $\omega=\omega_0$ this function is infinite. If $G(s)$ has right half-plane zeros, then the phase will be more negative than is indicated by formula (6.36).

Another important constraint is the requirement for tolerance to uncertainties. In practice this puts severe limitations on the frequency range over which the loop gain can be large. For a better understanding of its implications let us analyze the following simple example. Consider a unit feedback system with a nominal open-loop transfer function $G(j\omega)$, as depicted in Fig. 6.37, which is assumed to have a constant relative error of 20% throughout the entire frequency range. Let $A=G(j\omega_A)$, $B=G(j\omega_B)$, etc.; the circles centred at these points represent the admissible regions for the 'true' values $G'(j\omega_A)$, $G'(j\omega_B)$, etc. Clearly their radius decreases with decreasing G. Although a constant accuracy was assumed, the picture reveals that stability can be maintained for larger values of uncertainty, provided the present accuracy changes little in the vicinity of B.

In other words, what we need for stability of the closed-loop system is a good (accurate) open-loop model in the frequency range around the crossover frequency. More specifically, the smaller the least value of $|1+G(j\omega)|$, $\omega \in (0,\infty)$, the larger the required modelling accuracy; Quite naturally $(1+G)^{-1}$ is known as the 'sensitivity' of the closed-loop system, as defined in the following. In Fig. 6.38 we show the relative position of $(1+G)$ and G.

Definition: The *sensitivity* of a transfer function T to one of its parameters p, S_p^T, is defined as the ratio of percent change in T to percent change in p:

$$S_p^T = \frac{dT/T}{dp/p} = \frac{dl_nT}{d\, l_np} \tag{6.39}$$

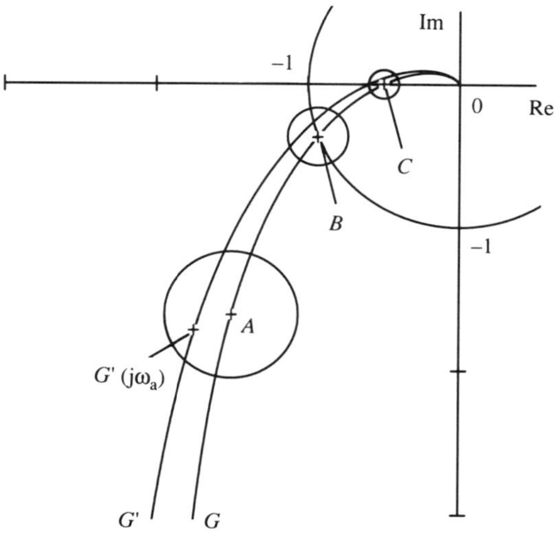

Figure 6.37 Nominal, $G(j\omega)$, and possible true system, $G'(j\omega)$.

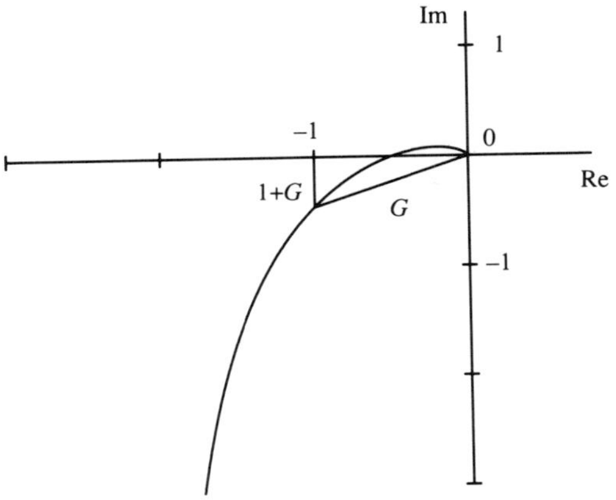

Figure 6.38 (1+*G*) as a measure of sensitivity to modelling errors.

or equivalently

$$S_p^T = \frac{\mathrm{d}T}{\mathrm{d}p} = \frac{p}{T} \tag{6.40}$$

Let us then compute the sensitivity function of Y/R in Fig. 6.34, assuming for simplicity that $G_t(s)=1$. Setting $G=G_c\,G_p$ we can write

$$S_G^{Y/R} = \frac{G}{Y/R}\,\frac{\mathrm{d}(Y/R)}{dG} \tag{6.41}$$

Because

$$\frac{\mathrm{d}(Y/R)}{\mathrm{d}G} = \frac{(1+G)-G}{(1+G)^2} = \frac{1}{(1+G)^2} \tag{6.42}$$

we have

$$S_G^{Y/R} = (1+G)^{-1} \tag{6.43}$$

as expected.

Exercise 6.13 Compute the sensitivity functions

$$S_{G_c}^{Y/R},\ S_{G_p}^{Y/R},\ S_{G_cG_p}^{Y/R},\ S_{G_t}^{Y/R}$$

in Fig 6.34.

Answers:

$$S_{G_c}^{Y/R} = S_{G_p}^{Y/R} = S_{G_cG_p}^{Y/R} = \frac{1}{1+G_cG_pG_t}$$

$$S_{G_t}^{Y/R} = \frac{-G_t}{1 + G_c G_p G_t}$$

If we denote by ΔH_{cl} and ΔH_{ol} the relative errors of the closed- and open-loop transfer functions, caused by a small change ΔG_p, in the plant G_p, we can write, from the definition of sensitivity,

$$\Delta H_{cl} \simeq (1 + G_c G_p G_t)^{-1} \Delta H_{ol} \qquad (6.44)$$

This is in agreement with our previous analysis, namely that feedback can only reduce the effects of plant uncertainty when $(1+G_cG_pG_t)$ is large.

In practice we find that plant uncertainty increases with frequency; this is due to unmodelled dynamics such as pure transport delays, mechanical resonances, etc. Therefore the loop gain must be small at those (high) frequencies in order to keep the closed-loop system stable. In particular this means that compensation can only increase the bandwidth of the system if the model remains relatively accurate; any attempt to broaden it, to an extent that encompasses a region of large uncertainty, will lead to undesirable performance or even instability.

What has just been said is summarized in Fig. 6.39, which depicts the desirable shape of the loop frequency response curve. In short, the purpose of frequency-domain compensation is to *reshape* the loop response in order to achieve the following:

1. High loop gain within the control bandwidth for control performance.
2. Well-behaved crossover for good stability properties (say, a slope close to 20 dB/decade over a reasonably broad frequency band, centred at crossover).
3. Low loop gain outside the control bandwidth for desensitization to modelling errors and noise rejection.

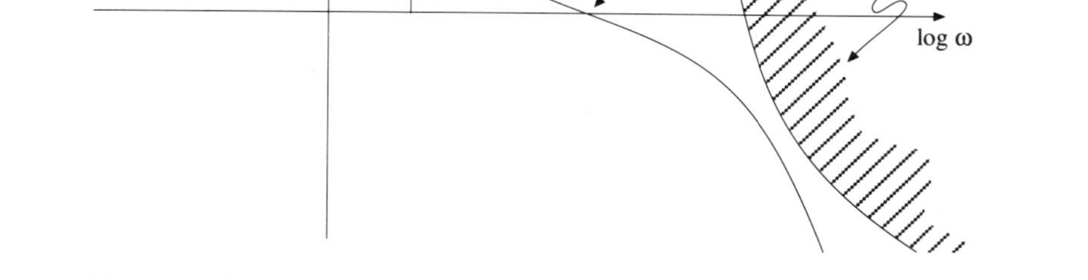

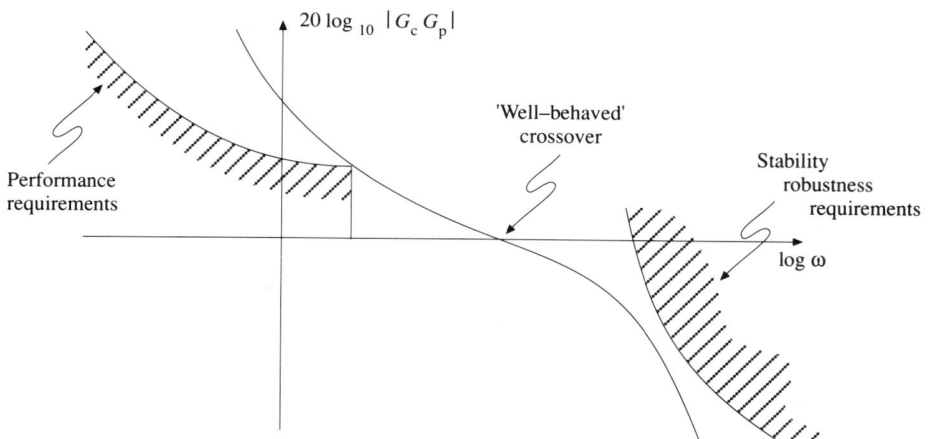

Figure 6.39 Frequency-domain requirements for feedback design.

Exercise 6.14 In the light of the previous discussion, can you see any disadvantage of the ideal PID controller when compared with the lag–lead network?
Answer: the ideal PID controller has the disadvantage of an increasing gain with increasing frequency. This is why in practice the derivative is usually implemented as $(sT_D)/(1+sT_1)$, $T_D >> T_1$, as considered later in the chapter. Besides the lag–lead network seems better tailored to reshape the system frequency response curve, as can be seen from Fig. 6.14.

Exercise 6.15 Show that, with the exception of the low frequency range, feedback actually 'magnifies' open-loop modelling uncertainties. Then illustrate this fact by plotting the open- and closed-loop relative errors in the unit feedback system where the true open-loop transfer function

$$G_2(s) = \frac{1}{s(s + 1.2)}$$

was modelled as

$$G_1(s) = \frac{1}{s(s + 1)}$$

Answer: from (6.44) and Fig. 6.38 we conclude that the magnitude of $(1+G)$, where G is the open-loop transfer function, eventually becomes smaller than 1 as frequency increases, leading to a magnification of the open-loop error.

Computing the magnitudes of the open- and closed-loop relative errors we get

$$\left| \frac{G_1 - G_2}{G_1} \right| = \frac{0.2}{\sqrt{((1.2)^2 + \omega^2)}}$$

and

$$\frac{\left| \dfrac{G_1}{1 + G_1} - \dfrac{G_2}{1 + G_2} \right|}{\left| \dfrac{G_1}{1 + G_1} \right|} = \frac{0.2\omega}{\sqrt{[(1 - \omega^2)^2 + (1.2)^2 \omega^2]}}$$

A plot of these functions, for $\omega \leq 2$ rad/s is shown in Fig. 6.40. The crossover frequency for G_1 is 0.79 rad/s, and the closed-loop error becomes larger for $\omega > 0.71$ rad/s. As expected, the closed-loop error eventually becomes greater as frequency increases.

In practice there are departures from the above-mentioned standard form of specifications. For example, in the face of a resonant plant, one may wish to have a large loop gain in a frequency range below and above the resonant frequency, and a low loop gain around it. As such, the magnitude curve will not be monotonic decreasing. This case will be treated in the next section, by the time we address the notch filter compensation, for a resonant plant with a conjugate pair of zeros.

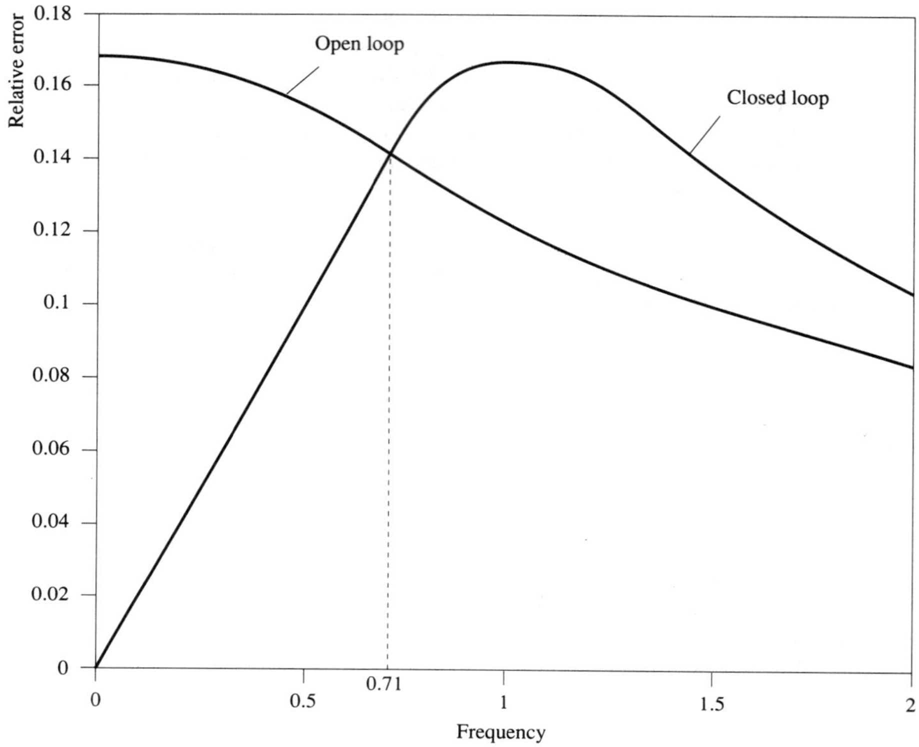

Figure 6.40 Plot of absolute and relative errors in exercise 6.15.

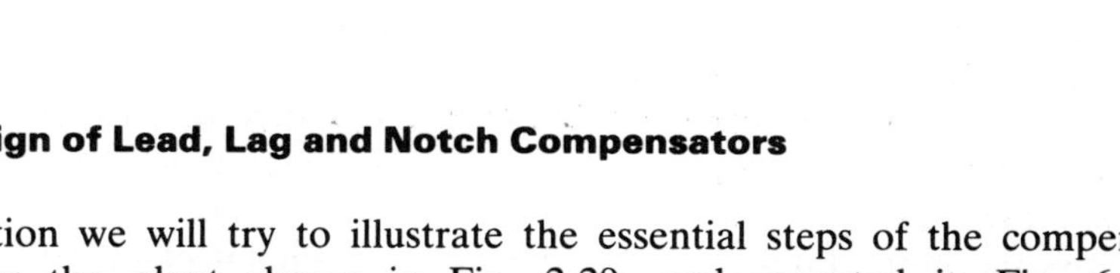

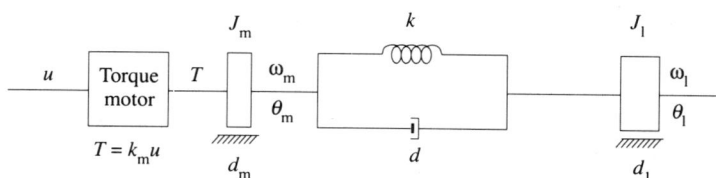

Figure 6.41 Torque motor coupled to an inertial load via an elastic shaft.

6.4.2 Design of Lead, Lag and Notch Compensators

In this section we will try to illustrate the essential steps of the compensation process, for the plant shown in Fig. 2.29, and repeated in Fig. 6.41 for convenience; recall that the relevant transfer functions were already computed in Chapter 3, namely Ω_m/T and Ω_l/T. In the case of rigid coupling ($K=\infty$) for a given set of parameters, namely $d=0$, in particular leads to

$$G_{p1}(s) = \frac{\theta_m(s)}{U(s)} = \frac{\theta_l(s)}{U(s)} = \frac{500}{s(s + 1)} \tag{6.45}$$

whose Bode plots are shown in Fig. 6.42. Assume we wish to compensate this plant to obtain

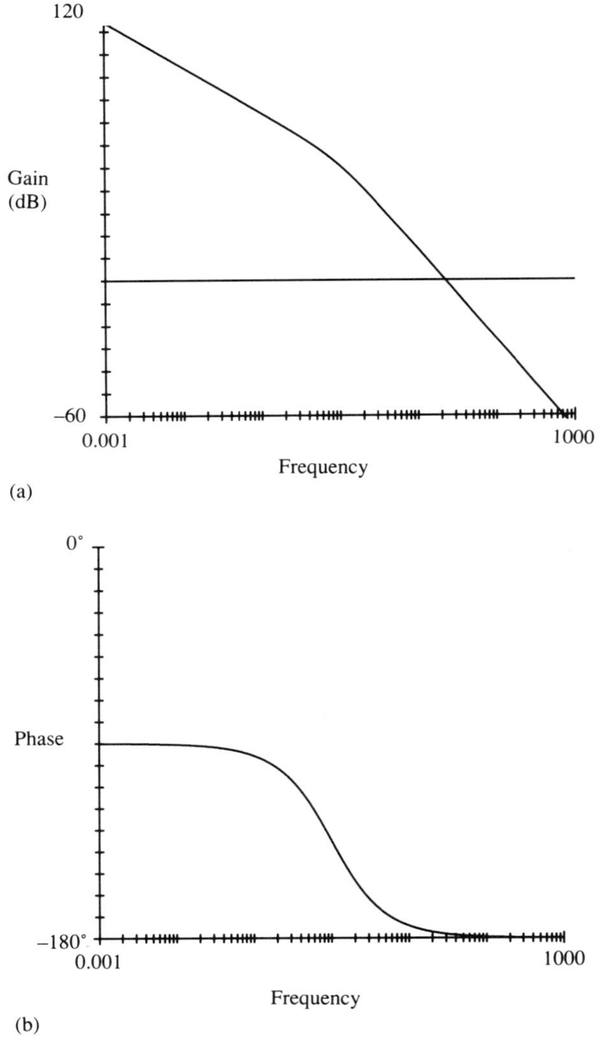

Figure 6.42 Bode plots for $500/(s(s+1))$. At $\omega=34$ rad/s we have gain $=-7.28$ dB and phase $=-178.3°$.

$$\text{phase margin} = 65° \tag{6.46}$$

and crossover frequency

$$\omega_c = 35 \text{ rad/s} \tag{6.47}$$

These specifications correspond to a closed-loop second-order system, with

$$\zeta = 0.65$$

$$\omega_n = 50 \tag{6.48}$$

or equivalently, to an open-loop transfer function

$$\frac{2500}{s(s + 65)} \tag{6.49}$$

Since at 35 rad/s the plant has only $-(180°-178°.4)=-1.6°$ of phase we must perform a lead compensation.

6.4.2.1 Lead compensation

The lead compensator, $G_{ld}(s)=K_c(s-z_c)/(s-p_c)$, must therefore contribute, at $\omega=35$ rad/s, with approximately 8 dB and +65°. From section 6.2 we know that a maximum phase contribution of 65° implies for the lead compensator a zero-pole ratio of $\alpha=0.05$.

If we set $\omega_{max}=35$ rad/s, we have

$$z_c = -\omega_{max}\sqrt{\alpha} = -7.83$$

and

$$p_c = -156.5$$

For simplicity let us set $z_c=-8$ and $p_c=-160$. For $K_c=11.4$ we have

$$|G_c\,(j35)| = 8 \text{ dB}$$

therefore

$$G_{ld}(s) = 11.4\frac{s + 8}{s + 160} \tag{6.50}$$

is the desired lead compensator.

The Bode plots for

$$G_{ld}\,G_{p1} = \frac{11.4 \times 500 \times (s + 8)}{s(s + 1)(s + 160)} \tag{6.51}$$

are depicted in Fig. 6.43, showing that our compensated system has a phase margin equal to 66.4° and a gain crossover frequency equal to 35.68 rad/s, well

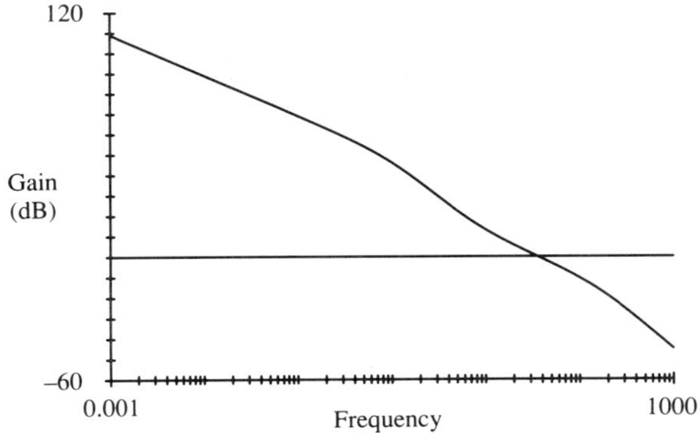

(a)

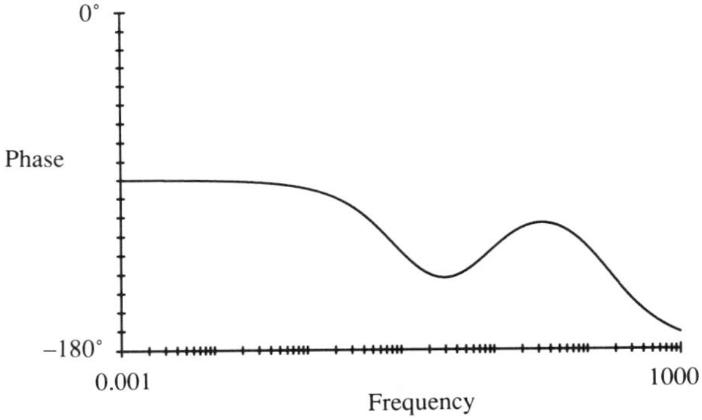

(b)

Figure 6.43 Bode plots for $(11.4\times500(s+8))/(s(s+1)(s+160))$. At 0 dB we have $-113.6°$ and $\omega=35.68$ rad/s.

within the specification. Notice the effect of the lead compensator, easing the slope of the magnitude curve when crossing the 0 dB line, and the simultaneous phase increase, in accordance with the Bode formula (6.36).

However, the speed error constant is now smaller: before compensation $K_v=500$, and after lead compensation $K_v=285$. If we wish to restore the initial value we must design a lag compensator.

6.4.2.2 Lag compensation

Following the accepted design heuristics (Hang, 1989) for the lag compensator

$$G_{lg}(s) = \frac{s - z_c}{s - p_c} \tag{6.52}$$

we choose $|z_c|$ from 0.5 to 1 decade below the gain crossover frequency or the closed-loop bandwidth. Therefore we set

$$G_{lg}(s) = \frac{s + 3.5}{s + 2} \tag{6.53}$$

Besides restoring the original velocity error constant, this compensator has virtually no effect on the Bode plot above 7 rad/s.

The lag filter can also be used to increase the system phase margin, although at the expense of a reduction in bandwidth, as shown in the following example.

Example 6.1 Given the plant

$$G_p(s) = \frac{20}{s(s + 1)(s + 2)}$$

design a series lag compensator such that the velocity error constant remains unchanged and the phase margin becomes 45°.

We have

$$G_p(\text{j}2.43 \text{ rad/s}) = \text{e}^{-\text{j}208.1°}$$

The closed-loop (unit feedback) is therefore unstable.

Given the lag compensator transfer function

$$G_c(s) = K_c \frac{s + z_c}{s + p_c}$$

we conclude that $K_c(z_c/z_p)=1$ in order to keep the velocity error constant unchanged.

Assumption: In the design we are going to place the zero of the compensator one decade below the required ω_c. In this fashion, at ω_c the phase contribution of the compensator can be neglected (hopefully!), and $|G_c|$ is approximately K_c. Since

$$G_p(\text{j}0.56) = 15\ \text{e}^{-\text{j}135°}$$

we conclude that the new crossover frequency must be ω_c=0.56 rad/s. Consequently

$$z_c = \frac{0.56}{10} = 0.056$$

The required attenuation at ω_c is approximately

$$K_c = (15)^{-1}$$

which in turn implies

$$p_c = K_c\, z_c = 0.003\ 73$$

Therefore

$$G_c = \frac{s + 0.056}{15(s + 0.003\ 73)}$$

At $\omega=0.562$ rad/s we have for the open-loop transfer function G_cG_p

$$|G_cG_p| = 1$$

$$\angle G_cG_p = -140.3°$$

showing that the phase margin is only 40°. If we wish to recover the missing 5°, then we must find a new lower ω_c where $\angle G_p(j\omega_c)\simeq 50°$ and repeat the above process.

So far we have assumed a rigid coupling in Fig. 6.41, i.e. $K=\infty$. For the same set of parameters as above, namely $d=0$, and for a given (finite) value of K we obtain

$$G_{p2}(s) = \frac{\theta_l(s)}{U(s)} = \frac{6\,052\,000}{s(s+1)(s+2\pm j110)} \tag{6.54}$$

We may then ask what is the effect of this resonant pole pair on the performance of the system? To gain insight into this question we have plotted in Fig. 6.44 the Bode diagram of

$$G_{p2}G_{ld} = \frac{11.4\times(s+8)\times 6\,052\,000}{s(s+1)(s+160)(s+2\pm j110)} \tag{6.55}$$

i.e. the just lead-compensated plant with the inclusion of the previously neglected resonant mode. For ease of comparison the curves for (6.51) are also included.

Exercise 6.16 Sketch the roots-loci of the lead compensated resonant plant (6.55), i.e.

$$KG_{p2}G_{ld} = K\frac{11.4(s+8)(6\,052\,000)}{s(s+1)(s+160)(s^2+4s+12\,104)}$$

and compute the value of K on the limit of stability.
Answer: $K=1/6=0.1667$.

We no longer have an infinite gain margin nor a monotone decreasing gain curve; besides, the phase curve exhibits a dangerous abrupt phase loss just after 100 rad/s; the magnitude curve intersects the 0 dB line at three points, whose frequencies and phases are

(40.75 rad/s; −114.9°)

(88.5 rad/s; −128.2°)

(122.1 rad/s; −300.7°).

Besides the phase curve intersects the −180° line at (108.5 rad/s; 15.57 dB). Therefore the closed-loop system is now unstable.

In order to stabilize it we have several alternatives. One is a simple gain reduction; since

$$G_{ld}\,G_{p2} = (j180.5\text{ rad/s}) = 6.003\, e^{-j180°}$$

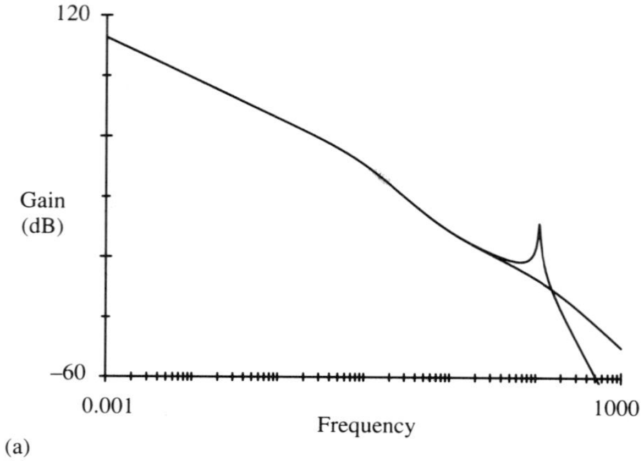

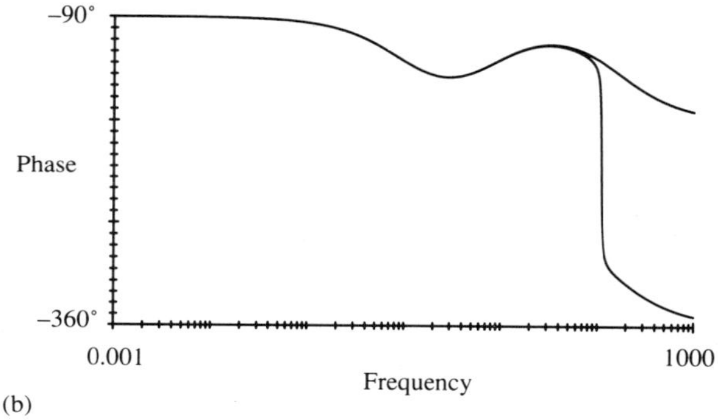

Figure 6.44 Bode plots for $(11.4(s+8)\ 6\ 052\ 000)/(s(s+1)(s+160)(s+2\pm j110))$. The curves of previous figure are also drawn for ease of comparison.

i.e. the open-loop polar plot intersects the negative real axis at the point -6.003, the closed-loop system becomes stable for a gain reduction greater than 15.57 dB ($20\log_{10}6.003=15.57$). For example, a gain reduction of 7 leads to the crossover frequency $\omega_c=7.435$ rad/s (at $-132.2°$) and the (minute) gain margin of 1.334 dB.

Another possibility is to use a lag compensator in series with $G_{ld}G_{p2}$. With such a device, we can attenuate high frequencies and leave the low frequency range unchanged. Since we are aiming at a crossover frequency $\omega_c=35$ rad/s, we choose the compensator zero one decade below ω_c, i.e. $z_c=3.5$. On the other hand, a high frequency attenuation of 10 (20 dB) appears a good compromise between the need to compensate for the phase loss induced by the lag compensator and the need to keep the bandwidth as broad as possible. Therefore we set

$$G_{lg}(s) = \frac{1}{10}\,\frac{s + 3.5}{s + 0.35}$$

This procedure yields a phase margin of 20°, a crossover frequency ω_c=6.42 rad/s and a gain margin of 4.44 dB. Despite its smaller phase margin, the lag compensation has produced a more robust design: comparing the polar plots of $1/7\ G_{ld}G_{p2}$ and $G_{lg}G_{ld}G_{p2}$ we see the former gets closer to the $(-1+j0)$ point than the latter.

However, if we wish to maintain the initial bandwidth then we must use a more complex form of compensation, namely a notch filter.

6.4.2.3 Notch-filter compensator

Going back to Fig. 6.44 we conclude that the action of notch filter is beneficial if it attenuates the spike in the magnitude curve around 110 rad/s and adds sufficient positive phase. On the basis of what we have learned, in section 6.2, about this filter we conclude a reasonable choice might be

$$G_n(s) = \frac{25(s + 2 \pm j110)}{(s + 550)^2} \tag{6.56}$$

Its phase and magnitude curves are shown in Fig. 6.45. The block diagram of the compensated system is depicted in Fig. 6.46 and the corresponding Bode plots in Fig.6.47 (case 1).

The gain crossover frequency is now ω_c=35.48 rad/s, the phase margin 59°, and the gain margin 19.3 dB. This is quite remarkable since the notch filter has not only stabilized the system but also (almost) brought it back to the initial specification of the non-resonant plant.

Table 6.1 Notch compensator design for the system of Fig. 6.46

Case	*Gain crossover frequency (rad/s)*	*Phase margin (degrees)*	*Gain margin (dB)*	*Phase crossover frequency (rad/s)*
$1(\omega_n=110)$	35.48	59	19.3	189.6
$2(\omega_n=100)$	113.2	51.3	16.96	190.1
$3(\omega_n=120)$	36.14	58.7	0.59	107.6

Unfortunately our notch filter design is not robust, i.e. it is very sensitive to any mismatch between the filter zeros and the resonant plant poles. In Fig. 6.47 we also depict such a situation, by placing the notch at 100 rad/s (case 2) and 120 rad/s (case 3). The results are summarized in Table 6.1. Although the magnitude curves have spikes in cases 2 and 3, the situation is desperate in case 3 (notch at 120) with the closed-loop system on the verge of instability: in this case the phase

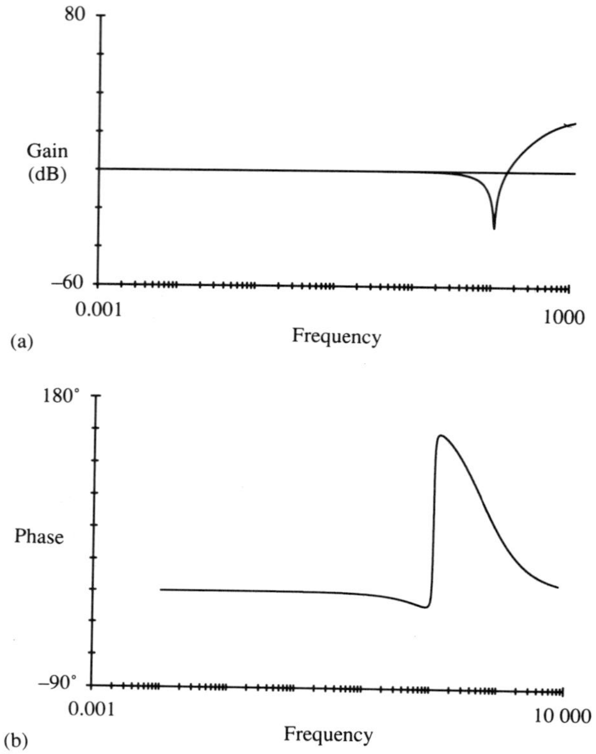

Figure 6.45 Bode plots for notch filter $(25(s+2\pm j110))/(s+550)^2$.

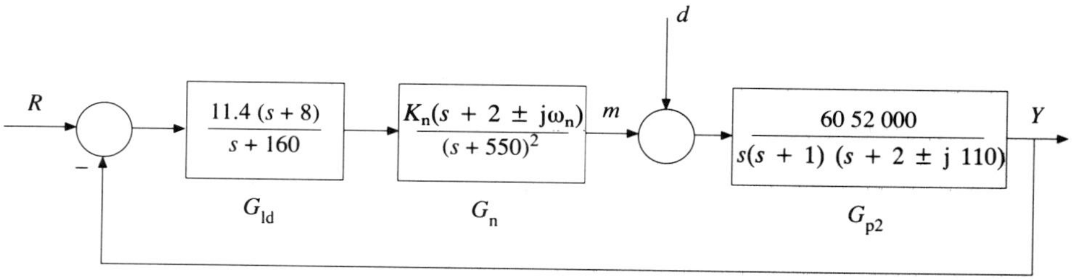

Figure 6.46 Block diagram of the resonant plant, compensated with a lead and a notch filter.

falls sharply to −270° in the vicinity of the gain crossover frequency.

The step responses, plotted in Fig. 6.48, confirm what has just been said.

Exercise 6.17 Based on the Bode plots of Fig. 6.47, and on the data of Table 6.1, sketch the Nyquist plots in situations 2 (ω_c=100 rad/s), and 3 (ω_c=120 rad/s).

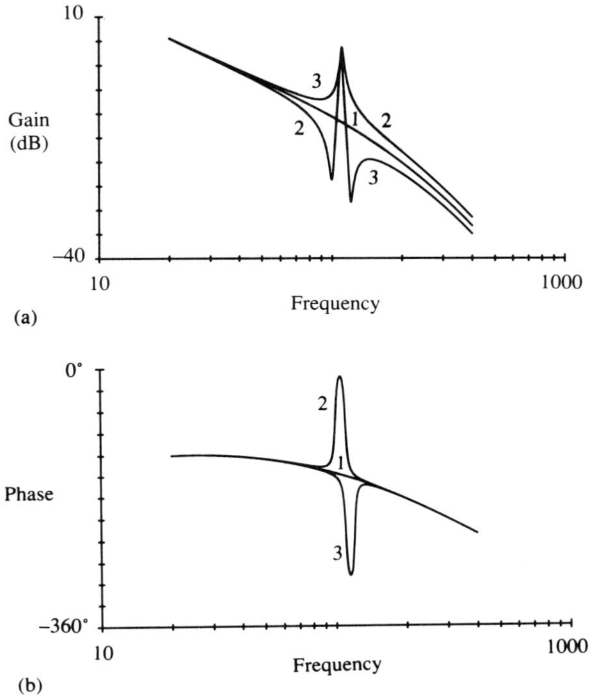

Figure 6.47 Bode plots of the system of previous picture, with $G_n=(K_n(s+2\pm j\omega_n))/(s+550)^2$ and 1, $K_n=25$, $\omega_n=110$; 2, $K_n=30.23$, $\omega_n=100$; 3, $K_n=21$, $\omega_n=120$.

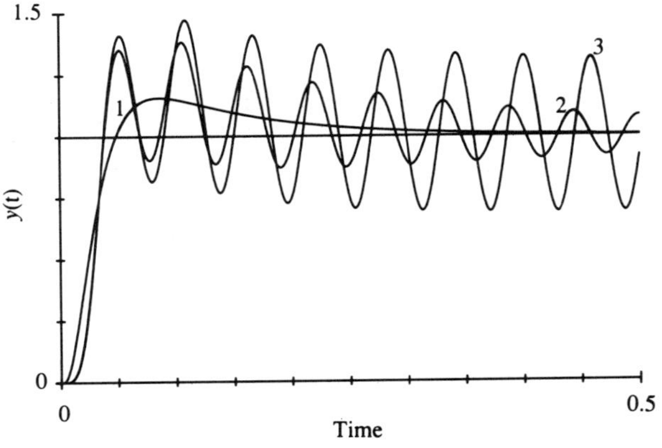

Figure 6.48 Responses, for a unit-step reference, of system of Figure 6.46, with: 1, $K_n=25$, $\omega_n=110$; 2, $K_n=30.23$, $\omega_n=100$; 3, $K_n=21$, $\omega_n=120$.

Since the notch filter compensation is rather sensitive to modelling errors in cases like this, we must take precautions when the plant is time-variant, in particular, designing the filter with a broader notch, at the expense of compensation efficiency, or update at regular intervals the position of the notch by means of a real-time identification procedure.

The same set of parameters for the flexible coupling, which led to G_{p2} in (6.54), also yields

$$G_{p3}(s) = \frac{\theta_m(s)}{U(s)} = \frac{3782.216(s + 0.35 \pm j40)}{s(s + 1)(s + 2 \pm j110)} \tag{6.57}$$

This corresponds to the case of the measurements being taken at the motor shaft. Although in practice we would like to measure the load position (θ_l), the collocation of the sensor at the motor shaft leads to a system with much better gain and phase margin, as shown in the following.

In Fig. 6.49 we have the Bode plots for

$$G_{p3}G_{ld} = \frac{11.4(s + 8)3782.216(s + 0.35 \pm j40)}{s(s + 1)(s + 160)(s + 2 \pm j110)} \tag{6.58}$$

We conclude that the resonant system is now stable given the beneficial positive phase contribution from the conjugate zero-pole pair. The gain margin is infinite, the phase margin is 36° and the gain crossover frequency has reached 212 rad/s. The addition of the notch filter G_n, defined in (6.56), even improves this state of affairs, as shown in Fig. 6.50: the phase margin is increased 74.5° and the crossover frequency becomes ω_c=870.7 rad/s. The addition of the notch filter has flattened the spike around 110 rad/s, in the magnitude curve, without any phase loss; consequently the step response becomes less oscillatory as depicted in Fig. 6.51. This increase in performance is achieved at expense of larger values of the control value m (cf. Fig. 6.46): when $r(t)$ is a unit-step at time $t=0$, $m(t)$ reaches, at $t=0^+$, 11.4 and (11.4×25)=285 before and after the notch filter addition, respectively.

Also of importance is the fact the compensated system for the plant with zeros (G_{p3}) is far more robust. For example, the closed-loop system remains stable in the presence of mismatches of the notch filter zeros and the plant poles, as shown by the Bode plots of Fig. 6.52. Again, an overestimation of the plant resonant frequency is more detrimental than an underestimation.

6.5 COMPENSATION OF TIME-DELAY SYSTEMS

The design techniques discussed so far are not suitable for plants containing non-negligible pure transport delays, because we no longer have a rational transfer function. A practical way to bypass this problem is to approximate the delay block

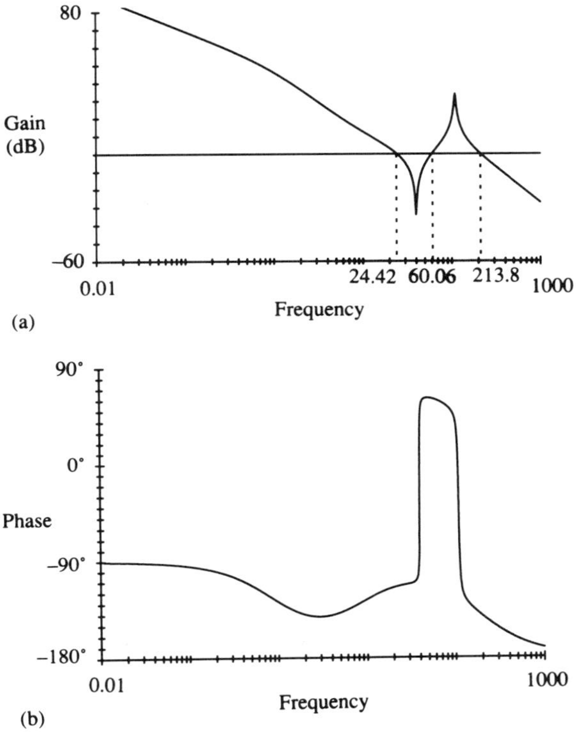

Figure 6.49 Bode plots for

$$G_{\mathrm{ld}}\,G_{\mathrm{p3}} = \frac{11.4(s+8)}{(s+160)} \times \frac{3782.216(s+0.35 \pm \mathrm{j}40)}{s(s+1)(s+2 \pm \mathrm{j}110)}.$$

by a rational transfer function, namely

$$e^{-Ts} \rightarrow \frac{(2/T) - s}{(2/T) + s} \tag{6.59}$$

This approximation does not change the magnitude Bode plot and in the low frequency range the phase curves are virtually indistinguishable, as shown in Fig. 6.53.

The above approximation is known as the first-order Padé approximant for $\exp(-sT)$. The second (n=2) and third-order (n=3) approximants are, respectively,

$$\frac{1 - (sT/2) + ((sT)^2/12)}{1 + (sT/2) + ((sT)^2/12)} \tag{6.60}$$

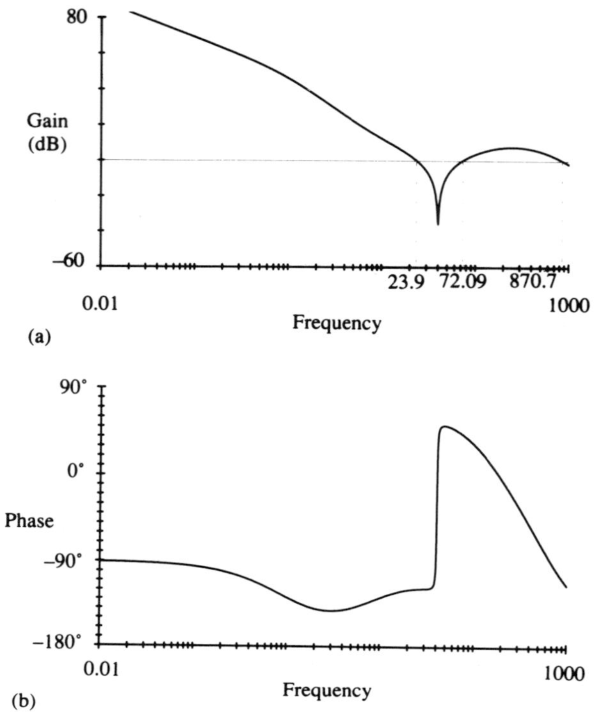

Figure 6.50 Bode plots for

$$G_{ld}\,G_n\,G_{p3} = \frac{11.4(s+8)}{(s+160)} \times \frac{25(s+2\pm j110)}{(s+550)^2} \times \frac{(s+0.35\pm j40)}{s(s+1)(s+2\pm j110)}.$$

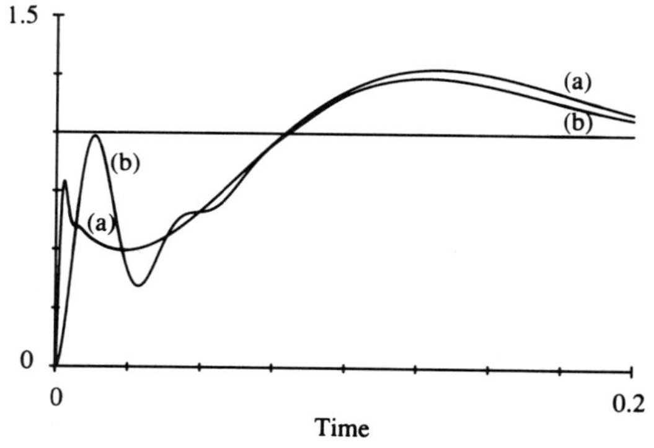

Figure 6.51 Unit-step reference responses of the unit feedback systems with loop transfer functions: (a) $G_{ld}\,G_{p3}$; (b) $G_{ld}\,G_n\,G_{p3}$.

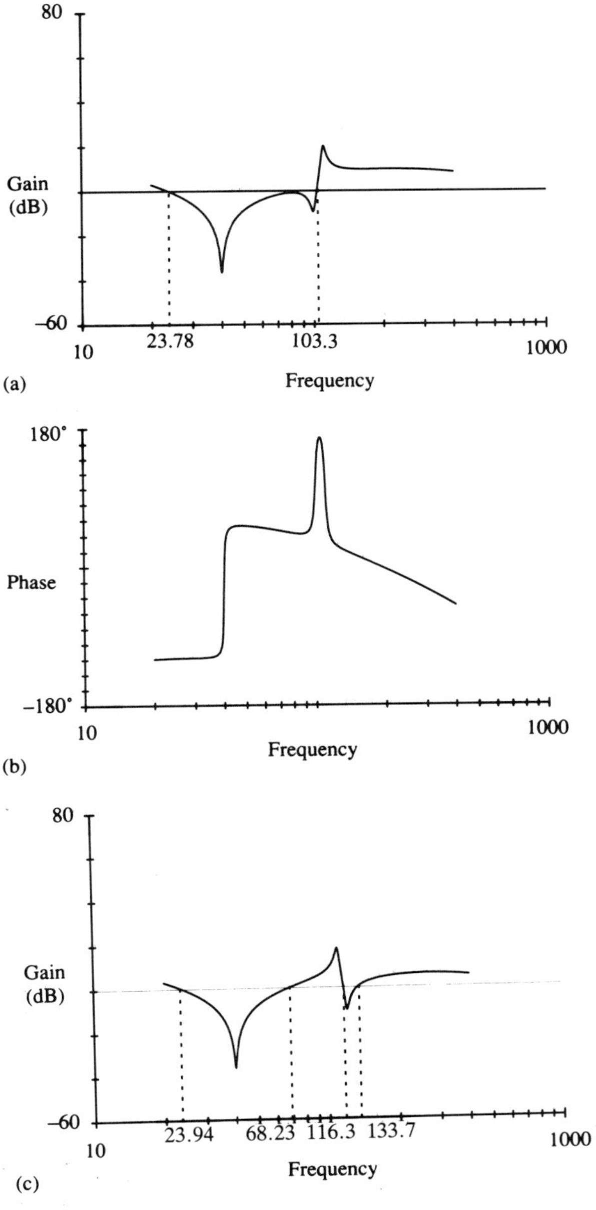

Figure 6.52 Bode plots for $G_{ld}\,G_n\,G_{p3}$, with G_{ld} and G_{p3} as before and: (a), (b) $G_n(s) = (30.23(s+2\pm j100))/(s+550)^2$; (c), (d) $G_n(s) = (21(s+2\pm j120))/(s+550)^2$.

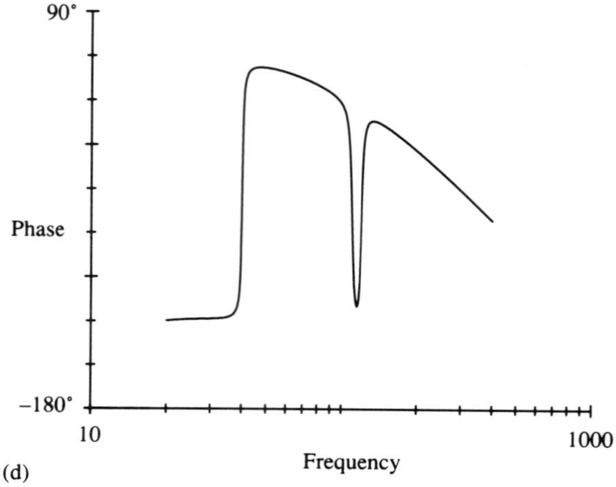

Figure 6.52 (cont.)

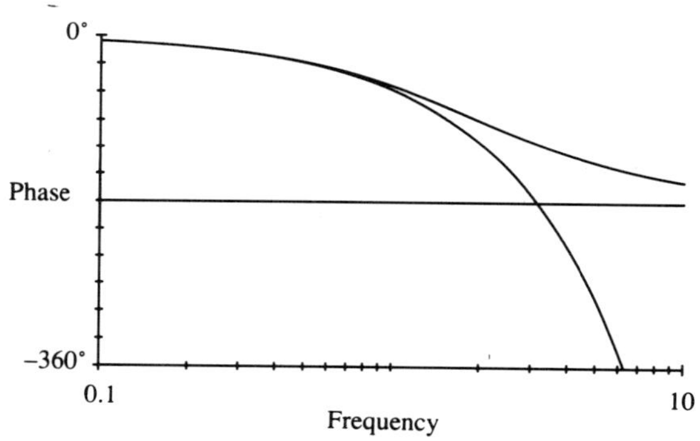

Figure 6.53 Phase Bode plots for exp $(-j\omega)$ and $(2-s)/(2+s)$.

and

$$\frac{1-((sT)/2)+((sT)^2/10)-((sT)^3/120)}{1+((sT)/2)+((sT)^2/10)+((sT)^3/120)} \tag{6.61}$$

These approximants have the property that the coefficients of their long division expansion agree with the Taylor series expansion of $\exp(-Ts)$ as far as the $(2n)$th term.

For example

$$\frac{1-((sT)/2)}{1+((sT)/2)} = 1 - sT + \frac{(sT)^2}{2} - \frac{(sT)^3}{4} + \ldots$$

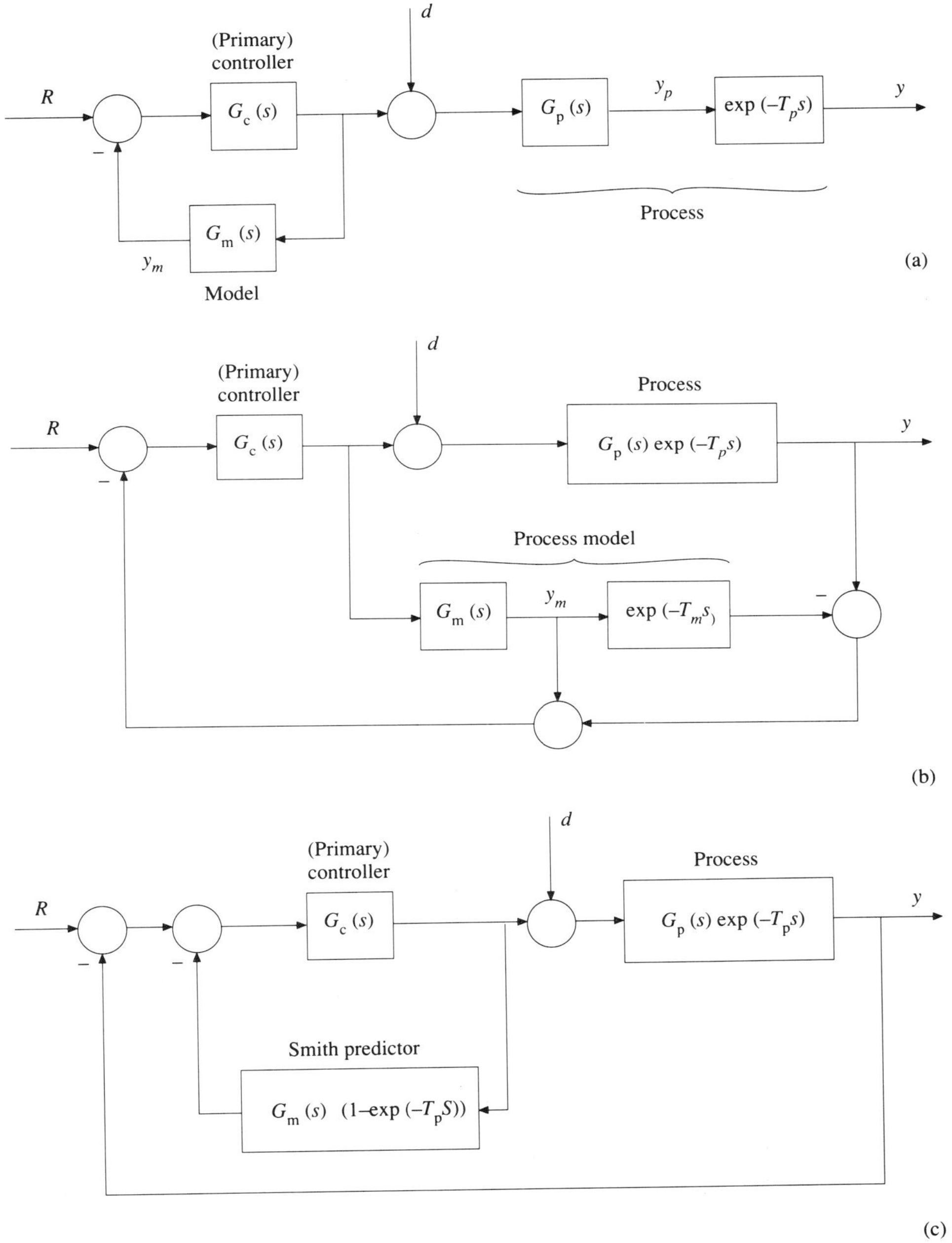

Figure 6.54 Smith predictor control scheme: (a) ideal; (b) practical; (c) equivalent form of (b).

and the Taylor series of $\exp(-Ts)$ is

$$1 - sT + \frac{(sT)^2}{2} - \frac{(sT)^3}{6} + \frac{(sT)^4}{24} - \ldots$$

It can be shown that Padé approximants have the form $N(s)/D(s)$ where $N(s)$ and $D(s)$ are polynomials of the same degree. Furthermore $N(s)=D(-s)$; this property ensures a unit gain at all frequencies, since the zeros are the mirror image points of the poles (with respect to the imaginary axis).

A more elaborate method to deal with time-delay plants is the Smith predictor or dead time compensator (Smith, 1957). The essence of the method is to move the pure delay outside the feedback loop, by means of a process model, thereby allowing the design of controllers as for delay-free plants. The method is summarized in Fig. 6.54(a): if we know $G_p(s)$ and T_p then we can make y_m to equal the physically unavailable variable y_p (assuming $d=0$); then the controller $G_c(s)$ can be designed as for a delay-free plant. In particular this means that the loss of performance created by the pure delay has been eliminated. In order to accommodate load disturbances, d, and inaccuracies in the process model, the set-up (b) must be used in practice as an alternative to (a).

Exercise 6.18

(a) show that in Fig. 6.54(b):

$$\frac{Y(s)}{R(s)} = \frac{G_c(s)\, G_p(s) \exp(-T_p s)}{1 + G_c(s)\, G_m(s)\, (1 - \exp(-T_m s)) + G_c(s)\, G_p(s) \exp(-T_p s)}$$

(b) Show that, when the process model is correct, i.e. $G_p=G_m$ and $T_p=T_m$,

$$\frac{Y(s)}{R(s)} = \frac{G_c(s)\, G_m(s)}{1 + G_c(s)\, G_m(s)} \exp(-T_p s)$$

In this case the Smith controller moves the pure delay outside the feedback loop.

Unfortunately there are situations where the Smith controller only works in the case of perfect model matching; i.e. the closed-loop system is stable *only* when there is no mismatch. Such systems were coined as *practically unstable* or infinitely sensitive (Palmor, 1980). As shown in exercise 6.18, the closed-loop transfer function $Y(s)/R(s)$ has only a finite number of poles in case of perfect matching. In case of mismatch, the poles are given by the zeros of

$$P(s) = 1 + G_c(s)G_m(s)\, (1 - \exp(-T_m s)) + G_c(s)G_p(s) \exp(-T_p s) \tag{6.62}$$

which is the denominator of $Y(s)/R(s)$.

If we denote by S the set of real parts of all the zeros of $P(s)$, i.e.

$$S = \{\sigma\colon \sigma = \operatorname{Re} z,\ P(z) = 0\}$$

we can define the stability of the system as follows:

Definition The closed-loop system is said to be stable if S has a negative upper bound.

With this definition, we rule out the possibility of an infinite chain of closed-loop poles asymptotic to the imaginary axis.

The following example (Yamanaka and Shimemura, 1987) will help to clarify these issues. Assume a process with $T_p=1$ and

$$G_p(s) = \frac{\mu s^2 + \mu s + 1}{(\mu s + 1)(s + 1)(2s + 1)} \tag{6.63}$$

with μ strictly positive, but very small, and

$$G_m(s) = \frac{1}{(s + 1)(2s + 1)} \tag{6.64}$$

With this choice, the relative error of our approximation is

$$\frac{G_p - G_m}{G_m} = \frac{\mu s^2}{\mu s + 1}$$

If we select as primary controller the PID controller

$$G_c(s) = \frac{1.25}{s} + 3.75 + 2.50s \tag{6.65}$$

based on the assumption that

$$\frac{e^{-s}}{(s + 1)(2s + 1)} \tag{6.66}$$

is the correct process model, we are led to the (wrong) conclusion that the closed-loop is stable, with transfer function

$$\frac{Y(s)}{R(s)} = \frac{e^{-s}}{0.8s + 1} \tag{6.67}$$

In fact the actual system has not only an infinite number of poles because of mismatch, but also an infinite number of them in the right half-plane no matter how small μ is ($\mu>0$). Only when $\mu=0$ (perfect matching) is the closed loop stable. Therefore the Smith controller of this example could not be used in a practical situation.

A design exercise is of no practical value if the closed loop cannot withstand small perturbations. When a Smith predictor controller system can cope with infinitesimal perturbations in the plant dynamics we call it *practically stable*.

This concept can be made precise as follows:

Definition (Yamanaka and Shimemura, 1987) – The closed-loop system is said to be *practically stable* if there exist positive numbers ω_M, δ_T and δ_G such that the Smith predictor controlled system is stable for every plant, satisfying

$$|T_p - T_m| < \delta_T \tag{6.68}$$

$$\left| \frac{G_p(j\omega)}{G_m(j\omega)} - 1 \right| < \delta_G, \; 0 < \omega < \omega_M \tag{6.69}$$

Notice that, when the system is stable, the steady-state behaviour is not affected by the time delay estimation error $|T_p - T_m|$; furthermore, in steady-state, all we need in order to make our Smith predictor controlled system insensitive to mismatch, is to ensure (besides stability) that

$$\lim_{s \to 0} \left| G_m(s) - G_p(s) \right| = 0 \tag{6.70}$$

In the following we give necessary and sufficient conditions for the stability of the closed-loop system.

This theorem was derived under the following assumptions: $G_m(s)$ is proper with no poles or zeros in the closed right half-plane (RHP). All of $G_p(s)$ under consideration have no poles in the closed RHP. $G_c(s)$ is chosen so that

$$\frac{G_m(s)\, G_c(s)}{1 + G_m(s)\, G_c(s)} \tag{6.71}$$

has no poles in the closed RHP.

Theorem (Yamanaka and Shimemura, 1987): assume that for some $K \geqslant 0$, $b \geqslant 0$ and some non-negative integer λ,

$$\left| \frac{G_p(j\omega) - G_m(j\omega)}{G_m(j\omega)} \right| \leqslant K\omega^{\lambda}, \; \omega > b \tag{6.72}$$

then, the closed-loop system is practically stable if and only if

$$\lim_{\omega \to 0} (K\omega^{\lambda} + 2) \left| \frac{G_m(j\omega)\, G_c(j\omega)}{1 + G_m(j\omega)\, G_c(j\omega)} \right| < 1 \tag{6.73}$$

Of interest is the fact that when the mismatch is only in the time delay, the condition of practical stability is simply

$$\lim_{\omega \to 0} \left| \frac{G_m(j\omega)\, G_c(j\omega)}{1 + G_m(j\omega)\, G_c(j\omega)} \right| < \frac{1}{2} \tag{6.74}$$

because K and b can then be set to zero. If $G_c(s)\, G_m(s)$ is strictly proper, (6.74) is automatically ensured. Also note that the upper bound $K\omega^{\lambda}$ naturally arises when high frequency poles are neglected, for example.

The theorem states that the Smith controller should be designed in such a way that the closed-loop gain decreases, at least as fast as the inverse of the rate of increase of model uncertainty. For example, if $G_c(s)$ is a PID controller, $G_p(s)$ a low-order transfer function and the rate of uncertainty growth is very large, it is most likely that practical stability will exclude the use of derivative and proportional action. As the rate of uncertainty growth decreases it may be possible to use a PI primary controller and if matters improve, a PID controller.

In the above-mentioned paper by Yamanaka and Shimemura we can find such an example, namely:

$$G_p(s) = \frac{1}{(s+1)(2s+1)} \tag{6.75}$$

and $T_p>0$. They show that the use of derivative action, in the primary controller, is restricted by the extent of uncertainty, i. e. the value of λ, as follows:

$\lambda = 2 \Rightarrow$ no derivative action

$\lambda = 1 \Rightarrow$ derivative time bounded above

$\lambda = 0 \Rightarrow$ derivative time unrestricted

Notice that the uncertainty decreases with decreasing λ.

Example 6.2 Let us go back to example 4.5 and design a Smith predictor controller with $G_c(s)$ of PID type. Recall that in example 4.5 the process transfer function is

$$5.464\,\frac{\exp(-0.88s)}{s+5.464} \tag{6.76}$$

and the (Cohen–Coon) PID settings were

$$\begin{cases} K = 0.551 \\ T_i = 1.11 \\ T_d = 0.166 \end{cases}$$

In our notation we have

$$G_m = G_P = \frac{5.464}{s+5.464} \tag{6.77}$$

$$T_p = 0.88$$

and K=0; b=0.

If

$$G_c(s) = \frac{c_0 + c_1 s + c_2 s^2}{s} \tag{6.78}$$

the ideal (perfect matching) closed-loop transfer function becomes:

$$\frac{G_c\,G_m}{1+G_c\,G_m} = \frac{5.464\,(c_0 + c_1 s + c_2 s^2)}{(1+5.464c_2)s^2 + 5.464(1+c_1)s + 5.464c_0} \tag{6.79}$$

which must be stable by the above theorem. From the Routh–Hurwitz criterion, stability is ensured if

$$c_0 > 0$$

$$c_1 > -1 \tag{6.80}$$

$c_2 > -0.1830$

The closed-loop system remains practically stable if

$$\lim_{\omega \to \infty} \left| \frac{G_c(j\omega)G_m(j\omega)}{1 + G_c(j\omega)\,G_m(j\omega)} \right| < \frac{1}{2} \tag{6.81}$$

This condition is equivalent to

$$\frac{5.464c_2}{1 + 5.464c_2} < \frac{1}{2} \text{ or } c_2 < \frac{1}{5.464} \tag{6.82}$$

Notice this condition is not a robustness result: it only guarantees that our closed-loop system will remain stable for 'infinitesimal' mismatches. How big such a mismatch must be in order to destabilize the system is another question.

Since the unit feedback system, with open-loop transfer function

$$5.464 \frac{\exp(-0.88s)}{s + 5.464}$$

has a pair of dominant poles near $(-0.12 \pm j2.98)$, we are going to design our Smith controller with a primary PID controller, such that the delay-free part of the closed-loop transfer function (perfect matching) has a double pole at (-3).

From (6.79), and setting $c_2 = 0.1$, we obtain $c_1 = 0.698$ and $c_0 = 2.547$, leading to

$$\frac{G_c\,G_m}{1 + G_cG_m} = \frac{0.3533s^2 + 2.466s + 9}{s^2 + 6s + 9} = \frac{0.3533(s + 3.49 \pm j3.6455)}{(s + 3)^2} \tag{6.83}$$

The resulting system is depicted in Fig. 6.55; the responses at m and y, following a unit-step reference change, at time $t=0$, and a step disturbance d, at time $t=5$, with magnitude 0.2 are plotted in Fig. 6.56.

To illustrate the robustness of our design against variations in the plant transport delay, we plot in Fig. 6.57 the output y, in response to a unit-step reference change, for $T_p=0.5$ and $T_p=1.2$. The response for the case of perfect matching is also shown, for comparison. Despite the relatively large parameter variations, the closed-loop remains stable, although an underestimation of the transport delay seems to be more detrimental.

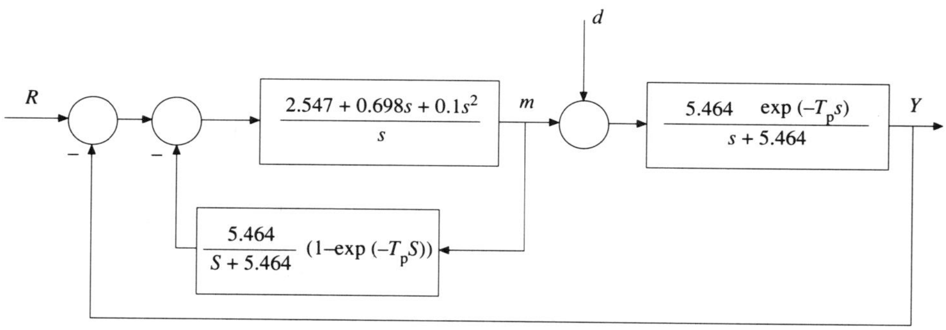

Figure 6.55 Block diagram of example 6.1.

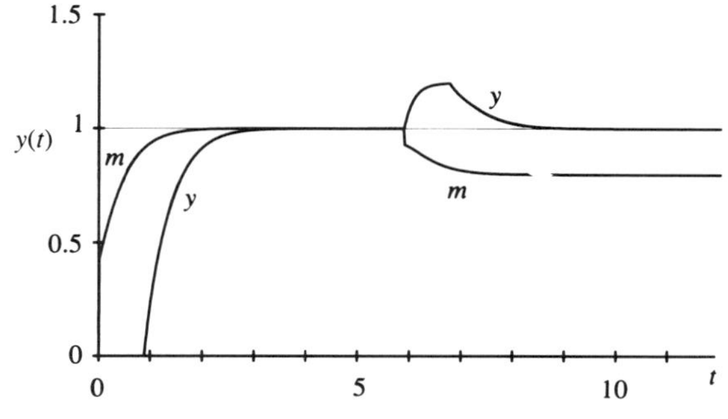

Figure 6.56 Response of system of Fig. 6.55 to a unit-step reference input at time $t=0$ and a step disturbance d, of magnitude 0.2, at time $t=5$, $T_p=T_m=0.88$.

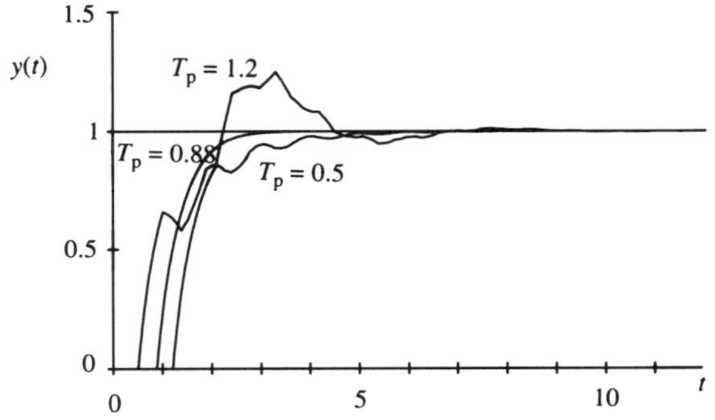

Figure 6.57 Unit-step responses of system of Fig. 6.55 with $T_m=0.88$ and $T_p=0.5$, 0.88, 1.2.

REFERENCES

Bode, H.W. (1945) *Network Analysis and Feedback Amplifier Design*, Van Nostrand, Princeton.

Hang, C.C (1989) 'The choice of controller zeros', *IEEE Control System Magazine*, January.

Lopes dos Santos, J.P. and J.L. Martins de Carvalho (1990) 'Automatic transfer function synthesis from a Bode plot', 29th IEEE Conference on Decision and Control, Honolulu, Hawaii, 5–7 December.

Machado, J.A.T. and J.L. Martins de Carvalho (1989) 'Engineering design of a multirate nonlinear controller for robot manipulators', *Journal of Robotic Systems*, **6**(1), 1–17.

Palmor, Z. (1980) 'Stability properties of Smith dead-time compensator controllers', *Int. J. Control*, **32**(6), 937–91.

Polak, E., D.Q. Mayne and D.M. Stimler (1984) 'Control system design via semi-infinite optimization: a review', *Proc. IEEE*, **72**(12), 1777–94.
Smith, O.J.M. (1957) 'A controller to overcome dead-time', *ISA Journal*, **6**(2), 28–33.
Yamanaka, K. and E.Shimemura (1987) 'Effects of mismatched Smith controller on stability in systems with time-delay', *Automatica*, **23**(6), 787–91.

7

State-space Analysis

In this chapter a detailed analysis of the internal mechanisms of a linear system – the mode generators – and the practical implications of the ways in which they can be coupled to the system inputs and outputs will be presented. This analysis also provides the natural setting for the study of controllability, observability and stability concepts, and the dynamical interpretation of poles and zeros of the transfer function.

7.1 MODAL ANALYSIS

Assume a system as described by the model:

$$\dot{\boldsymbol{x}} = \boldsymbol{A}\boldsymbol{x} + \boldsymbol{B}u$$

$$y = \boldsymbol{C}\boldsymbol{x} \qquad (7.1)$$

with $\boldsymbol{x} \in \mathbb{R}^n$; u, $y \in \mathbb{R}$; $\boldsymbol{A} \in \mathbb{R}^{n \times n}$. We are particularly interested in the evolution, or trajectory, of the state vector $\boldsymbol{x}(\mathrm{t})$ when the system is left on its own, i.e. when $u(t)=0$, $\forall t$. As will be shown below, the properties of this trajectory are determined by the eigenvalues of the matrix $\boldsymbol{A}$ and a simple geometric interpretation arises if a suitable basis is selected for the state-space. We start by dealing with the case of distinct eigenvalues before moving into the general case. This is a topic of increasing importance not only in traditional control studies but also in vibration analysis, particularly in active vibration control for flexible structures (Raps and Schmidt, 1987).

7.1.1 Distinct Eigenvalues

Result 7.1 If a matrix $A \in \mathbb{R}^{n \times n}$ has n distinct eigenvalues then it has n linearly independent eigenvectors.

This is a standard result in linear algebra and can be shown as follows: denote by $\lambda_1,\ldots,\lambda_n$ *the eigenvalues of* A and let $\{e_1,e_2,\ldots,e_n\}$ be a set of corresponding eigenvectors. The proof will be by contradiction. Assume than that $e_1,\ldots,e_n$ are linearly dependent, i.e. that there exists a set of scalars $\{\alpha_1,\ldots,\alpha_n\}$, not *all* zero, such that:

$$\sum_{i=1}^{n} \alpha_i \, e_i = 0$$

Assume also, without any loss of generality, that $\alpha_1 \neq 0$. If we multiply both sides of the above equation on the left, by

$$(A - \lambda_2 \mathbf{I})\,(A - \lambda_3 \mathbf{I}) \ldots (A - \lambda_n \mathbf{I})$$

we get

$$\alpha_1(\lambda_1 - \lambda_2)\,(\lambda_1 - \lambda_3) \ldots (\lambda_1 - \lambda_n)\, e_1 = 0$$

because $Ae_i = \lambda_i e_i$, $i=1,2,\ldots,n$. But this equation implies $e_1=0$, which is a contradiction of the definition of eigenvalue. Therefore we have to assume that the eigenvectors are linearly independent.

7.1.1.1 Diagonal Representation

We are now in a position to analyze the question stated at the beginning of the section, namely the study of the trajectories of $x(t)$ in (7.2) when $u(t)=0$. In this case x will be governed by the equation $\dot{x}=Ax$, or equivalently $x(t)=\exp(At)x(0)$. Of special interest are the trajectories starting on an eigenvector, i.e. $x(0)=a_i e_i$, $a_i \in$. In this case

$$\begin{aligned} x(t) &= a_i \exp(At) e_i \\ &= a_i(\mathbf{I} + At + A^2t^2/2! + \ldots)e_i \\ &= a_i(1 + \lambda_i t + \lambda_i^2 t^2/2! + \ldots)e_i \\ &= a_i \exp(\lambda_i t) e_i \end{aligned} \tag{7.2}$$

which shows that $x(t)$ remains in the straight line defined by the eigenvector e_i; the components of the eigenvector give the relative distribution of the associated natural mode between the components of the state. Furthermore, the other modes $\exp(\lambda_j t)$, $j \neq i$, are *not* present in $x(t)$.

In general we will have

$$x(0) = \sum_{k=1}^{n} a_k \, e_k$$

Then $x(t)$ will be the result of a combination of trajectories along the eigenvectors because

$$x(t) = e^{At} \sum_{k=1}^{n} a_k\, e_k = \sum_{k=1}^{n} a_k\, e^{At}\, e_k = \sum_{k=1}^{n} a_k\, e^{\lambda_k t}\, e_k \tag{7.3}$$

This result is illustrated in Fig. 7.1 for n=2.

Let $d_k(t)$ denote the coefficient a_k exp $(\lambda_k t)$ in (7.3), i.e.

$$d_k(t) = a_k \exp(\lambda_k t),\ k = 1,2,\ldots,n \tag{7.4}$$

Then (7.3) can be expressed as

$$x(t) = W\, d(t) \tag{7.5}$$

where $d=(d_1,d_2,\ldots,d_n)$ and W is a square matrix whose columns are the eigenvectors $e_1,e_2,\ldots,e_n$. On the other hand, (7.4) shows that d is a vector whose components evolve independently and can be generated as shown in Fig. 7.2; here $a_1,a_2,\ldots,a_n$ represent the initial conditions of the integrators. In matrix form Fig. 7.2 is expressed as

$$\dot{d} = \Lambda\, d \tag{7.6}$$

where

$$\Lambda = \begin{bmatrix} \lambda_1 & & & \mathbf{0} \\ & \cdot & & \\ & & \cdot & \\ & & & \cdot \\ \mathbf{0} & & & \lambda_n \end{bmatrix} \tag{7.7}$$

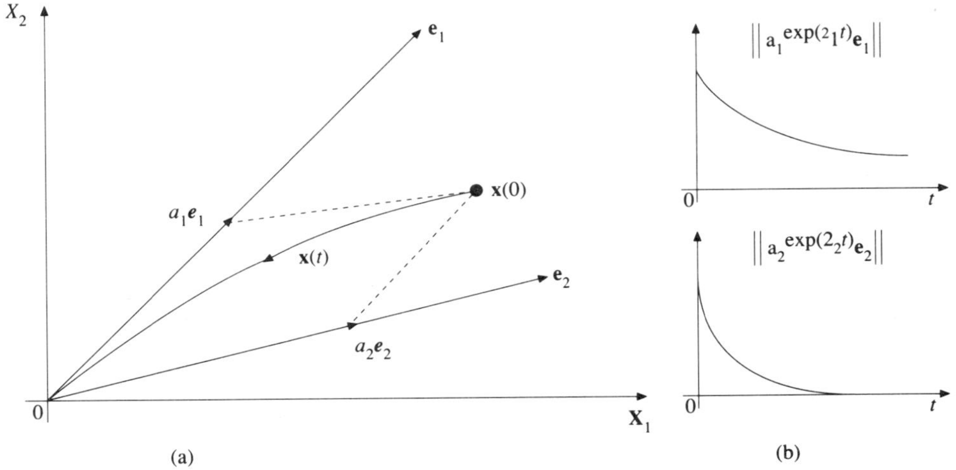

Figure 7.1 (a) State-space trajectory for $\dot{x}=Ax$, for n=2 and $|\lambda_1|<|\lambda_2|$; (b) Evolution of components along e_1 and e_2.

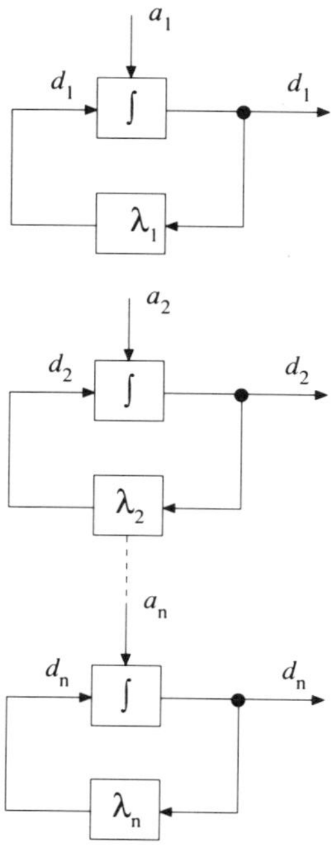

Figure 7.2 Generation of the modes of (7.1).

Representation (7.6) is called a diagonal representation of $\dot{\boldsymbol{x}}=\boldsymbol{A}\boldsymbol{x}$. The matrix $\boldsymbol{\Lambda}$ is related with $\boldsymbol{A}$ by means of the equation

$$\boldsymbol{\Lambda} = \boldsymbol{W}^{-1}\boldsymbol{A}\boldsymbol{W} \tag{7.8}$$

The proof of this is straighforward. In fact from (7.5) we have $\boldsymbol{W}\dot{\boldsymbol{d}}=\boldsymbol{A}\boldsymbol{W}\boldsymbol{d}$. Because $\boldsymbol{W}$ is invertible, we can write $\dot{\boldsymbol{d}}=\boldsymbol{W}^{-1}\boldsymbol{A}\ \boldsymbol{W}\ \boldsymbol{d}$ and (7.8) follows.

Although the components of $\boldsymbol{x}$ are linear combinations of the natural modes of the system, multiplication of $\boldsymbol{x}$ by $\boldsymbol{W}^{-1}$ transforms $\boldsymbol{x}$ into a vector where each component consists of a single mode *only* i.e. $\boldsymbol{d}=\boldsymbol{W}^{-1}\ \boldsymbol{x}$. This is why $\boldsymbol{W}^{-1}$ is also known as a *mode filter*.

The above results still hold if some eigenvalues are complex. However, it is desirable to have representations with real coefficients only. For example, with reference to Fig. 7.2 some integrators would have complex coefficients in the feedback loop which is undesirable for system analysis or simulation, for example. Next we give an interpretation of the modes associated with a pair of complex eigenvalues to introduce the procedure to obtain a 'diagonal' representation with real coefficients.

7.1.1.2 Complex modes

Let λ and λ^* be a pair of complex conjugate eigenvalues of $A \in \mathbb{R}^{n\times n}$ and $\boldsymbol{v}$ and $\boldsymbol{v}^*$ the corresponding eigenvectors.

> **Exercise 7.1** Show that the eigenvalues of a matrix with real entries occur in conjugate pairs. *Hint*: The eigenvalues are the roots of the polynomial det $[\lambda \boldsymbol{I} - \boldsymbol{A}]$ which is a polynomial with *real* coefficients.

Let

$$\lambda = \alpha + j\omega;\ \lambda^* = \alpha - j\omega$$

$$\boldsymbol{v} = \boldsymbol{v}' + j\boldsymbol{v}'';\ \boldsymbol{v}^* = \boldsymbol{v}' - j\boldsymbol{v}''$$

Notice that $\boldsymbol{v}'$ and $\boldsymbol{v}''$ are not only *real* vectors but also linearly independent vectors. In fact if we assume $\boldsymbol{v}' = \alpha \boldsymbol{v}''$, then we would have $\boldsymbol{v} = (\alpha + j)\boldsymbol{v}'$ and $\boldsymbol{v}^* = (\alpha - j)\boldsymbol{v}'$. But this is not possible because distinct eigenvalues must have linearly independent eigenvectors (result 7.1).

Assume again that $u(t) = 0$ in (7.1) and that $\boldsymbol{x}(0) = a\boldsymbol{v}'$, i.e. the initial condition lies on the real part of the conjugate eigenvectors. $\boldsymbol{x}(0)$ can also be expressed as

$$\boldsymbol{x}(0) = a/2\ (\boldsymbol{v} + \boldsymbol{v}^*)$$

Then from (7.3) we can write

$$\boldsymbol{x}(t) = a/2\ e^{\lambda t}\ \boldsymbol{v} + a/2\ e^{\lambda^* t}\ \boldsymbol{v}^*$$

Because the terms on the right hand side are a conjugate pair we have

$$\begin{aligned}\boldsymbol{x}(t) &= a\ \mathrm{Re}\{e^{\lambda t}\ \boldsymbol{v}\} \\ &= a\ \mathrm{Re}\ \{e^{\alpha t}\ e^{j\omega t}\ (\boldsymbol{v}' + j\boldsymbol{v}'')\} \\ &= a\ e^{\alpha t} \cos \omega t\ \boldsymbol{v}' - a\ e^{\alpha t} \sin \omega t\ \boldsymbol{v}'' \end{aligned} \tag{7.9}$$

This trajectory is a logarithmic spiral in the plane defined by the vectors $\boldsymbol{v}'$ and $\boldsymbol{v}''$ and shown in Fig. 7.3. This result is stated below as result 7.2.

> **Exercise 7.2** Show that $\boldsymbol{x}(t) = a\ e^{\alpha t} \sin \omega t\ \boldsymbol{v}' + a\ e^{\alpha t} \cos \omega t\ \boldsymbol{v}''$ if $\boldsymbol{x}(0) = a\ \boldsymbol{v}''$.

> **Result 7.2** Associated with each pair of complex-conjugate eigenvalues, λ and λ^*, of a matrix $\boldsymbol{A}$ there is a plane defined by the real and imaginary parts of the associated eigenvectors, such that any solution of $\dot{\boldsymbol{x}} = \boldsymbol{A}\boldsymbol{x}$ starting in this plane remains there. The trajectory is a logarithmic spiral if $\mathrm{Re}(\lambda) \neq 0$ and an ellipse if $\mathrm{Re}(\lambda) = 0$.

> **Exercise 7.3** Show that
>
> $$\boldsymbol{x}(t) = e^{\alpha t}\ \sqrt{(\alpha_{re}^2 + \alpha_{Im}^2)}\ \{\sin(\omega t + \varphi)\ \boldsymbol{v}' + \cos\ (\omega t + \varphi)\boldsymbol{v}''\}$$

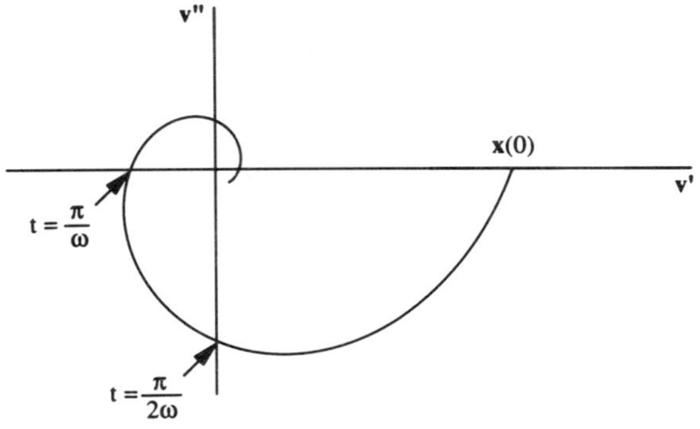

Figure 7.3 State-space trajectory associated with a pair of complex conjugate eigenvalues of (7.1) with negative real parts and $u(t)=0$.

where

$$\varphi = \text{arc cos } [\alpha_{\text{Im}}/\surd(\alpha_{\text{re}}^2 + \alpha_{\text{Im}}^2)]$$

if $x(0)=\alpha_{\text{re}}v'+\alpha_{\text{Im}}\,v''$.

7.1.1.3 'Diagonal' representation with real coefficients

The previous discussion suggests that in cases with complex eigenvalues it may be preferable to include in the basis for the state-space the real and imaginary components of each pair of conjugate eigenvectors instead of the eigenvectors themselves. Let us assume, for the sake of simplicity of exposition, that A has only one pair of complex eigenvalues, say λ_1 and λ_2. Denoting this pair by $\alpha\pm j\omega$, we have in this case:

$$\Lambda = \left[\begin{array}{cc|cccc} \alpha+j\omega & 0 & & & \mathbf{0} & \\ 0 & \alpha+j\omega & & & & \\ \hline & & \lambda_3 & & & \\ & \mathbf{0} & & \cdot & & \\ & & & & \cdot & \\ & & & & & \lambda_n \end{array}\right]$$

Going back to (7.5) remember that W transforms the coordinates with respect to the basis of eigenvectors $D=\{e_1,e_2,\ldots,e_n\}$ into the coordinates with respect to the natural basis $N=\{1,0,0,\ldots,0);\ (0,1,0,\ldots,0);\ (0,\ldots,0,1)\}$. If the basis D is replaced by D_{m} where $D_{\text{m}}=\{\text{Re}(e_1),\ \text{Im}(e_1),\ e_3,\ e_4,\ldots,e_n\}$ then only the 2×2 upper left block of Λ above will be changed. If P denotes the coordinate transformation matrix $D_m \rightarrow D$ then

$$P = \left[\begin{array}{cc|c} 1/2 & -j/2 & \\ 1/2 & +j/2 & 0 \\ \hline & 0 & 1 \end{array}\right] \tag{7.10}$$

because $\mathrm{Re}(\boldsymbol{e}_1)=1/2\ (\boldsymbol{e}_1+\boldsymbol{e}_1^*)$ and $\mathrm{Im}(\boldsymbol{e}_1)=-j/2(\boldsymbol{e}_1-\boldsymbol{e}_1^*)$. Similarly for the matrix $\boldsymbol{Q}$ of the inverse transformation $D \rightarrow D_m$

$$\boldsymbol{Q} = \left[\begin{array}{cc|cccc} 1 & 1 & & & & \\ j & -j & & 0 & & \\ \hline & & 1 & & & 0 \\ & 0 & & \cdot & & \\ & & & & \cdot & \\ & & 0 & & & 1 \end{array}\right] = \boldsymbol{P}^{-1}$$

Then in the basis D_m the transformation is represented $\tilde{\Lambda}$ where

$$\tilde{\Lambda} = \boldsymbol{Q}\Lambda\boldsymbol{P} = \left[\begin{array}{cc|cccc} \alpha & \omega & & & & \\ -\omega & \alpha & & 0 & & \\ \hline & & \lambda_3 & & & 0 \\ & 0 & & \cdot & & \\ & & & & \cdot & \\ & & 0 & & & \lambda_n \end{array}\right]$$

Notice that the above formula has an intuitive interpretation: $\boldsymbol{P}$ transforms the coordinates from D_m to D; then we apply Λ and the result is transformed back to D_m by $\boldsymbol{Q}$. If $\tilde{\boldsymbol{d}}$ represents the state vector in basis D_m then

$$d/dt\ \tilde{\boldsymbol{d}} = \tilde{\Lambda}\, \tilde{\boldsymbol{d}} \tag{7.12}$$

whose diagram is shown in Fig. 7.4 for $n=3$. From (7.9) we can write

$$\tilde{\boldsymbol{d}}(t) = \left[\begin{array}{c} a\, e^{\alpha t}\cos\omega t \\ -a\, e^{\alpha t}\sin\omega t \\ e^{\lambda_3 t} \end{array}\right]$$

for $\tilde{\boldsymbol{d}}\ (0)=(a,0,1)^T$.

Exercise 7.4

(a) Compute the eigenvalues of the matrix

$$\left[\begin{array}{cc} \alpha & \omega \\ -\omega & \alpha \end{array}\right]$$

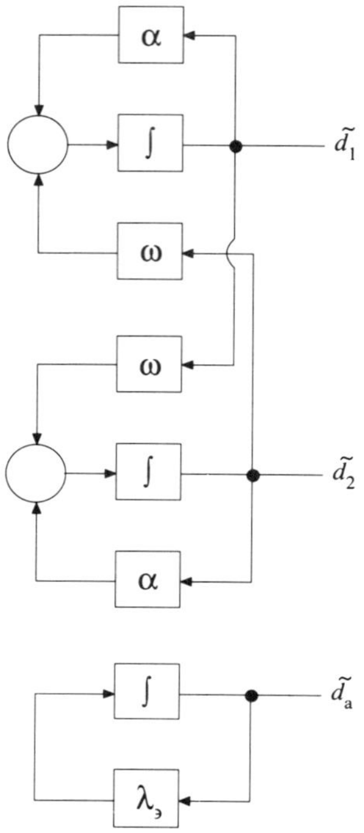

Figure 7.4 Mode generator, with real coefficients for system (7.1) with a pair of complex eigenvalues and n=3.

(b) Compute exp($\boldsymbol{A}t$) where $\boldsymbol{A}$ is the above matrix. *Hint*: Use (7.9) and exercise 7.2.

Answers:

(a) $\alpha \pm j\omega$

(b) $\begin{bmatrix} e^{\alpha t}\cos\omega t & e^{\alpha t}\sin\omega t \\ -e^{\alpha t}\sin\omega t & e^{\alpha t}\cos\omega t \end{bmatrix}$

Example 7.1 Consider the mechanical system depicted in Fig. 7.5 with M_1=100 kg, M_2=50 kg, K=100 N/m and $c_1=c_2$=1 N s/m. Let F denote the force exerted by the spring, i.e. $F=K(y_1-y_2)$.

The system equations are

$$F + M_1\ddot{y}_1 + C_1\dot{y}_1 = 0$$

$$-F + M_2\ddot{y}_2 + C_2\dot{y}_2 = 0$$

$$F = K(y_1 - y_2)$$

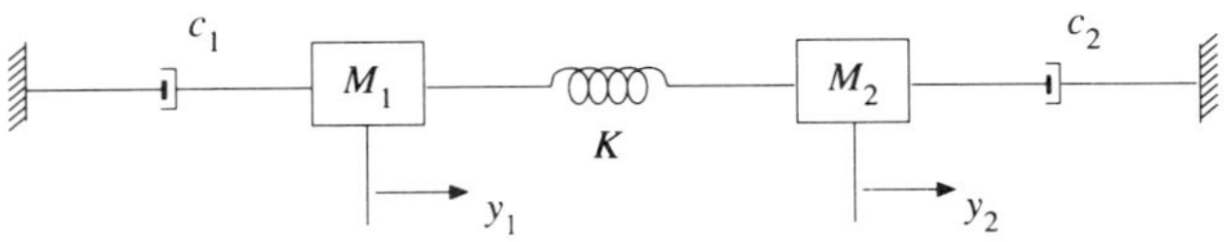

Figure 7.5 System for example 7.1.

Defining $x_1=\dot{y}_1$, $x_2=\dot{y}_2$ and $x_3=F$ we can write

$$\begin{bmatrix} \dot{x}_1 \\ \dot{x}_2 \\ \dot{x}_3 \end{bmatrix} = \begin{bmatrix} -0.01 & 0 & -0.01 \\ 0 & -0.02 & 0.02 \\ 100 & -100 & 0 \end{bmatrix} \begin{bmatrix} x_1 \\ x_2 \\ x_3 \end{bmatrix}$$

The system eigenvalues are

$$\lambda_1 = -1.33 \times 10^{-2}$$

$$\lambda_2 = -8.333 \times 10^{-3} + j1.732$$

$$\lambda_3 = \lambda_2^*$$

with associated eigenvectors

$$\boldsymbol{e}_1 = \begin{bmatrix} 6.882 \\ 6.882 \\ 2.294 \end{bmatrix} ; \boldsymbol{e}_2 = \begin{bmatrix} 0.7439 \\ -0.7621 \\ 6.833 \end{bmatrix} + j \begin{bmatrix} 1.025 \\ -7.085 \\ -0.9594 \end{bmatrix} ; \boldsymbol{e}_3 = \boldsymbol{e}_2^*$$

Let $\boldsymbol{x}(0)=a_1\boldsymbol{e}_1+a_2\text{Re}(\boldsymbol{e}_2)+a_3\text{Im}(\boldsymbol{e}_2)$. Then from the previous discussion and exercise 7.3 we can write:

$$\boldsymbol{x}(t) = a_1 \exp(-1.333 \times 10^{-2}t) \begin{bmatrix} 6.882 \\ 6.882 \\ 2.294 \end{bmatrix} + \sqrt{(a_2^2 + a_3^2)}$$

$$\times \exp(-8.333 \times 10^{-3}t)\{\sin(1.732t + \varphi) \begin{bmatrix} 0.7439 \\ -0.7621 \\ 6.833 \end{bmatrix}$$

$$+ \cos(1.732t + \varphi) \begin{bmatrix} 1.025 \\ -7.085 \\ -0.9594 \end{bmatrix} \}$$

with $\varphi=\text{arc cos}\,(a_3/\sqrt{(a_2^2+a_3^2)})$.

Figure 7.6(a) represents the first component, $x_1(t)$, of the state vector $\boldsymbol{x}(t)$ when $\boldsymbol{x}(0)=\boldsymbol{e}_1$; as expected, the oscillatory mode is not present. Figure 7.6(b) shows $x_1(t)$ for an initial condition on the plane defined by $\text{Re}(\boldsymbol{e}_2)$ and $\text{Im}(\boldsymbol{e}_2)$, namely $a_1=0$, $a_2=a_3=1$. The system exhibits a damped oscillation of frequency 1.732 rad/s; only the complex mode is present.

Finally, Fig. 7.6(c) shows $\boldsymbol{x}_1(t)$ for $\boldsymbol{x}(0)=\boldsymbol{e}_1+\text{Re}(\boldsymbol{e}_2)+\text{I}_m(\boldsymbol{e}_2)$; now $x_1(t)$ is a combination of all the modes, which increases the complexity of the identification process.

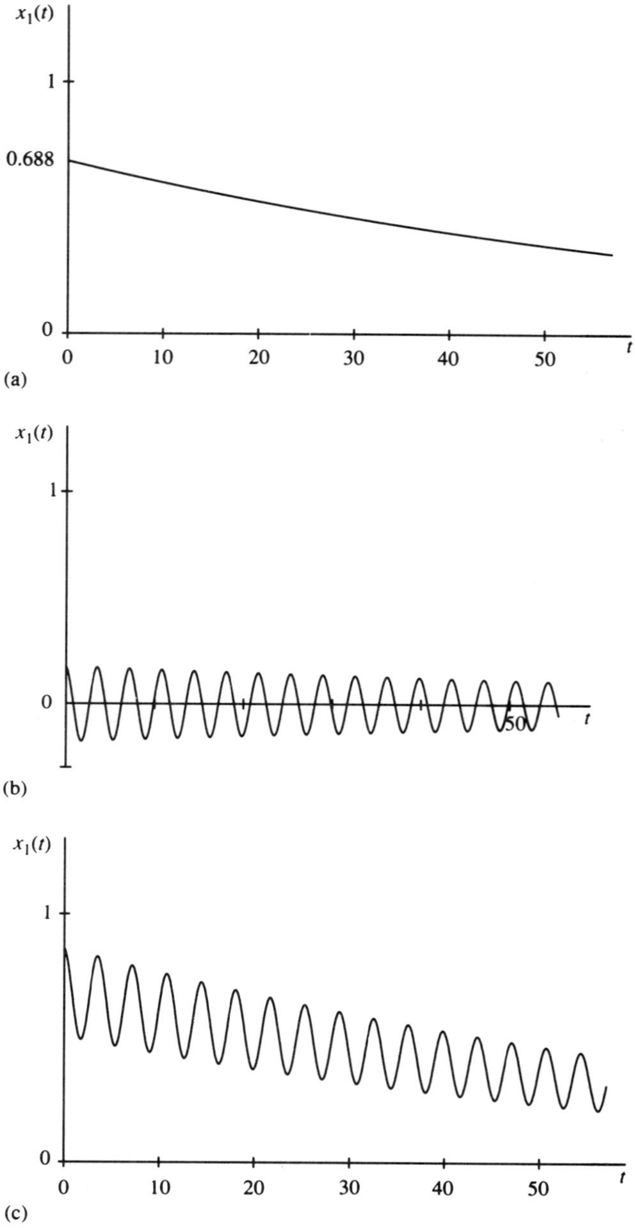

Figure 7.6 Modes of system of Fig. 7.5: (a) real mode; (b) complex mode; (c) real and complex modes superimposed.

Conclusions: The discussion so far has revealed that the system (7.39), with $u(t)=0$, may look like a first- or second-order system in special circumstances: if $\boldsymbol{x}(0)$ lies on an eigenvector associated with a real eigenvalue λ, then the components of $\boldsymbol{x}$ evolve as the impulse response of a first-order system with pole λ; if $\boldsymbol{x}(0)$ lies on the plane defined by the real and imaginary parts of a complex eigenvector associated with a complex eigenvalue $\lambda=\alpha+j\omega$, the components of $\boldsymbol{x}$ evolve as the impulse response of a second order system with a pair of conjugate poles $\alpha\pm j\omega$.

Before looking at the next example, it is convenient to restate the result of exercise 7.3 with respect to the natural basis N of the state-space as shown below.

Result 7.3 Given $\dot{\boldsymbol{x}}=\boldsymbol{A}\mathbf{x}$ and an eigenvector $\boldsymbol{v}$ of $\boldsymbol{A}$

$$\boldsymbol{v} = \boldsymbol{v}' + j\boldsymbol{v}'' = \begin{bmatrix} 1 \\ r_2 \exp(j\theta_2) \\ \cdot \\ \cdot \\ \cdot \\ r_n \exp(j\theta_n) \end{bmatrix}$$

with eigenvalue $\lambda=\alpha+j\omega$; then $\boldsymbol{x}(0)=\alpha_{re}\boldsymbol{v}'+\alpha_{Im}\boldsymbol{v}'' \Rightarrow$

$$\boldsymbol{x}(t) = \frac{r}{2}\exp(\alpha t) \begin{bmatrix} \cos(\omega t-\theta) \\ r_2 \cos(\omega t+\theta_2-\theta) \\ \cdot \\ \cdot \\ \cdot \\ r_n \cos(\omega t+\theta_n-\theta) \end{bmatrix}$$

where $\alpha_{re}+j\alpha_{Im}=Re^{j\theta}$.

This result constitutes a generalization of (7.2) in cases where $\mathbf{e}_i$ is a complex vector; it also shows that the magnitude of the components, $1, r_2, \ldots, r_n$ of a complex eigenvector correspond with the amplitude distribution of the associated oscillatory mode between the states in the system, and their arguments $0, \theta_2, \theta_3, \ldots, \theta_n$, with the relative phase shift between the states.

Proof: First we show that $\boldsymbol{x}(0)$ can be expressed as $r/2(\exp(-j\theta)\boldsymbol{v}+\exp(j\theta)\,\boldsymbol{v}^*)$. In fact

$$\begin{aligned} \boldsymbol{x}(0) &= \alpha_{re}/2\ (\boldsymbol{v} + \boldsymbol{v}^*) - j\alpha_{Im}/2\ (\boldsymbol{v} - \boldsymbol{v}^*) \\ &= 1/2(\alpha_r - j\alpha_{Im})\ \boldsymbol{v} + 1/2\ (\alpha_{re} + j\alpha_{Im})\ \boldsymbol{v}^* \\ &= r/2[\exp(-j\theta)\ \boldsymbol{v} + \exp(j\theta)\boldsymbol{v}^*] \end{aligned}$$

Then, for $t>0$,

$$\begin{aligned} \boldsymbol{x}(t) &= r/2\{\exp(-j\theta)\exp((\alpha + j\omega)t)\ \boldsymbol{v} + \exp(j\theta)\exp((\alpha - j\omega)t)\ \boldsymbol{v}^*\} \\ &= r\ \mathrm{Re}\{\exp(-j\theta)\exp((\alpha + j\omega)t)\ \boldsymbol{v}\} \end{aligned}$$

$$= r\exp(\alpha t)\,\mathrm{Re}\{\exp(j(\omega t-\theta))\begin{bmatrix}1\\ r_2\exp(j\theta_2)\\ \cdot\\ \cdot\\ \cdot\\ r_n\exp(j\theta_n)\end{bmatrix}\}$$

$$= r\exp(\alpha t)\begin{bmatrix}\cos(\omega t-\theta)\\ r_2\cos(\omega t+\theta_2-\theta)\\ \cdot\\ \cdot\\ \cdot\\ r_n\cos(\omega t+\theta_n-\theta)\end{bmatrix}$$

QED.

Example 7.2 The model represented in Fig. 7.7 is frequently used in vibration analysis. Writing the equations for each of the masses we get

$$M_1\ddot{x}_1 + k_{12}(x_1 - x_2) = 0$$

$$M_2\ddot{x}_2 + k_{12}(x_2 - x_1) + k_{20}\,x_2 + k_{23}\,(x_2 - x_3) = 0$$

$$M_3\ddot{x}_3 + k_{23}(x_3 - x_2) = 0$$

or in matrix form

$$\begin{bmatrix}M_1 & 0 & 0\\ 0 & M_2 & 0\\ 0 & 0 & M_3\end{bmatrix}\begin{bmatrix}\ddot{x}_1\\ \ddot{x}_2\\ \ddot{x}_3\end{bmatrix} + \begin{bmatrix}k_{12} & -k_{12} & 0\\ -k_{12} & (k_{12}+k_{20}+k_{23}) & -k_{23}\\ 0 & -k_{23} & k_{23}\end{bmatrix}\begin{bmatrix}x_1\\ x_2\\ x_3\end{bmatrix} = 0$$

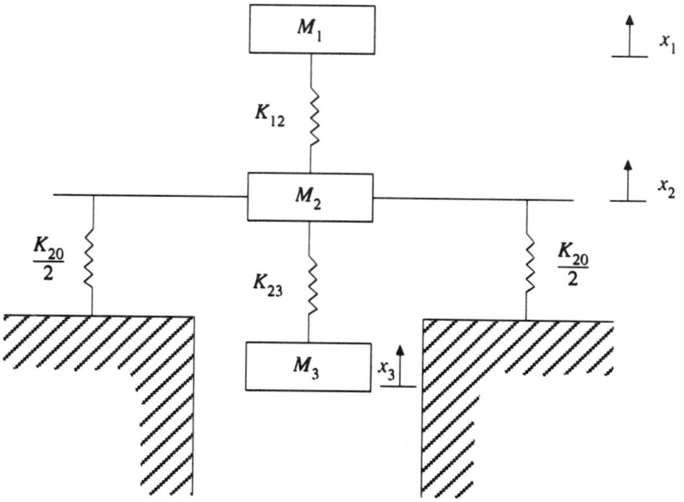

Figure 7.7 Spring–mass system for example 7.2.

Defining an additional set of variables: $x_4=\dot{x}_1$, $x_5=\dot{x}_2$ and $x_6=\dot{x}_3$, the above system is transformed in the following system of first order equations:

$$\begin{bmatrix} \dot{x}_1 \\ \dot{x}_2 \\ \dot{x}_3 \\ \dot{x}_4 \\ \dot{x}_5 \\ \dot{x}_6 \end{bmatrix} = \begin{bmatrix} 0 & 0 & 0 & 1 & 0 & 0 \\ 0 & 0 & 0 & 0 & 1 & 0 \\ 0 & 0 & 0 & 0 & 0 & 1 \\ & & & 0 & 0 & 0 \\ & -M^{-1}K & & 0 & 0 & 0 \\ & & & 0 & 0 & 0 \end{bmatrix} \begin{bmatrix} x_1 \\ x_2 \\ x_3 \\ x_4 \\ x_5 \\ x_6 \end{bmatrix} \Leftrightarrow \dot{\boldsymbol{x}}=\boldsymbol{A}\boldsymbol{x}$$

where

$$M = \begin{bmatrix} M_1 & 0 & 0 \\ 0 & M_2 & 0 \\ 0 & 0 & M_3 \end{bmatrix}$$

and

$$\boldsymbol{K} = \begin{bmatrix} k_{12} & -k_{12} & 0 \\ -k_{12} & (k_{12}+k_{20}+k_{23}) & -k_{23} \\ 0 & -k_{23} & k_{23} \end{bmatrix}$$

From our previous study we expected this system to have three pairs of conjugate eigenvalues with *zero* real parts because of the absence of damping, and thus to exhibit three undamped oscillatory modes.

Although $\boldsymbol{A}$ is a 6×6 matrix, its eigenvalues and eigenvectors are defined by the 3×3 matrix $-\boldsymbol{M}^{-1}\boldsymbol{K}$. The observation that

$$\boldsymbol{A}^2 = \left[\begin{array}{c|c} -\boldsymbol{M}^{-1}\boldsymbol{K} & \boldsymbol{0} \\ \hline \boldsymbol{0} & -\boldsymbol{M}^{-1}\boldsymbol{K} \end{array}\right]$$

shows the plausibility of the following result:

Result 7.4

(a) If λ_1, λ_2 and λ_3 are the eigenvalues of $(-\boldsymbol{M}^{-1}\boldsymbol{K})$, then $\pm\sqrt{\lambda_1}$, $\pm\sqrt{\lambda_2}$ and $\pm\sqrt{\lambda_3}$ are the eigenvalues of $\boldsymbol{A}$.

(b) If $\boldsymbol{e}$ is an eigenvector of $-\boldsymbol{M}^{-1}\boldsymbol{K}$ with eigenvalue λ, then $(\boldsymbol{e}^{\mathrm{T}}, \sqrt{\lambda}\,\boldsymbol{e}^{\mathrm{T}})^{\mathrm{T}}$ and $(\boldsymbol{e}^{\mathrm{T}}, -\sqrt{\lambda}\,\boldsymbol{e}^{\mathrm{T}})^{\mathrm{T}}$ are eigenvectors of $\boldsymbol{A}$.

Let us now prove these statements. The first is an immediate consequence of the fact that $\det(\boldsymbol{A}^2)=(\det\boldsymbol{A})^2=(\det(-\boldsymbol{M}^{-1}\boldsymbol{K}))^2$, and the structure of A^2, which means that each eigenvalue of $-\boldsymbol{M}^{-1}\boldsymbol{K}$ is an eigenvalue of $\boldsymbol{A}^2$ with multiplicity two; hence the result.

In order to prove the second property let $\boldsymbol{u}$ and $\boldsymbol{v}$ be eigenvectors of $\boldsymbol{A}$ with eigenvalues γ and $-\gamma$ respectively.

$$\boldsymbol{u} = \begin{bmatrix} \boldsymbol{u}_1 \\ \text{- -} \\ \boldsymbol{u}_2 \end{bmatrix}, \quad \boldsymbol{v} = \begin{bmatrix} \boldsymbol{v}_1 \\ \text{- -} \\ \boldsymbol{v}_2 \end{bmatrix}$$

where u_1, u_2, v_1 and v_2 are three-dimensional vectors. Then $Au=\gamma u$ implies $u_2=\gamma u_1$ and

$$(-M^{-1}K)u_1 = \gamma^2 u_2 \ (*)$$

For v we have similarly $v_2=-\gamma v_1$ and

$$(-M^{-1}K)v_1 = \gamma^2 v_1 \ (**)$$

Because γ^2 is an eigenvalue of $-M^{-1}K$ we conclude that (*) and (**) are equivalent equations and the result follows.

Let us resume example 7.2.

If

$$M_1 = 250 \times 10^3 \text{ kg} \qquad k_{12} = 0.75 \times 10^9 \text{ N m}$$

$$M_2 = 30 \times 10^3 \text{ kg} \qquad k_{23} = 15 \times 10^9 \text{ N m}$$

$$M_3 = 135\times 10^3 \text{ kg} \qquad k_{20} = 1.25 \times 10^9 \text{ N m}$$

then the eigenvalues of the 3×3 matrix $(-M^{-1}K)$ are:

$$\lambda_1 = -1.690 \times 10^3,\ \lambda_2 = -12.33 \times 10^3,\ \lambda_3 = -666.8 \times 10^3$$

with associated eigenvectors

$$\begin{bmatrix} 1 \\ 0.4018 \\ 0.4503 \end{bmatrix} \begin{bmatrix} 1 \\ -3.108 \\ -3.496 \end{bmatrix} \begin{bmatrix} 1 \\ -217.8 \\ 43.49 \end{bmatrix}$$

respectively. From result 7.4 above it follows that A has eigenvalues ±j41.11, ±j111.0, ±j816.6 with associated eigenvectors:

$$\begin{bmatrix} 1 \\ 0.402 \\ 0.450 \\ \pm j41.11 \\ \pm j16.52 \\ \pm j18.51 \end{bmatrix} = \begin{bmatrix} e^{j0} \\ 0.402\ e^{j0} \\ 0.450\ e^{j0} \\ 41.11\ e^{\pm j\pi/2} \\ 16.52\ e^{\pm j\pi/2} \\ 18.51\ e^{\pm j\pi/2} \end{bmatrix}$$

$$\begin{bmatrix} 1 \\ -3.108 \\ -3.496 \\ \pm j111.0 \\ \mp j345.1 \\ \mp j388.2 \end{bmatrix} = \begin{bmatrix} e^{j0} \\ 3.108\ e^{j\pi} \\ 3.496\ e^{j\pi} \\ 111.0\ e^{\pm j\pi/2} \\ 345.1\ e^{\mp j\pi/2} \\ 388.2\ e^{\mp j\pi/2} \end{bmatrix}$$

$$\begin{bmatrix} 1 \\ -217.8 \\ 43.49 \\ \pm j816.6 \\ \mp j177.9\times 10^3 \\ \pm j35.52\times 10^3 \end{bmatrix} = \begin{bmatrix} e^{j0} \\ 217.8\ e^{j\pi} \\ 43.49\ e^{j0} \\ 816.6\ e^{\pm j\pi/2} \\ 177.9\times 10^3\ e^{\mp j\pi/2} \\ 35.52\times 10^3\ e^{\pm j\pi/2} \end{bmatrix}$$

Mode shapes can now be easily analyzed with the help of result 7.3. Because we are only interested in relative phase shifts and amplitudes, we assume $r=1$ and $\theta=0$. Although the system is of order six its behaviour may be fully characterized by the first three components of each of the first three modes.

The first mode, with oscillating frequency 41.11 rad/s, is given by

$$\begin{bmatrix} x_1(t) \\ x_2(t) \\ x_3(t) \end{bmatrix} = \begin{bmatrix} \cos(41.11t) \\ 0.4018\cos(41.11t) \\ 0.4503\cos(41.11t) \end{bmatrix}$$

All masses oscillate in phase (see Fig. 7.8), with mass M_1 having an amplitude of about 2.5 times the amplitudes of masses M_2 and M_3.

The second mode has an oscillating frequency of 111.0 rad/s and equation

$$\begin{bmatrix} x_1(t) \\ x_2(t) \\ x_3(t) \end{bmatrix} = \begin{bmatrix} \cos(111.0t) \\ 3.108\cos(111.0t+\pi) \\ 3.496\cos(111.0t+\pi) \end{bmatrix}$$

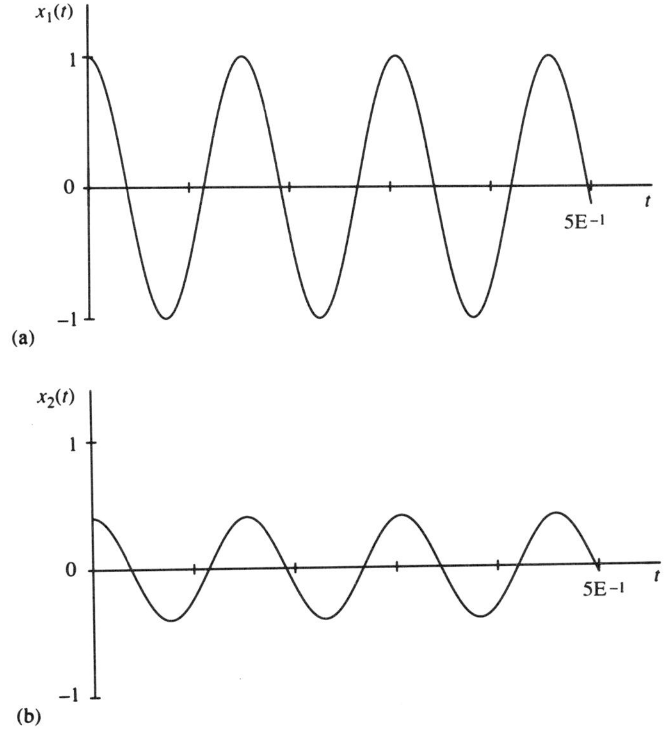

Figure 7.8 Mode shape for frequency 41.11 rad/s in example 7.2.

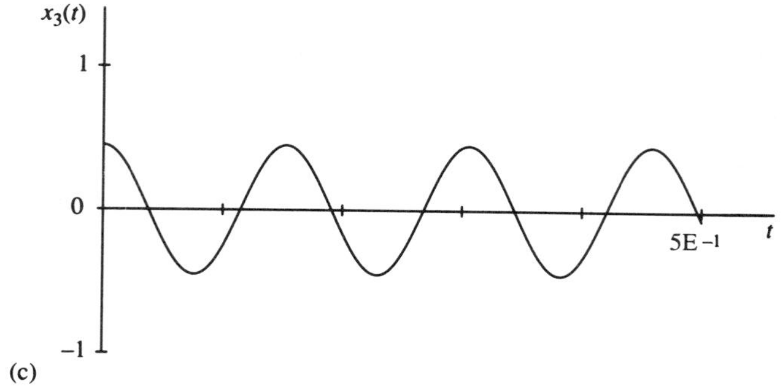

Figure 7.8 (cont.)

and is shown in Fig. 7.9. M_1 is oscillating π radians out of phase relative to M_2 and M_3. The amplitude of M_1 is about 1/3 of the amplitude of M_2 and M_3.

Mode three has a frequency of 816.6 rad/s and is described by the equation:

$$\begin{bmatrix} x_1(t) \\ x_2(t) \\ x_3(t) \end{bmatrix} = \begin{bmatrix} \cos(816.6t) \\ 217.8\cos(816.6t+\pi) \\ 43.49\cos(816.6t) \end{bmatrix}$$

The amplitude of oscillation of mass M_1 is almost negligible, in comparison with the other two amplitudes, which is not surprising given the large magnitude of mass M_1 and the high value of the frequency. This mode is represented in Fig. 7.10. Masses M_1 and M_2 oscillate in counter phase.

7.1.2 Repeated Eigenvalues

In Section 3.4 we saw that a pole, $-p_k$, of multiplicity m_k of a transfer function $W(s)$ contributes to the impulse response $w(t)$ with a set of modes of the form (cf. (3.41)):

$$\exp(-p_k t),\ t\exp(-p_k t)/1!,\ \ldots,\ t^{(m_k-1)}\exp(-p_k t)/(m_k-1)!$$

which can be generated as shown in Fig. 3.10. This figure also shows that these modes are not generated independently, but with only a set of m_k *coupled* integrators. As a result, a diagonal representation of the type (7.6) is no longer possible. In fact the state space representation of Fig. 3.10 is

$$\begin{bmatrix} \dot{d}_1 \\ \dot{d}_2 \\ \dot{d}_3 \\ \cdot \\ \cdot \\ \cdot \\ \dot{d}_{mk} \end{bmatrix} = \begin{bmatrix} -p_k & 1 & & & \\ & -p_k & 1 & & 0 \\ & & -p_k & \cdot & \\ & & & \ddots & \\ & 0 & & & \ddots \\ & & & \cdot & 1 \\ & & & & -p_k \end{bmatrix} \begin{bmatrix} d_1 \\ d_2 \\ d_3 \\ \cdot \\ \cdot \\ \cdot \\ d_{mk} \end{bmatrix} \Leftrightarrow \dot{\mathbf{d}} = \mathbf{\Lambda}_r\mathbf{d} \qquad (7.13)$$

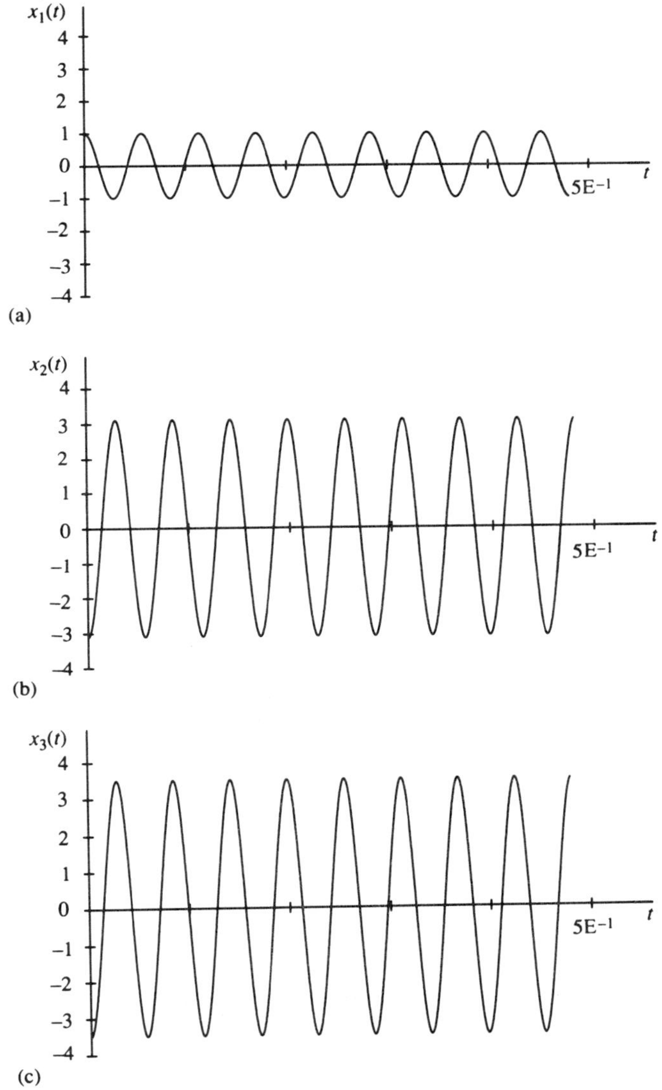

Figure 7.9 Mode shape for frequency 111.0 rad/s, in example 7.2.

where the series connection of the integrators is responsible for the appearance of ones above the main diagonal. Therefore (7.13) is the nearest we can get to diagonalizing a block associated with this repeated eigenvalue. Besides having a single eigenvalue, of multiplicity m_k, matrix $\mathbf{\Lambda}_r$ above has only one eigenvector, namely $(1,0,\ldots,0)^T$.

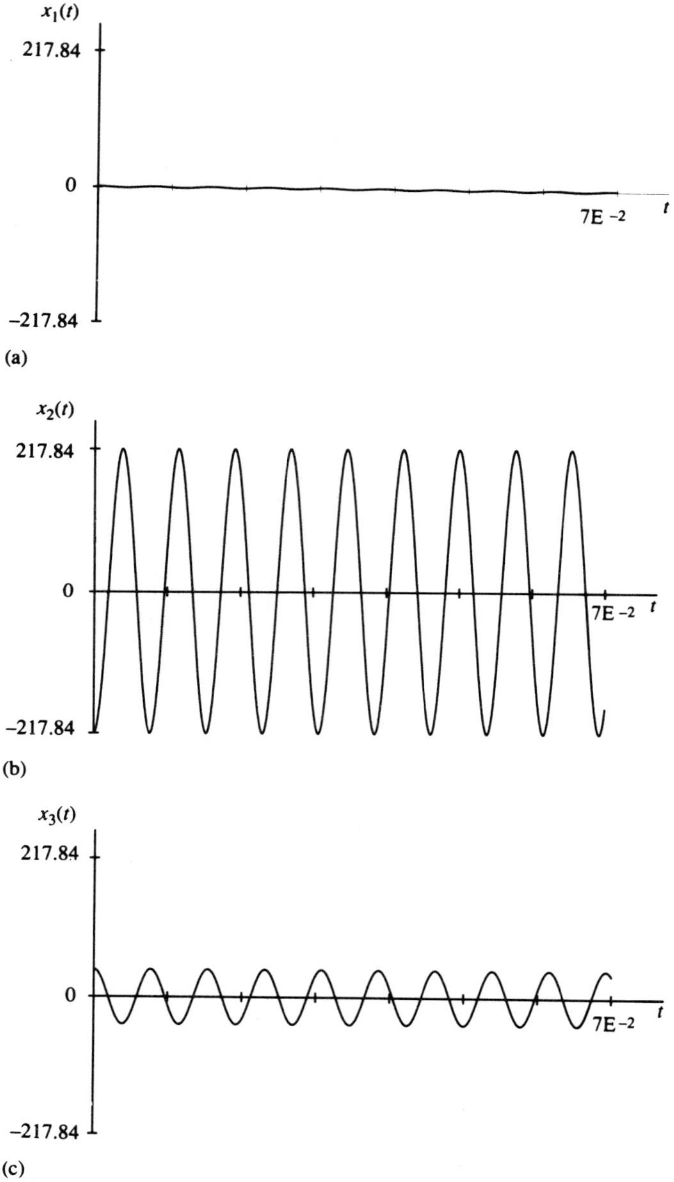

Figure 7.10 Mode shape for frequency 816.6 rad/s in example 7.2.

Exercise 7.5

(a) Check that $(\alpha, 0, 0, \ldots, 0)^T$ are the only eigenvectors of $\boldsymbol{\Lambda}_r$ in (7.13), $\forall\ \alpha \in \mathbb{R}$.

Hint: If $\boldsymbol{v}=(v_1,v_2,\ldots,v_{mk})^T$ denotes an eigenvector of $\boldsymbol{\Lambda}_r$, then $(\boldsymbol{A}+p_k\mathbf{I})\ \boldsymbol{v}=\mathbf{0}$; on the other hand

$$(\boldsymbol{A} + p_k\mathbf{I}) = \begin{bmatrix} 0\,1\,0\,\cdot & & \mathbf{0} \\ & 0\,1\,\cdot\,\cdot & \\ & & \cdot\,\cdot\,\cdot\,0 \\ \mathbf{0} & & \cdot\,\cdot\,1 \\ & & \cdot\,0 \end{bmatrix}$$

then

(b) Assuming $m_k=4$ verify that

$$(\boldsymbol{A} + p_k\mathbf{I})^2 = \begin{bmatrix} 0&0&1&0 \\ 0&0&0&1 \\ 0&0&0&0 \\ 0&0&0&0 \end{bmatrix};\ (\boldsymbol{A} + p_k\mathbf{I})^3 = \begin{bmatrix} 0&0&0&1 \\ 0&0&0&0 \\ 0&0&0&0 \\ 0&0&0&0 \end{bmatrix};\ (\boldsymbol{A} + p_k\mathbf{I})^4 = [\mathbf{0}]$$

Exercise 7.6

(a) Show that if

$$\boldsymbol{\Lambda}_r = \begin{bmatrix} \lambda & 1 \\ 0 & \lambda \end{bmatrix}$$

then

$$\exp(\boldsymbol{\Lambda}_r t) = \begin{bmatrix} \exp(\lambda t) & t\exp(\lambda t) \\ 0 & \exp(\lambda t) \end{bmatrix}$$

Hint: use the definition of matrix exponential.

(b) Show that

$$\exp\boldsymbol{\Lambda}_r t = \begin{bmatrix} \exp(-p_k t) & t\exp(-p_k t) & \dfrac{t^2}{2!}\exp(-p_k t) & \ldots & \dfrac{t^{m_k-1}}{(m_k-1)!}\exp(-p_k t) \\ 0 & \exp(-p_k t) & t\exp(-p_k t) & \ldots & \dfrac{t^{m_k-2}}{(m_k-2)!}\exp(-p_k t) \\ \cdot & \cdot & \cdot & & \cdot \\ \cdot & \cdot & \cdot & & \cdot \\ \cdot & \cdot & \cdot & & \cdot \\ 0 & 0 & 0 & \ldots & t\exp(-p_k t) \\ 0 & 0 & 0 & \ldots & \exp(-p_k t) \end{bmatrix}$$

if $\boldsymbol{\Lambda}_r$ is the matrix in (7.13).

Note: if $\boldsymbol{d}(0)=(0,0,\ldots,0,1)^T$ in (7.13) then $\exp(\boldsymbol{\Lambda}_r t)\ \boldsymbol{d}(0)$ generates the modes shown in Fig. 3.10, as expected.

The question now is how to select a suitable basis for the state-space to 'diagonalize' a representation $\dot{\boldsymbol{x}}=\boldsymbol{A}\boldsymbol{x}$ in case of multiple eigenvalues. We are generally short of eigenvectors because an eigenvalue with multiplicity m does not necessarily have m associated linearly independent eigenvectors. How can we then find the extra basis vectors? A technique to obtain them is discussed below and we will call them *generalized eigenvectors*.

For simplicity, let us assume for the moment that the system has only one eigenvector associated with the eigenvalue λ of multiplicity n. Let $D_e=\{\boldsymbol{b}_1,\boldsymbol{b}_2,\ldots,\boldsymbol{b}_n\}$ denote the basis we are looking for, i.e. with respect to this basis, the system $\dot{\boldsymbol{x}}=\boldsymbol{A}\boldsymbol{x}$ is expressed as in (7.13). Let $\boldsymbol{W}$ be a matrix whose columns are the elements of D_e, i.e. $\boldsymbol{W}=[\boldsymbol{b}_1|\boldsymbol{b}_2|\ldots|\boldsymbol{b}_n]$. Then, as shown in the previous section we can write

$$\boldsymbol{\Lambda}_r = \boldsymbol{W}^{-1}\boldsymbol{A}\boldsymbol{W} \tag{7.14}$$

where $\boldsymbol{\Lambda}_r$ is the matrix in (7.13). If we rewrite (7.14) as $\boldsymbol{AW}=\boldsymbol{W\Lambda}_r$ and equate the columns of the matrices on both sides we get:

$$\begin{aligned} \boldsymbol{A}\boldsymbol{b}_1 &= \lambda\boldsymbol{b}_1 \\ \boldsymbol{A}\boldsymbol{b}_2 &= \lambda\boldsymbol{b}_2 + \boldsymbol{b}_1 \\ \boldsymbol{A}\boldsymbol{b}_3 &= \ldots \lambda\boldsymbol{b}_3 + \boldsymbol{b}_2 \\ \boldsymbol{A}\boldsymbol{b}_n &= \lambda\boldsymbol{b}_n + \boldsymbol{b}_{n-1} \end{aligned} \tag{7.15}$$

The first equation shows that $\boldsymbol{b}_1$ is an eigenvector of $\boldsymbol{A}$ associated with the eigenvalue λ. The remainder of the elements of D_e can then be computed recursively from (7.15). The vectors $\boldsymbol{b}_2$, $\boldsymbol{b}_3,\ldots,\boldsymbol{b}_n$ are the generalized eigenvectors of $\boldsymbol{A}$.

Exercise 7.7 Show that $S=\text{span}\{\boldsymbol{b}_1,\ldots,\boldsymbol{b}_n\}$ is an invariant sub-space under $\boldsymbol{A}$ i.e. if $\boldsymbol{v}\in S$ then $\boldsymbol{Av}\in S$. *Hint*: if $\boldsymbol{v}\in S$ then $\boldsymbol{v}=\alpha_1\boldsymbol{b}_1+\ldots+\alpha_{n_k}\boldsymbol{b}_{n_k}$, *with* $\alpha_1,\alpha_2,\ldots,\alpha_{n_k}\in\mathbb{R}$. Then apply defining equation (7.15).

Exercise 7.8

(a) Assume $n=2$ in equations (7.15). Then show $\{\boldsymbol{b}_1,\boldsymbol{b}_2\}$ is a basis for $N(\boldsymbol{A}-\lambda\boldsymbol{I})^2$ where $N(\boldsymbol{M})$ denotes the null space of matrix $\boldsymbol{M}$. *Hint*: the eigenvector $\boldsymbol{b}_1$ is a basis for $N(\boldsymbol{A}-\lambda\boldsymbol{I})\subset N((\boldsymbol{A}-\lambda\boldsymbol{I})^2)$.

(b) Generalize the result in (a) for $n>2$.

Answer: $\{\boldsymbol{b}_1, \boldsymbol{b}_2,\ldots,\boldsymbol{b}_n\}$ is a basis for $N(\boldsymbol{A}-\lambda\boldsymbol{I})^n$.

Assuming now that the only eigenvalue (of multiplicity n) has two associated eigenvectors, say $\boldsymbol{b}_1$ and $\boldsymbol{b}_n$, and that the remaining $(n-2)$ generalized eigenvectors $\boldsymbol{b}_2,\ldots,\boldsymbol{b}_{n-1}$ can be computed from $\boldsymbol{b}_1$ in the same fashion as in (7.15), it can be shown that these n vectors constitute a basis for the n-dimensional space $N((\boldsymbol{A}-\lambda\boldsymbol{I})^{n-1})$. Then $\boldsymbol{\Lambda}_r$ will look like

$$\Lambda_r = \left[\begin{array}{cccccc|c} \lambda & 1 & & & & & \\ & \lambda & 1 & & 0 & & \\ & & \lambda & 1 & . & & \\ & & & \cdot & \cdot & & 0 \\ & & & & \cdot & \cdot & \\ & 0 & & & . & \lambda \; 1 & \\ & & & & & \lambda & \\ \hline & & 0 & & & & \lambda \end{array}\right]$$

In general the diagonalization of a real square matrix consists of finding the largest possible number of subspaces S_i, $i=1,2,\ldots,s$, $s \leq n$, *invariant* under $\boldsymbol{A}$, such that $\mathbb{R}^n$ is the direct sum of them, i.e. for any vector $\boldsymbol{v} \in \mathbb{R}^n$ we have a *unique* decomposition

$$\boldsymbol{v} = \boldsymbol{v}_1 + \boldsymbol{v}_2 + \ldots + \boldsymbol{v}_s, \text{ with } \boldsymbol{v}_i \in S_i, \; i = 1,2, \ldots, s$$

Ideally we would like to have all of these subspaces with dimension one to get perfect diagonalization, as in the case of distinct eigenvalues. Recall that an eigenvector of $\boldsymbol{A}$ lies in a subspace of dimension one and *invariant* under $\boldsymbol{A}$. In case of a repeated eigenvalue λ of multiplicity m, $m>1$, there may not exist m linearly independent eigenvectors associated with λ. In this case the '*smallest*' invariant subspace under $\boldsymbol{A}$ containing the eigenvectors of $\boldsymbol{A}$ associated with λ and having dimension m is $N((\boldsymbol{A}-\lambda\boldsymbol{I})^k)$ for some $k \leq m$. The proof of this result can be found in Desoer (1970). The following example illustrates an application of the above properties.

Example 7.3 In this example we are going to 'diagonalize' the system $\dot{\boldsymbol{x}}=\boldsymbol{A}\boldsymbol{x}$ where

$$\boldsymbol{A} = \begin{bmatrix} 1 & 1 & 0 & 0 & 0 \\ -1 & 3 & 0 & 0 & 0 \\ 0 & 0 & 3 & 2 & -1 \\ 0 & 0 & 3 & 5 & -3 \\ 0 & 0 & 7 & 8 & -5 \end{bmatrix}$$

This matrix has two distinct eigenvalues: $\lambda_1=2$ with multiplicity four and $\lambda_2=-1$. The eigenvalue 2 is associated with the eigenvectors $\boldsymbol{v}_1=(1,1,0,0,0,)^T$ and $\boldsymbol{v}_3=(0,0,1,0,1)^T$ and the eigenvalue -1 with the eigenvector $\boldsymbol{v}_5=(0,0,0,1,2)^T$. Because there are no more eigenvectors, perfect diagonalization is not possible. Therefore we must look for the smallest subspace, invariant under $\boldsymbol{A}$, containing the eigenvectors associated with the eigenvalue 2. In other words, we need to compute two generalized eigenvectors, which can be done by means of equations (7.15). The equation

$$\boldsymbol{A}\boldsymbol{v}_2 = 2\boldsymbol{v}_2 + \boldsymbol{v}_1 \tag{7.17}$$

yields $\boldsymbol{v}_2=(0,1,0,0,0)^T$, i.e. $\boldsymbol{v}_2$ is a generalized eigenvector associated with eigenvector $\boldsymbol{v}_1$. Similarly, for the generalized eigenvector $\boldsymbol{v}_4$ associated with $\boldsymbol{v}_3$, we can write:

$$\boldsymbol{A}\boldsymbol{v}_4 = 2\boldsymbol{v}_4 + \boldsymbol{v}_3 \tag{7.18}$$

which yields the solution $v_4=(0,0,0,1,1)^T$.

Exercise 7.9 With $v_1, v_2, \ldots, v_5$ as above, show that the subspaces span$\{v_1,v_2\}$ and span$\{v_3,v_4\}$ are invariant under A. With respect to the basis $D_e=\{v_1,v_2,v_3,v_4,v_5\}$ the representation $\dot{x}=Ax$ becomes $\dot{d}=\Lambda_r d$ where

$$\Lambda_r = \begin{bmatrix} 2 & 1 & 0 & 0 & 0 \\ 0 & 2 & 0 & 0 & 0 \\ 0 & 0 & 2 & 1 & 0 \\ 0 & 0 & 0 & 2 & 0 \\ 0 & 0 & 0 & 0 & -1 \end{bmatrix} \tag{7.19}$$

and $x=Wd$. W is a matrix whose columns are the vectors v_1, v_2, v_3, v_4, v_5, i.e. the elements of D_e. Notice that the blocks, 'the Jordan Blocks', in Λ_r

$$\begin{bmatrix} 2 & 1 \\ 0 & 2 \end{bmatrix}, \begin{bmatrix} 2 & 1 \\ 0 & 2 \end{bmatrix}, [-1]$$

are associated with the invariant subspaces span$=\{v_1,v_2\}$, span$\{v_3,v_4\}$ and span$\{v_5\}$ respectively.

Exercise 7.10

(a) Compute $(A-2I)^2$ and $(A-2I)^3$.
(b) Show that v_5 does not belong to $N(A-2I)^2$.
(c) Check that $N((A-2I)^2)=N((A-2I)^3)=N((A-2I)^4)=\ldots$
(d) Show that $\{v_1,v_2,v_3,v_4\}$ is a basis for $N(A-2I)^2)$.

Answer:

$$(A-2I)^2 = \begin{bmatrix} 0 & 0 & 0 & 0 & 0 \\ 0 & 0 & 0 & 0 & 0 \\ 0 & 0 & 0 & 0 & 0 \\ 0 & 0 & -9 & -9 & 9 \\ 0 & 0 & -18 & -18 & +18 \end{bmatrix}; (A-2I)^3 = \begin{bmatrix} 0 & 0 & 0 & 0 & 0 \\ 0 & 0 & 0 & 0 & 0 \\ 0 & 0 & 0 & 0 & 0 \\ 0 & 0 & -81 & -81 & 81 \\ 0 & 0 & -162 & -162 & +162 \end{bmatrix}$$

From the Cayley–Hamilton theorem we know that A verifies its own characteristic equation, i.e. if $p(\lambda)$ denotes the characteristic polynomial of A then $p(A)=0$. However, there may exist a polynomial l of lesser degree than p such that $l(A)=0$. In the previous example the characteristic polynomial of A is $p(\lambda)=(\lambda-2)^4\,(\lambda+1)$. However, the matrix A also satisfies the polynomial $l(\lambda)=(\lambda-2)^2\,(\lambda+1)$. It can be shown that l is *the* minimal polynomial of A.

Exercise 7.11 Verify that $l(A)=0$.

The multiplicity of the eigenvalue, as a zero of the minimal polynomial, gives the size of the largest associated Jordan block (Desoer, 1970). Looking at (7.19) we see that although the eigenvalue 2 has multiplicity 4, the associated Jordan blocks have size 2. In particular this means that the system $\dot{d}=\Lambda_r d$ with Λ_r as in (7.19) will generate only the modes of e^{2t} and $t\,e^{2t}$ as shown in Fig. 7.11. With a Jordan block of size four the modes $t^2e^{2t}/2!$ and $t^3e^{2t}/3!$ would also be present because all the integrators would then be coupled in series as in Fig. 3.10.

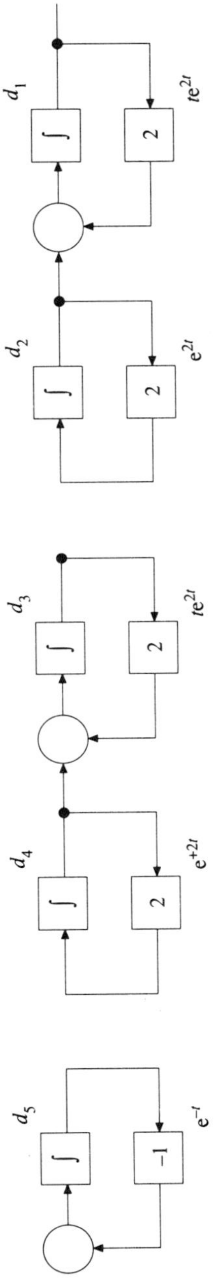

Figure 7.11 Block diagram of $\dot{\boldsymbol{d}}=\Lambda_r\boldsymbol{d}$ defined by (7.19).

The results of sections 7.1.1 and 7.1.2 can be summarized in the Jordan form theorem whose proof can be found in Chen (1984). The gist of the theorem is as follows. Given a matrix $\boldsymbol{A} \in \mathbb{R}^{n \times n}$, with d distinct eigenvalues, we can always find a collection of *invariant* subspaces under $\boldsymbol{A}, \{N_1, N_2, \ldots, N_d\}$,such that

$$\mathbb{R}^n = N_1 \oplus N_2 \oplus \ldots \oplus N_d$$

i.e. for any $\boldsymbol{v} \in \mathbb{R}^n$

$$\boldsymbol{v} = \boldsymbol{v}_1 + \boldsymbol{v}_2 + \ldots \boldsymbol{v}_d, \ \boldsymbol{v}_i \in N_i, \ i = 1,2, \ldots, d$$

and this decomposition is *unique*. If we now select a basis for $\mathbb{R}^n$ as the union of the bases of $N_1, N_2, \ldots, N_d$ then, with respect to this basis, $\boldsymbol{A}$ will have the 'diagonal' form

$$\boldsymbol{A} = \begin{bmatrix} \boldsymbol{A}_1 & & & & & \\ & \boldsymbol{A}_2 & & & & \\ & & \cdot & & & \\ & & & \cdot & & \\ & & & & \cdot & \\ & & & & & \boldsymbol{A}_d \end{bmatrix} \tag{7.20}$$

N_i, i=1,2,. . .,d, is the null space of $(\boldsymbol{A} - \lambda_i \boldsymbol{I})^{m_i}$ and λ_i is an eigenvalue of $\boldsymbol{A}$ with multiplicity m_i as a zero of the *minimal* polynomial of $\boldsymbol{A}$. The Jordan form theorem guarantees the existence of a basis for N_i that leads to a simple form of $\boldsymbol{A}_i$ consisting of as many Jordan blocks as there are linearly independent eigenvectors associated with λ_i (cf. (7.19)). This basis can be constructed by means of equations (7.15).

7.2 STABILITY

The stability concepts presented in this section are related to the type of system description studied so far, namely the external or input–output description and the internal or state-variable description. The concept of bounded-input, bounded-output (b.i.b.o.) stability will be presented in connection with the former. Two concepts of stability – stability in the sense of Liapunov (i.s.L.) and asymptotic stability – will be introduced in terms of the state-space representation. For simplicity of exposition, only single-input, single-output (s.i.s.o.) will be considered, but this by no means implies any loss of generality; the results can be easily extended to the multivariable case; see Desoer (1970).

7.2.1 Bounded-input, Bounded-output Stability

Given a linear time invariant dynamical system described by the input–output relation (convolution integral):

$$y(t) = \int_0^t w(t - t')u(t')\, dt' \tag{7.21}$$

or equivalently by its transfer function $W(s)$, we say it is *b.i.b.o. stable* if and only if any bounded-input produces a bounded-output.

Definition: A function $f(t)$, $t \in \mathbb{R}$, is bounded if there exists a finite number M such that $|f(t)| \leqslant M$, $\forall\ t \in \mathbb{R}$. If $f(t)$ is a vector-valued function, i.e. $f(t) = (f_1(t), f_2(t), \ldots, f_n(t))^T$, then we use instead $\|f(t)\| \leqslant M$, $\forall\ t \in \mathbb{R}$, where $\|.\|$ denotes a norm in $\mathbb{R}^n$.

Theorem: The system described by (7.21) is b.i.b.o. stable if and only if its impulse response is absolutely integrable, i.e. there exists a finite L such that

$$\int_0^t |w(t - t')|\, dt' \leqslant L,\ \forall\ t \in \mathbb{R}^+$$

Proof: We start by showing the condition is sufficient. From (7.21) and bearing in mind that $|u(t)| \leqslant K$, $\forall t$, we have

$$|y| = |\int_0^t w(t - t'))\ u(t')\ dt'| \leqslant \int_0^t |w(t - t')\ u(t')| dt'$$

$$\leqslant K \int_0^t |w(t - t')| dt' \leqslant KL$$

and the result follows.

We now prove by contradiction that the condition is also necessary. Suppose

$$\int_0^t |w(t - t')|\ dt'$$

was unbounded; then given $l \in \mathbb{R}$, no matter how large, there would exist t_l such that

$$\int_0^{t_l} |w(t_l - t')| dt' > l$$

Selecting a bounded input $u_l(t')$ such that $u_l(t') = \text{sgn}\ w(t_l - t')$, $0 \leqslant t' \leqslant t_l$ the corresponding output at time t_l would be precisely

$$\int_0^{t_l} |w(t_l - t')| dt' > l$$

showing that the system is not b.i.b.o, stable; therefore the condition is also necessary.

Corollary Let $W(s)$ denote the Laplace transform of the impulse response $w(t)$ of the system defined in (7.21), and assume

$$W(s) = N(s)/D(s) \tag{7.22}$$

where $N(s)$ and $D(s)$ are polynomials in s with degrees a and b, respectively, and $a<b$. Then the system is b.i.b.o. stable if and only if the zeros of $D(s)$ have negative real parts.

Proof: A partial fraction decomposition of W(s), as shown in equation (3.41), reveals that the impulse response will be a linear combination of terms of the form

$$t^l \exp(p_i t)/l!, \; 0 \leqslant l \leqslant n_i - 1 \tag{7.23}$$

where p_i is a pole of $W(s)$ with multiplicity n_i. Notice that we do not have either the Dirac function $\delta(t)$ because it is assumed that the degree of $N(s)$ is strictly less than the degree of $D(s)$. Therefore

$$\int_0^t |w(t')| \, dt'$$

is bounded if the terms of the form (7.23) are absolutely integrable. Now integrating by parts we get

$$\int_0^t x^l \, e^{ax} \, dx = 1/a \, t^l \, e^{at} - l/a \int_0^t x^{l-1} \, e^{ax} \, dx \tag{7.24}$$

and these terms are absolutely integrable if, for l=0,1,. . .,n_i−1, $|t^l \exp(p_i t)|$ and

$$\int_0^t |\exp(p_i x)| dx$$

are bounded functions of t. This is satisfied if and only if Re(p_i)<0. This completes the proof.

Exercise 7.12 Verify the correctness of formula (7.24). *Hint*: Compute the derivative of both sides.

Exercise 7.13 Prove the above corollary by first establishing the conditions to be satisfied by a *rational Laplace transform* in order to have a bounded inverse transform. *Hint*: use (3.41); this will reveal that the poles must have negative real parts or lie on the imaginary axis provided they are simple. Then, think of an input signal sharing a pole on the imaginary axis with the transfer function.

7.2.2 Asymptotic Stability and Stability in the Sense of Liapunov

The concepts of asymptotic stability and stability in the sense of Liapunov are applicable to the zero input solutions of $\dot{\boldsymbol{x}}=\boldsymbol{Ax}+\boldsymbol{Bu}$, i.e. to the system

$$\dot{\boldsymbol{x}} = \boldsymbol{A}\mathbf{x}, \; \boldsymbol{x} \in \mathbb{R}^n \tag{7.25}$$

Given a general, possibly non-linear, dynamical system

$$\dot{\boldsymbol{x}} = f(\boldsymbol{x}) \tag{7.26}$$

most of the time stability is not a property of the system but is defined for each particular solution.

Definition: A solution $\boldsymbol{x}^o(t)$ of (7.26) starting at $\boldsymbol{x}(0)$ is *stable in the sense of Liapunov* if an arbitrarily small perturbation of the initial state results in an arbitrarily small perturbation of the corresponding solution. Or more precisely: given ε>0 there exists δ(ε)>0 such that

$$||\boldsymbol{x}^o(0) - \boldsymbol{x}(0)|| < \delta \Rightarrow ||\boldsymbol{x}(t) - \boldsymbol{x}^o(t)|| < \varepsilon, \; \forall t > 0$$

Definition: A solution $x^o(t)$ of (7.26) is *asymptotically stable* if it is stable in the sense of Liapunov and there exists $\delta>0$ such that

$$\|x(0) - x^o(0)\| < \delta \Rightarrow \| x(t) - x^o(t)\| \to 0 \text{ as } t \to \infty$$

For a linear system, if a solution is stable then all solutions are also stable. Stability thus becomes a global property, i.e. a property of the system and not of particular solutions. In fact, given a solution of $\dot{x}=Ax$, say $\mathbf{x}(t)$, if the initial condition $x(0)$ is perturbed we get another solution $x_p(t)$ with $x_p(0)=x(0)+\Delta x_0$; defining $\bar{x}=x_p-x$, we have for $t\geq 0$

$$d\bar{x}(t)/dt = A\,\bar{x}(t), \text{ with } \bar{x}(0) = \Delta x_0$$

Therefore all solutions are stable if and only if the zero solution is stable.

Because the stability properties of a system are independent of the basis for the state-space, the characterization of the stability of $\dot{x}=Ax$ may follow immediately from the diagonalization and modal analysis results presented in the previous section. Because $\mathbf{x}(t)=e^{At}\,x(0)$ we have $\dot{x}=Ax$ is asymptotically stable if and only if all eigenvalues of A have negative real parts. $\dot{x}=Ax$ is stable in the sense of Liapunov if and only if:

1. A has all eigenvalues with real parts less than or equal to zero.
2. Any eigenvalue of A on the imaginary axis has multiplicity one as a zero of the minimal polynomial of A.

The last condition shows that it is possible for the system $\dot{x}=Ax$ to be stable in the sense of Liapunov and to have an eigenvalue λ of multiplicity m greater than 1. However, being a simple zero of the minimal polynomial, the associated Jordan blocks have dimension one, which means that the m integrators associated with λ are uncoupled, therefore generating only the mode $e^{\lambda t}$; or equivalently that λ has m linearly independent eigenvectors.

7.2.3 Relationships Between Stability Concepts

The stability concepts applied to the system:

$$\dot{x} = Ax + Bu$$

$$v = Cx$$

are related, as shown in Fig. 7.12. Because the poles of the system transfer function are eigenvalues of the system matrix A we have

asymptotic stability $\Rightarrow$ b.i.b.o. stability

The ideal integrator, having transfer function $1/s$, shows that

stability i.s.L. $\not\Rightarrow$ b.i.b.o stability

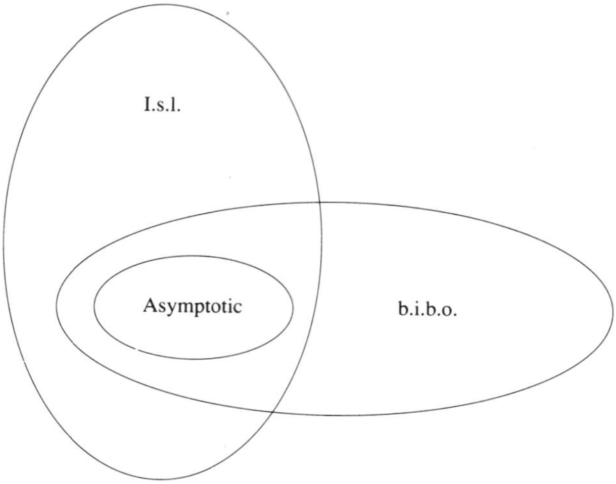

Figure 7.12 Relationship between stability concepts.

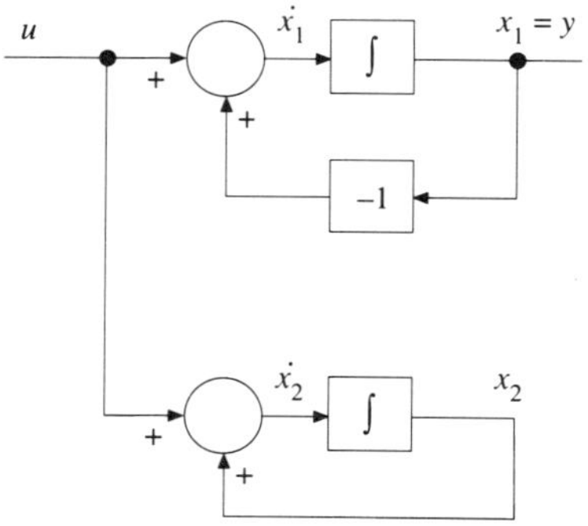

Figure 7.13 An example of a b.i.b.o. stable system and unstable i.s.L.

The system depicted in Fig 7.13 shows that

b.i.b.o. stability $\not\Rightarrow$ stability i.s.L.

Exercise 7.14 Consider the one-dimensional motion of a ball rolling on a surface under the influence of gravity, as shown in Fig. 7.14. Regarding the ball as a dynamical system whose state is the position and velocity, we may characterize the trajectories starting at the equilibrium point assuming:
(a) A frictionless surface.
(b) A frictional force proportional to velocity.
Answers: (a) stable; (b); asymptotic stable.

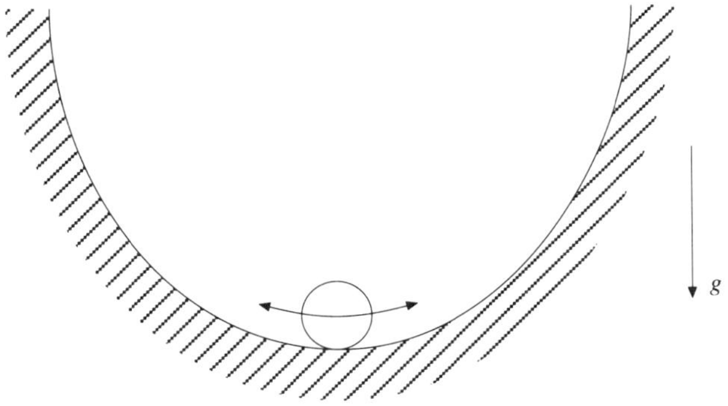

Figure 7.14 System for exercise 7.14.

7.3 CONTROLLABILITY AND OBSERVABILITY

The concepts of controllability and observability are very important in system analysis and design, particularly in realization theory. As defined by Kalman (1963), the problem of *realization* consists in identifying a state-space representation that generates an experimentally observed impulse response. The word realization is justified because such representation can be 'concretely realized', e.g. on an analog (or digital) computer. The solution, when it exists, is not unique; in fact for a given impulse response one can always find a realization with dimension as large as desired, and an infinite number of realizations with the same dimension. In general the most interesting ones are those whose state-space is of minimal dimension, i.e. the so-called *irreducible realizations*. It is shown in the above-mentioned paper that a realization is irreducible if and only if it is controllable and observable. This follows from the 'canonical structure theorem', which we shall also illustrate at the end of this section. The characterization of controllability and observability is basically a problem of determining the dimension of the range and null space of a linear transformation as we shall see presently.

7.3.1 Controllability

Consider the time-invariant state-space representation

$$\begin{aligned} \dot{\boldsymbol{x}} &= \boldsymbol{A}\boldsymbol{x} + \boldsymbol{B}\boldsymbol{u} \\ \boldsymbol{y} &= \boldsymbol{C}\boldsymbol{x} + \boldsymbol{D}\boldsymbol{u} \end{aligned} \tag{7.27}$$

with $\boldsymbol{x} \in \mathbb{R}^n$, $\boldsymbol{u} \in \mathbb{R}^m$ and $\boldsymbol{y} \in \mathbb{R}^p$, and where $\boldsymbol{A}$, $\boldsymbol{B}$, $\boldsymbol{C}$ and $\boldsymbol{D}$ have compatible dimensions.

Definition: The state $\boldsymbol{x}$ of (7.27) is said to be *controllable* if there exists an input $\boldsymbol{u}$ that steers $\boldsymbol{x}$ to the origin in a finite time interval.

Exercise 7.15 Show if state $\boldsymbol{x}_0$ is controllable then all states on the line defined by $\boldsymbol{x}_0$ and the origin of the state-space are also controllable. *Hint*: any point $\boldsymbol{x}$ on this line is of the form $\boldsymbol{x}=\lambda\boldsymbol{x}_0$, $\lambda\in\mathbb{R}$. Then use (3.62).

Definition: The representation (7.27) is said to be *completely controllable* if all states are controllable.

Hereafter, complete controllability will be simply referred to as controllability. For example, the system depicted in Fig. 7.15 is not controllable because the second component x_2 of the state vector $\boldsymbol{x}$ is not affected by the input. Furthermore, given an initial non-zero value for x_2, it will vanish only after an *infinite* time interval.

Theorem: The state-space representation (7.27) is (completely) controllable if and only if the $(n\times nm)$ matrix

$$\boldsymbol{Q} = [\boldsymbol{B} \mid \boldsymbol{AB} \mid \ldots \mid \boldsymbol{A}^{n-1}\boldsymbol{B}] \tag{7.28}$$

has rank n.

Notice the characterization of controllability is independent of the matrix $\boldsymbol{C}$. In our previous example we have

$$\boldsymbol{A} = \begin{bmatrix} 1 & 1 \\ 0 & -1 \end{bmatrix} ; \boldsymbol{B} = \begin{bmatrix} 1 \\ 0 \end{bmatrix} ; \boldsymbol{Q} = \begin{bmatrix} 1 & 1 \\ 0 & 0 \end{bmatrix}$$

Because $\boldsymbol{Q}$ is not full rank we conclude the system is not controllable. For a proof of this theorem see Desoer (1970).

There are other useful ways of defining controllability; the following are equivalent to our definition:

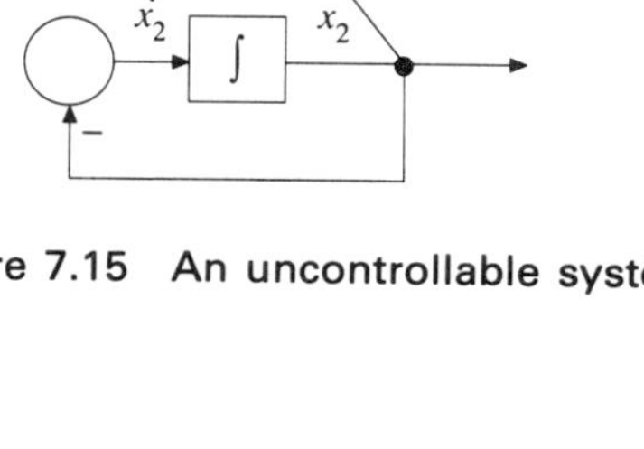

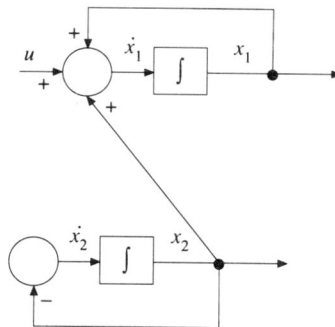

Figure 7.15 An uncontrollable system.

1. The zero state can be transferred to any state, in a finite time, by some input.
2. Any state $\boldsymbol{x}_0$ can be transferred to any state $\boldsymbol{x}_1$ in a finite time interval by some input.
3. The range of the map

$$\boldsymbol{u}(t) \rightarrow \int_0^{t_1} \exp \boldsymbol{A}(t_1 - t')\boldsymbol{B}\boldsymbol{u}(t')\mathrm{d}t'$$

with $t \in [0, t_1]$, is $\mathbb{R}^n$.

The proof of equivalence for the first two definitions is obvious. The last is an immediate consequence of (3.62). It will also be shown that controllability is equivalent to the fact that matrix $\boldsymbol{M}$ defined by (7.36) is non-singular. Recalling the definition of exp $(\boldsymbol{A}t)$ and the Cayley–Hamilton theorem, we can write: $\exp(\boldsymbol{A}t) = \alpha_0(t)\,\boldsymbol{I} + \alpha_1(t)\,\boldsymbol{A} + \ldots + \alpha_{n-1}(t)\,\boldsymbol{A}^{n-1}$. This, together with (3.62), shows that the set of controllable states is, at most, span $\{\boldsymbol{B}, \boldsymbol{AB}, \ldots, \boldsymbol{A}^{n-1}\boldsymbol{B}\}$ i.e. the range of the matrix $\boldsymbol{Q}$ defined in (7.28).

Exercise 7.16 Show that the set of controllable states of (7.27) is a subspace of $\mathbb{R}^n$. Recall a set S is a subspace if and only if: (a) $x \in S \Rightarrow \alpha x \in S$, $\forall \alpha \in \mathbb{R}$; (b) if x_1 and x_2 belong to S, then $(\alpha_1 x_1 + \alpha_2 x_2) \in S$, for all α_1 and α_2 in $\mathbb{R}$. *Hint*: let $\boldsymbol{u}_1$ and $\boldsymbol{u}_2$ denote the controls that steer $\boldsymbol{x}_1$ and $\boldsymbol{x}_2$ to the origin, respectively; then using (3.62) show that $(\boldsymbol{u}_1 + \boldsymbol{u}_2)$ steers $(\boldsymbol{x}_1 + \boldsymbol{x}_2)$ to the origin.

Theorem: The range of $\boldsymbol{Q}$, defined in (7.28) is the subspace of controllable states. *Proof*: See Desoer (1970).

Example 7.4 Consider the familiar stick balancing problem shown in Fig. 7.16. For simplicity, let us assume that the motion is restricted to a plane. Complex balancing problems are frequently modelled as inverted pendulums, particularly in robotics, where robot manipulators can be regarded as a series of concatenated pendulums. The linearized equation around $\theta = 0$, assuming no friction, is

$$(\mathrm{I} + mL^2)\,\ddot{\theta} + mL\,\ddot{u} - mgL = 0 \tag{7.29}$$

or equivalently

$$4/3\,L\,\ddot{\theta} - g\,\theta + \ddot{u} = 0 \tag{7.30}$$

where g is the gravitational constant, $2L$ the length of the stick, m its mass and $\mathrm{I} = 1/3mL^2$ the moment of inertia with respect to the centre of gravity. Using the formulae given with (3.13) in chapter 3, we can write (7.29) in state-space form

$$\dot{\boldsymbol{x}} = \begin{bmatrix} 0 & 1 \\ \frac{3g}{4L} & 0 \end{bmatrix} \boldsymbol{x} + \begin{bmatrix} 0 \\ 1 \end{bmatrix} u \tag{7.31}$$

$$y = \left(-\frac{9g}{16L^2}\,;\, 0 \right) \boldsymbol{x} - \frac{3}{4L}\,u$$

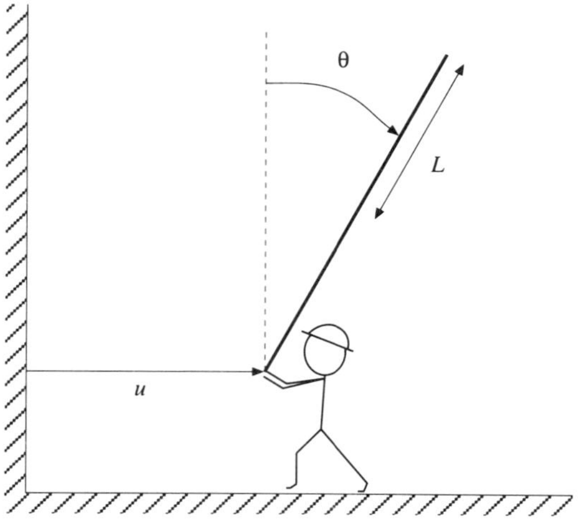

Figure 7.16 Stick-balancing problem.

where $y=\theta$. Notice the system matrix has eigenvalues on the real axis, $\pm\sqrt{(3g)/(4L)}$, showing the system is unstable. On the other hand, from (7.31) we have

$$Q = [\boldsymbol{BAB}] = \begin{bmatrix} 0 & 1 \\ 1 & 0 \end{bmatrix}$$

which shows the system is controllable.

Exercise 7.17

(a) Check the transfer functions of (7.29) and (7.31) are the same.

(b) Knowing that the linearized equation around $\theta=\pi$, of the system in Fig. 7.16, is

$$\frac{4L}{3}\ddot{\alpha} + g\alpha - \ddot{u} = 0$$

verify that the system remains controllable, with eigenvalues $\pm\, j\,\sqrt{(3g/(4L))}$, as expected. Notice that this is in agreement with the well-known result for the ideal simple pendulum, when the angle described, α, is small, namely that the frequency of oscillation is proportional to $\sqrt{(g/(2L))}$, and independent of its mass.

Controllability has been defined in terms of the state, but it can also be interpreted in terms of system modes. In fact controllability is invariant under change of coordinates, as shown in the next exercise.

Exercise 7.18 If $\boldsymbol{x}=\boldsymbol{Tz}$, det $\boldsymbol{T}\neq 0$, show (7.27) is controllable if and only if

$$\dot{z} = \boldsymbol{T}^{-1}\boldsymbol{A}\,\boldsymbol{T}\,z + \boldsymbol{T}^{-1}\boldsymbol{B}\,\boldsymbol{u}$$

$$\boldsymbol{y} = \boldsymbol{CT}\,z + \boldsymbol{Du} \tag{7.32}$$

is controllable. *Hint*: compute the matrix $\boldsymbol{Q}$, (7.28), for the above system.

Now we show that if the representation is not controllable there are natural modes that cannot be excited by the input and vice versa. The proof is straightforward if we use the system canonical (or diagonal) representation. For simplicity, let us start by assuming we have a scalar input system and a representation with distinct eigenvalues. Then the representation is not controllable if and only if the vector $T^{-1}B$, in the diagonal representation, has some of its elements equal to zero. In fact the components of this vector control the excitation to the natural mode generators and if any element is zero, the associated mode cannot be excited by the input, i.e. the representation is not controllable. On the other hand, the matrix Q for the diagonal representation can be written as

$$Q = [T^{-1}B|\Lambda T^{-1}B| \dots |\Lambda^{n-1}T^{-1}B] \tag{7.33}$$

Because Λ is diagonal, with distinct elements, Q is full rank if $T^{-1}B$ has non-zero elements.

Exercise 7.19 Show Q in (7.33) is full rank if and only if Λ has distinct diagonal elements and the vector $T^{-1}B$ has non-zero elements. *Hint*: let $T^{-1}B=(\alpha_1,\alpha_2,\dots,\alpha_n)^T$ and start by showing

$$Q = \begin{bmatrix} \alpha_1 & & & & \\ & \alpha_2 & & \mathbf{0} & \\ & & \cdot & & \\ & & & \cdot & \\ & \mathbf{0} & & & \cdot \\ & & & & \alpha_n \end{bmatrix} \begin{bmatrix} 1 & \lambda_1 & \dots & \lambda_1^{n-1} \\ 1 & \lambda_2 & \dots & \lambda_2^{n-1} \\ \cdot & \cdot & & \cdot \\ \cdot & \cdot & & \cdot \\ \cdot & \cdot & & \cdot \\ 1 & \lambda_n & \dots & \lambda_n^{n-1} \end{bmatrix} \tag{7.34}$$

where $\lambda_1,\lambda_2,\dots,\lambda_n$, are the diagonal elements of Λ, and then use the fact that the determinant of the product of two matrices is the product of their determinants. Furthermore the matrix on the right hand side of (7.34) is the well-known Vandermonde matrix whose determinant is non-zero if and only if $\lambda_1, \lambda_2,\dots, \lambda_n$ are distinct.

Exercise 7.20 Show that the determinant of the Vandermonde matrix in (7.34) is non-zero if and only if $\lambda_i \neq \lambda_j$ for $i \neq j$. *Hint*: we know, from Section 7.1.1, that an $(n \times n)$ matrix with n distinct eigenvalues has n linearly independent eigenvectors. Then, using exercise 3.3, show that if $\lambda_1,\lambda_2,\dots,\lambda_n$ are the eigenvalues of the companion matrix

$$\begin{bmatrix} 0 & 1 & & & \\ & & 1 & \mathbf{0} & \\ & & & \cdot & \\ & \mathbf{0} & & & \cdot \\ & & & & 1 \\ -a_n & -a_{n-1} & -a_{n-2} & \dots & -a_1 \end{bmatrix}$$

then $(1, \lambda_i, \lambda_i^2,\dots,\lambda_i^{n-1}), i=1,2,\dots,n$ are the associated eigenvectors.

In cases of repeated eigenvalues, a representation, in order to be controllable, must have at least as many inputs as the maximum number of Jordan blocks associated with a single eigenvalue. For example, the representation $\dot{\boldsymbol{x}}=\boldsymbol{A}\boldsymbol{x}+\boldsymbol{B}\boldsymbol{u}$ with

$$\boldsymbol{A} = \begin{bmatrix} \lambda & 0 \\ 0 & \lambda \end{bmatrix}$$

has two Jordan blocks and is not controllable with a single input. However, if $\boldsymbol{B}=(b_1,b_2)^{\mathrm{T}}$ and

$$\boldsymbol{A} = \begin{bmatrix} \lambda & 1 \\ 0 & \lambda \end{bmatrix}$$

the representation has a single Jordan block and is controllable provided $b_2 \neq 0$.

In general the subsystem associated with a Jordan block is controllable, if and only if the first integrator, in the chain associated with the block, is linked to a system input.

Exercise 7.21 Show the last mentioned statement, i.e. if

$$\boldsymbol{A} = \begin{bmatrix} \lambda & 1 & & \mathbf{0} \\ & \ddots & \ddots & \\ & & \lambda & 1 \\ \mathbf{0} & & & \lambda \end{bmatrix}, \quad \boldsymbol{b} = \begin{bmatrix} b_1 \\ b_2 \\ \vdots \\ b_l \end{bmatrix}$$

then $\dot{\boldsymbol{x}}=\boldsymbol{A}\boldsymbol{x}+\boldsymbol{b}u$ is controllable if and only if $b_l \neq 0$. *Hint*: notice that the controllability matrix has the form

$$[\boldsymbol{b}\ \boldsymbol{A}\boldsymbol{b}\ \ldots\ \boldsymbol{A}^{l-1}\ \boldsymbol{b}] = \begin{bmatrix} b_1 & b_2 & b_3 & \ldots & b_l \\ b_2 & b_3 & & \cdot & \\ b_3 & & \cdot & & \\ \vdots & \cdot & & & \mathbf{0} \\ b_l & & & & \end{bmatrix} \begin{bmatrix} 1 & \lambda & \lambda^2 & \lambda^3 & \ldots & \lambda^{l-1} \\ & 1 & 2\lambda & 3\lambda^2 & & \vdots \\ & & 1 & 3\lambda & \ddots & \vdots \\ & & & 1 & \ddots & (l-1)\lambda^2 \\ & \mathbf{0} & & & \ddots & (l-1)\lambda \\ & & & & & 1 \end{bmatrix}$$

If $\boldsymbol{A}=\boldsymbol{\Lambda}_{\mathbf{r}}$, as defined by (7.19), and $\boldsymbol{B}=(b_1,\ldots,b_5)^{\mathrm{T}}$, the controllability matrix has the form

$$Q = \begin{bmatrix} x & x & x & x & x \\ b_2 & 2b_2 & 2^2b_2 & 2^3b_2 & 2^4b_2 \\ x & x & x & x & x \\ b_4 & 2b_4 & 2^2b_4 & 2^3b_4 & 2^4b_4 \\ x & x & x & x & x \end{bmatrix}$$

which implies that the representation is not controllable. Because each Jordan block associated with an eigenvalue λ is also associated with a row in the system matrix with λ in the main diagonal and the remaining elements equal to zero (cf. (7.19)), we find that each eigenvalue will give rise to as many linearly dependent rows in Q as the number of associated Jordan blocks. Therefore this rank deficiency must be compensated for by an increase in the columns of B, i.e. in the number of system inputs. If B is replaced by a 5×2 matrix in the previous example, Q will have full rank *provided* the second and fourth rows of B are linearly independent. An interpretation of this result in terms of block diagrams of the type of Figs 3.10 and 3.11 means that controllability requires, for each eigenvalue, that each of the first integrators in the chain associated with a Jordan block must be connected to a separate input. A detailed treatment of this situation can be found in Chen (1984).

The modes associated with the controllable subsystem are called the controllable modes (or controllable eigenvalues). The modes in the diagonal representation that cannot be excited by the input are called the uncontrollable modes. When all the uncontrollable modes are stable (i.e. they lie in the left half of the complex plane) we say the system is *stabilizable*. The reason, as seen later in the book, lies in the fact that it is possible to shift the controllable modes arbitrarily, by state-feedback.

An alternative method to analyze the controllability of an eigenvalue is by means of the Hautus test (Hautus, 1969):

Theorem: An eigenvalue λ_i, $i=1,2,\ldots,r$, of the representation $\dot{\boldsymbol{x}}=\boldsymbol{A}\boldsymbol{x}+\boldsymbol{B}\boldsymbol{u}$, $x\in\mathbb{R}^n$, $r\leq n$, is controllable if and only if rank $[\mathbf{A}-\lambda_i\mathbf{I}\ \mathbf{B}]=n$.

The proof of this result is straightforward, in the case of single-input, single-output systems with distinct eigenvalues. Because the controllability and the eigenvalues of a system are invariant under change of coordinates, we can assume $\boldsymbol{A}$ is diagonal. Then rank $[\boldsymbol{A}-\lambda_i\mathbf{I}\ \boldsymbol{B}]$ equals the rank of

$$\begin{bmatrix} (\lambda_1-\lambda_i) & 0 & \ldots & 0 & b_1 \\ 0 & (\lambda_2-\lambda_i) & \ldots & 0 & b_2 \\ \ldots & \ldots & \ldots & \ldots & \ldots \\ 0 & 0 & \ldots & (\lambda_n-\lambda_i) & b_n \end{bmatrix}$$

which is equal to n if and only if $b_i\neq0$. But $b_i\neq0$ means that the mode generator $\exp(\lambda_i t)$ has an input; therefore this mode is controllable (cf. Fig. 7.20, for example).

The following result shows how to compute a control $\boldsymbol{u}$ that steers the state from a given $\boldsymbol{x}_0$ to $\boldsymbol{x}_1$.

Fact: If representation (7.27) is controllable then the input

$$\boldsymbol{u}(t) = -\boldsymbol{B}^T\exp(\boldsymbol{A}^T(t_0 - t))\boldsymbol{M}^{-1}(t_0,t_1)\,(\boldsymbol{x}_0 - \exp(\boldsymbol{A}(t_0 - t_1))\boldsymbol{x}_1) \tag{7.35}$$

steers $\boldsymbol{x}_0$, at time t_0, to x_1 at time t_1, with $\boldsymbol{M}$ defined by

$$\boldsymbol{M}(t_0,t_1) = \int_{t_0}^{t_1} \exp(\boldsymbol{A}(t_0 - z))\boldsymbol{B}\boldsymbol{B}^T \exp(\boldsymbol{A}^{\mathrm{T}}(t_0 - z))\,\mathrm{d}z \tag{7.36}$$

Notice that this fact states that controllability is *equivalent* to the existence of $\boldsymbol{M}^{-1}(t_0,t_1)$.

Proof: From (3.62) we have

$$\boldsymbol{x}(t) = \exp \boldsymbol{A}(t - t_0)\,\{\boldsymbol{x}_0 + \int_{t_0}^{t} \exp(\boldsymbol{A}(t_0 - z))\,\boldsymbol{B}\,\boldsymbol{u}(z)\,\mathrm{d}z\} \tag{7.37}$$

because

$$\exp \boldsymbol{A}(t_1 + t_2) = \exp \boldsymbol{A}t_1 \exp \boldsymbol{A}t_2$$

Then inserting (7.35) into (7.37) we get

$$\begin{aligned}\boldsymbol{x}(t_1) &= \exp[\boldsymbol{A}(t_1 - t_0)]\{\boldsymbol{x}_0 - \int_{t_0}^{t_1} \exp[\boldsymbol{A}(t_0 - z)]\boldsymbol{B}\boldsymbol{B}^{\mathrm{T}}\exp[\boldsymbol{A}^{\mathrm{T}}(t_0 - z)]\mathrm{d}z\\ &\qquad \times\boldsymbol{M}^{-1}(t_0,t_1)\,\{\boldsymbol{x}_0 - \exp[\boldsymbol{A}(t_0 - t_1)]\boldsymbol{x}_1\}\}\\ &= \exp[\boldsymbol{A}(t_1 - t_0)]\{\boldsymbol{x}_0 - \boldsymbol{M}(t_0,t_1)\boldsymbol{M}^{-1}(t_0,t_1)\,\{\boldsymbol{x}_0 - \exp[\boldsymbol{A}(t_0 - t_1)]\boldsymbol{x}_1\}\}\\ &= \boldsymbol{x}_1\end{aligned}$$

QED.

Example 7.5 Given the system

$$\dot{\boldsymbol{x}} = \begin{bmatrix} 0 & 1 \\ 0 & 0 \end{bmatrix} \boldsymbol{x} + \begin{bmatrix} 0 \\ 1 \end{bmatrix} u$$

$$y = (1,0)\boldsymbol{x} \tag{7.38}$$

let us consider the problem of transferring the origin of the state-space to $(1,0)^{\mathrm{T}}$ in the time interval $[0,t_1]$. From (7.36) we have

$$\begin{aligned}\boldsymbol{M}(0,t_1) &= \int_0^{t_1} \begin{bmatrix} 1 & -z \\ 0 & 1 \end{bmatrix} \begin{bmatrix} 0 \\ 1 \end{bmatrix} (0,1) \begin{bmatrix} 1 & 0 \\ -z & 1 \end{bmatrix} \mathrm{d}z\\ &= \int_0^{t_1} \begin{bmatrix} z^2 & -z \\ -z & 1 \end{bmatrix} \mathrm{d}z = \begin{bmatrix} t_1^3/3 & -t_1^2/2 \\ -t_1^2/2 & t_1 \end{bmatrix}\end{aligned}$$

The required control can now be computed from (7.35):

$$\begin{aligned}u(t) &= -\,12/(t_1^4)\,(0,-1) \begin{bmatrix} 1 & 0 \\ -t & 1 \end{bmatrix} \begin{bmatrix} t_1 & t_1^2/2 \\ t_1^2/2 & t_1^3/3 \end{bmatrix} \begin{bmatrix} 1 & -t_1 \\ 0 & 1 \end{bmatrix} \begin{bmatrix} 1 \\ 0 \end{bmatrix}\\ &= 6/t_1^2 - 12t/t_1^3\end{aligned}$$

where

$$\boldsymbol{M}^{-1}(0,t_1) = 12/(t_1^4) \begin{bmatrix} t_1 & t_1^2/2 \\ t_1^2/2 & t_1^3/3 \end{bmatrix}$$

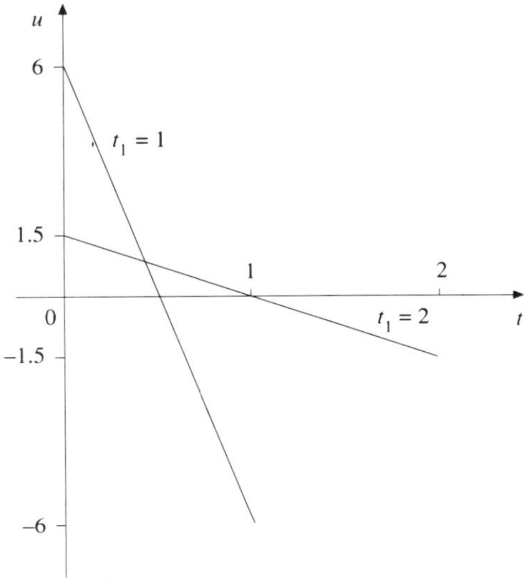

Figure 7.17 Inputs of system (7.48) for a given state transition in time intervals [0,1] and [0,2].

Figure 7.17 shows this control for $t_1=2$ and $t_1=1$. As expected, the larger the time interval, the smaller the control effort. Although we can make the transition in an arbitrarily small time interval, provided the representation is controllable, this example shows that this is achieved at the expense of an input with very large magnitude. This means that in practice there is always a lower bound for this time interval because physical variables are always amplitude-constrained.

7.3.2 Observability

Observability is concerned with the zero input response of (7.27).

Definition: A state $\boldsymbol{x}_0 \neq 0$ is said to be *unobservable* if for $\boldsymbol{u}(t)=0\ \forall t$, we have $y(t)=0$ on some non-zero time interval.

Example 7.6 Figure 7.18 shows an unobservable system; the second component of the state does not contribute to the system output.

Definition: The representation (7.27) is said to be *completely observable* if all the states are observable.

Hereafter complete observability will be simply referred to as observability. From (7.38) we conclude that if $\boldsymbol{x}_0$ is an unobservable state then $\boldsymbol{C} \exp(\boldsymbol{A}t)\, \boldsymbol{x}_0$ is identically zero. Therefore the set of unobservable states is a sub-space of $\mathbb{R}^n$.

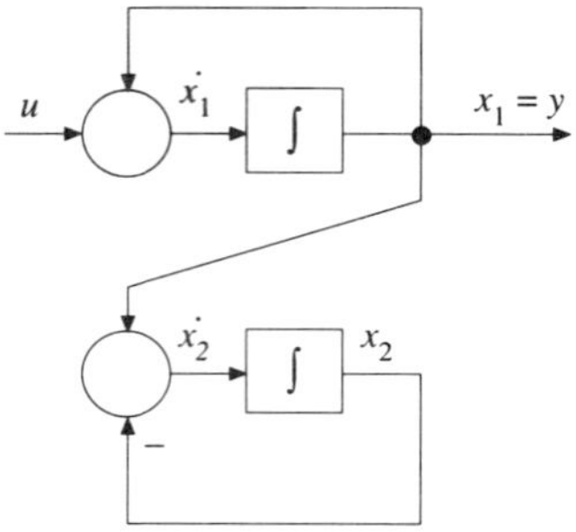

Figure 7.18 A non-observable system.

Theorem: The state-space representation (7.27) is observable if and only if the $(np \times n)$ matrix

$$R = \begin{bmatrix} C \\ \hdashline CA \\ \hdashline \cdot \\ \cdot \\ \cdot \\ CA^{n-1} \end{bmatrix} \qquad (7.39)$$

has rank n. For a proof see Chen (1984). As expected the characterization of observability is independent of B.

Example 7.7 For the system of Fig. 7.18 we have

$$R = \begin{bmatrix} 1\,0 \\ 1\,0 \end{bmatrix}$$

Because R has rank one the representation is unobservable.

Theorem: The set of unobservable states of (7.27) is the null space of R, $N(R)$.

This theorem states that an unobservable state x_0 is orthogonal to the linear sub-space span$\{C^T, A^TC^T, \ldots, (A^T)^{n-1}C^T\}$. For a detailed proof see Chen (1984). Here we shall only prove the condition is sufficient: Because the zero input response is $C\exp(At)x_0$ and

$$C\exp(At) = [\alpha_0(t)C + \alpha_1(t)CA + \ldots + \alpha_{n-1}(t)CA^{n-1}]$$

we conclude that if $x_0 \in N(R)$ then

$$C\exp(At)x_0 = 0$$

for all values of t. Therefore the set of unobservable states must contain $N(R)$.

We can also interpret observability in terms of system modes. A representation is not observable if and only if there are natural modes that cannot contribute to the system output; in fact the rank deficiency of R is the number of unobservable modes. This is obvious from the canonical representation, and the discussion above for mode controllability applies verbatim to mode observability. Example 7.8 illustrates these ideas.

Exercise 7.22 Show that observability is invariant under change of coordinates.

Given a transfer function

$$W(s) = \frac{b_1 s^{n-1} + \ldots + b_{n-1}s + b_n}{s^n + a_1 s^{n-1} + \ldots + a_{n-1}s + a_n} \tag{7.40}$$

we can obtain, by simple inspection, two useful state-space realizations of $W(s)$ with the least number of coefficients, as given below for n=3.

1. **Controllable canonical form:**

$$\begin{bmatrix} \dot{x}_1 \\ \dot{x}_2 \\ \dot{x}_3 \end{bmatrix} = \begin{bmatrix} 0 & 1 & 0 \\ 0 & 0 & 1 \\ -a_3 & -a_2 & -a_1 \end{bmatrix} \begin{bmatrix} x_1 \\ x_2 \\ x_3 \end{bmatrix} + \begin{bmatrix} 0 \\ 0 \\ 1 \end{bmatrix} u$$

$$y = (b_3, b_2, b_1)\boldsymbol{x} \tag{7.41}$$

Exercise 7.23
(a) Show the above realization is controllable.
(b) Using the result of exercise (7.3), compute its transfer function.

2. **Observable canonical form**

$$\begin{bmatrix} \dot{x}_1 \\ \dot{x}_2 \\ \dot{x}_3 \end{bmatrix} = \begin{bmatrix} 0 & 1 & -a_3 \\ 1 & 0 & -a_2 \\ 0 & 0 & -a_1 \end{bmatrix} \begin{bmatrix} x_1 \\ x_2 \\ x_3 \end{bmatrix} + \begin{bmatrix} b_3 \\ b_2 \\ b_1 \end{bmatrix} u$$

$$y = (0,0,1)\boldsymbol{x} \tag{7.42}$$

Exercise 7.24
(a) Show the above realization is observable.
(b) Show that the characteristic polynomials of a matrix A and of its transpose $\boldsymbol{A}^{\mathrm{T}}$ are the same.
(c) Using the results of exercise 3.2, compute the transfer function of this realization.

Exercise 7.25 Given the partial fraction expansion

$$W(s) = \sum_{i=1}^{n} \sum_{j=1}^{m_i} \frac{k_{ij}}{(s + p_i)^j}$$

can you obtain a straightforward state-space realization? Is it controllable? And observable? *Hint*: use the results of section 3.4, particularly Figs 3.10 and 3.11.

7.3.3 Kalman's Canonical Structure Theorem

This theorem rests on the following result:

> **Fact:** $N(R)$, the sub-space of unobservable states, and $R(Q)$, the sub-space of controllable states, are invariant under A.

> **Exercise 7.26** Prove the above fact. *Hint*: use the definition of invariance and the Cayley–Hamilton theorem.

Kalman's theorem states that it is possible to find a basis for the state-space $\mathbb{R}^n$ such that:

$$\mathbb{R}^n = \Sigma co \oplus \Sigma c\not{o} \oplus \Sigma \not{c}o \oplus \Sigma \not{c}\not{o} \tag{7.43}$$

where $\oplus$ denotes direct sum. This means that any state $\boldsymbol{x} \in \mathbb{R}^n$ has a *unique* decomposition

$$\boldsymbol{x} = \boldsymbol{x}_1 + \boldsymbol{x}_2 + \boldsymbol{x}_3 + \boldsymbol{x}_4, \text{ with } \boldsymbol{x}_1 \in \Sigma co,\ \boldsymbol{x}_2 \in \Sigma c\not{o},\ \boldsymbol{x}_3 \in \Sigma \not{c}o \text{ and } \boldsymbol{x}_4 \in \Sigma \not{c}\not{o}$$

$\Sigma c\not{o}$ is *the* sub-space of controllable and unobservable states; $\Sigma \not{c}o$ is *a* sub-space of observable and not controllable states and finally $\Sigma \not{c}\not{o}$ is *a* sub-space of uncontrollable and unobservable states. Furthermore

$$R(\mathbf{Q}) = \Sigma c\not{o} \oplus \Sigma co$$

and

$$N(\mathbf{R}) = \Sigma c\not{o} \oplus \Sigma \not{c}\not{o}$$

These results are now illustrated by means of an example.

Example 7.8 Consider the representation

$$\dot{\boldsymbol{x}} = \begin{bmatrix} 0 & 0 & 0 \\ 1 & 0 & -2 \\ 0 & 1 & -3 \end{bmatrix} \boldsymbol{x} + \begin{bmatrix} 2 \\ 4 \\ 2 \end{bmatrix} u$$

$$y = (-\ 1/2, 1, -\ 1)\boldsymbol{x} \tag{7.44}$$

The system matrix has eigenvalues 0, -1 and -2, with eigenvectors $\boldsymbol{v}_1 = (2,3,1)^{\mathrm{T}}$, $\boldsymbol{v}_2 = (0,2,1)^{\mathrm{T}}$ and $\boldsymbol{v}_3 = (0,1,1)^{\mathrm{T}}$ respectively. Let us use the eigenvectors as the new coordinate system, i.e. $\boldsymbol{x} = \boldsymbol{Tz}$ where

$$\boldsymbol{T} = \begin{bmatrix} 2 & 0 & 0 \\ 3 & 2 & 1 \\ 1 & 1 & 1 \end{bmatrix} \tag{7.45}$$

The new representation then becomes

$$\dot{\boldsymbol{z}} = \boldsymbol{T}^{-1}\boldsymbol{A}\ \boldsymbol{Tz} + \boldsymbol{T}^{-1}\boldsymbol{B}\ u$$

$$y = \boldsymbol{CTz} \tag{7.46}$$

with

$$\Lambda = T^{-1}AT = \begin{bmatrix} 0 & 0 & 0 \\ 0 & -1 & 0 \\ 0 & 0 & -2 \end{bmatrix} ; T^{-1}B = \begin{bmatrix} 1 \\ 0 \\ 1 \end{bmatrix} ; \mathbf{CT} = (1,1,0)$$

The diagrams of these two (algebraically) equivalent representations are shown in Figs 7.19 and 7.20. For representation (7.44) we have

$$Q = \begin{bmatrix} 2 & 0 & 0 \\ 4 & -2 & 4 \\ 2 & -2 & 4 \end{bmatrix}$$

and

$$R = \begin{bmatrix} -0.5 & 1 & -1 \\ 1 & -1 & 1 \\ -1 & 1 & -1 \end{bmatrix}$$

Because rank Q=rank R=2 we see that the representation is neither controllable nor observable. On the other hand we have:

$$N(R) = \text{span}\,\{(0,1,1)^T\} = \text{span}\{v_3\}$$

and

$$R(Q) = \text{span}\{(1,2,1)^T, (0,1,1)^T\} = \text{span}\{v_1, v_3\}$$

therefore

$$\Sigma c\emptyset = \text{span}\{(0,1,1)\}$$

The transfer function $W(s)$ is invariant under change of coordinates; therefore we conclude from Fig. 7.21 that $W(s)=1/s$. This is also confirmed by the formula:

$$W(s) = C\,(sI - A)^{-1}B = \frac{(s+1)(s+2)}{s(s+1)(s+2)} = \frac{1}{s}$$

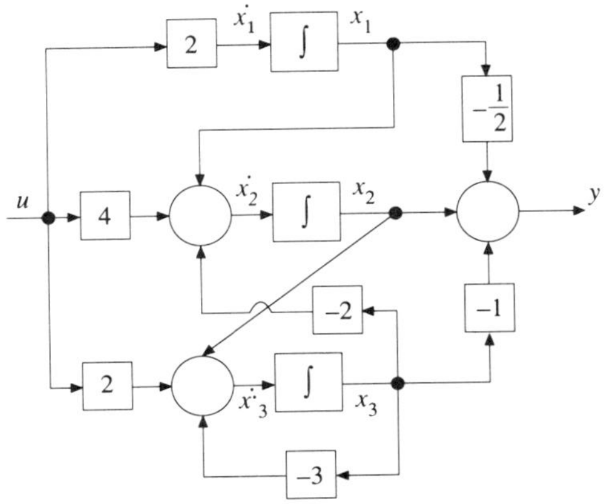

Figure 7.19 Simulation diagram of system (7.44).

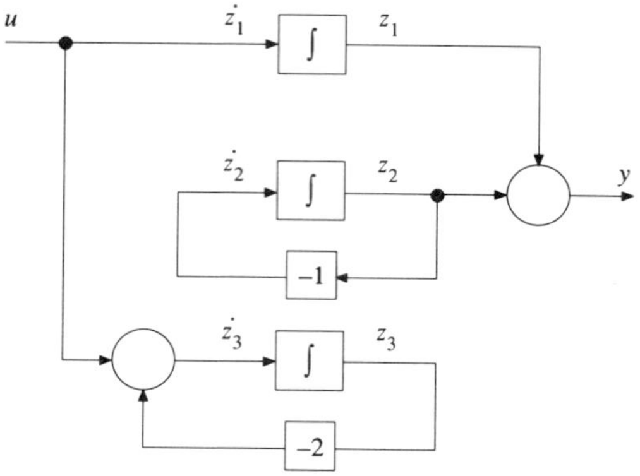

Figure 7.20 Simulation diagram of system (7.46).

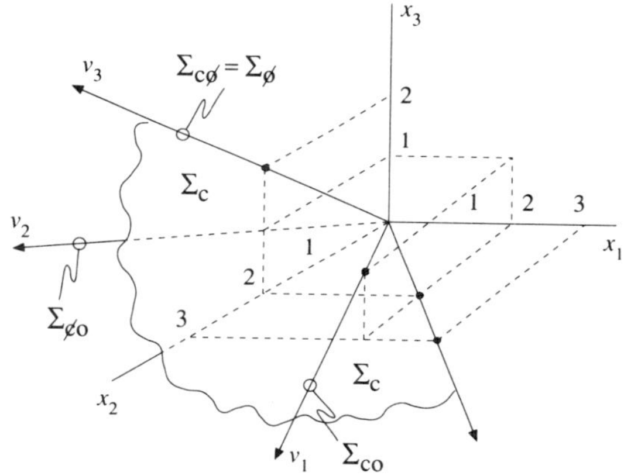

Figure 7.21 Kalman's canonical decomposition for representation (7.44).

which reveals the cancellation of two modes. Such cancellations always occur when the realization is not minimal, i.e. controllable and observable. Sub-spaces for the decomposition (7.43) can now be easily found, and are shown in Fig. 7.21. The only sub-space that is uniquely defined is $\Sigma c\not{o}=\text{span}\{\boldsymbol{v}_3\}$. A possible choice for the other sub-spaces is: $\Sigma co=\text{span}\{\boldsymbol{v}_1\}$, $\Sigma\not{c}o=\text{span}\{\boldsymbol{v}_2\}$ and $\Sigma\not{c}\not{o}=\{0\}$. Notice, for example, that any vector on the plane $\text{span}\{\boldsymbol{v}_1,\boldsymbol{v}_3\}$ containing the origin, with the exception of $\boldsymbol{v}_3$, is also *a* sub-space of controllable and observable states.

Exercise 7.27 Show that the set of observable states and the set of uncontrollable states are not sub-spaces. Identify these subsets in Fig. 7.21.

Exercise 7.28 With respect to the equation $\dot{\boldsymbol{x}}=\boldsymbol{A}\boldsymbol{x}+\boldsymbol{B}u$ show that any solution $\boldsymbol{x}(t)$, $t\geqslant 0$, such that $\boldsymbol{x}(0)$ belongs to the sub-space of controllable states, Σc, remains

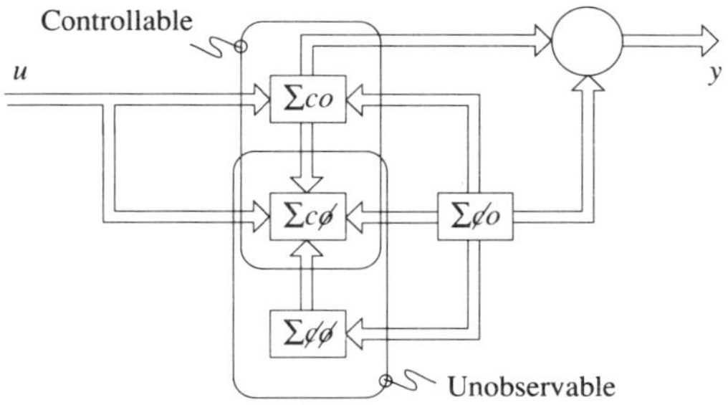

Figure 7.22 Canonical structure subsystems.

there ever after, no matter what the value of the input **u**. *Hint*: use the definition of controllable state and reason by contradiction.

Let us now take a closer look at the diagram in Fig. 7.22 which illustrates decomposition (7.43):

1. The zero state response is only affected by Σco. This means in our example that any trajectory starting on span$\{\boldsymbol{v}_3\}$ will evolve on the plane span$\{\boldsymbol{v}_1,\boldsymbol{v}_3\}$ and only the component along $\boldsymbol{v}_1$ will contribute to the output. Therefore a one-dimensional representation is enough to realize the transfer function of (7.38), which is confirmed by Fig. 7.20.
2. Because $\Sigma c\not{o}$ only has arrows pointing inwards we can see that, in the case of zero input, a trajectory starting on $\Sigma c\not{o}$ remains there ever after, and will not contribute to the output. In our example we have span$\{\boldsymbol{v}_3\}=\Sigma c\not{o}$; therefore the trajectory will stay in span$\{\boldsymbol{v}_3\}$ if $\boldsymbol{u}$ remains equal to zero.
3. $\Sigma \not{c}o$ only has arrows pointing outwards. This means we cannot freely change the course of trajectories starting on span$\{\boldsymbol{v}_2\}$. Any input $\boldsymbol{u}$ will give rise to a non-zero state component on the plane span$\{\boldsymbol{v}_1, \boldsymbol{v}_3\}$, but there is nothing we can do about the component along span$\{\boldsymbol{v}_2\}$. Furthermore, such state trajectory will never cross or touch the plane span$\{\boldsymbol{v}_1,\boldsymbol{v}_2\}$.

7.4 DYNAMICAL INTERPRETATION OF ZEROS

So far we have only addressed the role of system poles. As shown later in the book, the zeros play an important part in the regulator design problem, where the input is not zero. For the moment we can say that if z is a zero, then there exists an initial state, $\boldsymbol{x}(0)$, such that the output vanishes identically when the system input is $\exp(zt)$, $t \geqslant 0$. The details are given below.

Fact (Desoer, 1970): assume the system described by

$$\begin{cases} \dot{\boldsymbol{x}} = \boldsymbol{A}\boldsymbol{x} + \boldsymbol{B}u \\ y = \boldsymbol{C}\boldsymbol{x} + du \end{cases} \tag{7.47}$$

with transfer function $H(s)$. If $\boldsymbol{\nu}$ is not an eigenvalue of $\boldsymbol{A}$, then there exists an initial state $\boldsymbol{x}(0)$, such that the response to $u(t)=e^{\nu t}$, $t \geqslant 0$, is $y(t)=H(\boldsymbol{\nu})\ e^{\nu t}$. In particular, if ν is a zero of the transfer function, then this response is identically zero.

Proof: Let

$$\boldsymbol{x}(0) = (\nu \boldsymbol{I} - \boldsymbol{A})^{-1}\boldsymbol{B} \tag{7.48}$$

and

$$U(s) = \frac{1}{s - \nu}$$

Since

$$Y(s) = \boldsymbol{C}\ (s\boldsymbol{I} - \boldsymbol{A})^{-1}\boldsymbol{x}(0) + [\boldsymbol{C}(s\boldsymbol{I} - A)^{-1}\ \boldsymbol{B} + d]\ U(s) \tag{7.49}$$

we can write

$$Y(s) = \boldsymbol{C}\ (s\mathbf{I} - \boldsymbol{A})^{-1}(\nu\boldsymbol{I} - \boldsymbol{A})^{-1}\boldsymbol{B} + \frac{1}{s - \nu}\boldsymbol{C}\ (s\boldsymbol{I} - \boldsymbol{A})^{-1}\boldsymbol{B} + \frac{d}{s - \nu} \tag{7.50}$$

Using the resolvent identity

$$(s\boldsymbol{I} - \boldsymbol{A})^{-1} = (\nu I - \boldsymbol{A})^{-1} - (s - \nu)(s\boldsymbol{I} - \boldsymbol{A})^{-1}(\nu\boldsymbol{I} - \mathbf{A})^{-1} \tag{7.51}$$

we arrive at

$$Y(s) = (\boldsymbol{C}(\nu\boldsymbol{I} - \boldsymbol{A})^{-1}\boldsymbol{B} + d)\frac{1}{s - \nu}$$

Recalling that

$$H(s) = \boldsymbol{C}(s\boldsymbol{I} - \boldsymbol{A})^{-1}\boldsymbol{B} + d \tag{7.52}$$

the result follows.

Exercise 7.29 Bearing in mind the above fact, show that the zeros of the transfer function are the zeros of the polynomial

$$\det \left[\begin{array}{c|c} z\boldsymbol{I}_n - \boldsymbol{A} & -\boldsymbol{B} \\ \hline \boldsymbol{C} & d \end{array}\right] \tag{7.53}$$

[cf. the transfer function representation (3.17)].
Answer: Proof:

(A) z_0 is a zero of $H(s) \Rightarrow z_0$ is a zero of (7.53).

If we write

$$x(0) = (z_0\boldsymbol{I}_n - \mathrm{A})^{-1} \text{ and } y(t) = \exp(z_0 t)$$

then, from the above fact,

$$C\,x(0) + d = 0$$

and

$$[z_0 I_n - A]x(0) - B = \mathbf{0}$$

Both equations can be condensed in matrix form as

$$\begin{bmatrix} z_0 I_n - A & -B \\ C & d \end{bmatrix} \begin{bmatrix} x(0) \\ 1 \end{bmatrix} = \mathbf{0}$$

But this implies (7.53) must be zero, since the vector $(x^T(0),1) \neq 0$.

(B) Now we show the reverse implication ($\Leftarrow$).
If z_0 is a zero of (7.53), and z_0 is not an eigenvalue of A then there exists $x(0) \neq \mathbf{0}$ such that

$$\begin{bmatrix} z_0 I_n - A & -B \\ C & d \end{bmatrix} \begin{bmatrix} x(0) \\ 1 \end{bmatrix} = 0$$

From this we can write

$$((z_0 I_n - A)x(0) - B)\exp(z_0 t) = 0$$

$$(C\,x(0) + d)\exp(z_0 t) = 0$$

But this shows that

$$x(t) = x(0)\exp(z_0 t),\ y(t) = 0$$

is the solution of (7.47), when

$$u(t) = \exp(z_0 t)$$

and

$$x(0) = (z_0 I - A)^{-1} B$$

which ends the proof.

7.5 CONCLUSIONS

It has been shown that the dynamics of an nth order dynamical system is generated by the interconnection of two types of elementary sub-systems, consisting of either a pair of cross-coupled integrators or an integrator with a simple feedback loop. While the former always exhibits oscillatory behaviour, the latter is characterized by a monotone response. These sub-systems constitute the mode generators in state-space representations. The geometric equivalent of this analysis is the fact that the state-space trajectory is a composition of elementary trajectories – the *system modes* – evolving independently in sub-spaces whose direct sum is the entire state-space. These sub-spaces are defined by the real and imaginary parts of the (possibly generalized) system eigenvectors and the trajectories characterized by the corresponding eigenvalues.

When the system is regarded as a 'black box', we can describe it as an operator acting upon the system inputs by means of the transfer function or the convolution integral. With this type of input–output description we may not be able to describe all of the internal mechanisms of the system: only those states which are both controllable and observable affect the input–output relation. This is not surprising if we bear in mind that controllability and observability are related to the number and position of actuators and sensors in a real system.

Also of interest is the fact that the input e^{zt}, $t \geqslant 0$, where z is a zero of the transfer function, has the ability to counteract the system-free response for an appropriate initial condition, leading to an output vanishing identically on $t \geqslant 0$.

REFERENCES

Chen, C.T. (1984) *Linear System Theory and Design*, Holt, Rinehart and Winston.

Desoer, C.A. (1970) *Notes for a Second Course on Linear Systems*, Van Nostrand, Reinhold.

Kalman, R.E. (1963) 'Mathematical descriptions of linear dynamical systems', *J.S.I.A.M. Control, Ser. A*, **1**(2), pp. 152–92.

Raps, F. and G. Schmidt (1987) 'Modelling, simulation and experiments in active vibration control for flexible structures', IMACS/IFAC, Hiroshima, October.

Hautus, M. L. J. (1969) 'Controllability and observability conditions of linear autonomous systems', *Proc. Kon. Ned. Akad. Wetensch., Ser. A*, **72**, 445–8.

8

State-space and Algebraic Control Systems Design

8.1 STATE-SPACE DESIGN

In the root-locus and frequency response design methods we use output feedback and, most of the time, compensators with two design parameters. Consequently the number of closed-loop poles under direct design control was restricted to two. Compensators with more design parameters proved difficult to use because of the difficulty of relating the compensator parameters with the desired system properties.

However, if the plant is modelled in state-space it is sensible to regard the state vector as a possible input to the compensator since the state encompasses *all* the available information about the plant, up to that moment. As shown below, this approach turns out to be rewarding because it allows, under mild assumptions, *all* the plant eigenvalues to be shifted to any desired locations; furthermore the relation between the design parameter and the closed-loop eigenvalues is very simple, regardless of the dimension of the system.

8.1.1 Eigenvalue Assignment

8.1.1.1 Plant model in controllable canonical form

If the state-space representation of the plant is in controllable canonical form, namely

$$\dot{\mathbf{x}} = \begin{bmatrix} 0 & 1 & & & \\ & & 1 & & \\ & & & & \\ & & & & 1 \\ -a_n & & -a_{n-1} & \cdots & -a_1 \end{bmatrix} \boldsymbol{x} + \begin{bmatrix} 0 \\ \cdot \\ \cdot \\ \cdot \\ 0 \\ 1 \end{bmatrix} u \qquad (8.1)$$

$$y = \boldsymbol{C}\boldsymbol{x}$$

the characteristic polynomial is defined by the last row of A, namely

$$\det(sI - A) = a(s) = s^n + a_1 s^{n-1} + \ldots + a_{n-1} s + a_n \tag{8.2}$$

Let $\lambda_1, \lambda_2, \ldots, \lambda_n$ denote the eigenvalues of A. Assume

$$\alpha(s) = s^n + \alpha_1 s^{n-1} + \ldots + \alpha_{n-1} s + \alpha_n \tag{8.3}$$

is the new desired characteristic polynomial, with roots $\mu_1, \mu_2, \ldots, \mu_n$. Then, defining v as the new external input, and setting the plant input as

$$u = \mathrm{v} - Kx \tag{8.4}$$

where

$$K = (k_1, k_2, \ldots, k_n)$$

we obtain the configuration shown in Fig. 8.1, and the system equation becomes

$$\begin{cases} \dot{\mathbf{x}} = (\mathbf{A} - \mathbf{BK})\,\mathbf{x} + b\mathrm{v} \\ y = Cx \end{cases} \tag{8.5}$$

If we set

$$K = (\alpha_n - a_n, \alpha_{n-1} - a_{n-1}, \ldots, \alpha_1 - a_1) \tag{8.6}$$

then det $[s\mathbf{I}-(\mathbf{A}-\mathbf{BK})]=\alpha(s)$, since

$$\mathbf{BK} = \begin{bmatrix} \mathbf{0} \\ \hline k_1\, k_2 \ldots k_n \end{bmatrix} \tag{8.7}$$

In this fashion we have solved the eigenvalue assignment problem in a straightforward manner.

Exercise 8.1 Show that state feedback does not change the system zeros.
Answer: The numerator of the transfer function of system (8.5) is

$$\det \left[\begin{array}{c|c} s\mathbf{I}-\mathbf{A}+\mathbf{BK} & -\mathbf{B} \\ \hline \mathbf{C} & 0 \end{array}\right] = \det \left[\begin{array}{c|c} s\mathbf{I}-\mathbf{A} & -\mathbf{B} \\ \hline \mathbf{C} & 0 \end{array}\right]$$

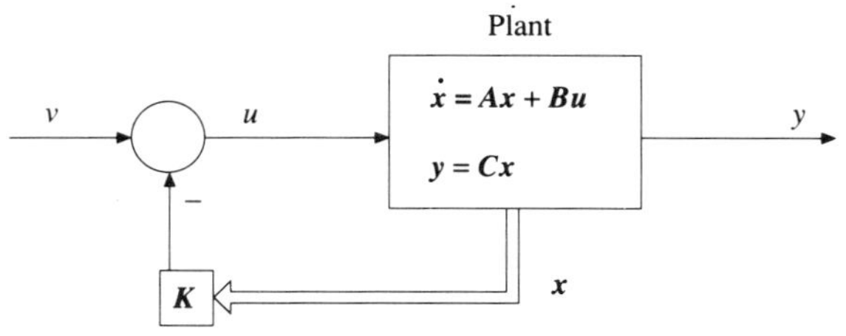

Figure 8.1 State-variable feedback.

(we have added to the first column block the product of the last column by **K**). But the right hand side is the numerator of the transfer function of the system before compensation. The result follows.

8.1.1.2 General case

Arbitrary eigenvalue assignment is possible if and only if the representation is controllable. In fact, in the case of non-controllability there are natural modes that cannot be excited by the input, as we have shown in section 7.3. On the other hand if the representation is controllable we can always find a basis of $\mathbb{R}^n$ with respect to which the representation has the form as in (8.1). The reader is invited to prove this result in exercise 8.2.

Exercise 8.2 Given a controllable representation

$$\begin{cases} \dot{x} = \mathbf{A}\mathbf{x} + \mathbf{B}u \\ y = \mathbf{C}\mathbf{x} \end{cases}$$

compute a matrix $\mathbf{T}_c$ such that

$$\begin{cases} \dot{\bar{\mathbf{x}}} = \bar{\mathbf{A}}\,\bar{\mathbf{x}} + \bar{\mathbf{B}}\,u \\ y = \bar{C}\,\bar{\mathbf{x}} \end{cases}$$

is in controllable canonical form, where $\mathbf{x} = \mathbf{T}_c\,\bar{\mathbf{x}}$.

Answer: Let

$$\mathbf{Q} = [\mathbf{B} \mid \mathbf{A}\mathbf{B} \mid \ldots \mid \mathbf{A}^{n-1}\mathbf{B}]$$

$$\bar{\mathbf{Q}} = [\bar{\mathbf{B}} \mid \bar{\mathbf{A}}\bar{\mathbf{B}} \mid \ldots \mid \bar{\mathbf{A}}^{n-1}\bar{\mathbf{B}}]$$

$$= [\mathbf{T}_c^{-1}\,\mathbf{B} \mid \mathbf{T}_c^{-1}\,\mathbf{A}\mathbf{T}_c\mathbf{T}_c^{-1}\mathbf{B} \mid \ldots \mid \mathbf{T}_c^{-1}\,\mathbf{A}^{n-1}\,\mathbf{B}]$$

$$= \mathbf{T}_c^{-1}\,[\mathbf{B} \mid \mathbf{A}\mathbf{B} \mid \ldots \mid \mathbf{A}^{n-1}\mathbf{B}] = \mathbf{T}_c^{-1}\mathbf{Q}$$

By assumption $\bar{\mathbf{Q}}$ is invertible. Therefore

$$\mathbf{T}_c = \mathbf{Q}(\bar{\mathbf{Q}})^{-1}$$

The columns of $\mathbf{T}_c$ constitute the basis of $(\bar{A}, \bar{B}, \bar{C})$.

Therefore the only question that is left is the computation of the feedback gain $\boldsymbol{K}$ in the case of a controllable, non-canonical realization.

8.1.1.3 Formulae for the feedback gain

Formula 1

Let the columns of $\mathbf{T}_c \in \mathbb{R}^{n \times n}$ represent the basis of $\mathbb{R}^n$ with respect to which the state representation is in controllable canonical form (cf. exercise 8.2). Then, if $\bar{\boldsymbol{K}}$ denotes the feedback gain vector for the canonical form, the feedback gain $\boldsymbol{K}$ for the (controllable) representation $(\boldsymbol{A}, \boldsymbol{B}, \boldsymbol{C})$ is

$$K = \bar{K}\,\mathbf{T}_c^{-1} \tag{8.8}$$

where

$$\mathbf{T}_c = Q(\bar{Q})^{-1} \tag{8.9}$$

$$Q = [\mathbf{b} \mid \mathbf{AB} \mid \mathbf{A}^2\mathbf{B} \mid \dots \mid \mathbf{A}^{n-1}\mathbf{B}] \tag{8.10}$$

$$\bar{\mathbf{Q}} = [\bar{\mathbf{B}} \mid \bar{\mathbf{A}}\,\bar{\mathbf{B}} \mid \bar{\mathbf{A}}^2\bar{\mathbf{B}} \mid \dots \mid \bar{\mathbf{A}}^{n-1}\bar{\mathbf{B}}] \tag{8.11}$$

$(\bar{\mathbf{A}},\bar{\mathbf{B}},\bar{\mathbf{C}})$ is an algebraically equivalent (controllable) canonical form of the controllable representation $(\mathbf{A},\mathbf{B},\mathbf{C})$.

Proof:

$$\dot{\bar{x}} = (\bar{\mathbf{A}} - \bar{\mathbf{B}}\,\bar{\mathbf{K}})\,\bar{\mathbf{x}} + \bar{\mathbf{B}}u$$

Since $x = T_c\,\bar{x}$ we can write

$$\mathbf{T}_c^{-1}\,\dot{x} = (\bar{A} - \bar{B}\,\bar{K})\,\mathbf{T}_c^{-1}\,x + \bar{B}\,u$$

or equivalently

$$\dot{x} = (\mathbf{T}_c\,\bar{A}\,\mathbf{T}_c^{-1} - (\mathbf{T}_c\,\bar{B})\,(\bar{K}\,\mathbf{T}_c^{-1}))\,x + \mathbf{T}_c\,\bar{B}u$$

Also

$$\mathbf{T}_c\,\bar{A}\,\mathbf{T}_c^{-1} = A$$

$$\mathbf{T}_c\,\bar{B} = B$$

Therefore $\bar{K}\,\mathbf{T}_c^{-1} = K$.

Formula 2: Ackermann's formula

With the same notation as above

$$K = (0, \dots, 0,1)\,\mathbf{Q}^{-1}\,\alpha(\mathbf{A}) \tag{8.12}$$

where $\alpha(.)$ is the desired characteristic polynomial. (For other formulae, the reader is referred to Kailath, 1980.)

Proof: With our notation we now have

$$\bar{K} = \alpha_n - a_n,\ \alpha_{n-1} - a_{n-1}, \dots, \alpha_1 - a_1)$$

$\bar{A}$ as in (8.1). From (8.2) we can write

$$a(\bar{A}) = \bar{A}^n + a_1\,A^{n-1} + \dots + a_{n-1}\,\bar{A} + a_n\,\mathbf{I} = 0$$

because a matrix satisfies its own characteristic polynomial. Also from (8.3),

$$\alpha(\bar{A}) = \bar{A}^n + \alpha_1 A^{n-1} + \dots + \alpha_{n-1}\,\bar{A} + \alpha_n\,I$$

$$= (\alpha_1 - a_1)\,\bar{A}^{n-1} + (\alpha_2 - a_2)\,\bar{A}^{n-1} + \dots + (\alpha_{n-1} - a_{n-1})\,\bar{A} + (\alpha_n - a_n)\,I$$

Defining

$$e_1^{\mathrm{T}} = (1,0,\dots,0)$$

we conclude that

$$e_1^T\bar{A} = (0,1,0. . .,0) = e_2^T$$
$$(e_1^T\bar{A})\bar{A} = (0,0,1,0,. . .,0) = e_3^T$$
$$\vdots$$
$$e_1^T(\bar{A})^{n-1} = (0,0,. . .,0,1) = e_n^T$$

Therefore

$$e_1^T\ \alpha(\bar{A}) = \bar{K}$$

But

$$\begin{aligned} K = \bar{K}\ T_c^{-1} &= e_1^T\ \alpha(\bar{A})\ T_c^{-1} \\ &= e_1^T\ \alpha(T_c^{-1}\ A\ T_c)\ T_c^{-1} \\ &= e_1^T\ T_c^{-1}\ \alpha(A) = e_1^T\ \bar{Q}\ Q^{-1}\alpha(A) \end{aligned}$$

Since the first row of $\bar{Q}$ is

$$e_n^T = (0,0,. . .,0,1)$$

we can write

$$K = (0,0,. . .,0,1)\ Q^{-1}\alpha(A)$$

which is Ackermann's formula.

8.1.2 State Estimation

The procedure discussed so far assumes that all the components of the state vector can be measured. Unfortunately this is not the case in the majority of situations. What we do in practice is to produce an estimate $\hat{\mathbf{x}}$, of the true state $\mathbf{x}$, by some mechanism known as observer or estimator, and to replace the true state by its estimate in the control law $u=-Kx$. As shown in the following, this is a feasible procedure provided the system representation is *observable*.

If the representation is observable we can compute the state from the derivates of the system input and output (cf. exercise 8.3). However, for systems of order greater than two, this is not a recommended practical procedure given its high sensitivity to noise.

Exercise 8.3 Given the observable state-space representation

$$\begin{cases} \dot{x} = Ax + Bu \\ y = Cx \end{cases}$$

$x \in \mathbb{R}^n$, show that

$$x(t) = R^{-1}[Y(t) - T\ U\ (t)]$$

where

$$Y(t) \equiv \left(y, \frac{dy}{dt}, \frac{d^2y}{dt^2}, \ldots, \frac{d^{n-1}}{dt^{n-1}} \right)^T$$

$$U(t) \equiv \left(u, \frac{du}{dt}, \ldots, \frac{d^{n-1}y}{dt^{n-1}} \right)^T$$

$\boldsymbol{T}$ a Toeplitz matrix with first column $(0, \boldsymbol{CB}, \boldsymbol{CAB}, \ldots, \boldsymbol{CA}^{n-2}\boldsymbol{B})^{\mathrm{T}}$.
Answer: The result is obvious since

$$y = \boldsymbol{Cx}$$

$$\dot{y} = \boldsymbol{C\dot{x}} = \boldsymbol{CAx} + \boldsymbol{CB}u$$

$$\ddot{y} = \boldsymbol{CA\dot{x}} + \boldsymbol{CB}\dot{u} = \boldsymbol{CA}^2\boldsymbol{x} + \boldsymbol{CAB}u + \boldsymbol{CB}\dot{u}$$

$$\vdots$$

Therefore

$$\boldsymbol{Y} = \begin{bmatrix} \boldsymbol{C} \\ \hline \boldsymbol{CA} \\ \hline \boldsymbol{CA}^2 \\ \hline \vdots \end{bmatrix} x + \left[\begin{array}{c:c:c:c:c} 0 & 0 & & 0 & 0 \\ \boldsymbol{CB} & 0 & & \vdots & \vdots \\ & & \cdots & & \\ \boldsymbol{CAB} & \boldsymbol{CB} & & 0 & \\ \vdots & & & \boldsymbol{CB} & 0 \end{array}\right] \boldsymbol{U}$$

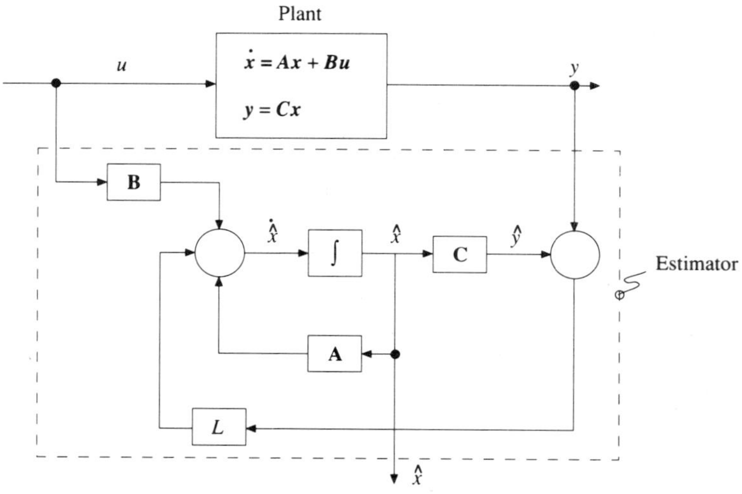

Figure 8.2 A state estimator.

An alternative is shown in Fig. 8.2. Note the inputs to the estimator are measurable variables, namely the input and output of the plant.

Defining the estimation error $\tilde{\mathbf{x}}$ as

$$\tilde{\mathbf{x}} = \mathbf{x} - \hat{\mathbf{x}} \tag{8.13}$$

we can write, from the plant and estimator equations,

$$\dot{\tilde{\mathbf{x}}} = (\mathbf{A} - \mathbf{LC})\tilde{\mathbf{x}} \tag{8.14}$$

Therefore, if the eigenvalues of $(\mathbf{A}-\mathbf{LC})$ lie in the interior of the left half-plane, $\hat{\mathbf{x}}$ will converge to the true state $\mathbf{x}$, no matter what the value of the input v.

> *Fact 8.1*: The eigenvalues of (8.14) can be arbitrarily assigned, by choice of $\mathbf{L}$, if $(\mathbf{A},\mathbf{B},\mathbf{C})$ is an observable representation.

The proof is obvious if $(\mathbf{A},\mathbf{B},\mathbf{C})$ is in observable canonical form. If not, the arguments we have used about $(\mathbf{A}-\mathbf{BK})$ (controller design) can be applied here verbatim, the only change being the substitution of 'controllability' by 'observability'.

Exercise 8.4 Given an observable state-space representation $(\mathbf{A},\mathbf{B},\mathbf{C})$ find a basis of $\mathbb{R}^n$ with respect to which this representation is algebraically equivalent to the observable canonical form.

Answer: let

$$\begin{cases} \dot{\bar{\mathbf{x}}} = \bar{\mathbf{A}}\,\bar{\mathbf{x}} + \bar{\boldsymbol{B}}u \\ y = \bar{\mathbf{C}}\,\bar{\mathbf{x}} \end{cases}$$

denote the equivalent state-space representation in observable canonical form; define the observability matrices

$$\mathbf{R} = \begin{bmatrix} \mathbf{C} \\ --- \\ \mathbf{CA} \\ --- \\ \cdot \\ \cdot \\ \cdot \\ --- \\ \mathbf{CA}^{n-1} \end{bmatrix}$$

$$\bar{\mathbf{R}} = \begin{bmatrix} \bar{\mathbf{C}} \\ --- \\ \bar{\mathbf{C}}\,\bar{\mathbf{A}} \\ --- \\ \cdot \\ \cdot \\ \cdot \\ --- \\ \bar{\mathbf{C}}\,\bar{\mathbf{A}}^{n-1} \end{bmatrix}$$

which are invertible matrices, by assumption.

Let $\mathbf{x}=\mathbf{T}_0\,\bar{\mathbf{x}}$. Then

$$\bar{\mathbf{R}} = \begin{bmatrix} \mathbf{C}\mathbf{T}_0 \\ --- \\ \mathbf{C}\mathbf{T}_0\,\mathbf{T}_0^{-1}\,\mathbf{A}\mathbf{T}_0 \\ --- \\ \cdot \\ \cdot \\ \cdot \\ --- \\ \mathbf{C}\,\mathbf{A}^{n-1} \end{bmatrix} = \begin{bmatrix} \mathbf{C} \\ --- \\ \mathbf{C}\,\mathbf{A} \\ --- \\ \cdot \\ \cdot \\ \cdot \\ --- \\ \mathbf{C}\,\mathbf{A}^{n-1} \end{bmatrix} \mathbf{T}_0 = \mathbf{R}\,\mathbf{T}_0$$

Consequently $\mathbf{T}_0=\mathbf{R}^{-1}\,\bar{\mathbf{R}}$. The columns of $\mathbf{T}_0$ constitute the desired basis.

Again the estimator gains, i.e. the components of vector $\mathbf{L}$ can be computed by simple inspection for the observer canonical form. The formulae for the estimator gains in the general case, i.e. non-canonical observable representation, can be obtained as in the controller design. Let $\beta(s)$ denote the desired observer polynomial, i.e.

$$\beta(s) = \det\,(s\mathbf{I} - \mathbf{A} + \mathbf{L}\mathbf{C}) \tag{8.15}$$

8.1.2.1 Formulae for the observer gain L

Formula 1

Assume $(\bar{\mathbf{A}}, \bar{\mathbf{B}}, \bar{\mathbf{C}})$ is an observable canonical form and $(\mathbf{A},\mathbf{B},\mathbf{C})$ is an algebraically equivalent (observable) representation, i.e.

$$\mathbf{x} = \mathbf{T}_0\,\bar{\mathbf{x}} \tag{8.16}$$

where

$$\mathbf{T}_0 \equiv \mathbf{R}^{-1}\,\bar{\mathbf{R}} \tag{8.17}$$

and

$$\mathbf{R} = \begin{bmatrix} \mathbf{C} \\ --- \\ \mathbf{C}\mathbf{A} \\ --- \\ \cdot \\ \cdot \\ \cdot \\ --- \\ \mathbf{C}\mathbf{A}^{n-1} \end{bmatrix} \tag{8.18}$$

$$\bar{\mathbf{R}} = \begin{bmatrix} \bar{\mathbf{C}} \\ --- \\ \bar{\mathbf{C}}\,\bar{\mathbf{A}} \\ --- \\ \cdot \\ \cdot \\ \cdot \\ --- \\ \bar{\mathbf{C}}\,\bar{\mathbf{A}}^{n-1} \end{bmatrix} \tag{8.19}$$

If $\bar{\mathbf{L}}$ denotes the estimator gain vector for the observable canonical form, then the estimator gain $\mathbf{L}$ for the algebraically equivalent (observable) representation $(\mathbf{A},\mathbf{B},\mathbf{C})$ is

$$\mathbf{L} = \mathbf{T}_0\,\bar{\mathbf{L}} \tag{8.20}$$

Exercise 8.5 Prove formula (8.20).

Answer: with the plant and estimator in canonical form the equation error is

$$\dot{\tilde{x}} = (\bar{A} - \bar{L}\,\bar{C})\,\tilde{x}\ (*)$$

where

$$\tilde{x} \equiv \bar{\mathbf{x}} - \hat{\bar{\mathbf{x}}}$$

Since

$$\mathbf{x} = \mathbf{T}_0\,\bar{\mathbf{x}}$$

we have

$$A = \mathbf{T}_0\,\bar{\mathbf{A}}\,\mathbf{T}_0^{-1}$$

$$\bar{\mathbf{C}} = \mathbf{C}\mathbf{T}_0$$

$$\tilde{\mathbf{x}} = \mathbf{T}_0\,\tilde{x}$$

Equation (*) then becomes

$$\dot{\tilde{\mathbf{x}}} = \mathbf{T}_0\,(\bar{\mathbf{A}} - \bar{\mathbf{L}}\,\bar{\mathbf{C}})\,\mathbf{T}_0^{-1}\,\tilde{\mathbf{x}}$$

or

$$\dot{\tilde{\mathbf{x}}} = (\mathbf{A} - \mathbf{T}_0\,\bar{L}\,C)\,\tilde{\mathbf{x}} \Rightarrow \mathbf{L} = \mathbf{T}_0\,\bar{\mathbf{L}}$$

Formula 2: Ackermann's formula

In the controller problem we know how to compute $\boldsymbol{K}$ such that

$$\det\,(s\mathbf{I} - \mathbf{A} + \mathbf{B}\mathbf{K}) = \alpha(s)$$

If we rewrite the observer polynomial as

$$\det\,(s\mathbf{I} - \mathbf{A}^{\mathrm{T}} + \mathbf{C}^{\mathrm{T}}\,\mathbf{L}^{\mathrm{T}}) = \beta(s)$$

we can use the solution of the controller problem. All we have to do is to replace **A**, **B** and **K**, by A^T, C^T and L^T, respectively. These, in turn, imply that

$$\mathbf{Q} = [\mathbf{B} \;\vdots\; \mathbf{AB} \;\vdots\; \ldots \;\vdots\; \mathbf{A}^{n-1}\,\mathbf{B}]$$

is replaced by

$$[\mathbf{C}^T \;\vdots\; \mathbf{A}^T\,\mathbf{C}^T \;\vdots\; \ldots \;\vdots\; (\mathbf{A}^{n-1})^T\,\mathbf{C}^T = \mathbf{R}^T$$

Therefore, Ackermann's formula in estimator form becomes

$$L = \beta(A)\, R^{-1} \begin{bmatrix} 0 \\ \cdot \\ \cdot \\ \cdot \\ 0 \\ 1 \end{bmatrix} \tag{8.21}$$

8.1.3 Combined Eigenvalue Assignment and State Estimation

If we feed back the state estimate $\hat{x}$ as shown in Fig. 8.3, we get a $2n$-dimensional system, with state vector

$$\begin{bmatrix} x \\ \cdots \\ \hat{x} \end{bmatrix} \tag{8.22}$$

and state-space representation

$$\begin{bmatrix} \dot{x} \\ \cdots \\ \dot{\hat{x}} \end{bmatrix} = \left[\begin{array}{c|c} A & -BK \\ \hline LC & A-BK-LC \end{array}\right] \begin{bmatrix} x \\ \cdots \\ \hat{x} \end{bmatrix} + \begin{bmatrix} B \\ \cdots \\ B \end{bmatrix} \mathrm{v}$$

$$y = (C : \mathbf{0}) \begin{bmatrix} x \\ \cdots \\ \hat{x} \end{bmatrix} \tag{8.23}$$

The characteristic polynomial is

$$\det(s\mathbf{I} - A + BK)\,\det(s\mathbf{I} - A + LC) \tag{8.24}$$

since

$$\det \left[\begin{array}{c|c} sI-A & BK \\ \hline -LC & sI-A+BK+LC \end{array}\right] = \det \left[\begin{array}{c|c} sI-A & BK \\ \hline -sI+A-LC & sI-A+LC \end{array}\right]$$

$$= \det \left[\begin{array}{c|c} sI-A+BK & BK \\ \hline \mathbf{0} & sI-A+LC \end{array}\right] = \det[sI - A + BK]\,\det[sI - A + LC]$$

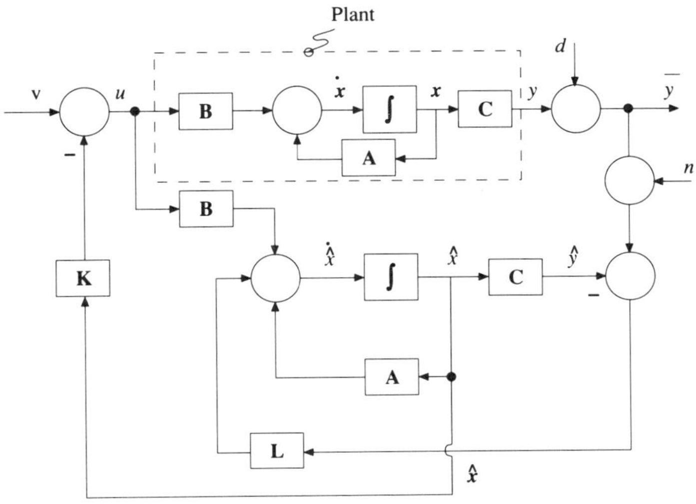

Figure 8.3 Eigenvalue assignment combined with state estimation. d and n represent disturbances.

To compute the transfer function $Y(s)/V(s)$ recall formula (3.18) in Chapter 3. The numerator polynomial is then

$$\det\left[\begin{array}{c|c|c} sI_n-A & BK & -B \\ \hline -LC & sI_n-A+BK+LC & -B \\ \hline C & 0 & 0 \end{array}\right]$$

$$= \det\left[\begin{array}{c|c|c} sI_n-A & BK & -B \\ \hline C & 0 & 0 \\ \hline -LC & sI_n-A+BK+LC & -B \end{array}\right]$$

$$= \det\left[\begin{array}{c|c|c} s\mathbf{I}_n-A & 0 & -B \\ \hline C & 0 & 0 \\ \hline -LC & sI_n-A+LC & -B \end{array}\right]$$

$$= \det\left[\begin{array}{c|c|c} s\mathbf{I}_n-A & -B & 0 \\ \hline C & 0 & 0 \\ \hline -LC & -B & sI_n-A+LC \end{array}\right]$$

$$= B_p(s)\det(s\boldsymbol{I}_n - \boldsymbol{A} + \boldsymbol{LC})$$

$B_p(s)$ is the numerator of the plant transfer function

$$\frac{Y(s)}{U(s)} = \frac{B_p(s)}{A_p(s)} \tag{8.25}$$

Therefore, the overall transfer function in Fig. 8.3 is

$$\frac{Y(s)}{V(s)} = \frac{B_p(s)\det(s\mathbf{I}_n - \mathbf{A} + \mathbf{LC})}{\det(s\mathbf{I}_n - \mathbf{A} + \mathbf{BK})\det(s\mathbf{I}_n - \mathbf{A} + \mathbf{LC})}$$

$$= \frac{B_p(s)}{\det(s\mathbf{I}_n - \mathbf{A} + \mathbf{BK})} \tag{8.26}$$

8.1.3.1 Conclusions

1. The use of the state estimate $\hat{\mathbf{x}}$ instead of $\mathbf{x}$, in the feedback law $u=\mathrm{v}-\mathbf{Kx}$, has not changed the transfer function of the compensated system. The cancellation of the observer polynomial reveals that the state vector of the estimator, $\hat{\mathbf{x}}$, is *not* controllable from v. In fact if the system is at rest at time $t=0$, i.e. $\mathbf{x}(0)=\hat{\mathbf{x}}(0)=\mathbf{0}$, then $\hat{\mathbf{x}}(t)=\mathbf{x}(t)$, $\forall\ t\geq 0$, whatever $\mathrm{v}(t)$.
2. As expected, the plant zeros are also zeros of the compensated system.

In general, $\mathbf{x}(0)\neq\hat{\mathbf{x}}(0)$; therefore the observer modes will also be present in $y(t)$. The observer dynamics can also be excited by disturbance and noise inputs, d and n, shown in Fig. 8.3.

In fact, with the exception of $Y(s)/V(s)$, the observer poles are not cancelled in the transfer functions $U(s)/D(s)$, $U(s)/N(s)$, $Y(s)/D(s)$ and $Y(s)/N(s)$. Therefore the observer dynamics must be faster than the prescribed poles for the compensated system. In general we set the observer dynamics three to four times faster than the closed loop system.

One would think at this stage that the faster the observer the better, because that would bring us closer to the ideal situation of Fig. 8.1. But here lies the crux of the problem: the faster the observer poles the wider the estimator bandwidth. Consequently the bandwidth of transfer functions U/N and U/D increases, allowing high frequency signals to pass on the actuator input. These signals, above the working range may not be noticeable at the plant output, but may have undesirable consequences on the actuator.

Let us assume, for example, that the plant is an inertial load driven by an electric motor and that a lot of noise is present in n; although the load inertia will suppress such variations, the motor windings will dissipate more heat given the presence of larger currents.

The next example will help to clarify these issues.

Example 8.1 Consider a plant with a state-space representation

$$\begin{cases} \dot{\boldsymbol{x}} = \begin{bmatrix} 0 & 1 \\ -1 & -2 \end{bmatrix} \boldsymbol{x} + \begin{bmatrix} 0 \\ 1 \end{bmatrix} u \\ y = (1,0)\,\boldsymbol{x} \end{cases} \tag{8.27}$$

with eigenvalues at (-1) and transfer function

$$\frac{Y(s)}{U(s)} = \frac{1}{(s+1)^2}$$

We wish to shift the eigenvalues to (-2), by state feedback (criteria for the choice of controller poles will be dealt with in section 8.2). Since

$$(s+1)^2 = s^2 + 2s + 1$$

$$(s+2)^2 = s^2 + 4s + 4$$

and the representation is in control canonical form, by simple inspection we can compute

$$\boldsymbol{K} = (k_1, k_2) = (3,2)$$

From (8.26) we have

$$\frac{Y(s)}{V(s)} = \frac{1}{(s+2)^2}$$

The eigenvalues of the estimator are given by

$$\det\,[s\boldsymbol{I} - \boldsymbol{A} + \boldsymbol{LC}] = \det \begin{bmatrix} s+l_1 & -1 \\ 1+l_2 & s+2 \end{bmatrix} = s^2 + s(l_1+2) + (2l_1 + l_2 + 1) \tag{8.28}$$

where $\boldsymbol{L}=(l_1,l_2)^{\mathrm{T}}$.

If we specify an observer with a double pole at p_o, we can write from (8.28)

$$\boldsymbol{L} = \begin{bmatrix} -2(p_o+1) \\ \text{- - - - - - -} \\ p_o^2+4p_o+3 \end{bmatrix} = \begin{bmatrix} l_1 \\ \\ l_2 \end{bmatrix} \tag{8.29}$$

In order to analyze the effects of the noise at n, as a function of p_o, we compute the transfer functions $U(s)/N(s)$ and $Y(s)/N(s)$. To simplify the computations we redraw Fig. 8.3 in the equivalent form shown Fig. 8.4. Then we can write

$$\begin{bmatrix} \dot{\boldsymbol{x}} \\ \text{- - -} \\ \dot{\hat{\boldsymbol{x}}} \end{bmatrix} = \left[\begin{array}{c|c} \boldsymbol{A} & -\boldsymbol{BK} \\ \hline \boldsymbol{LC} & \boldsymbol{A}-\boldsymbol{BK}-\boldsymbol{LC} \end{array}\right] \begin{bmatrix} \boldsymbol{x} \\ \text{- - -} \\ \hat{\boldsymbol{x}} \end{bmatrix} + \begin{bmatrix} \boldsymbol{0} \\ \text{- - -} \\ \boldsymbol{L} \end{bmatrix} n$$

$$u = (\boldsymbol{0},\boldsymbol{K}) \begin{bmatrix} \boldsymbol{x} \\ \text{- - -} \\ \hat{\boldsymbol{x}} \end{bmatrix} \tag{8.30}$$

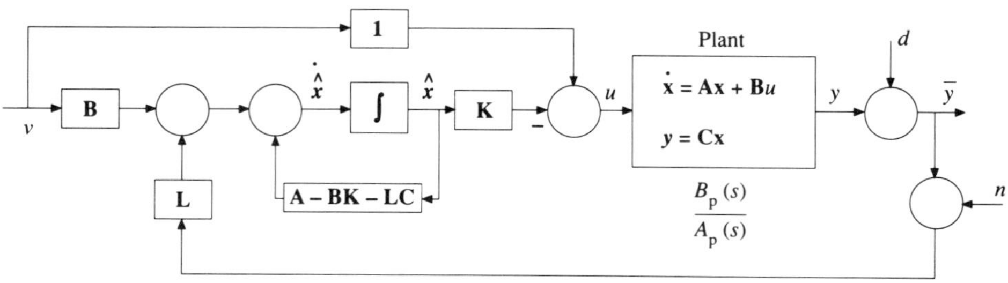

Figure 8.4 Equivalent diagram of Fig. 8.3.

The numerator of $U(s)/N(s)$ is

$$\det \left[\begin{array}{c|c|c} s\mathbf{I}_n - A & BK & \mathbf{0} \\ \hline -LC & s\mathbf{I}_n - A + BK + LC & -L \\ \hline \mathbf{0} & K & 0 \end{array}\right]$$

$$= \det \left[\begin{array}{c|c|c} s\mathbf{I}_n - A & BK & \mathbf{0} \\ \hline -LC & sI_n - A + BK + LC & -L \\ \hline \mathbf{0} & K & 0 \end{array}\right]$$

$$= \left[\begin{array}{c|c|c} sI_n - A & \mathbf{0} & \mathbf{0} \\ \hline -LC & sI_n - A + BK + LC & -L \\ \hline \mathbf{0} & K & 0 \end{array}\right]$$

$$= \det\,[sI_n - A]\,\det \left[\begin{array}{c|c} sI_n - A + BK + LC & -L \\ \hline K & 0 \end{array}\right] \tag{8.31}$$

Substituting **A**, **B**, **C**, **K** and **L** we have

$$\det \left[\begin{array}{c|c} s\mathbf{I}_n - A + BK + LC & -L \\ \hline K & 0 \end{array}\right] = -\, l_1k_2 + 2k_1l_1 + k_1l_2 + s\,(k_1l_1 + k_2l_2) \tag{8.32}$$

Therefore

$$\frac{U(s)}{N(s)} = -\frac{[l_1k_2 + 2k_1l_1 + k_1l_2 + s(k_1l_1 + k_2l_2)](s+1)^2}{(s - p_0)^2(s+2)^2} \tag{8.33}$$

$$\frac{Y(s)}{N(s)} = \frac{U(s)}{N(s)} \times \frac{B_p(s)}{A_p(s)} = -\frac{[l_1k_2 + 2k_1l_1 + k_1l_2 + s(k_1l_1 + k_2l_2)]}{(s - p_0)^2(s+2)^2} \tag{8.34}$$

Table 8.1 shows $U(s)/N(s)$, assuming the feedback gain $\boldsymbol{K}=(3,2)$ and observer polynomial $(s-p_0)^2$. As $|p_0|\to\infty$ the (simple) numerator zero converges to -1.5 and the steady-state gain converges to -0.75.

Figure 8.5 shows the magnitude Bode plots for the transfer functions $1/(s+2)^2=Y/V$ and U/N when $p_0=-4$ and $p_0=-8$. It is apparent from these plots that, as the speed of response of the estimator increases, the gain of the transfer function U/N also increases.

In the limit ($p_0\to\infty$) we have the curve depicted in Fig. 8.6: the compensator greatly amplifies the noise, in a wide frequency band well above the desired closed-loop system bandwidth, which is defined by curve (a) in Fig. 8.5.

Therefore, the choice of observer poles is dictated by the shape and magnitude of the spectrum of the sensor noise.

For the sake of completeness we also depict the magnitude Bode plots for Y/N in Fig. 8.7. In comparison with the $|U(s)/N(s)|$ curves they have a negligible effect, except in the low frequency range, where they exhibit similar gains.

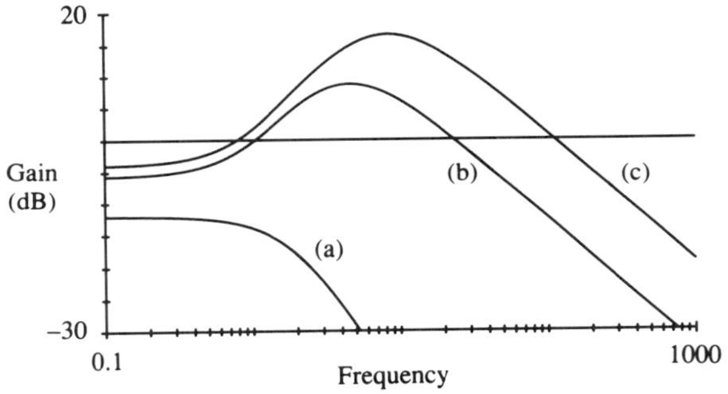

Figure 8.5 Magnitude Bode plots for: (a) $1/(s+2)^2$; (b) $24(s+1.375)(s+1)^2/((s+4)^2(s+2)^2)$; (c) $112(s+1.4375)(s+1)^2/((s+8)^2(s+2)^2)$.

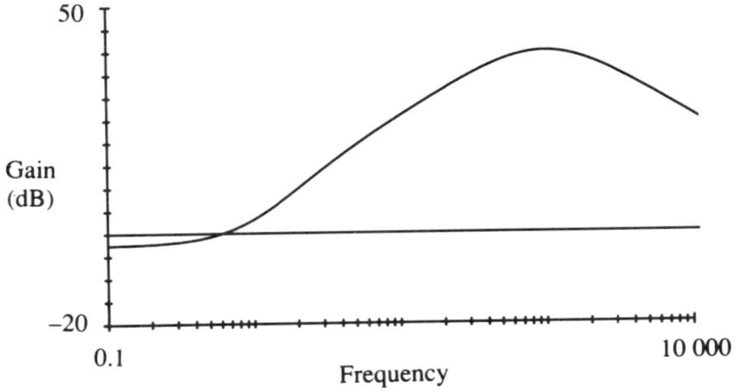

Figure 8.6 Magnitude Bode plot for $19\,800(s+1.495)(s+1)^2/((s+100)^2(s+2)^2)$.

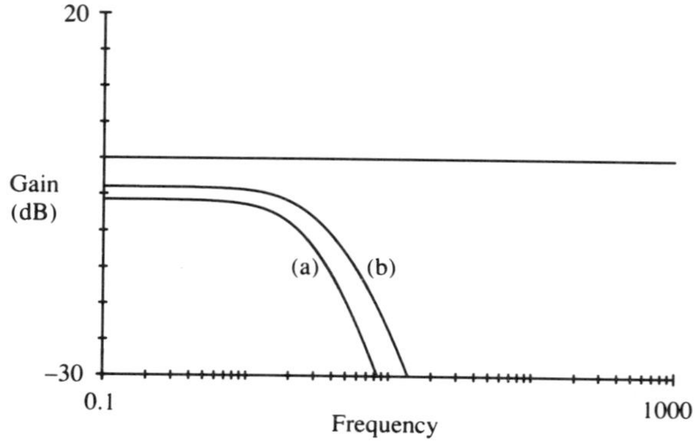

Figure 8.7 Bode plot for: (a) $24(s+1.375)/((s+4)^2(s+2)^2)$; (b) $112(s+1.4375)/((s+8)^2(s+2)^2)$.

However, the noise effects at y are still very significant when compared with the response to reference inputs, because the compensated system exhibits a lower steady-state gain, namely 0.25, from the reference v to the output y.

Table 8.1 Effects of the noise as a function of observer poles in example 8.1

p_o	L	Transfer function $U(s)/N(s)$	*Steady-state gain*
−4	$\begin{pmatrix}6\\3\end{pmatrix}$	$\dfrac{-24(s+1.375)(s+1)^2}{(s+4)^2(s+2)^2}$	−0.5156
−8	$\begin{pmatrix}14\\35\end{pmatrix}$	$\dfrac{-112(s+1.4375)(s+1)^2}{(s+8)^2(s+2)^2}$	−0.6289
−100	$\begin{pmatrix}198\\9603\end{pmatrix}$	$\dfrac{-19\,800(s+1.495)(s+1)^2}{(s+100)^2(s+2)^2}$	−0.74

Table 8.1 can also be interpreted in the light of the discussion about the effect of zeros on the step response (cf. section 3.6): the zeros of U/N remain practically unchanged as $p_0 \to -\infty$. Therefore the 'speed' of p_0, as seen from the zeros, increases and the system behaves approximately as a fast second-order system with a zero near the origin.

8.1.4 The Tracking Problem

So far we have addressed the *regulator problem*, i.e. we have only prescribed the closed-loop poles; this is sufficient when the reference input is assumed to be identically zero. However, in the *servo* or *tracking problem* we wish to specify the response to certain types of reference inputs. Since such responses are also dictated by the zeros of the closed-loop transfer function, and given that our previous design method allows no control upon such zeros we must attempt some modifications in order to achieve this goal.

Note that the configuration of Fig. 8.4 does not even allow control of the closed-loop steady-state gain $Y(0)/V(0))$ as we found in example 8.1.

Since the system zeros cannot be changed by feedback, the alternative is to introduce compensators in the forward paths. Consequently we replace, in the configuration of Fig. 8.4, the forward blocks '$\boldsymbol{B}$' and '1' with adjustable gain blocks $\boldsymbol{M}$, $\boldsymbol{M} \in \mathbb{R}^n$, and N, $N \in \mathbb{R}$, as shown in Fig. 8.8. The state-space representation then becomes

$$\begin{bmatrix} \dot{\boldsymbol{x}} \\ \dot{\hat{\boldsymbol{x}}} \end{bmatrix} = \left[\begin{array}{c|c} \boldsymbol{A} & -\boldsymbol{BK} \\ \hline \boldsymbol{LC} & \boldsymbol{A}-\boldsymbol{BK}-\boldsymbol{LC} \end{array}\right] \begin{bmatrix} \boldsymbol{x} \\ \hat{\boldsymbol{x}} \end{bmatrix} + \begin{bmatrix} N\boldsymbol{B} \\ \boldsymbol{M} \end{bmatrix} \mathrm{v}$$

$$y = (\boldsymbol{C}\ \mathbf{O}) \begin{bmatrix} \boldsymbol{x} \\ \hat{\boldsymbol{x}} \end{bmatrix} \tag{8.35}$$

The system matrix remains unchanged. Therefore the denominator of the transfer function $Y(s)/V(s)$ is the same as in (8.23), namely

$$\det(s\mathbf{I} - \boldsymbol{A} + \boldsymbol{BK}) \det(s\mathbf{I} - \boldsymbol{A} + \boldsymbol{LC}) = \alpha_c(s)\, \alpha_o(s)$$

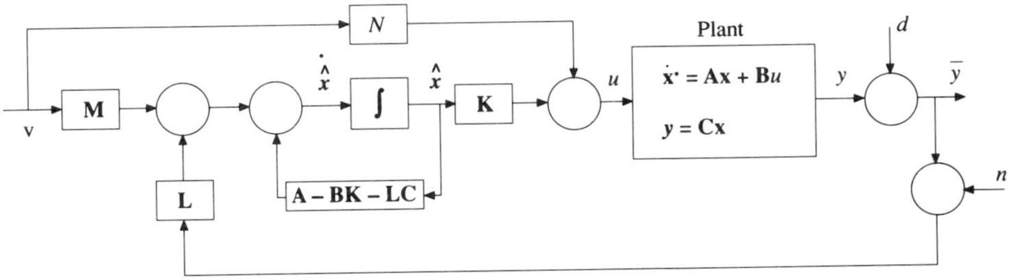

Figure 8.8 Configuration for the tracking problem.

As for the numerator we have

$$N \det \begin{bmatrix} sI_n - A & | & B \\ ---- & | & --- \\ C & | & 0 \end{bmatrix} \det \begin{bmatrix} sI_n - A + BK + LC - \frac{1}{N} MK \end{bmatrix}$$

$$= N\, B_p(s)\, \gamma(s) \tag{8.36}$$

N determines the steady-state gain, and the zeros of polynomial $\gamma(s)$ are dictated by $\boldsymbol{M}$.

8.2 ALGEBRAIC DESIGN

8.2.1 Introduction

The design methods discussed so far have only addressed the question of closed-loop stability, i.e. the location of closed-loop poles. The position of the zeros of the compensated system was not explicitly addressed and their location could not be controlled by the design method. Because the zeros of a transfer function are of no interest to zero-input responses, the previous design methods can only explicitly address *regulation problems*. However, if the responses to given inputs are also specified, as happens in tracking or *servo problems*, we need design procedures that are also capable of controlling the locations of the zeros. In other words, we need a design method capable of synthesizing a desired closed-loop transfer function B_m/A_m.

The approach to be described here consists of two stages. In the first one we select an overall closed-loop transfer function B_m/A_m that meets the specifications, and in the second stage we compute the required compensators. The latter is also known as the '*model matching*' problem and the question is to know under what conditions matching will be possible. Before addressing this problem we recall a few definitions.

Let $N(s)$ and $D(s)$ denote polynomials in s. If $G(s)=N(s)/D(s)$ we say $G(s)$ is a *rational* transfer function. Furthermore $G(s)$ is said to be strictly proper, proper or improper if degree N<degree D, degree N $\leqslant$ degree D or degree N>degree D, respectively. If the polynomials $N(s)$ and $D(s)$ have no roots in common then they are said to be *coprime*. The difference between the number of poles and zeros of G is known as the pole-excess of G.

Mathematically, the model matching problem can be solved in many different ways. However, not all of them are feasible in practice. Assume, for example, that we have a plant with proper transfer function

$$\frac{B_p(s)}{A_p(s)} = \frac{b_0 s^n + b_1 s^{n-1} + \ldots + b_n}{s^n + a_1 s^{n-1} + \ldots + a_n} \tag{8.37}$$

and we would like to compensate it in order to have an overall transfer function

$$\frac{B_m(s)}{A_m(s)} = \frac{b_0' s^n + b_1' s^{n-1} + \ldots + b_n'}{s^n + a_1' s^{n-1} + \ldots + a_n'} \tag{8.38}$$

If the state vector can be measured, then we can think of feedback and feedthrough compensation as shown in Fig, 8.9. Notice that the compensators cancel the plant dynamics and add the new dynamics. For example, at the output y the contribution of the state component x_n, before compensation, was $b_1 x_n$;

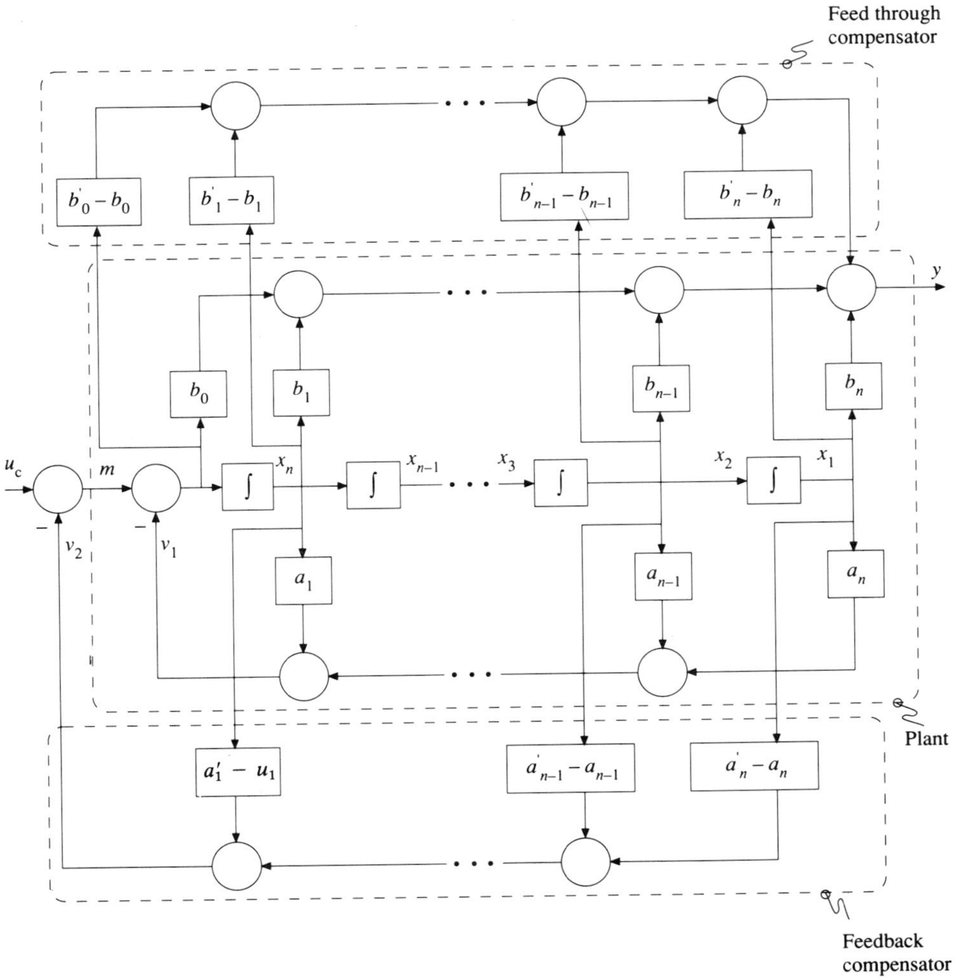

Figure 8.9 An ideal model matching procedure.

after compensation it becomes $b_1x_n+(b_1'-b_1)x_n=b_1'x_n$. This procedure also shows that the poles can be arbitrarily assigned by state feedback.

If state feedback is a feasible practical procedure as we have already seen, state feedthrough is impossible to implement because the forward paths do not pass through the plant. This is known as 'plant leakage'. In fact, we cannot act directly upon the plant output y except via its input m; for example, if the plant is an electric furnace, the way we manipulate its temperature is by acting upon the electric current. Otherwise changes would have to be made in the plant. This is why the forward compensator, as shown in Fig. 8.9, is not a feasible solution.

An appealing alternative is to place in series with the plant a compensator with transfer function

$$G_c = \frac{B_m A_p}{A_m B_p} \tag{8.39}$$

This is a pure feedforward solution where the compensator cancels the process dynamics and adds the desired dynamics. Unfortunately, the possible presence of right half-plane zeros (also known as *non-minimum phase* zeros) in B_p/A_p would render G_c unstable. On the other hand the absence of feedback means that the compensated system would always exhibit errors in the presence of disturbance or modelling inaccuracies. Therefore it is imperative to establish the practical constraints that a given configuration with an overall function B_m/A_m must satisfy.

8.2.2 Implementable Transfer Functions

No matter what configuration is used to implement a given overall transfer function B_m/A_m, the transfer function of every possible input–output pair of the resulting configuration must be proper and stable. In particular this applies to the transfer function $M(s)/U_c(s)$, where U_c denotes the reference input and M the plant input, and which is given by

$$\frac{M}{U_c} = \frac{B_m/A_m}{B/A} = \frac{B_m A}{A_m B} \tag{8.40}$$

as depicted in Fig. 8.10. It can be shown (Chen, 1987) that B_m/A_m is *implementable* if and only if B_m/A_m is stable and M/U_c is stable and proper. It is

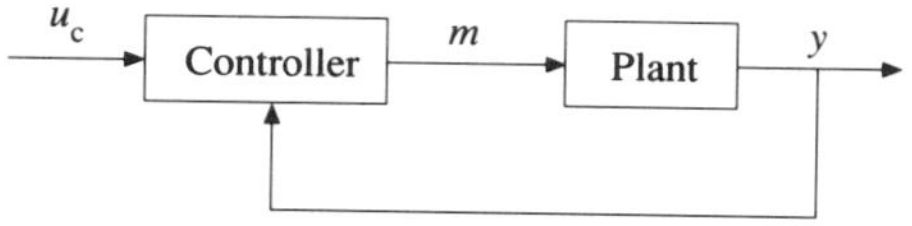

Figure 8.10 General control configuration.

assumed that the compensators have proper transfer functions, but that they are not necessarily stable. The same applies to the plant transfer function B/A.

From this result we have the following two important corollaries:

Fact 8.2: The properness of M/U_c implies

$$\text{deg. } A_m - \text{deg. } B_m \geq \text{deg. } A - \text{deg. } B \tag{8.41}$$

i.e. the pole excess of the prescribed model cannot be less than the pole excess of the plant. This means that the gain of the compensated system, at large frequencies, must decrease at least at the same rate as in the plant. In the discrete-time case this means that the delay of the compensated system must be greater than or equal to the plant delay.

Fact 8.3: In order to preserve the stability of M/U_c, the right half-plane plant zeros must remain zeros of the compensated system: if B has one such zero, then it must be cancelled by B_m , since A and B are coprime; otherwise M/U_c is unstable (cf. (8.40)).

8.2.3 Pole-zero Assignment

In this section we show how any transfer function B_m/A_m, satisfying the two just stated facts, can be implemented by means of proper compensators, with the structure shown in Fig. 8.11. In essence the desired poles will be generated in the feedback system, whose transfer function is $Y/\bar{U}_c$; the compensator T/R will provide the new zeros and will cancel any additional poles and zeros introduced by $Y/\bar{U}_c$.

Let us now analyze the solution of the problem in some detail. From Fig. 8.11 we have

$$\frac{Y}{\bar{U}_c} = \frac{BT}{AR + BS} \tag{8.42}$$

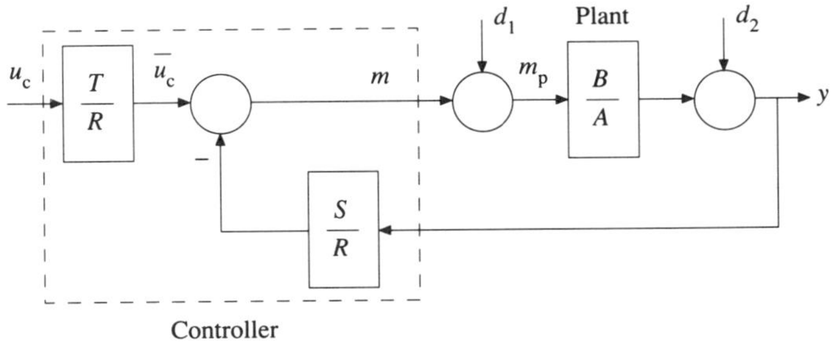

Figure 8.11 General two-parameter configuration.

$$\frac{Y}{U_c} = \frac{BR}{AR + BS} \tag{8.43}$$

Let

$$B = B^+ B^- \tag{8.44}$$

where B^- denotes the zeros in common with B_m (in particular the RHP zeros); then

$$B_m = B^- B'_m \tag{8.45}$$

Since Y/U_c does not contain the zeros of B^+, $(AR+BS)$ must necessarily have B^+ as a factor. Therefore R must factor as

$$R = B^+R' \tag{8.46}$$

because A and B are coprime by assumption. Consequently

$$\frac{Y}{U_c} = \frac{BR'}{AR' + B^- S} = \frac{BR'}{A_oA_m} \tag{8.47}$$

where A_o denotes the additional poles introduced by the compensation. Then we can write

$$AR' + B^- S = A_oA_m \tag{8.48}$$

or equivalently

$$AR + BS = B^+ A_oA_m \tag{8.49}$$

If we set

$$\mathrm{T} = B'_mA_0 \tag{8.50}$$

then

$$\frac{T}{R} = \frac{B'_mA_o}{R} \tag{8.51}$$

and the desired overall transfer function B_m/A_m is obtained.

The role of polynomial A_o is to ensure the existence of a solution R in (8.49), of sufficiently high degree in order to render S/R and T/R proper transfer functions. From what has been said we conclude that our design problem has been reduced to the solution of polynomial equation (8.49), with unknowns R and S, subject to the constraint of properness of transfer functions S/R and T/R.

Exercise 8.6 Check that R, S, T satisfying (8.49) and (8.51) yield the desired overall transfer function B_m/A_m.

Exercise 8.7 Discuss the amplitude of the control signal m, in Fig. 8.11, when

$$\frac{B}{A} = \frac{1}{s+1} \text{ and } \frac{B_m}{A_m} = \frac{100}{s+100}$$

Notice the difference in 'speed' between the plant and the compensated system.
Answer: since

$$\frac{M}{U_c} = \frac{100(s+1)}{(s+100)}$$

and assuming u_c is a step at time $t=0$, then $m(t)$ will assume large values at time $t=0^+$ given 'the speed of the pole as seen from the zero'.

Exercise 8.8 Given the polynomial equation

$$AX + BY = C$$

where A, B, C are known polynomials:

(a) Show that if a solution (X,Y) exists then any common factor of A and B also factors C.
(b) Assuming $\bar{X}$ and $\bar{Y}$ is a solution, show that

$$\begin{cases} X = \bar{X} + MB \\ Y = \bar{Y} - MA \end{cases}$$

is also a solution, whatever polynomial M.
(c) Show that if a solution (X,Y) exists, then we can find another solution (X,Y) such that $\deg.Y < \deg.A$.

Exercise 8.9 Assuming

$$\frac{B}{A} = \frac{(5-s)(s+4)}{(s+1)(s+2)(s+3)}$$

state when the design conditions are violated if:

(a) $\dfrac{B_m}{A_m} = \dfrac{(5-s)}{(s+3)}$

(b) $\dfrac{B_m}{A_m} = \dfrac{(5-s)}{(s+5)^2}$

(c) $\dfrac{B_m}{A_m} = \dfrac{(5-s)(s+10)}{(s+1)(s+2)(s+3)}$

Answers: (a) not implementable; (b) implementable; (c) implementable.

We are now in a better position to understand why the root-locus and frequency-domain methods are essentially pole-assignment methods. Only one compensator is used, either in the feedback path or in series with the plant as shown in Fig. 8.12. Denoting the compensator transfer function by S/R, we have the overall transfer function

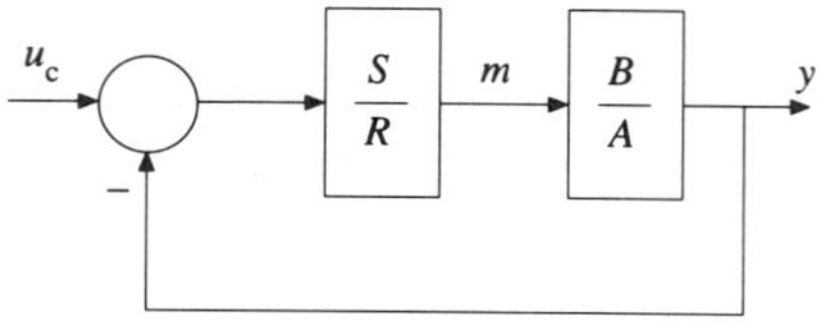

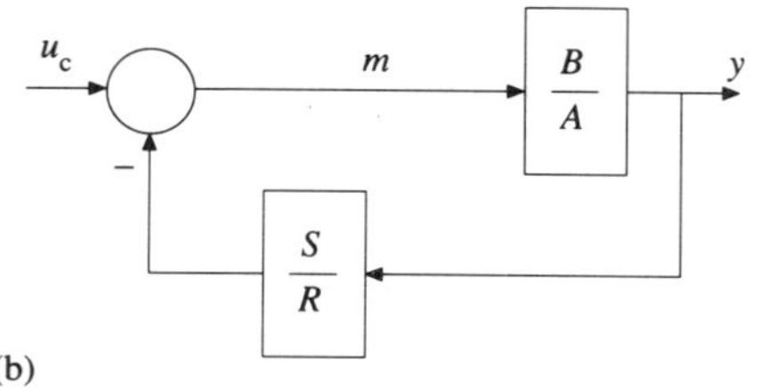

Figure 8.12 Series (a) and feedback (b) compensation.

$$\frac{BR}{AR + BS} \tag{8.52}$$

for the feedback compensation, and

$$\frac{BS}{AR + BS} \tag{8.53}$$

for the series compensation.

The desired poles can be arbitrarily assigned since the denominators are the same as before. However, once A_o and A_m have been selected in equation (8.49) we have *no control* over the locations of the zeros of R and S. Even if we set $R = B^+R'$ all we get is

$$\frac{BR'}{A_oA_m} \tag{8.47}$$

and

$$\frac{B^-S}{A_oA_m} \tag{8.54}$$

for each of the closed-loop transfer functions.

Coming back to the configuration of Fig. 8.11, known as a *two-parameter* configuration, we see that A_o is not present in the overall transfer function Y/U_c, since it is cancelled by T. As such, its modes will not be excited by command inputs u_c – the modes of A_o are *uncontrollable* from u_c. However they can be excited by disturbances such as d_1, since

$$\frac{Y}{D_1} = \frac{Y}{\bar{U}_c} = \frac{BR'}{A_o A_m} \tag{8.47}$$

Therefore we suggest that the zeros of A_o are chosen to be at least four times faster than the desired closed-loop poles.

8.2.4 Implementation of Compensators

Given an nth order proper rational transfer function such as (8.37), we know how to realize it with the help of n integrators as shown in Fig. 2.1. An alternative realization is depicted in Fig. 8.13.

Exercise 8.10 Show that the diagram of Fig. 8.13 has the transfer function (8.37). *Hint*: notice that (8.37) can be rewritten as

$$Y = b_0 M + \frac{1}{s}(b_1 M - a_1 Y) + \ldots + \frac{1}{s^{n-1}}(b_{n-1} M - a_{n-1} Y) + \frac{1}{s^n}(b_n M - a_n Y)$$

When we have a two-input, one-output compensator, as shown in Fig. 8.11, there are a number of reasons showing that T/R and S/R should not be implemented as two separate units, namely:

1. If R has unstable roots, the transfer function $\bar{U}_c/U_c$ will be unstable, therefore violating the implementability conditions; furthermore, the unstable poles T/R and S/R will never be exactly equal in practice; therefore a perfect pole-zero cancellation between $\bar{U}_c/U_c$ and $Y/\bar{U}_c$ is impossible and the result would be an unstable overall transfer function Y/U_c.

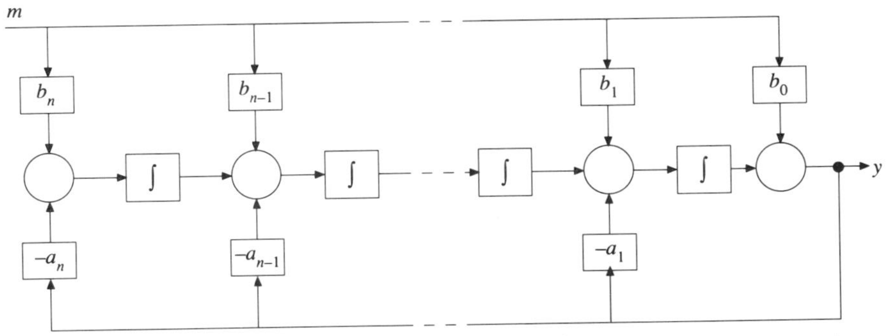

Figure 8.13 An analog simulation diagram for transfer function (8.37).

2. A separate implementation of T/R and S/R would require a number of integrators equal to 2 deg. R. Such a procedure would not take advantage of the fact that T/R and S/R share a common denominator. However, if we implement the transfer function $1/R$ as shown in Fig. 8.13 or Fig. 2.1, and then construct the transfer functions

$$\frac{T}{R} = \frac{(l_0 s^n + l_1 s^{n-1} + \ldots + l_n}{(s^n + a_1 s^{n-1} + \ldots + a_n)}$$

and

$$\frac{S}{R} = \frac{(m_0 s^n + m_1 s^{n-1} + \ldots + m_n)}{(s^n + a_1 s^{n-1} + \ldots + a_n)}$$

by inserting the coefficients of T and S in the forward paths, as depicted in Fig. 8.14, we have the desired result.

8.2.5 Solution of the Polynomial Equation

Recall that our design problem relies upon the solution of the polynomial equations

$$AR + BS = B^+ A_o A_m \tag{8.49}$$

$$T = B'_m A_o \tag{8.50}$$

in the unknowns R, S and T, subject to the constraint of properness of T/R and S/R, and knowing that:

1. A and B are coprime, and B/A is proper.

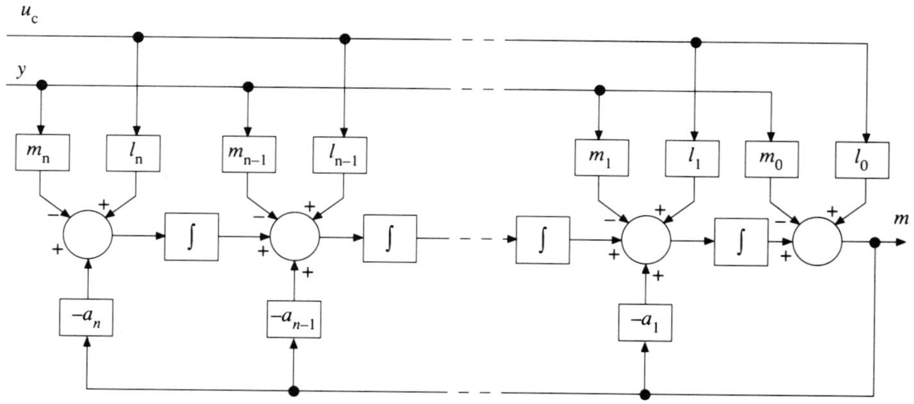

Figure 8.14 Realization of the compensator of Fig. 8.11 with only n integration.

2. A, B, A_m and B_m satisfy Facts 8.2 and 8.3.
3. $B = B^+B^-$.
4. $B_m = B^-B'_m$.
5. $R = B^+R'$.

It is now shown that a solution exists to the problem just stated, provided the degree of A_o *is sufficiently large.*

Fact 8.4: There exist solutions R, S and T to (8.49) and (8.50) with T/R and S/R proper, provided B/A is a proper irreducible transfer function, Facts 8.2 and 8.3 are verified and

$$\text{deg. } A_o \geqslant 2 \text{ deg. } A - \text{deg. } A_m - \text{deg. } B^+ \tag{8.55}$$

8.2.5.1 Proof of properness of *S/R*

Let R and S denote a solution of (8.49) such that deg. $S<$deg. A_p. It can be shown (Aström and Wittenmark, 1984) that such a solution exists because A and B are coprime. If A_o is such that (8.55) is satisfied, we have

$$\text{deg. } (AR + BS) = \text{deg. } AR = \text{deg. } (B^+A_oA_m) \geqslant 2 \text{ deg. } A \tag{8.56}$$

Therefore S/R is proper.

8.2.5.2 Proof of properness of *T/R*

Because the already computed solutions S and R satisfy (8.56), we can write

$$\text{deg. } R = \text{deg. } A_m + \text{deg. } A_o + \text{deg. } B^+ - \text{deg. } A \tag{8.57}$$

From (8.41) we have

$$\text{deg. } A_m - \text{deg. } B^- - \text{deg. } B'_m \geqslant \text{deg. } A - \text{deg. } B^+ - \text{deg. } B^- \tag{8.58}$$

or equivalently

$$\text{deg. } A_m + \text{deg. } B^+ + \text{deg. } A_o \geqslant \text{deg. } A + \text{deg. } B'_m + \text{deg. } A_o \tag{8.59}$$

The combination of (8.57) and (8.59) leads to

$$\text{deg. } R \geqslant \text{deg. } B'_m + \text{deg. } A_o = \text{deg. } T \tag{8.60}$$

which completes the proof of Fact 8.4.

At this point we would like to draw attention to the fact that equation (8.49) will be poorly conditioned when B and A have roots which are close together (Aström, 1980). Such transfer functions normally occur when one tries to fit a high order model to a given set of data. In such cases one must resort to robust identification procedures that produce lower-order models and still retain the essential underlying dynamics (Lopes dos Santos and Martins de Carvalho, 1990).

Exercise 8.11 Show that when deg. B<deg. A, i.e. B/A is strictly proper, then (8.55) can be relaxed to

$$\text{deg. } A_o \geq 2 \text{ deg. } A - \text{deg. } A_m - \text{deg. } B^+ - 1 \tag{8.61}$$

Answer: in this case deg. $(BS) \leq 2n-2$. From (8.61)

$$\text{deg. } (B^+ A_o A_m) \geq 2 \text{ deg. } A - 1$$

From (8.56)

$$\text{deg. } (AR + BS) = \text{deg. } AR \geq 2 \text{ deg. } A - 1$$

showing that deg. $R \geq (n-1)$. Therefore S/R is proper.

The proof for the properness of T/R remains unaltered.

8.2.6 Disturbance Rejection

A system is said to achieve *step* disturbance rejection if the output goes to zero, as $t \to \infty$, following a step disturbance at time $t = t_0$. With reference to Fig. 8.11 we have

$$\frac{Y(s)}{D_1(s)} = \frac{BR}{AR + BS} \text{ and } \frac{Y(s)}{D_2(s)} = \frac{AR}{AR + BS}$$

Therefore, if $R(0)=0$ the system achieves step disturbance rejection. In principle, the step disturbance rejection problem can be solved if, in equation (8.49), we set

$$R(s) = s\,\bar{R}(s)$$

However we must show that a solution (S,R), with S/R proper, continues to exist.

The proof we gave earlier for the solution of (8.49) assumed unconstrained zeros for S and R. If we define $\bar{A}$ as $\bar{A} = s\,A$, we know a feasible solution, say $(\bar{R},\bar{S})$, exists for

$$\bar{A}\,\bar{R} + B\bar{S} = B^+ A_o A_m$$

because we are not constraining $\bar{R}$ and $\bar{S}$. Rewriting this equation as

$$s\,A\bar{R} + B\bar{S} = B^+ A_o A_m$$

we conclude that $R = s\bar{R}$ and $\bar{S}$ constitute feasible solutions to our problem.

Let us now illustrate these issues by means of a few examples.

Example 8.2 Assume that we have a plant with transfer function

$$\frac{B}{A} = \frac{1}{(s+1)^2}$$

and we wish to design a controller such that the overall transfer function becomes

$$\frac{B_m}{A_m} = \frac{4}{(s+2)^2}$$

In this case we have

$$A = (s + 1)^2,\ B = B^+ = B^- = 1$$

$$A_m = (s + 2)^2,\ B_m = B'_m = 4$$

From (8.55)

$$\text{deg. } A_o \geqslant 4 - 2 - 0 = 2$$

As suggested in section 8.2.3 we set

$$A_o = (s + 8)^2$$

Equation (8.49) assumes the form

$$(s + 1)^2 R + S = (s + 8)^2 (s + 2)^2 \qquad (*)$$

Since we know there exists a proper solution S/R such that deg. S<deg. A, we can set

$$S = s_1 s + s_0$$
$$R = r_2 s^2 + r_1 s + r_0$$

Substituting into equation (*) above we get

$$S = 112\, s + 161 = 112\, (s + 1.44)$$

$$R = s^2 + 18\, s + 95 = (s + 9 \pm j3.74)$$

and from (8.50)

$$T = 4\, (s + 8)^2 = 4\, (s^2 + 16\, s + 64)$$

The reader is invited to compute the transfer function $BT/(AR+BS)$ in order to verify the result.

Example 8.3 In this example we solve the pole-zero allocation problem with a lower-order observer. Since B/A is strictly proper, we can lower the degree of polynomial A_o, as described in exercise 8.11. If we then set

$$A_o = (s + 8)$$

equation (8.49) becomes

$$(s + 1)^2 R + S = (s + 8)\, (s + 2)^2 \qquad (**)$$

and naturally deg. R=deg. S=1. Then

$$S = \bar{s}_1 s + \bar{s}_o$$

$$R = r_1 s + r_0$$

Substituting into equation (**) and equating equal powers of s, we get

$$S = 15s + 22$$

$$R = s + 10$$

Also

$$T = B'_m A_o = 4(s + 8)$$

With these choices, the overall transfer function becomes

$$\frac{BT}{AR + BS} = \frac{4(s + 8)}{s^3 + 12s^2 + 36s + 32} = \frac{4(s + 8)}{(s + 2)^2 (s + 8)} = \frac{4}{(s + 2)}$$

as expected.

Analyzing the result of this compensation, in a frequency-domain or root-locus perspective, we conclude that we have performed a lead feedback compensation. In fact we can see from the root-locus of the feedback block shown in Fig. 8.15 that the plant poles at (-1)were shifted to the left by the action of the lead compensator

$$\frac{S}{R} = \frac{15(s + 1.467)}{s + 10}$$

and that the poles of Y/U_c are -2, -2 and -8.

Example 8.4 A simple computation reveals that, in examples 8.2 and 8.3, $y(t) \to 0.371$ and $y(t) \to 0.313$ as t goes to infinity, respectively, when d is a unit-step disturbance. In order to achieve step disturbance rejection, we repeat the pole-zero allocation design of these examples with the added restriction of $R(0)=0$.

Set

$$R(s) = r_2 s^2 + r_1 s$$

$$S(s) = s_1 s + s_0$$

Then deg. $(AR+BS)=4$, and consequently deg. $A_o=2$.

The solution of (8.49) gives rise to five equations; but our choice of R and S produces only four unknowns: s_0, s_1, r_1 and r_2. We invite the reader to check that no solution exists in this case. Consequently we set

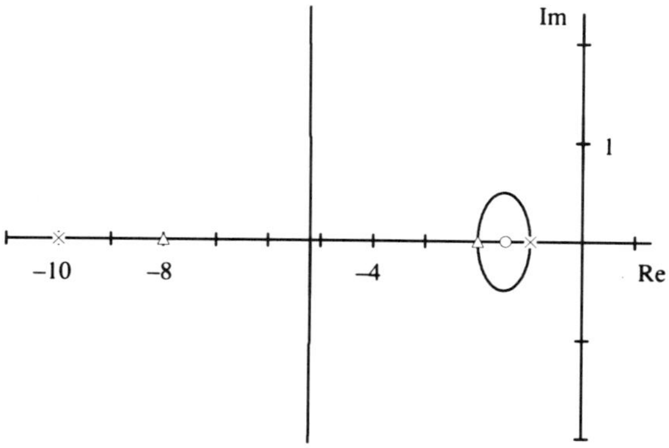

Figure 8.15 Root-loci of example 8.3.

$$R(s) = r_2 s^2 + r_1 s$$

$$S(s) = s_2 s^2 + s_1 s + s_0$$

$$AR + BS = r_2 s^4 + (2r_2 + r_1)s^3 + (r_2 + 2r_1 + s_2)s^2 + (r_1 + s_1)s + s_0$$

and

$$B^+ A_o A_m = (s + 8)^2 (s + 2)^2 = s^4 + 20s^3 + 132s^2 + 320s + 256$$

Equating equal powers of s in

$$AR + BS = B^+ A_o A_m \tag{8.49}$$

we arrive at

$$\begin{cases} r_1 = 18 \\ r_2 = 1 \end{cases} \qquad \begin{cases} s_0 = 256 \\ s_1 = 302 \\ s_2 = 95 \end{cases}$$

The solution is

$$R(s) = s(s + 18)$$

$$S(s) = 95s^2 + 302s + 256 = 95(s + 1.589 \pm j0.4103)$$

$$T(s) = 4(s + 8)^2$$

It is of interest to analyze the roots-loci of the feedback block with transfer function Y/U_c: these are plotted in Fig. 8.16 showing the closed-loop poles at the break-away points of the loci. The poles at -8 are very fast poles, as seen from the zeros of S, i.e from $-1.589 \pm j0.4103$. Consequently any transfer function, having S as a factor in the numerator, will exhibit a large overshoot in its step response. This is the case of

$$\frac{M(s)}{D_2(s)} = \frac{-AS}{AR + BS} = -\frac{95s^4 + 492s^3 + 955s^2 + 814s + 256}{s^4 + 20s^3 + 132s^2 + 320s + 256}$$

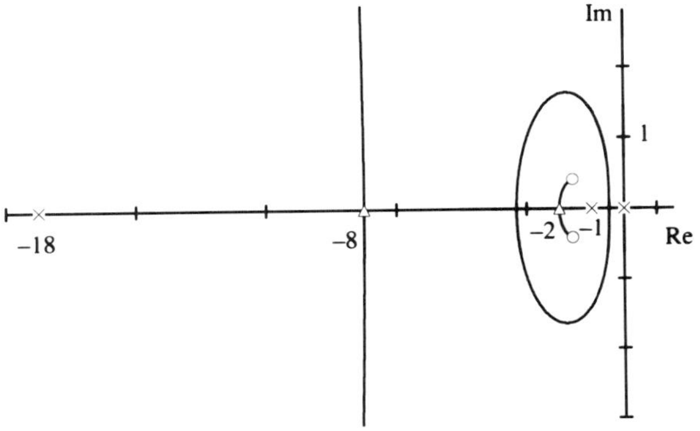

Figure 8.16 Root-loci of example 8.4.

whose unit-step response assumes the magnitude of 95 at time $t=0^+$. Therefore large disturbances may cause actuator saturation. This is not unexpected since we have prescribed the loop poles eight times faster than the plant poles, and that can be achieved only at the expense of a large control effort. We will came back to this question in the next example.

The response $y(t)$ to unit-step disturbances d_1 and d_2 are shown in Fig. 8.17: as prescribed, the output goes to zero, because

$$\frac{Y}{D_1} = \frac{BR}{AR + BS} = \frac{s(s + 18)}{(s + 8)^2 (s + 2)^2}$$

$$\frac{Y}{D_2} = \frac{AR}{AR + BS} = \frac{s(s + 1)^2(s + 18)}{(s + 8)^2 (s + 2)^2}$$

These responses are dominated by the modes $t\,e^{-2t}$ and e^{-2t}, as t increases. In fact, if we expand the transform of the step response of Y/D_2 into partial fractions, after inversion, we get

$$y(t) = 13.61t\,e^{-8t} + 2.009\,e^{-8t} + 0.444\,t\,e^{-2t} - 1.009\,e^{-2t}$$

Figure 8.18 depicts $y(t)$ and $(0.444t\,e^{-2t} - 1.009\,e^{-2t})$: it is clear that for $t>1$ they are indistinguishable.

At this point one may ask if we can achieve a faster disturbance rejection and still retain the overall transfer, i.e. can we handle separately the tracking requirements and the regulation requirements? The answer is in the affirmative and is the topic of the next section.

8.2.7 Tracking and Regulation with Different Specifications

Assume we wish to assign our controlled system a set of modes to disturbance responses and another, possibly disjoint, set of modes to the responses to

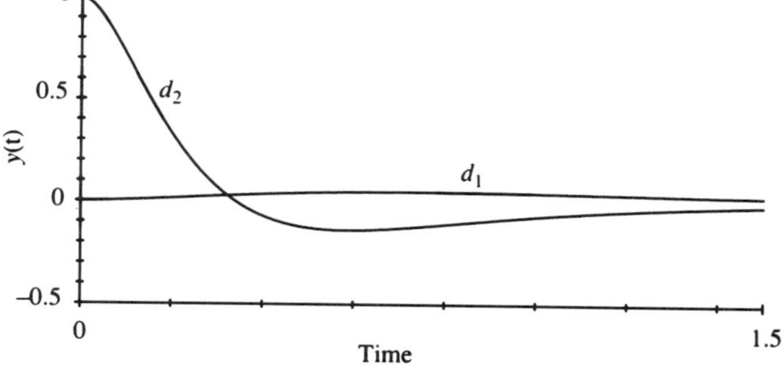

Figure 8.17 Responses for unit-step disturbances at d_1 and d_2, Fig. 8.11, in example 8.4.

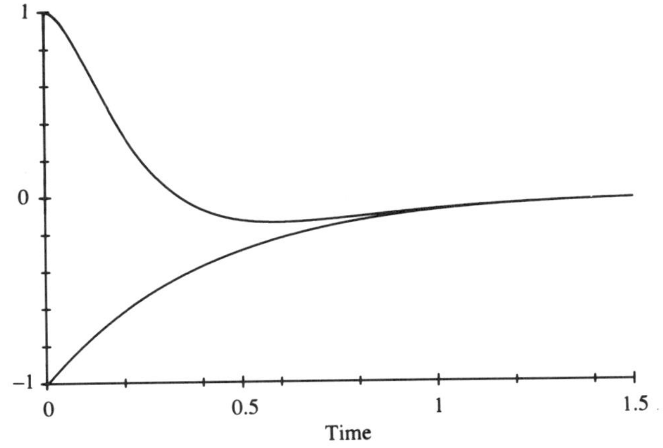

Figure 8.18 Contribution of the dominant modes of the response to the unit-step disturbance at d_2 in example 8.4.

command inputs. This can be achieved with the procedure discussed so far, with minor alterations, as depicted in Fig. 8.19.

In essence, we begin by implementing a transfer function with the same pole-excess as B/A, namely

$$\frac{Y}{U_c} = \frac{B^-}{(s - p_1) \ldots (s - p_e)} \tag{8.62}$$

where e=deg. A−deg. B^+. $p_1, p_2, \ldots p_e$ are prescribed poles for the disturbance transfer functions Y/D_1 and Y/D_2; recall their denominator is

$$A_o\,(s - p_1) \ldots (s - p_e)$$

Then we place in series with the feedback system a pre-compensator with transfer function

$$\frac{B'_m(s - p_1) \ldots (s - p_e)}{A_m} \tag{8.63}$$

Notice this is a proper transfer function, since we have assumed

$$\text{deg. } A_m - \text{deg. } B'_m \geqslant \text{deg. } A - \text{deg. } B^+$$

and consequently

$$\text{deg. } A_m \geqslant \text{deg. } B'_m + e$$

The following example illustrates the application of this method.

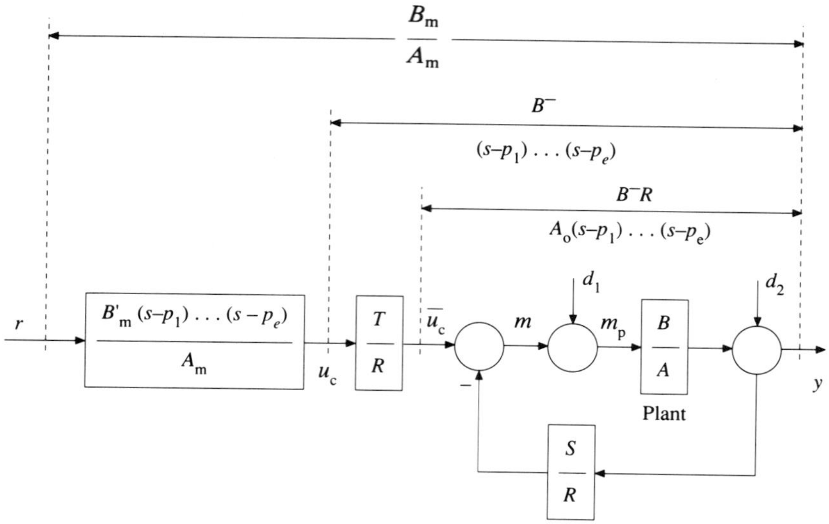

Figure 8.19 Tracking and regulation with different specifications.

Example 8.5 Assume we have again the same plant

$$\frac{B}{A} = \frac{1}{(s+1)^2}$$

and the same desired transfer function from reference to output

$$\frac{B_m}{A_m} = \frac{4}{(s+2)^2}$$

Furthermore we want, besides step disturbance rejection, the poles of the disturbance to output transfer functions, Y/D_1 and Y/D_2, equal to (-8).

From the algorithm just described we have $e=2$, $p_1=p_2=-8$. Again we set

$$R(s) = r_2 s^2 + r_1 s$$

$$S(s) = s_2 s^2 + s_1 s + s_0$$

therefore

$$AR + BS = (s+8)^4$$
$$= (s^2 + 16s + 64)(s^2 + 16s + 64)$$
$$= s^4 + 32s^3 + 384s^2 + 2048s + 4096$$

From example 8.4 we know that

$$AR + BS = r_2 s^4 + (2r_2 + r_1)s^3 + (r_2 + 2r_1 + s_2)s^2 + (r_1 + s_1)s + s_0$$

Equating equal powers of s we obtain

$$\begin{cases} r_1 = 30 \\ r_2 = 1 \end{cases} \qquad \begin{cases} s_0 = 4096 \\ s_1 = 2018 \\ s_2 = 323 \end{cases}$$

The solution is

$$R(s) = s^2 + 30s$$

$$S(s) = 323s^2 + 2018s + 4096 = 323\ (s + 3.124 \pm \mathrm{j}1.74)$$

$$T = (s + 8)^2$$

$$B'_{\mathrm{m}} = 4$$

leading to

$$\frac{Y}{U_{\mathrm{c}}} = \frac{Y}{D_1} = \frac{s(s + 30)}{(s + 8)^4}$$

$$\frac{T}{R} = \frac{(s + 8)^2}{s(s + 30)}$$

$$\frac{B'_{\mathrm{m}}(s - p_1)\ (s - p_2)}{A_{\mathrm{m}}} = \frac{4\ (s + 8)^2}{(s + 2)^2}$$

and

$$\frac{Y}{R} = \frac{4}{(s + 2)^2}$$

Let us now analyze in some detail the results of this compensation. To take into consideration practical limitations, we have constrained the plant input to the range [–5,5] during the simulation. This is shown in Fig. 8.20 and is the result of a unit step applied at r, at time t=0, and step disturbances at d_1 and d_2, at times 3 and 5, with amplitudes 1 and 0.2 respectively. The system performs as expected for the step changes at r and d_2, clearly showing the faster dynamics for the input from d_2.

Furthermore d_1 has virtually no effect upon y. Things are different with respect to d_2, despite the fact that it has an amplitude five times smaller. We conclude the compensated system is now highly sensitive to output disturbances. This is confirmed by the transfer function

$$\frac{M_{\mathrm{p}}(s)}{D_2(s)} = -\frac{AS}{AR + BS} = -\frac{323s^4 + 2664s^3 + 8455s^2 + 10\ 210s + 4096}{s^4 + 32s^3 + 384s^2 + 2048s + 4096}$$

In fact for a unit step at d_2, at time 0, $m_{\mathrm{p}}(t)$ will reach -323 at time $t=0^+$. Consequently as m_{p} saturates the transient period becomes longer. In our case, m_{p} remains equal to -5 for about 0.16 time units as shown in the figure. We can also see the extent of its impact on y is considerably stronger than the influence of d_1.

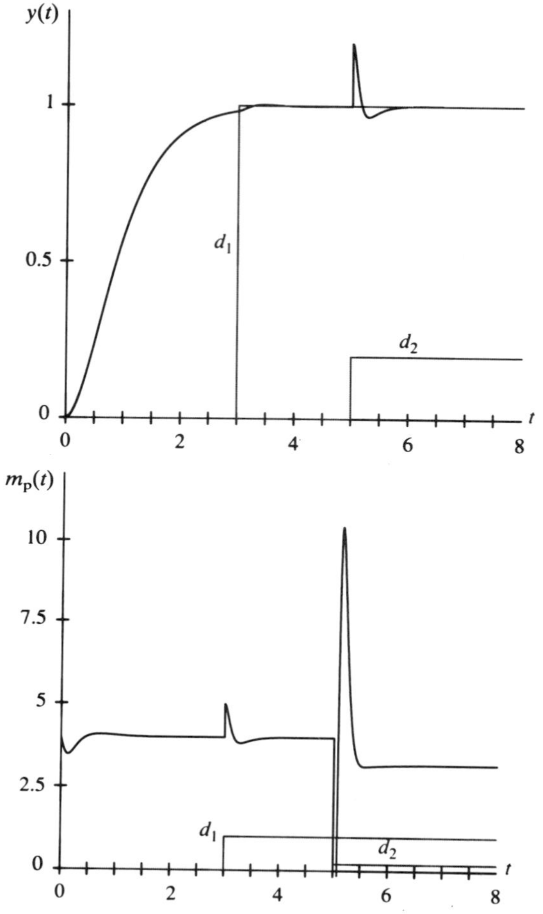

Figure 8.20 Evolution of the control effort, m_p, and plant output y, following step disturbances at r, d_1 and d_2, at times 0, 3 and 5, and magnitudes 1, 1 and 0.2, respectively.

8.2.8 Transfer Function Selection

Given the desired response, either in the time or in the frequency domains, for the overall closed-loop system, it is not straightforward to find the transfer function, if any, that matches those requirements, except in the case of second-order systems. The 'best' transfer function depends upon the selected performance criteria. The most popular ones are the concept of dominant poles and the optimization of performance measures (ITAE, quadratic performance index, etc.). Unfortunately, when the actuating signal $m_p(t)$ is, quite naturally, amplitude-constrained

$$|m_p(t)| \leqslant M$$

none of these methods is capable of incorporating this constraint. The only way to check if it is not violated is by simulation. Therefore the selection of the overall transfer function is essentially an iterative, computer-aided procedure.

In the following we look at some scalar performance measures for a control system.

8.2.8.1 Integral of time multiplied by absolute error (ITAE)

The performance index J is defined as

$$J = \int_0^\infty t\,|e(t)|\,\mathrm{d}t \tag{8.64}$$

where

$$e(t) = y(t) - u(t)$$

and y and u denote the system output and input.

The multiplication of the error magnitude by t has the advantage of putting less weight on the (larger) errors at the start of the transient period.

Graham and Lathrop (1953) have obtained, by analog computation, sets of optimizing transfer functions of J, assuming $u(t)$ is a step function, some of which are shown in Table 8.2. Unfortunately no prototype transfer functions with right half-plane zeros are available.

Table 8.2 Optimal ITAE transfer functions

Transfer function	*n*	*Denominator*
$\dfrac{\alpha_0}{P(s)}$	1	$s + \omega_0$
	2	$s^2 + 1.4\,\omega_0 s + \omega_0^2$
	3	$s^3 + 1.75\,\omega_0 s^2 + 2.15\,\omega_0^2 s + \omega_0^3$
	4	$s^4 + 2.1\,\omega_0 s^3 + 3.4\,\omega_0^2 s^2 + 2.7\,\omega_0^3 s + \omega_0^4$
$\dfrac{\alpha_1 + \alpha_0}{P(s)}$	2	$s^2 + 3.2\,\omega_0 s + \omega_0^2$
	3	$s^3 + 1.75\,\omega_0 s^2 + 3.25\,\omega_0^2 s + \omega_0^3$
	4	$s^4 + 2.41\,\omega_0 s^3 + 4.93\,\omega_0^2 s^2 + 5.14\,\omega_0^3 s + \omega_0^4$
$\dfrac{\alpha_2 s^2 + \alpha_1 s + \alpha_0}{P(s)}$	3	$s^3 + 2.97\,\omega_0 s^2 + 4.94\,\omega_0^2 s + \omega_0^3$
	4	$s^4 + 3.71\,\omega_0 s^3 + 7.88\,\omega_0^2 s^2 + 5.93\,\omega_0^3 s + \omega_0^4$

$$P(s) = s^n = \alpha^{n-1}\, s^{n-1} + \ldots + \alpha_2 s^2 + \alpha_1 s + \alpha_0$$

Exercise 8.12 If we do not restrict ourselves to strictly proper transfer functions, what would be the result of optimizing J in (8.64)?
Answer: a transfer function of unit.

Notice that the parameter ω_0 in Table 8.2 is not fixed because the optimal transfer function is independent of the time-scale. ω_0 determines the system bandwidth and can be determined from the amplitude constraint on the control effort. Also of interest is the fact that equal powers of s, in the numerator and denominator, have the same coefficient. This property ensures zero steady-state errors to constant, ramp and parabola inputs as shown in exercise 8.13.

Exercise 8.13

(a) Compute the expression for the steady-state error of the system with transfer function

$$T(s) = \frac{K\,(s^m + \beta_{m-1}s^{m-1} + \ldots + \beta_1 s + \beta_0)}{s^n + \alpha_{n-1}s^{n-1} + \ldots + \alpha_1 s + \alpha_0} \tag{8.65}$$

and the conditions for zero steady-state error to constant, ramp and parabola inputs.

(b) If $(-p_i)$ and $(-z_i)$ denote the poles and zeros of $T(s)$, show that zero steady-state error to a step input is ensured if

$$\pi(-\, z_i) = \pi(-\, p_i)$$

Hint: recall formula relating the zeros and the coefficients of a polynomial, presented in section 3.5.

(c) If $T(s)$ has zero steady-state error to step inputs, show that

$$\frac{1}{K_v} = \sum \frac{1}{z_i} - \sum \frac{1}{p_i} \tag{8.66}$$

Answers: (a) If $y(t)$ and $u(t)$ denote the system output and input, respectively, the error $e(t)$ becomes

$$e(t) = y(t) - u(t)$$

and

$$E(s) = [T(s) - 1]\; U(s) = \frac{\ldots + (k\beta_2 - \alpha_2)s^2 + (k\beta_1 - \alpha_1)s + (k\beta_0 - \alpha_0)}{s^n + \alpha^{n-1}s^{n-1} + \ldots + \alpha_1 s + \alpha_0}\, U(s)$$

Since

$$e_{ss} = \lim_{t\to\infty} e(t) = \lim_{s\to 0} s\; E\,(s)$$

we can write

$$e_{ss} = \frac{1}{\alpha_0}\left[\left(\lim_{t\to\infty} u(t)\right)(k\beta_0 - \alpha_0) + \left(\lim_{t\to\infty} \frac{du}{dt}\,(k\beta_1 - \alpha_1)\right) + \ldots\right]$$

Therefore:

(i) $(k\beta_0 - \alpha_0) = 0$ ensures zero steady-state error for step inputs.

(ii) $(k\beta_0 - \alpha_0) = 0$ and $(k\beta_1 - \alpha_1) = 0$ ensure zero steady-state error for step and ramp inputs.

(c) Assume $n=3$. Then

$$\frac{1}{p_1} + \frac{1}{p_2} + \frac{1}{p_3} = \frac{p_2p_3 + p_1p_3 + p_1p_2}{p_1p_2p_3} = \frac{\alpha_1}{\alpha_0}$$

It is obvious this result holds for any value of n, bearing in mind formula (3.43). Therefore

$$\sum \frac{1}{p_i} = \frac{\alpha_1}{\alpha_0}$$

Similarly,

$$\sum \frac{1}{z_i} = \frac{\beta_1}{\beta_0}$$

By assumption $k\beta_0 = \alpha_0$

$$\begin{aligned} e_{ss}\ (\text{ramp}) &= \frac{1}{k_v} = \frac{k\beta_1 - \alpha_1}{\alpha_0} \\ &= \frac{1}{\alpha_0}(k\ \beta_1 - \alpha_1) \\ &= \frac{1}{\alpha_0}\left(k\ \beta_0 \sum \frac{1}{z_i} - \alpha_0 \sum \frac{1}{p_i}\right) \\ &= \sum \frac{1}{z_i} - \alpha_0 \sum \frac{1}{p_i} \end{aligned}$$

Exercise 8.14

(a) Show that the non-minimum phase system

$$\frac{1 - s}{s^2 + s + 1}$$

has non-zero steady-state error to ramp inputs.

(b) Compute a and b in order that

$$\frac{(1 - s)\ (as + b)}{s^3 + s^2 + s + 1}$$

has zero velocity error.

Answer: $a=2$; $b=1$.

Example 8.6 We have seen, in example 8.2, that

$$|m_p(t)| \leqslant 4$$

for a unit-step reference input. Let us now compute the optimal ITAE transfer function, assuming the same plant and the same amplitude constraint.

In order to satisfy the pole-excess condition we select from Table 8.2 the overall transfer function

$$G_o = \frac{\omega_0^2}{s^2 + 1.4\,\omega_0 s + \omega_0^2}$$

Then

$$M_p(s) = \frac{G_o}{B/A}\frac{1}{s} = \frac{\omega_0^2(s^2 + 2s + 1)}{s(s^2 + 1.4\,\omega_0 s + \omega_0^2)}$$

Since the maximum occurs at time $t=0^+$ we have

$$\lim_{t\to 0^+} m_p(t) = \lim_{s\to\infty} sM_p(s) = \omega_0^2$$

Therefore $\omega_0=2$. The damping factor for G_0 is then $\zeta=0.7$.

This contrasts with the design of example 8.2 by having an overshoot greater than zero and a faster rise time.

8.2.8.2 Quadratic regulator theory

Given a system described by the controllable and observable representation

$$\begin{cases} \dot{\boldsymbol{x}} = \boldsymbol{A}\boldsymbol{x} + \boldsymbol{B}u \\ y = \boldsymbol{C}\boldsymbol{x} \end{cases}, \ \boldsymbol{x}\in\mathbb{R}^n \tag{8.67}$$

with transfer function $B(s)/A(s)$, B and A coprime polynomials, and the performance index

$$J = \int_0^\infty [y^2(t) + ru^2(t)]\mathrm{d}t \tag{8.68}$$

with $r>0$, it can be shown (Kailath, 1980) that J is minimized by the feedback law

$$u(t) = -\boldsymbol{K}\boldsymbol{x}(t).$$

The vector $\boldsymbol{K}$ is such that the eigenvalues of the system

$$\dot{\boldsymbol{x}} = (\boldsymbol{A} - \boldsymbol{B}\boldsymbol{K})\boldsymbol{x}$$

are the left half-plane roots of the $2n$-degree polynomial equation

$$A_p(s)\,A_p(-s) + r^{-1}\,B_p(s)\,B_p(-s) = 0 \tag{8.69}$$

Note we can apply our well known roots-loci method to analyze the variation of the roots of (8.69) with respect to r. In fact this equation is the characteristic equation of the unit feedback system with open-loop transfer function

$$r^{-1}\frac{B_p(s)\,B_p(-s)}{A_p(s)\,A_p(-s)} \tag{8.70}$$

Rewriting (8.69) as

$$r^{-1}\frac{\pi(s-z_i)\,(s+z_i)\,(-1)^m}{\pi(s-p_i)\,(s+p_i)\,(-1)^n}=-1 \tag{8.71}$$

where

$$B_p(s)=\prod_{i=1}^{m}(s-z_i) \text{ and } A_p(s)=\prod_{i=1}^{n}(s-p_i)$$

we see that we have 180° loci if $(n-m)$ is even and 0° loci if $(n-m)$ is odd. Also note these loci never intersect the imaginary axis; this is obvious from (8.69) since A and B are coprime and r is strictly positive.

In comparison with the ITAE criterion, the quadratic performance index above has the advantage of taking the plant characteristics into consideration since it weighs the control effort u. Quite naturally, if r is large, u must remain small. In practical terms this means that the dynamic properties, i.e. the poles, of the feedback system must remain close to the original poles. If we wish a significant reduction in the settling time, then we have to accept large values of u, and consequently r must be small.

This criterion gives no indication about the 'optimal' zeros, since these are not involved in zero-input responses, and cannot be changed by state feedback either (see exercise 8.15).

Exercise 8.15 Given a controllable and observable representation

$$\begin{cases} \dot{\boldsymbol{x}} = \boldsymbol{A}\boldsymbol{x} + \boldsymbol{B}u \\ y = \boldsymbol{C}\boldsymbol{x} + Du \end{cases}$$

with transfer function $B_p(s)/A_p(s)$, show that the feedback law

$$u = -\,\boldsymbol{k}\,\boldsymbol{x} + r$$

does not change the zeros of the transfer functions.

Answer: The representation becomes

$$\begin{cases} \dot{\boldsymbol{x}} = (\boldsymbol{A} - \boldsymbol{B}\boldsymbol{K})\boldsymbol{x} + \boldsymbol{B}r \\ y = (\boldsymbol{C} - D\boldsymbol{K})\boldsymbol{x} + Dr \end{cases}$$

Let $N(s)/D(S)$ denote its transfer function. We know the transfer function zeros are the zeros of the polynomial

$$\det\left[\begin{array}{c|c} s\boldsymbol{I}-(\boldsymbol{A}-\boldsymbol{B}\boldsymbol{K}) & -\boldsymbol{B} \\ \hline (\boldsymbol{C}-D\boldsymbol{K}) & D \end{array}\right]=0 \tag{*}$$

If we right multiply the last column block by $\boldsymbol{K}$ and add it to the first column block, the determinant will remain unchanged. Therefore the zeros of the feedback system are the zeros of the equation

$$\det \left[\begin{array}{c|c} sI-A & -B \\ \hline C & D \end{array} \right] = 0$$

But this is also the equation for the zeros of the original system! Therefore state feedback has not changed the system zeros.

Once the poles have been selected, a natural way to select the zeros is by means of the specification of the steady-state response to step, ramp and parabola inputs, bearing in mind that we can have, at most, m zeros (cf. (8.71)).

Example 8.7 In this example we illustrate the application of the quadratic performance index with the plant of Example 8.6, namely

$$\frac{B_p(s)}{A_p(s)} = \frac{1}{(s+1)^2}$$

under the same plant input amplitude constraint

$$|u| \leqslant 4$$

when u_c is a unit step; see Fig. 8.21.

Since the plant has no zeros ($m=0$), non-zero steady-state error is only possible for step inputs. In this case, (8.69) becomes

$$(s^2 + 2s + 1)(s^2 - 2s + 1) + r^{-1} = s^4 - 2s^2 + (1 + r^{-1}) = 0 \tag{8.72}$$

We know the roots of this polynomial are symmetrically located with respect to the imaginary axis, with half of them in the left-half plane. This is confirmed by the roots-loci of (8.72) shown in Fig. 8.22. Therefore we can write

$$s^4 - 2s^2 + (1 + r^{-1}) = D_o(s)\, D_o(-s)$$

where

$$D_o(s) = d_0 + d_1 s + d_2 s^2$$

$$D_o(-s) = d_0 - d_1 s + d_2 s^2$$

and d_0, d_1, d_2 are strictly positive.

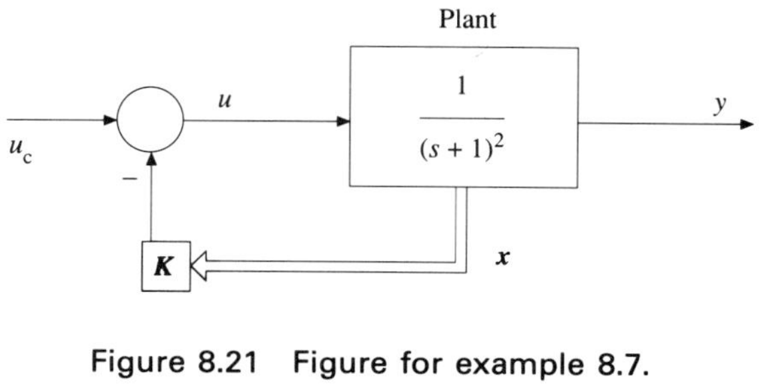

Figure 8.21 Figure for example 8.7.

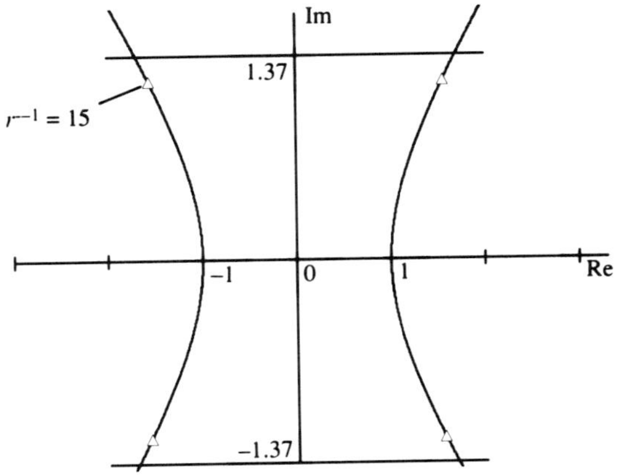

Figure 8.22 Root-loci for the equation $(s^2+2s+1)(s^2-2s+1) + r^{-1}=0$.

Consequently,

$$s^4 - 2s^2 + (1 + r^{-1}) = d_0^2 + (2d_0d_2 - d_1^2)s^2 + d_2^2\, s^4$$

By identifying coefficients of equal powers, we obtain

$$\begin{cases} d_0 = \sqrt{(1 + r^{-1})} \\ d_1 = \sqrt{(2 + 2\sqrt{(1 + r^{-1})})} \\ d_2 = 1 \end{cases}$$

The 'optimal' poles will then be again be given by the roots of the polynomial

$$D_o(s) = s^2 + \sqrt{(2 + 2\sqrt{(1 + r^{-1})})}\, s + \sqrt{(1 + r^{-1})}$$

If zero error is desired for step-reference inputs, the 'optimal' overall transfer function becomes

$$G_0(s) = \frac{Y(s)}{U_c(s)} = \frac{d_0}{d_0 + d_1s + d_2s^2}$$

We must now use the plant input amplitude constraint to select r. Recall that

$$\frac{U(s)}{U_c(s)} = \frac{G_0(s)}{B_p(s)/A_p(s)} = \frac{(s + 1)^2\sqrt{(1 + r^{-1})}}{s^2 + (\sqrt{[2 + 2\sqrt{(1 + r^{-1})}]})\, s + \sqrt{(1 + r^{-1})}}$$

when $u_c(t)$ is a unit step, at time $t=0$, the maximum of $u(t)$ occurs at time $t=0^+$, and is equal to

$$u(0^+) = \lim_{s\to\infty} \frac{(s + 1)^2\sqrt{(1 + r^{-1})}}{s^2 + (\sqrt{[2 + 2\sqrt{(1 + r^{-1})}]})\, s + \sqrt{(1 + r^{-1})}} = \sqrt{(1 + r^{-1})}$$

Then

$$\sqrt{(1 + r^{-1})} \leqslant 4 \Rightarrow r \geqslant \frac{1}{15}$$

With $r=1/15$, the 'optimal' poles are $-1.581\pm j1.225$, which corresponds to an undamped natural frequency $\omega_n=2$ and a damping factor $\zeta=0,791$.

Note that the ITAE criterion has produced, for the same problem, a lower damping factor (cf. example 8.6).

8.2.9 Relationships with State-space Methods

The state-space design method of section 8.1, applied to a plant with a controllable and observable representation, and transfer function $B_p(s)/A_p(s)$, led to a compensated system with the structure shown in Fig. 8.23. The state-space representation of this system is

$$\begin{bmatrix} \dot{x} \\ \hline \dot{\hat{x}} \end{bmatrix} = \left[\begin{array}{c|c} A & -BK \\ \hline LC & A-BK-LC \end{array}\right] \begin{bmatrix} x \\ \hline \hat{x} \end{bmatrix} + \begin{bmatrix} B\,N \\ \hline \mathbf{M} \end{bmatrix} \mathrm{r} \tag{8.73}$$

$$y = [C \;-\; DK] \begin{bmatrix} x \\ \hline \hat{x} \end{bmatrix} + DN\,\mathrm{r}$$

with characteristic polynomial

$$\det\,[s\mathbf{I} - A + BK]\,\det\,[s\mathbf{I} - A + LC] = \alpha_c(s)\,\alpha_0(s) \tag{8.74}$$

since

$$\det \left[\begin{array}{c|c} s\mathbf{I}-A & BK \\ \hline -LC & s\mathbf{I}-A+BK+LC \end{array}\right] = \det \left[\begin{array}{c|c} s\mathbf{I}-A & BK \\ \hline -s\mathbf{I}+A-LC & s\mathbf{I}-A+LC \end{array}\right]$$

$$= \det \left[\begin{array}{c|c} s\mathbf{I}-A+BK & BK \\ \hline \mathbf{0} & s\mathbf{I}-A+LC \end{array}\right]$$

and transfer function

$$\frac{Y(s)}{\mathrm{R}(s)} = \frac{N\det\left[s\mathbf{I} - A + \dfrac{1}{D}BC\right]\det\left[s\mathbf{I} - A + BK + LC - \dfrac{1}{N}MK\right]}{\det\,[s\mathbf{I} - A + BK]\,\det\,[s\mathbf{I} - A + LC]}$$

$$= N\frac{B_p(s)\gamma(s)}{\alpha_c(s)\alpha_o(s)} \tag{8.75}$$

This can be shown as follows: the zeros of the transfer function are the zeros of tl equation

$$\det \begin{bmatrix} sI-A & BK & -BN \\ -LC & sI-A+BK+LC & -M \\ C & -\mathbf{d}K & DN \end{bmatrix} = 0$$

By dividing the last column by N, and adding its product with K to the middle column block, the determinant remain unchanged.

Therefore an equivalent equation is

$$N \det \begin{bmatrix} sI-A & 0 & -B \\ -LC & sI-A+BK+LC-\frac{1}{N}MK & -\frac{1}{N}M \\ C & 0 & D \end{bmatrix} = 0$$

By interchanging the last two column blocks and then the last two row blocks we arrive at the equivalent form

$$N \det \begin{bmatrix} sI-A & -B & 0 \\ C & D & 0 \\ -LC & -\frac{1}{N}M & sI-A+BK+LC-\frac{1}{N}MK \end{bmatrix} = 0$$

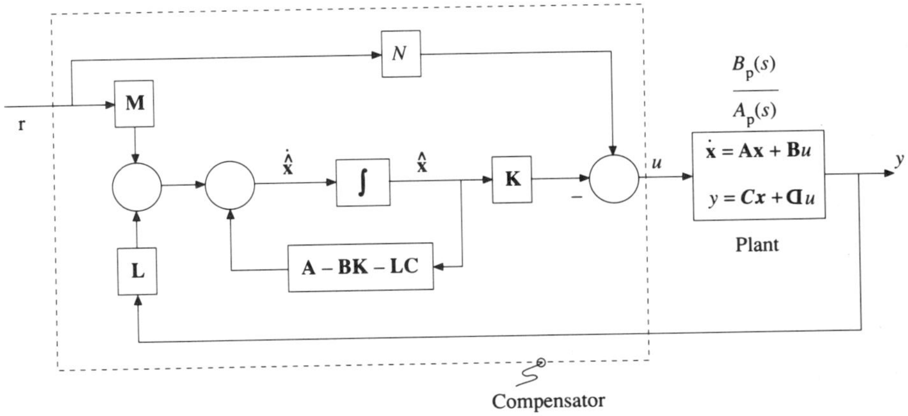

Figure 8.23 Compensated system via the state-space design method of section 8.1.

The left hand side of this equation is

$$N \det \begin{bmatrix} s\mathbf{I}-A & -B \\ C & D \end{bmatrix}$$

$$\times \det \left[s\mathbf{I}-A+BK+LC-\frac{1}{N}MK \right] = N \det \left[s\mathbf{I}-A+\frac{1}{D}BC \right]$$

$$\times \det \left[s\mathbf{I}-A+BK+LC-\frac{1}{N}MK \right]$$

which is the numerator of (8.75).

Notice that the zeros of det $[sI-A+(1/D)\,BC)$ are the plant zeros, i.e. the zeros of $B_p(s)$.

If we split the state vector $\hat{x}$ of the compensator into the components

$$\hat{x} = \dot{\hat{x}}_{ff} + \hat{x}_{fb} \tag{8.76}$$

the compensator can be redrawn as depicted in Fig. 8.24. Then we have

$$\frac{T}{R} = \frac{\det \begin{bmatrix} s\mathbf{I}-(A-BK-LC) & -M \\ -K & N \end{bmatrix}}{\det [s\mathbf{I} - (A - BK - LC)]}$$

$$= \frac{\det \begin{bmatrix} s\mathbf{I}-(A-BK-LC+\frac{1}{N}MK) & -\frac{1}{N}M \\ \mathbf{0} & 1 \end{bmatrix}}{\det [s\mathbf{I} - (A - BK - LC)]} \times \mathrm{N}$$

or

$$\frac{T}{R} = N \frac{\det \left[s\mathbf{I}-(A-BK-LC+\frac{1}{N}MK) \right]}{\det [sI-(A-BK-LC)]} \tag{8.77}$$

A similar procedure reveals that

$$\frac{S}{R} = \frac{\det \begin{bmatrix} sI-(A-BK-LC) & -L \\ K & 0 \end{bmatrix}}{\det [s\mathbf{I}-(A-BK-LC)]} \tag{8.78}$$

From (8.75) and (8.77) we conclude that

$$\frac{Y(s)}{U_{ff}(s)} = \frac{B_p(s)\ \det[sI - A + BK + LC]}{\alpha_c(s)\ \alpha_0(s)} \tag{8.79}$$

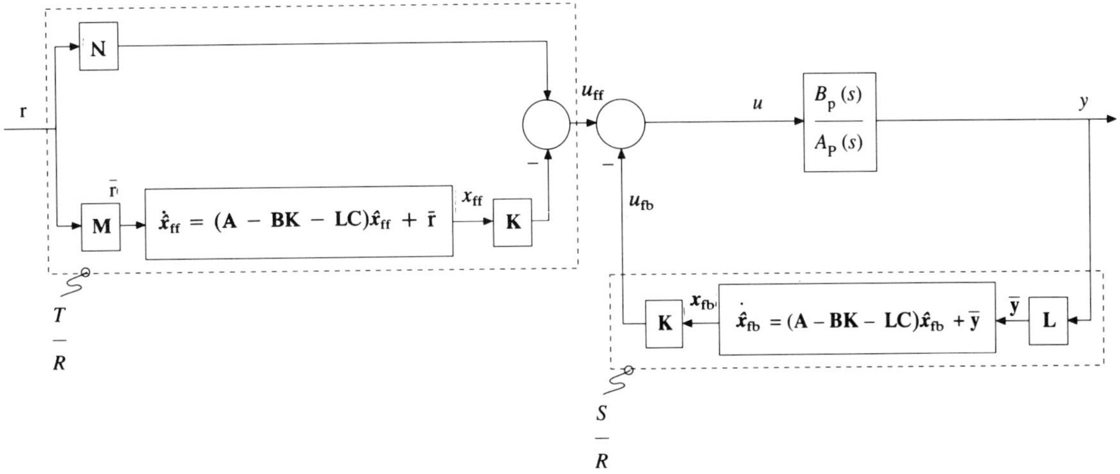

Figure 8.24 System of previous figure showing the contribution of each of the compensator inputs to the state vector $\hat{\mathbf{x}}=\hat{\mathbf{x}}_{ff}+\hat{\mathbf{x}}_{fb}$.

For ease of exposition let deg. $B_p=m$ and deg. $A_p=n$. The differences between state-space design and the algebraic design method are apparent from what has just been said.

In essence we have less freedom in the state-space method since the degree of the compensator is always n (or $(n-1)$ for a reduced order observer). Consequently we can not alter the pole-excess, and the maximum number of (stable) zeros under direct design control is n; when deg. $B_m>n$, then the compensated system, via state-space, shares (deg. B_m-n) zeros with the plant. These facts are summarized in Table 8.3.

Table 8.3 Algebraic and state-space design

	Pole excess, p_e, of the compensated system	*Dimension of the compensator*	*Maximum number of prescribed stable zeros under direct design control*
State-space design	$p_e = (n - m)$	n	n
Algebraic (transfer-function) design	$p_e \geqslant (n - m)$	deg. $R \geqslant n$	deg. B_m (no restriction)

For example if the prescribed transfer functions B_m/A_m is such that deg. $A_m=2n$, deg. $B_m=n+m$, then only n zeros of B_m can be allocated at will, by the state-space method; the remaining m must be the plant zeros.

8.3 CONTROLLER WINDUP

Actuators have a limited range of linear operation and most exhibit a saturation type input–output characteristic, as in Fig. 8.25. While saturation takes place, the

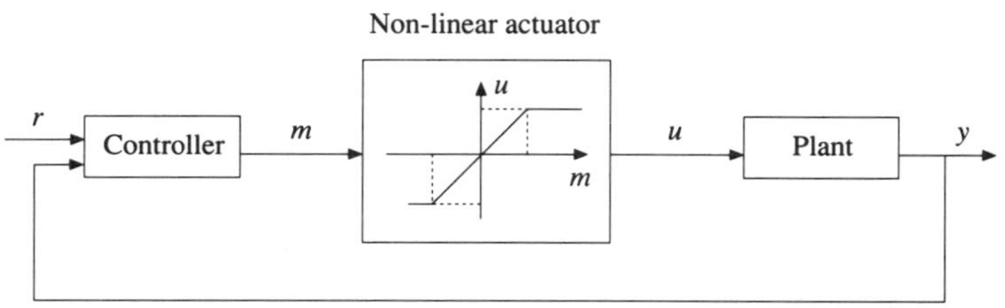

Figure 8.25 Control system with saturation-type non-linearity.

feedback loop is broken, since u remains constant no matter the value of y. Therefore, if the plant or the controller are unstable, their outputs may reach large values after saturation takes place, and in some cases the control system may not even regain stable operation.

In the following we analyze some practical ways to prevent controller windup. These techniques depend on how the controller is described, namely state-space or transfer-function form.

8.3.1 Regulator in State-space Form

In section 8.1 two 'equivalent' realizations were shown for the state feedback controller in Figs 8.3 and 8.4. Equivalent was written in inverted commas to stress that equivalence is no longer true if actuator saturation takes place, for example. The problem with the version of Fig.8.4 is that saturation breaks the loop into the plant and a system whose eigenvalues are the zeros of

$$\det [s\mathbf{I} - (\boldsymbol{A} - \boldsymbol{BK} - \boldsymbol{LC})] = 0 \tag{8.80}$$

which are not guaranteed to be stable. This corresponds to the situation of Fig. 8.26.

By contrast, in the version of Fig. 8.3 the plant input is also an explicit input to the controller. Naturally, if the controller has the actual plant input as one of its inputs, then the state estimate $\hat{\boldsymbol{x}}$ will continue to converge to the true plant state, $\boldsymbol{x}$, regardless of saturation. This situation is depicted in Fig. 8.27. The controller eigenvalues in this case remain the zeros of

$$\det [s\mathbf{I} - (\boldsymbol{A} + \boldsymbol{LC})] = 0 \tag{8.81}$$

which are stable by construction. Therefore windup will never take place.

8.3.2 Regulator in Transfer-function Form

Assume in Fig. 8.25 a controller with error feedback and transfer function

$$G_c(s) = \frac{N_c(s)}{D_c(s)}$$

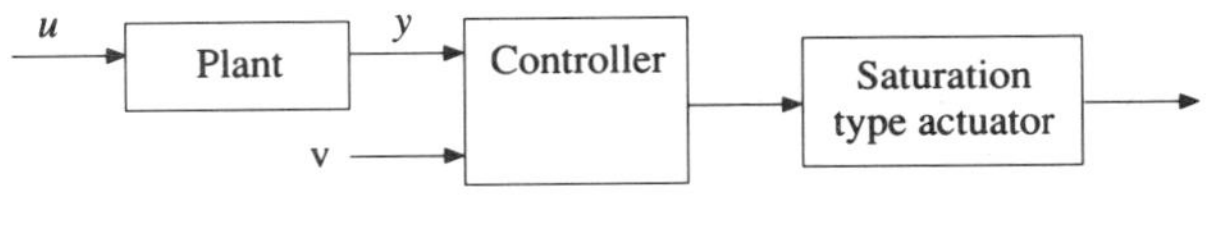

Figure 8.26 State-feedback controller with implicit observer in the presence of saturation.

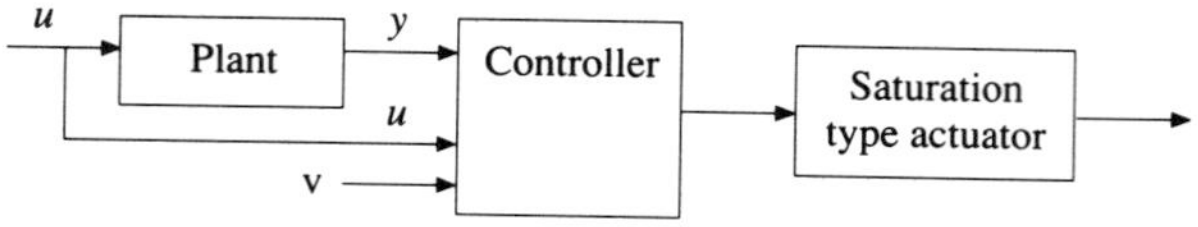

Figure 8.27 State-feedback controller with explicit observer in the presence of saturation.

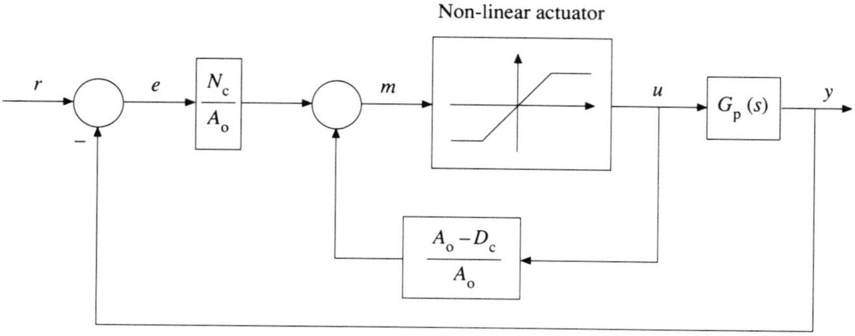

Figure 8.28 Practical implementation of the controller $G_c(s)=N_c(s)/D_c(s)$ that avoids the windup problem.

i.e.

$$M(s) = \frac{N_c(s)}{D_c(s)}(R(s) - Y(s)) \tag{8.82}$$

When $G_c(s)$ is unstable, the set-up of Fig. 8.28 avoids the controller windup problem, provided A_o is stable and of degree not less than the degree of D_c; this requirement is in order to ensure that $(A_o - D_c)/A_o$ is implementable.

Note that in Fig. 8.28

$$\frac{U}{E} = \frac{N_c}{A_o} \cdot \frac{1}{1 - (A_o - D_c)/A_o} = \frac{N_c}{D_c} \tag{8.83}$$

in the absence of saturation.

The above technique cannot be applied to an already existing PID controller, without hardware alterations, since it would require the elimination of the integral mode. In the case of PID controllers the standard procedure to avoid integrator windup is shown in Fig. 8.29.

In the absence of saturation, $m=u$ and consequently $U(s)/E(s)=1/s$. When saturation takes place, we may regard the system as having e and u as inputs, and

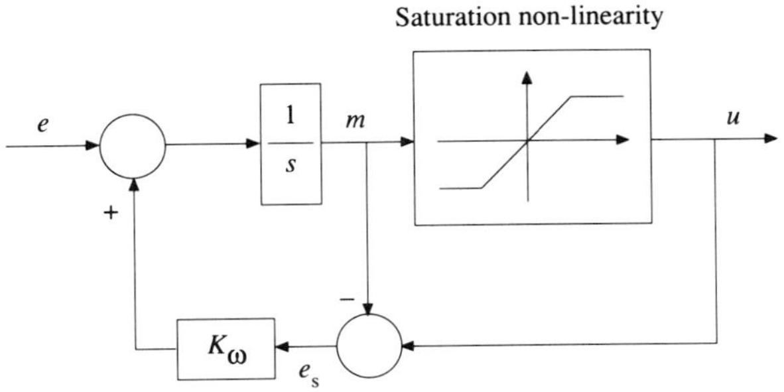

Figure 8.29 Anti-windup integrator.

output m; then we can write

$$M(s) = \frac{1}{s + K_\omega} E(s) + \frac{K_\omega}{s + K_\omega} U(s) \tag{8.84}$$

The larger K_ω, the faster $m(t)$ will converge to the saturation limit, i.e. the faster $e_s(t)=(u(t)-m(t))$ will go to zero.

Consequently $1/K_\omega$ is called the tracking time constant.

8.4 THE PID CONTROLLER REVISITED

The PID controller still constitutes the most frequently encountered control algorithms in industry. In Chapter 4 we analyzed it in some detail, although with a somewhat simplified structure. In this section, we present a slightly more complex version of the PID controller which follows naturally from the methods and requirements that we have been discussing. The PID controller was introduced by the equation

$$m(t) = K\left(e + \frac{1}{T_i} \int_0^t e(t')\, dt' - T_d \frac{dy}{dt}\right) \tag{8.85}$$

as in Fig. 8.30.

The first thing that draws our attention is the presence of the derivative term $T_d\, dy/dt$; it contradicts that we have said earlier about controller bandwidth and transfer function implementability. Such a term will amplify the high frequency measurement noise n, and may cause instability in the presence of modelling uncertainities. Besides, its transfer function is improper.

In most commercial controllers the pure derivative block is replaced by the

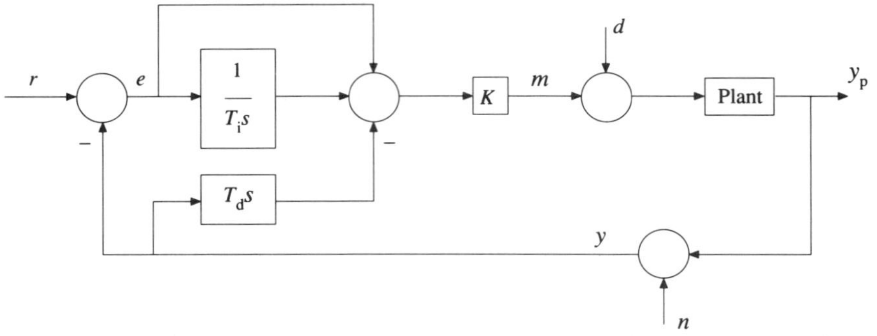

Figure 8.30 'Textbook' PID controller.

transfer function

$$\frac{T_d s}{1 + (T_d/N)\, s} \tag{8.86}$$

where N is typically in the range of 10–100. At low frequencies this block behaves as the ideal derivative, and at high frequencies the gain does not exceed N.

We have also learned how to tune this controller to achieve a satisfactory response to load disturbances. Unfortunately we were unable to control the response to step reference signals which, very often, exhibited significant overshoot. From what we have learned we know the reason lies in the fact that the zeros are not under direct design control.

Going back to section 8.2, we learned that the polynominal T, in the forward transfer function T/R, is responsible for the closed-loop zero allocation. Besides, the system response to load disturbances is independent of T. This suggests the introduction of an extra design parameter as in Fig. 8.31.

> **Exercise 6.19** Show that the transfer function $Y(s)/D(s)$, in Fig. 8.31, is independent of the parameter f.

By changing f, we can adjust the response to a reference step, for example, without changing the system characteristics to load disturbances.

Summing up, we have arrived at a PID controller with the following structure:

$$M(s) = P(s) + I(s) + D(s) \tag{8.87}$$

where

$$P(s) = K\,(fR(s) - Y(s)) \tag{8.88}$$

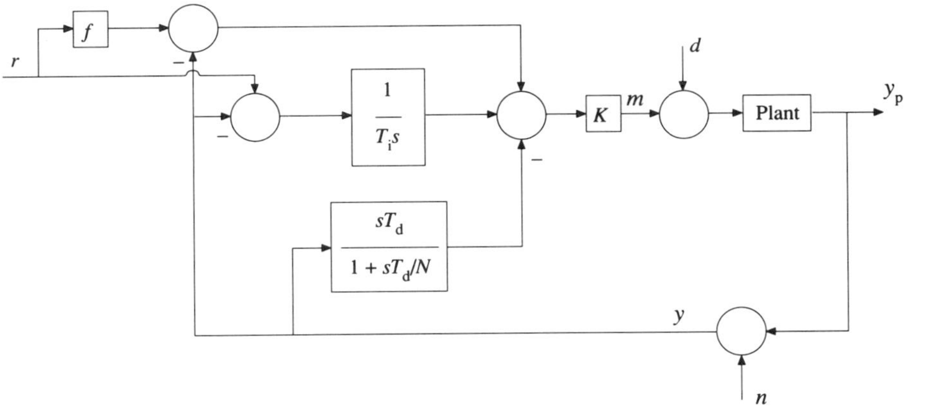

Figure 8.31 Commercial PID controller.

$$I(s) = \frac{K}{T_i s}(R(s) - Y(s)) \tag{8.89}$$

$$D(s) = -\frac{sT_d}{1 + sT_d/N} KY(s) \tag{8.90}$$

Note that the integral term still remains a function of $(r-y)$ in order to prevent steady-state errors in responses to constant command inputs.

For a thorough treatment of PID controllers, the reader is referred to Aström and Hägglund (1988).

Some commercial controllers implement the following algorithm, known as 2 PID:

$$m(t) = K\left[\left(e(t) + \frac{1}{T_i}\int e(t)\mathrm{d}t + T_\mathrm{d}\frac{\mathrm{d}\,e(t)}{\mathrm{d}t}\right) - \left(a\,r(t) + b\,T_\mathrm{d}\frac{\mathrm{d}\,r(t)}{\mathrm{d}t}\right)\right]$$

as depicted in Fig. 8.32.

As shown below, the closed-loop system poles are independent of a and b, and the closed-loop zeros are the plant zeros and the zeros of

$$((1 - b)T_iT_ds^2 + (1 - a)T_is + 1)$$

In this fashion, by changing a and b we can change the controller zeros, without changing the system response to load disturbances.

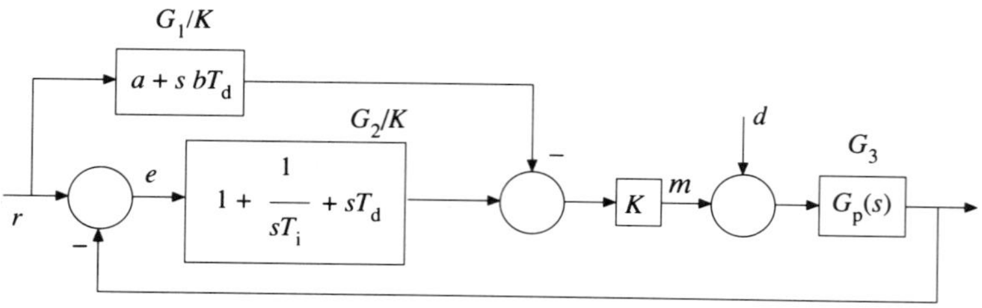

Figure 8.32 2 PID controller.

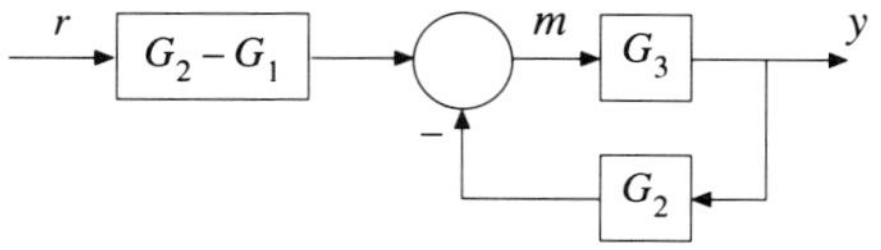

Figure 8.33 Equivalent diagram of Fig. 8.32.

First note that our control system can be written as in Fig. 8.33. Therefore

$$\frac{Y}{R} = \frac{G_3(G_2 - G_1)}{1 + G_2G_3}$$

Letting $G_i = N_i/D_i$, $i=1, 2, 3$, we can write:

$$\frac{Y}{R} = \frac{\frac{N_3}{D_3}\left(\frac{N_2}{D_2} - \frac{N_1}{D_1}\right)}{1 + \frac{N_2\,N_3}{D_2\,D_3}} = \frac{N_3(D_1N_2 - D_2N_1)}{D_1\,D_2\,D_3 + N_2\,N_3\,D_1}$$

In our case $D_1=1$, $D_2=sT_i$, $N_1=K(a+sbT_d)$ and $N_2=K(1+sT_i+s^2T_iT_d)$. Therefore the closed-loop poles are independent of a and b, and the zeros introduced by the controller are the roots of the polynomial

$$D_1N_2 - D_2N_1 = 1 + sT_i(1 - a) + s^2T_iT_d(1 - b)$$

REFERENCES

Aström, K.J. (1980), 'Robustness of a design method based on assignment of poles and zeros', *IEEE Trans. Auto. Control*, **AC–25**(3), pp. 588–91.

Aström, K.J. and B. Wittenmark (1984) *Computer Controlled Systems: Theory and Design*, Prentice Hall.

Aström, K.J. and T. Hägglund (1988) *Automatic Tuning of PID Controllers*, Instrument Society of America.

Chen, C.T. (1987) *Control System Design – Conventional, Algebraic and Optimal Methods*, Pond Woods Press.

Lopes dos Santos, P. and J.L. Martins de Carvalho (1990) 'Automatic transfer function synthesis from a Bode plot', 29th IEEE Conference on Decision and Control, 5–7 December, Honolulu, Hawaii.

Graham, D. and R.C. Lathrop (1953) 'The synthesis of optimum response: criteria and standard forms', *AIEE*, **72**, pt. II, 273–88.

Kailath, T. (1980) *Linear Systems*, Prentice Hall.

9

Discrete-time Systems Analysis

In this chapter we will develop the theory and the tools required to analyze (process) models when the controller is a digital computer, namely z-transforms, discrete-time models, controllability, observability and stability conditions. The development parallels that of Chapters 3 and 7 and many of the results obtained there apply, *mutatis mutandis*, to the discrete-time case.

9.1 COMPUTER-CONTROLLED SYSTEMS

As shown in Fig. 9.1, the structure of a computer-controlled system does not differ from the analog one. However, because a digital computer works intermittently, it is unable to monitor the process variables *continuously*. You would find yourself in a similar position if when cycling, for example, you let your eyes open only for a short time at regular intervals: while they are closed, (visual) feedback is suppressed and the faster you move or the more obstacles (disturbances) are present, the greater the need to *sample* the environment more frequently.

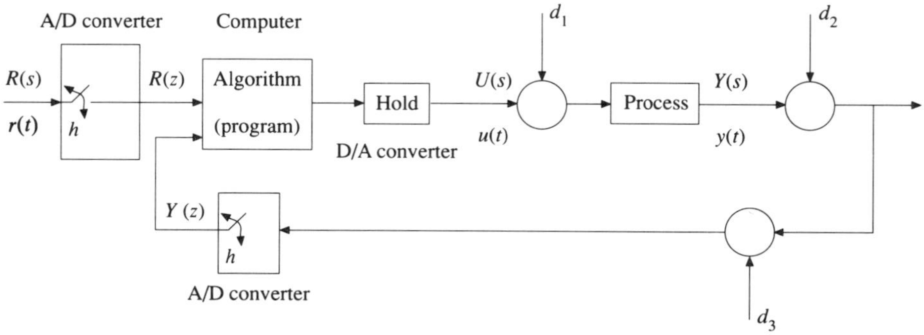

Figure 9.1 Structure of a computer-controlled process.

Although the lower theoretical limit for the sampling interval, h, is the period of the computer internal clock, in practice a lot of clock impulses are required to perform calculations. As a consequence, the frequency a computer can sample a process decreases with the amount of calculations to be performed between sampling instants. A computer-controlled system is also known as a *sampled-data system*, because it has both continuous and discrete (sampled) signals. As will be seen in Chapter 10 the mixture of continuous and discrete signals complicates the mathematical treatment and can cause ambiguities, mainly because sampling is a *time-varying* operation (see below). However, this problem can be avoided if we consider signals only at the sampling instants, or equivalently, if we regard the process as 'seen' from the computer; then we only have to deal with sequences of numbers, i.e. *discrete-time signals*.

Another characteristic of a computer-controlled system is the fact that the analog-to-digital (A/D) converter can only generate a finite set of equally spaced numbers. This means that the (continuous) range of variation of our process variables is going to be replaced in the computer by a finite set of values; such an operation is called *quantization* and represents a loss of accuracy that impedes the performance of the control system. The consequences of quantization are oscillations in the controlled variable (limit cycles), similar to those induced by on–off controllers as shown in Fig. 1.3, steady-state errors and noise. A signal which is both discrete and quantized is called a *digital signal*.

Although digital computers work with sequences of numbers it will be helpful, for a later frequency-domain interpretation of the sampling process (cf. section 10.2), to model them as sequences of Dirac impulses, h time units apart, whose areas are the values of the continuous-time signals at the corresponding instants, as shown in Fig. 9.2. This will cause no ambiguities, such as products of impulses at the same instant, provided the operations upon the signals in the computer are linear, as will be the case. Then the digital to analog (D/A)

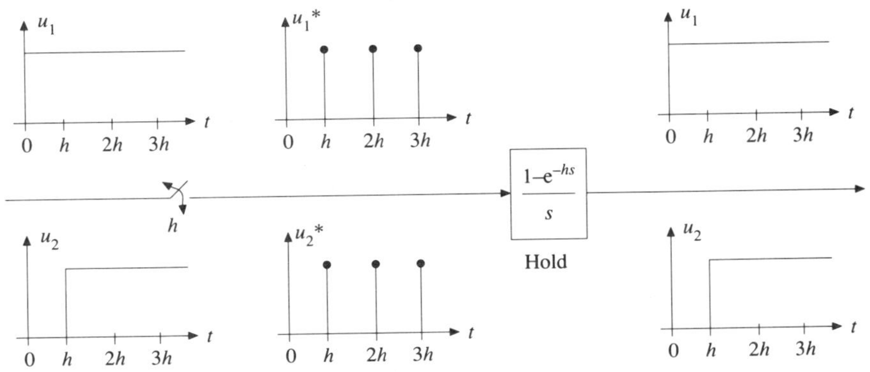

Figure 9.2 Illustrating the time-varying nature of the sampling process.

conversion can be modelled by the hold block of Fig. 9.2, whose transfer function is

$$\frac{1 - \exp(-hs)}{s}$$

In fact, the response of this transfer function to the Dirac impulse, with area A, applied at time $t=0$, is the rectangular pulse with height A and width h, starting at time=0. In Fig. 9.3 we construct such response as the result of the algebraic sum of the responses of blocks $1/s$ and $-e^{-hs}/s$.

This process of signal reconstruction is known as 'zero-order hold' if, during the sampling interval, the continuous-time signal is replaced by a constant, i.e. a zero degree polynomial. Figure 9.2 also illustrates the time-varying nature of the sampling process mentioned above: when the input is delayed by an amount d, $d \neq kh$, $k \in \mathbb{N}_0$, the output is not delayed by the same amount as it would be in a time-invariant system; however, when $d=kh$, the sampling process behaves as time-invariant, i.e. the sampling process is a periodic transformation with period h.

Exercise 9.1 Plot the response of system of Fig. 9.2 to a ramp input.

Exercise 9.2 Discuss what types of open-loop systems are obtained when the loop in Fig. 9.1 is cut:
(a) At the process input.
(b) After the A/D converter.
Recall that the process input and output are continuous-time signals, while the signals in the computer can only change at times $t=kh$, $k \in \mathbb{N}_0$.
Answers: (a) sampled-data system; (b) discrete-time system.

Exercise 9.3 What is the transfer function of the system of Fig. 9.2?
Answer: it does not exist because $U'_1(s)/U_1(s) \neq U'_2(s)/U_2(s)$.

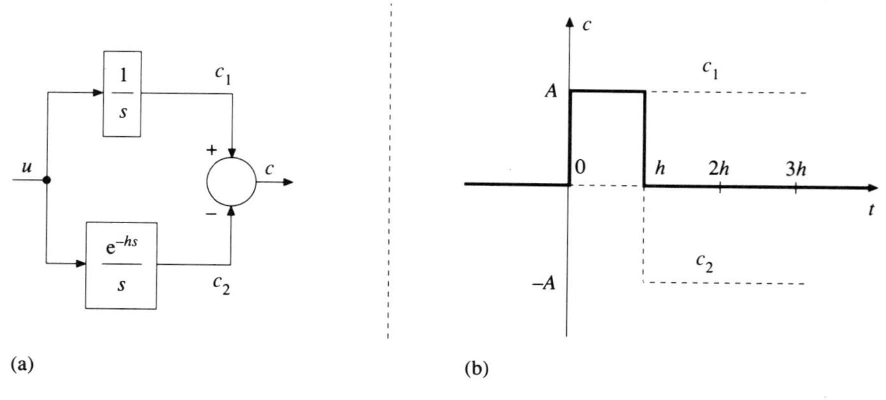

Figure 9.3 Computing the response of $1-e^{-hs}/s$ to the Dirac impulse of area A, at time $t=0$, by the superposition principle.

9.2 THE z-TRANSFORM

When a sampled (or discrete-time) signal is modelled as a sequence of *equally spaced* Dirac impulses, its Laplace transform takes a very simple and meaningful form as shown below.

If $u(t)$ is a continuous-time signal, such that $u(t)=0$ for $t<0$, and $u^*(t)$ denotes its sampled version as shown in Fig. 9.2 we have

$$u^*(t) = \sum_{k=0}^{\infty} u(kh)\ \delta(t - kh) \tag{9.1}$$

and

$$\mathcal{L}(u^*(t)) = \sum_{k=0}^{\infty} u(kh)\ e^{-khs} \tag{9.2}$$

because $\mathcal{L}(\delta(t-a))=e^{-as}$. With the change of variable

$$z = e^{hs} \tag{9.3}$$

equation (9.2) can be written as

$$U(z) = \sum_{k=0}^{\infty} u(kh)\ z^{-k} \tag{9.4}$$

which we will define as the *z-transform* of the continuous-time signal u and denote by $\mathcal{Z}(u)$. For example the z-transform of the signal

$$u(t) = \begin{cases} 0;\ t < 0 \\ \exp(-\ at);\ t \geqslant 0 \end{cases}$$

computed by definition (9.4), is as follows

$$\mathcal{Z}(u) = 1 + e^{-ah}\ z^{-1} + e^{-2ah}\ z^{-2} + \ldots = \frac{z}{z - e^{-ah}}$$

for $|z|>e^{-ah}$.

Exercise 9.4 Show the above result for the summation of the series. *Hint*: recall that $1+r+r^2+\ldots+r^n=1-r^{n+1}/(1-r)$.

Table 9.1 shows the z-transforms of the most common types of time functions; these are assumed zero for negative instants. For a complete list of z-transform pairs, the reader is referred to Smith (1987).

Table 9.1 Some time functions and corresponding z-transforms

Function	*z-Transform*
a^t	$\dfrac{z}{z - a^h}$
t	$\dfrac{h\, z}{(z - 1)^2}$
t^2	$\dfrac{h^2\, z\, (z + 1)}{(z - 1)^3}$
e^{-at}	$\dfrac{z}{z - e^{-ah}}$
$t\, e^{-at}$	$\dfrac{h\, z\, e^{-ah}}{(z - e^{-ah})^2}$
$\sin\omega t$	$\dfrac{z \sin(\omega h)}{z^2 - 2\, z \cos(\omega h) + 1}$
$e^{-at} \sin(\omega t)$	$\dfrac{z\, e^{-ah} \sin(\omega h)}{z^2 - 2\, z\, e^{-ah} \cos(\omega h) + e^{-2ah}}$
$\cos(\omega t)$	$\dfrac{z\, (z - \cos(\omega h))}{z^2 - 2\, z \cos(\omega h) + 1}$
$e^{-at} \cos(\omega t)$	$\dfrac{z^2 - z\, (e^{-ah} \cos(\omega h))}{z^2 - 2\, z\, e^{-ah} \cos(\omega h) + e^{-2ah}}$
$\sin(\omega t + \phi)$	$\dfrac{z^2 \sin \phi + z \sin(\omega h - \phi)}{z^2 - 2\, z \cos(\omega h) + 1}$
$e^{-at} \sin(\omega t + \phi)$	$\dfrac{z^2 \sin \phi + z\, e^{-ah} \sin(\omega h - \phi)}{z^2 - 2\, z\, e^{-ah} \cos(\omega h) + e^{-2ah}}$

9.2.1 Properties of the z-Transform

Analogously to the Laplace transform, the z-transform enjoys a set of useful properties, some of which are listed below. Recall that all time functions are assumed zero for $t<0$.

9.2.1.1 Linearity

$$\mathfrak{Z}(ax(t) + by(t)) = a\mathfrak{Z}(x(t)) + b\mathfrak{Z}(y(t))$$

9.2.1.2 Time shift

Let n denote a non-negative integer, $X(z)$ the z-transform of $x(t)$, and h the sampling period; then

1. $\mathfrak{Z}(y(t))=z^{-n} X(z)$, if $y(t)=x(t-nh)$, $\forall t$, since $x(t-nh)=0$ for $t<nh$.
2. $\mathfrak{Z}(y(t))=z^n X(z)-z^n x(0)-z^{n-1} x(h)- \dots -zx((n-1)h)$ if $y(t)=x(t+nh)$, $t\geqslant 0$ and $y(t)=0$ for $t<0$. (9.5)

We now prove (2) for $n=1$. From the definition we have:

$$\mathfrak{Z}(y(t)) = x(h) + z^{-1} x(2h) + z^{-2} x(3h) + \dots$$

which can be written as

$$\mathfrak{Z}(y(t)) = \mathfrak{Z}[x(0) + z^{-1} x(h) + z^{-2} x(2h) + \dots]- zx(0)$$

The sum in brackets on the right hand side is $X(z)$; therefore

$$\mathfrak{Z}(y(t)) = zX(z) - zx(0)$$

QED.

Exercise 9.5 Show, by finite induction, formula (9.5).

As we will see shortly, properties (1) and (2) above play a role similar to the Laplace transform of the time derivative of a signal, by transforming difference equations into algebraic ones.

9.2.1.3 Initial value theorem

$$x(0) = \lim_{z\to\infty} X(z) \tag{9.6}$$

This follows from the definition of $X(z)$ by letting $z\to\infty$.

9.2.1.4 Final value theorem

If $\lim_{t\to\infty} x(t)$ exists, then

$$\lim_{t\to\infty} x(t) = \lim_{z\to 1} [(z - 1) X(z)] \tag{9.7}$$

These two properties are very useful for computing the limiting values of $x(t)$ without going through the process of inverting the z-transform. However, the latter may lead to erroneous results when $x(t)$ has no limit, as $t \to \infty$. For example, for $x(t)$ such that, at the sampling instants, $x(kh)=(-1)^k$ we have

$$\lim_{z \to 1} (z - 1)\, X(z) = 0$$

where $X(z)=z/(z+1)$; however $\lim_{t \to \infty} x(t)$ does not exist!

Let us now show result (9.7). Because

$$\mathfrak{Z}(x(t)) = \sum_{k=0}^{\infty} x(kh) z^{-k}$$

we can write

$$(z - 1)\, X(z) - z\, x(0) = \sum_{k=0}^{\infty} x((k + 1)h) z^{-k} - \sum_{k=0}^{\infty} x(kh) z^{-k}$$

$$(z - 1)\, X(z) = zx(0) + \sum_{k=0}^{\infty} [x((k + 1)h) - x(kh)]\, z^{-k}$$

If $\lim_{t \to \infty} x(t)$ exists, the right hand side above becomes $z\, x(0) + \lim_{t \to \infty} x(t) - x(0)$ and the result follows with $z=1$.

9.2.1.5 Time multiplication

Another property which is often useful for the inversion of z-transforms, is the following:

$$\mathfrak{Z}(t\, x(t)) = -\, h\, z\, \mathrm{d}/\mathrm{d}z(X(z)) \tag{9.8}$$

This result is a straightforward consequence of the definition of the z-transform; in fact

$$\mathfrak{Z}(t\, x(t)) = \sum_{k=0}^{\infty} (kh)\, x(kh) z^{-k}$$

and

$$X(z) = \sum_{k=0}^{\infty} x(kh) z^{-k}$$

but

$$d/dz\ (X(z)) = \sum_{k=0}^{\infty} x(kh)\ (-\ k)z^{-k-1}$$

and the result follows.

9.2.1.6 Scaling in the *z*-plane

$$\mathcal{Z}(a^{-t}x(t)) = X(a^{h}z) \tag{9.9}$$

where $\mathcal{Z}(x(t))=X(z)$. This result can be shown by simple direct substitution:

$$\mathcal{Z}(a^{-t}x(t)) = \sum_{k=0}^{\infty} a^{-kh}x(kh)z^{-k}$$

$$= \sum_{k=0}^{\infty} x(kh)(a^{h}z)^{-k}$$

$$= X(a^{h}z)$$

q.e.d.

This property is useful, for example, in the derivation of *z*-transforms, as illustrated in the following.

Example 9.1 Compute $\mathcal{Z}(e^{-ct}\sin(\omega t))$ knowing that

$$\mathcal{Z}(\sin(\omega t)) = \frac{z\ \sin(\omega h)}{z^2 - 2z\cos(\omega h) + 1}$$

Letting $a=e^{c}$ in section 9.2.1.6 we get

$$\mathcal{Z}(e^{-ct}\sin(\omega t)) = \frac{ze^{ch}\sin(\omega h)}{z^2e^{2ch} - 2ze^{ch}\cos(\omega h) + 1}$$

$$= \frac{ze^{-ch}\sin(\omega h)}{z^2 - 2ze^{-ch}\cos(\omega h) + e^{-2ch}}$$

9.2.2 *z*-Transform Inversion

When the *z*-transform is a rational function of the complex variable *z*, we can expand it into partial fractions, as we did with Laplace transforms.

Example 9.2 Compute $(x(nh))n \in \mathbb{N}_0$ knowing that

$$X(z) = \frac{2z}{(z+1)(z-1)}$$

Expanding $X(z)/z$ into partial fractions we get

$$\frac{X(z)}{z} = \frac{-1}{z+1} + \frac{1}{z-1}$$

Then

$$X(z) = \frac{-z}{z+1} + \frac{z}{z-1}$$

From the z-transform table we have

$$\mathscr{Z}^{-1}\left(\frac{z}{z+1}\right) = ((-1)^n)_{n \in \mathbb{N}_0} \text{ and } \mathscr{Z}^{-1}\left(\frac{z}{z-1}\right) = 1$$

Therefore

$$\{x(nh)\}_{n \in \mathbb{N}_0} = \begin{cases} 2; & n \text{ odd} \\ 0; & n \text{ even} \end{cases}$$

Because the z-transforms appearing in the table have z as a factor in the numerator, it is convenient to make the partial fraction expansion of $X(z)/z$ instead of $X(z)$, as shown in the previous example.

The definition of z-transform, equation (9.4), provides another method for its inversion: if $X(z)$ is a rational fraction, we can expand it into an infinite power series by dividing the numerator and denominator polynomials. Although this has the disadvantage of not leading directly to a closed form solution, it is ideally suited to a digital computer. The previous example can be solved by this method as follows:

$$X(z) = \frac{2z}{(z+1)(z-1)} = \frac{2z}{z^2-1}$$

By long division we get

$$\begin{array}{l|l}
2z & z^2 - 1 \\ \cline{2-2}
-2z + 2z^{-1} & 2z^{-1} + 2z^{-3} + 2z^{-5} + \ldots \\ \hline
\quad 2z^{-1} & \\
\quad -2z^{-1} + 2z^{-3} & \\ \hline
\qquad 2z^{-3} & \\
\qquad -2z^{-3} + 2z^{-5} & \\ \hline
\qquad\quad +2z^{-5} & \\
\qquad\quad \vdots &
\end{array}$$

i.e.

$$X(z) = 2(z^{-1} + z^{-3} + z^{-5} + \ldots)$$

or equivalently $x(0)=0$, $x(h)=2$, $x(2h)=0$, $x(3h)=2$, etc., which coincides with the result obtained by the partial fraction expansion method.

As we shall see in the next section, a signal with z-transform $X(z)$ can be regarded as the unit pulse response of a system with transfer function $X(z)$. On the other hand it is straightforward to convert a transfer function into its equivalent difference equation. Consequently $\mathcal{Z}^{-1}X(z)$ can also be computed by solving recursively this difference equation for the unit pulse input, applied at time $k=0$. In the example just discussed, we see that the associated difference equation is

$$y(k + 2) - y(k) = 2u(k)$$

Then solving for the unit pulse input, with zero initial conditions, i.e. $y(k)=0$, $k \leqslant 0$, we get for $k=-1, 0, 1, 2, \ldots$

$$y(1) = 0$$

$$y(2) = 2$$

$$y(3) = 0$$

$$y(4) = 2$$

$$\ldots$$

which coincides with the result obtained by long division.

Another method of z-transform inversion parallels the inversion integral of Laplace transforms, and is of interest in numerical inversion problems. Recall that

$$\mathrm{f}(t) = \frac{1}{2\pi \mathrm{j}} \int_{c+\mathrm{j}\infty}^{c+\mathrm{j}\infty} \mathrm{F}(s)\, \mathrm{e}^{ts}\, \mathrm{d}s \tag{9.10}$$

where $\mathcal{L}(f(t))=F(s)$ and c is a real number greater than the real parts of all the singularities of the integrand $\mathrm{F}(s)$. It can be shown, e.g. Kuo (1980), that for the z-transforms we have a similar formula, namely

$$f(kh) = \frac{1}{2\pi \mathrm{j}} \oint_{\Gamma} F(z)\, z^{k-1}\, \mathrm{d}z$$

where Γ is a closed path enclosing all the singularities of $F(z)$; such a formula is not unexpected given the fact that the z-transform is a Laplace transform and that the path of integration in (9.10) is mapped into the z-plane circle $|z|=\mathrm{e}^{ch}$ by the transformation $z=\mathrm{e}^{sh}$.

Exercise 9.6

(a) Show that the transformation $z=\mathrm{e}^{hs}$ maps a straight line parallel to the imaginary axis with abscissa c, in the s-plane, into a circle of radius e^{ch} in the z-plane. Plot the result for $c=-1, 0, 1$.

(b) What is the image, by the same transformation, of a straight line parallel to the s-plane real axis?

Answer: a straight line starting at the origin of the z-plane.

9.3 DISCRETE-TIME MODELS

In Chapter 2 discrete-time models were introduced. These find wide application in areas such as economics, ecology (population models), biology, etc. where the variables are intrinsically discrete. A good catalog of examples can be found in Luenberger (1979). However, continuous-time phenomena can also give rise to discrete-time models if they are observed only at discrete instants, as illustrated by the following simple example.

Assume that at time $t=0$ the terminals of a charged capacitor are connected via a resistor, and that the voltage is not monitored continuously but (sampled) at times $t=kh$, $k \in \mathbb{N}_0$, as shown in Fig. 9.4. An elementary analysis reveals that voltage v is governed by the equation:

$$RC\, \mathrm{d}v/\mathrm{d}t + v = 0 \tag{9.11}$$

Denoting by $v(kh)$ the voltage at time $t=kh$, we have

$$v(t) = v(kh) \exp[-(t-kh)/(RC)],\ t \geq kh$$

Particularly for $t=(k+1)h$ this gives

$$v((k+1)h) = v(kh) \exp[-h/(RC)]$$

or in a more compact form

$$v(k+1) = a\, v(k) \tag{9.12}$$

with $a=\exp[-h/(RC)]$. A change in the observation process has changed the description of the circuit from a first-order homogeneous linear differential equation, (9.11), into a first-order homogeneous linear *difference* equation, (9.12).

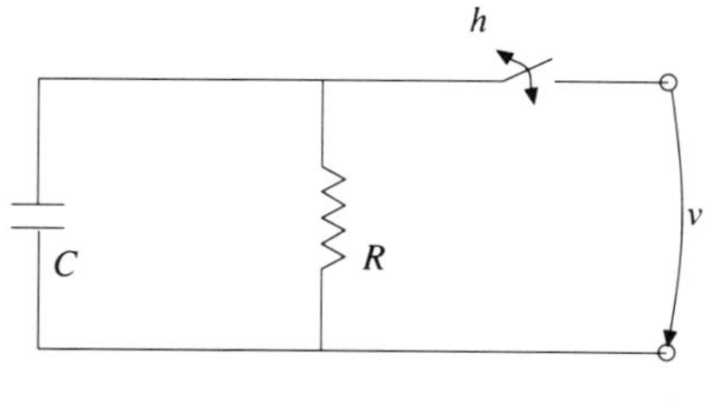

Figure 9.4 Sampling a continuous-time signal.

In general a linear discrete-time dynamical system with input $\{u(k)\}_{k\in_0}$ and output $\{y(k)\}_{k\in_0}$ can be described by a linear difference equation

$$y(k+n) + a_{n-1}y(k+n-1) + \ldots + a_0 y(k) = b_n u(k+n) + b_{n-1}u(k+n-1) + \ldots + b_0 u(k) \tag{9.13}$$

or by a state-space model

$$\begin{aligned} \boldsymbol{x}(k+1) &= \mathbf{G}\,\boldsymbol{x}(k) + \mathbf{H}\,u(k) \\ y(k) &= \boldsymbol{C}\,\boldsymbol{x}(k) + D\,u(k) \end{aligned} \tag{9.14}$$

with $\boldsymbol{x}\in\mathbb{R}^n$.

In the design of a computer-controlled system as in Fig. 9.1, a model of the type (9.13) or (9.14) will be required to describe the dynamical behaviour of the control loop between the input of the D/A converter and the output of the A/D converter. Such a model is derived below assuming the plant described by a (continuous-time) state-space model. Such a description is also known as the 'hold equivalent' because it provides an exact model, at the sampling times, when the plant input is a piecewise-constant signal. In Chapter 10 we will present other methods to discretize continuous-time models in connection with digital controller synthesis by translation of analog design.

9.3.1 Discretization of State-space Models

9.3.1.1 Delay-free case

Assume we have a (continuous-time) plant with m inputs and r outputs described by the model

$$\begin{aligned} \dot{\boldsymbol{x}} &= \boldsymbol{Ax} + \boldsymbol{Bu} \\ \boldsymbol{y} &= \boldsymbol{Cx} + \boldsymbol{Du} \end{aligned} \tag{9.15}$$

whose inputs change only at the sampling instants, as shown in Fig. 9.5. Then our

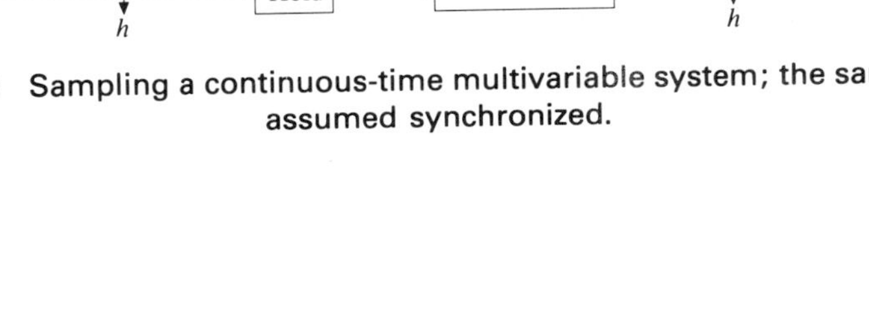

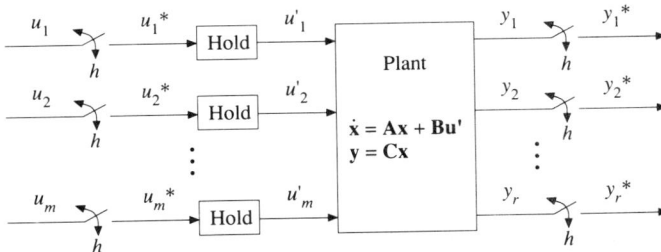

Figure 9.5 Sampling a continuous-time multivariable system; the samplers are assumed synchronized.

problem is to find a model of the type (9.14) that relates inputs and outputs at the sampling instants.

Bearing in mind the time-invariance of the plant and the fact that $u'(t)=\mathbf{u}(kh)$, $kh\leq t<(k+1)h$, we have from (3.60):

$$\boldsymbol{x}((k+1)h) = \exp(\mathbf{A}h)\,\boldsymbol{x}(kh) + \int_0^h \exp(\mathbf{A}(h-t'))\mathrm{d}t'\,\boldsymbol{B}\,\boldsymbol{u}(kh)$$

By the change of variable $t''=h-t'$ the above equation becomes

$$\boldsymbol{x}((k+1)h) = \exp(\mathbf{A}h)\,x(kh) + \int_0^h \exp(\mathbf{A}t'')\mathrm{d}t''\,\boldsymbol{B}\,u(kh) \tag{9.16}$$

which has the structure of (9.14) with

$$\boldsymbol{G}(h) = \exp(\mathbf{A}h) \tag{9.17}$$

$$\mathbf{H}(h) = \int_0^h \exp(\mathbf{A}t)\mathrm{d}t\,\boldsymbol{B} \tag{9.18}$$

Notice that $\boldsymbol{G}$ and $\boldsymbol{H}$ are functions of the sampling period. Therefore its choice is important in determining the dynamic properties of the discrete-time model; for example, controllability may be lost for certain choices of h as shown later.

Exercise 9.7
(a) Show that $\boldsymbol{G}(h)$ is never a singular matrix. *Hint*: use the fact that det $\boldsymbol{G}(h)=\exp[\text{trace }(\boldsymbol{A}h)]$.
(b) Verify that the calculation of $\boldsymbol{G}$ and $\boldsymbol{H}$ in (9.17) and (9.18) can be based on a single series calculation, namely

$$\boldsymbol{G} = I + \boldsymbol{A}\Psi$$

$$\boldsymbol{H} = \Psi\boldsymbol{B}$$

where

$$\Psi = \int_0^h \exp(\boldsymbol{A}t)\mathrm{d}t = Ih + \boldsymbol{A}h^2/2! + \boldsymbol{A}^2h^3/3! + \ldots + \boldsymbol{A}^l h^{l+1}/(l+1)! + \ldots$$

Exercise 9.8 Compute the discrete-time state-space model for the system in Fig. 9.5, assuming the plant described by

$$\dot{\boldsymbol{x}} = \begin{bmatrix} 0 & -2 \\ -1 & -1 \end{bmatrix} \boldsymbol{x} + \begin{bmatrix} 0 \\ 1 \end{bmatrix} u$$

$$y = (1,0)\,\boldsymbol{x}$$

Hint: Recall that the exponential of the above matrix was already computed in example 3.3 in section 3.7.

Solution of the discrete-time state equation

Difference equations are easier to solve on a digital computer than differential equations because the exact solution of the former can always be computed recursively. This is what we will do in the first method; in the second method, the solution is computed by means of the z-transform method, which has advantages when a closed form solution is desired.

Result 9.1: The solution of the discrete time equation

$$\boldsymbol{x}((k+1)h) = \boldsymbol{G}\,\boldsymbol{x}(kh) + \boldsymbol{H}\,\boldsymbol{u}(kh)$$

is given by

$$\boldsymbol{x}(kh) = \boldsymbol{G}^k\,\boldsymbol{x}(0) + \sum_{j=0}^{k-1} \boldsymbol{G}^{k-j-1}\,\boldsymbol{H}\,\boldsymbol{u}(jh),\ k \in \mathbb{N}_0 \tag{9.19}$$

where the second term on the right hand side is zero for $k=0$. The result is now shown by finite induction. For $k=1$ the result is obviously true. Now we prove that if it is true for $k=n$, $n>1$, then it is necessarily true for $k=n+1$. From our hypothesis and (9.14) we have

$$\boldsymbol{x}((n+1)h) = \boldsymbol{G}[\boldsymbol{G}^n\,\boldsymbol{x}(0) + \sum_{j=0}^{n-1} \boldsymbol{G}^{n-j-1}\,\boldsymbol{H}\,\boldsymbol{u}(jh)] + \boldsymbol{H}\,\boldsymbol{u}(nh)$$

$$= \boldsymbol{G}^{n+1}\,\boldsymbol{x}(0) + \sum_{j=0}^{n} \boldsymbol{G}^{n-j}\,\boldsymbol{H}\,\boldsymbol{u}(jh) + \boldsymbol{G}^0\,\boldsymbol{H}\,\boldsymbol{u}(nh)$$

$$= \boldsymbol{G}^{n+1}\,\boldsymbol{x}(0) + \sum_{j=0}^{n} \boldsymbol{G}^{n-j}\,\boldsymbol{H}\,\boldsymbol{u}(jh)$$

q.e.d.

Result 9.2: The solution of (9.14) is given by

$$\{\boldsymbol{x}(kh)\}_{k \in \mathbb{N}_0} = \mathfrak{Z}^{-1}((z\mathbf{I} - \boldsymbol{G})^{-1}z)\,\boldsymbol{x}(0) + \mathfrak{Z}^{-1}((z\mathbf{I} - \boldsymbol{G})^{-1}\boldsymbol{H}\boldsymbol{U}(z)) \tag{9.20}$$

where $\mathfrak{Z}^{-1}$ denotes the inverse z-transform operator.
Proof: Computing the z-transform of both sides of (9.14) we have

$$z\boldsymbol{X}(z) - z\boldsymbol{x}(0) = \boldsymbol{G}\boldsymbol{X}(z) + \boldsymbol{H}\,\boldsymbol{U}(z)$$

Rearranging terms, we obtain

$$(z\mathbf{I} - \boldsymbol{G})\,\boldsymbol{X}(z) = z\,\boldsymbol{x}(0) + \boldsymbol{H}\,\boldsymbol{U}(z)$$

or equivalently

$$\boldsymbol{X}(z) = (z\mathbf{I} - \boldsymbol{G})^{-1}z\,\boldsymbol{x}(0) + (z\mathbf{I} - \boldsymbol{G})^{-1}\,\boldsymbol{H}\,\boldsymbol{U}(z) \tag{9.21}$$

and the result follows.

At this point it is useful to take a closer look at equations (9.19) and (9.20) and compare them with their continuous-time counterparts, namely (3.62):

1. Just like the continuous-time solution of equation (9.15), (9.19) is formed by two terms: the first depends only on the initial condition and is called

the free response, while the second, which depends only on the system input, is the forced response.

2. Because (9.14) has a unique solution, the combination of (9.19) and (9.20) implies that

$$\mathfrak{Z}(\{\boldsymbol{G}^k\}_{k\in\mathbb{N}_0}) = (z\mathbf{I} - \boldsymbol{G})^{-1}z \tag{9.22}$$

and

$$\mathfrak{Z}(\{y'_k\}_{k\in\mathbb{N}_0}) = (z\mathbf{I} - \boldsymbol{G})^{-1}\,\boldsymbol{H}\,U(z) \tag{9.23}$$

where

$$y'_k = \begin{cases} 0, \; k = 0 \\ \sum_{j=0}^{k-1} \boldsymbol{G}^{k-j-1}\,\boldsymbol{H}\,u(jh), \; k \geqslant 1 \end{cases}$$

3. When $\boldsymbol{x}(0)=0$ and $u(kh)=0$ for $k\neq 0$, the response of (9.14), with $D=0$, is given by

$$y(0) = 0$$
$$y(k) = \boldsymbol{C}\,\boldsymbol{G}^{k-1}\,\boldsymbol{H}\,u(0), \; k \geqslant 1 \tag{9.24}$$

If in addition $u(0)=1$ then $\{y(k)\}_{k\in_0}$ is the unit impulse response of (9.14); therefore the second term on the right hand side of (9.19) is the discrete version of the *convolution* integral (3.20).

4. If $\boldsymbol{x}(0)=0$ we have from (9.21) and (9.15)

$$y(z) = (\boldsymbol{C}(z\mathbf{I} - \boldsymbol{G})^{-1}\,\boldsymbol{H} + \boldsymbol{D})\;U(z)$$

i.e.

$$\boldsymbol{C}(z\mathbf{I} - \boldsymbol{G})^{-1}\,\boldsymbol{H} + \boldsymbol{D} \tag{9.25}$$

is the *discrete transfer matrix* of the system of Fig. 9.5. Note that the continuous-time transfer matrix from $u(t)$ to $y(t)$, $t\in\mathbb{R}$, in Fig. 9.5 does not exist, given the time-varying nature of the sampling process.

5. (9.24) is the sampled version of the continuous-time impulse response of the hold and plant combination.

Exercise 9.9 If $m=r=1$ in Fig. 9.5, show that (9.25) is the z-transform of the continuous-time impulse response of the plant and hold combination.
Hint: show that (9.25) is the z-transform of (9.24), and recall that

$$(z\mathbf{I} - G)^{-1} = z^{-1}\,\mathbf{I} + z^{-2}\,\boldsymbol{G} + z^{-3}\,\boldsymbol{G}^2 + \ldots$$

9.3.1.2 Plants with pure transport delays

When we studied the Smith predictor algorithm, we found that many plants could

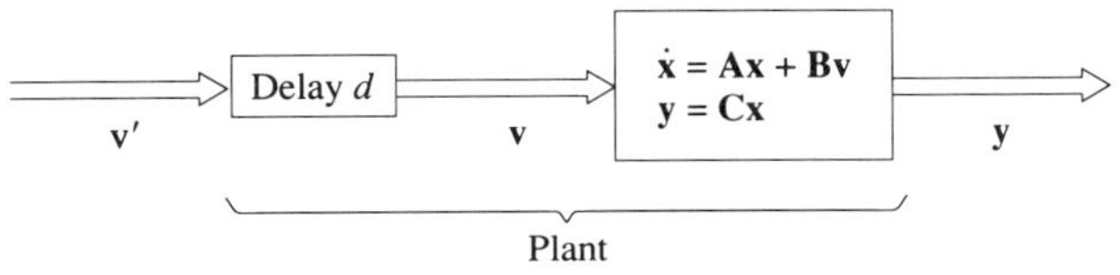

Figure 9.6 Plant modelled as a pure transport delay in series with a delay-free dynamical block. Note that v is inaccessible to measurement.

be satisfactorily described by a delay-free dynamic model in series with a pure transport delay, as shown in Fig. 9.6. This leads to a representation of the form

$$\dot{\boldsymbol{x}}(t) = \boldsymbol{A}\,\boldsymbol{x}(t) + \boldsymbol{B}\boldsymbol{v}'(t-d)$$
$$\boldsymbol{y}(t) = \boldsymbol{C}\,\boldsymbol{x}(t) \tag{9.26}$$

where d is the plant transport delay. (9.26) is an *infinite-dimensional* representation because the state is no longer the vector $\boldsymbol{x} \in \mathbb{R}^n$. In fact if we recall that the state of a system at time t_0 is all the information that, together with the input for $t \geqslant t_0$, is required to compute the output $\mathbf{y}(t)$, $t \geqslant t_0$, in (9.26) we need, besides $\mathbf{x}(t_0)$, the function $\mathbf{v}'(t)$ in the interval $[t_0-d, t_0]$. Because the space of real valued functions defined on a closed interval has infinite dimension, the result follows.

By contrast, the space of piecewise constant functions changing only at a finite set of (fixed) instants, constitutes a finite dimensional space. For example, the class of piecewise constant functions defined on the interval $[0,5h)$, h constant, with discontinuity points at the instants h, $2h$, $3h$ and $4h$, such as function v' of Fig. 9.7, constitutes a linear space of dimension 5; in fact

$$\mathrm{v}'(t) = \mathrm{v}(0)u_0(t) + \mathrm{v}(h)u_1(t) + \ldots + \mathrm{v}(4h)u_4(t)$$

$t \in [0,5)$, where $u_i(t)$, $i=0,1,\ldots,4$, is given by

$$u_i(t) = \begin{cases} 1; & ih \leqslant t < (i+1)h \\ 0; & \text{elsewhere} \end{cases}$$

Consequently a proper state-space representation is possible for sampled plants containing pure transport delays.

If the plant of Fig. 9.6 is sampled, as shown in Fig. 9.5, and $d=lh$, $l \in \mathbb{N}$, v will be a piecewise constant signal with discontinuity points *at the sampling instants*. Therefore the discrete-time model for the delay-free block of Fig. 9.6 becomes

$$\boldsymbol{x}((k+1)h) = \boldsymbol{G}\,\boldsymbol{x}(kh) + \boldsymbol{H}\,v(kh)$$
$$\boldsymbol{y}(kh) = \boldsymbol{C}\,\boldsymbol{x}(kh)$$

with $\boldsymbol{G}$ and $\boldsymbol{H}$ given by (9.17) and (9.18), respectively. However, the input to the plant is $\boldsymbol{v}'$ and not $\boldsymbol{v}$; because $\boldsymbol{v}(kh)=\boldsymbol{v}'((k-l)h)$, we need to store at time kh, besides $\boldsymbol{x}(kh)$, the vector $(\boldsymbol{v}'^{\mathrm{T}}(k-l)h),\ldots,\boldsymbol{v}'^{\mathrm{T}}((k-1)h))^{\mathrm{T}}$ in order to be able to compute $y(nh),n\geq h$; therefore the state of the plant becomes the vector $(\boldsymbol{x}^{\mathrm{T}}(kh),\boldsymbol{v}'^{\mathrm{T}}(k-l)h,\ldots,\boldsymbol{v}'^{\mathrm{T}}((k-1)h))^{\mathrm{T}}\in\mathbb{R}^{(n+lm)}$, where $\boldsymbol{v}'\in\mathbb{R}^m$.

Summing up, we see that the state-space model for the sampled plant of Fig.9.6, with $d=lh$, $l\in\mathbb{N}$, is

$$\begin{bmatrix} \boldsymbol{x}((k+1)h) \\ \boldsymbol{v}'((k+1-l)h) \\ \cdot \\ \cdot \\ \cdot \\ \\ \boldsymbol{v}'(kh) \end{bmatrix} = \begin{bmatrix} \mathbf{G} & \mathbf{H} & \mathbf{0} & \ldots & \mathbf{0} \\ \mathbf{0} & \mathbf{0} & \mathbf{I} & \ldots & \mathbf{0} \\ \cdot & \cdot & \cdot & & \cdot \\ \cdot & \cdot & \cdot & & \cdot \\ \cdot & \cdot & \cdot & & \cdot \\ & & & & \mathbf{I} \\ \mathbf{0} & \mathbf{0} & \mathbf{0} & \ldots & \mathbf{0} \end{bmatrix} \begin{bmatrix} \boldsymbol{x}(kh) \\ \boldsymbol{v}'((k-l)h) \\ \cdot \\ \cdot \\ \cdot \\ \\ \boldsymbol{v}'((k-1)h) \end{bmatrix} + \begin{bmatrix} \mathbf{0} \\ \mathbf{0} \\ \cdot \\ \cdot \\ \cdot \\ \mathbf{0} \\ \mathbf{I} \end{bmatrix} \boldsymbol{v}'(kh) \qquad (9.27)$$

$$y(kh) = (\boldsymbol{C},0,\ldots,0)(\boldsymbol{x}^{\mathrm{T}}(kh),\boldsymbol{v}'^{\mathrm{T}}(k-l)h,\ldots,\boldsymbol{v}'^{\mathrm{T}}((k-1)h))^{\mathrm{T}} = \boldsymbol{C}\,\boldsymbol{x}(kh)$$

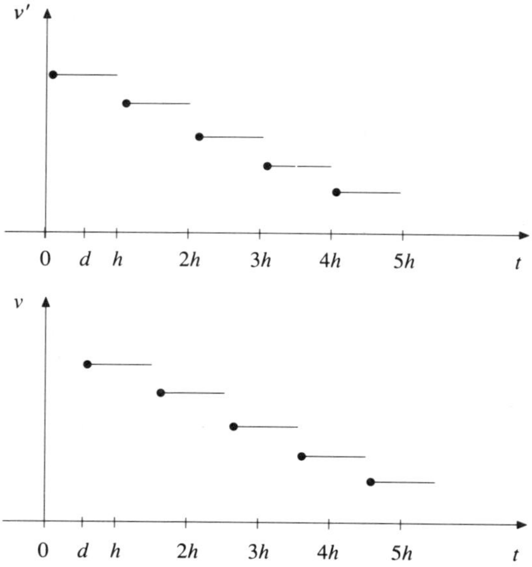

Figure 9.7 Illustration of the shift between sampling instants and inputs, when plant of Figure 9.6 is sampled.

with G and H defined by (9.17) and (9.18), respectively. Notice that the system matrix is now singular, in contrast with the delay-free case (cf. exercise 9.7); in fact it possesses, besides the eigenvalues of G, lm zero eigenvalues. On the other hand we see, from the time-shift property of z-transforms, that the transfer matrix of (9.27) is given by

$$z^{-l}(C(zI - G)^{-1}H) \tag{9.28}$$

where $(C(zI-G)^{-1}H)$ is the transfer matrix in the absence of delay, as given by (9.25).

Exercise 9.10

(a) Prove the last mentioned statement.

(b) Show that the characteristic polynomial of the system matrix in (9.27) is z^{lm} $\det(z\mathbf{I}-G)$.

(c) Check expression (9.28) by computing the transfer matrix directly from (9.27).

If $d \neq lh$, $l \in \mathbb{N}$, we can repeat the procedure as for (9.15), bearing in mind that $\boldsymbol{v}$ will now have discontinuities at times different from the sampling instants. Two cases must then be considered:

1. $0<d<h$.
2. $d>h$.

In the first case the input $\boldsymbol{v}$ to the delay-free block is piecewise constant changing at instants of the form $(kh+d)$, $k \in \mathbb{N}_0$, as shown in Fig. 9.7. Then from (3.60) we have:

$$\boldsymbol{x}((k+1)h) = \exp(\boldsymbol{A}h)\boldsymbol{x}(\mathrm{kh}) + \int_{kh}^{(k+1)h} exp(\boldsymbol{A}((k+1)h - t'))\boldsymbol{B}\boldsymbol{v}(t')\,\mathrm{d}t' \tag{9.29}$$

or equivalently

$$\boldsymbol{x}((k+1)h) = \exp(\boldsymbol{A}h)\boldsymbol{x}(kh) + \int_{kh}^{kh+d} \exp(\boldsymbol{A}((k+1)h - t'))\mathrm{d}t'\,\boldsymbol{B}\,\boldsymbol{v}'((k-1)h) + \int_{kh+d}^{(k+1)h} \exp(\boldsymbol{A}((k+1)h - t'))\boldsymbol{B}\,\mathrm{d}t'\boldsymbol{v}'(kh)$$

With the change of variable $t=(k+1)h-t'$ the above equation becomes

$$\boldsymbol{x}((k+1)h) = \exp(\boldsymbol{A}h)\boldsymbol{x}(kh) + \int_{h-d}^{h} \exp(\boldsymbol{A}t)\mathrm{d}t\boldsymbol{B}\boldsymbol{v}'((k-1)h)) + \int_{0}^{h-d} \exp(\boldsymbol{A}t)\mathrm{d}t\boldsymbol{B}\boldsymbol{v}'(kh)$$

With another obvious change of variable we get

$$\boldsymbol{x}((k+1)h) = \exp(\boldsymbol{A}h)\boldsymbol{x}(kh) + \exp(\boldsymbol{A}h)\int_{0}^{d} \exp(-\boldsymbol{A}t)\mathrm{d}t\,\boldsymbol{B}\boldsymbol{v}'((k-1)h) + \int_{0}^{h-d} \exp(\boldsymbol{A}t)\mathrm{d}t\,\boldsymbol{B}\boldsymbol{v}'(kh) \tag{9.30}$$

Inspecting this equation we conclude it has the following structure:

$$x((k+1)h) = G(h)x(kh) + H_1(h,d)v'((k-1)h) + H_0(h,d)v'(kh) \tag{9.31}$$

with $G(h)$ as in (9.17) and

$$H_1(h,d) = \exp(Ah)\int_0^d \exp(-At)\mathrm{d}t\,B \tag{9.32}$$

$$H_0(h,d) = \int_0^{h-d} \exp(At)\mathrm{d}t\,B \tag{9.33}$$

The input to the plant is now constant in the interval $((k-1)h,\ (k-1)h+d)$; because the space of constant functions on a fixed interval (d is assumed constant) is a one-dimensional space, we can now get a proper state-space representation for (9.26) by defining the new state vector at time kh as $(x^T(kh),\ v'^T((k-1)h)^T$; we then have

$$\begin{bmatrix} x((k+1)h) \\ v'(kh) \end{bmatrix} = \begin{bmatrix} G & H_1 \\ 0 & 0 \end{bmatrix} \begin{bmatrix} x(kh) \\ v'((k-1)h) \end{bmatrix} + \begin{bmatrix} H_0 \\ \mathrm{I} \end{bmatrix} v'(kh)$$

$$y(kh) = Cx(kh) \tag{9.34}$$

which is a state-space model for (9.26). Notice now that the dimension of the state vector is $(n+m)$, where m is the number of plant inputs, and the fact that the system matrix has m extra eigenvalues at the origin; although $d<h$, we conclude from (9.34) that the plant delay, as seen by the computer, is h. If $d=h$, then $H_0=0$ and $H_1=H$, H as in (9.18) and as expected.

The transfer matrix of (9.34) can be easily computed from equation (9.31): z-transforming both sides and assuming zero initial conditions we get

$$X(z) = C(z\mathrm{I} - G)^{-1}(H_1 z^{-1} + H_0)V'(z)$$

Therefore the transfer matrix of (9.34) is

$$C(zI - G)^{-1}(H_1 z^{-1} + H_0) \tag{9.35}$$

Exercise 9.11 Consider the process defined by (9.26) with $A=-a$, $a>0$, $B=C=1$ and $0<d<h$. Then compute a state-space representation of the type (9.34).
Answer: $G=\mathrm{e}^{-ah}$; $H_1=1/a\ \mathrm{e}^{-ah}\ (\mathrm{e}^{ad}-1)$; $H_0=1/a\ (1-\mathrm{e}^{-a(h-d)})$.

When $d>h$ we can write $d=lh+d'$, $l\in\mathbb{N}$, $0<d'<h$. Then (9.27) becomes

$$x((k+1)h) = \exp(Ah)\,x(kh) + \int_{kh}^{kh+d'} \exp(A(k+1)h - t')\mathrm{d}t'\,B\,v'((k-l-1)h) + \int_{kh+d'}^{(k+1)h} \exp(A((k+1)h - t'))B\,\mathrm{d}t'\,v'((k-l)h)$$

leading to

$$x((k+1)h) = G(h)x(kh) + H_1(h,d')v'((k-l-1)h) + H_0(h,d')\,v'((k-l)h) \tag{9.36}$$

Now the *memory* of (9.26) goes back further l sampling units. Given the piecewise constancy of the input this means that l extra values of the input vector must be added to the state of (9.34) to define the state at time kh. Then the state-space representation becomes

$$\begin{bmatrix} \boldsymbol{x}((k+1)h) \\ \boldsymbol{v}'((k-1)h) \\ \cdot \\ \cdot \\ \cdot \\ \boldsymbol{v}'(kh) \end{bmatrix} = \begin{bmatrix} \boldsymbol{G} & \boldsymbol{H}_1 & \boldsymbol{H}_0 & \mathbf{0} & \dots & \mathbf{0} \\ \mathbf{0} & \mathbf{0} & \mathbf{I} & \mathbf{0} & \dots & \mathbf{0} \\ \cdot & \cdot & \cdot & \cdot & & \cdot \\ \cdot & \cdot & \cdot & \cdot & & \cdot \\ \cdot & \cdot & \cdot & \cdot & & \cdot \\ \mathbf{0} & \mathbf{0} & \mathbf{0} & \mathbf{0} & \dots & \mathbf{I} \\ \mathbf{0} & \mathbf{0} & \mathbf{0} & \mathbf{0} & \dots & \mathbf{0} \end{bmatrix} \begin{bmatrix} \boldsymbol{x}(kh) \\ \boldsymbol{v}'((k-l-1)h) \\ \boldsymbol{v}'((k-1)h) \\ \cdot \\ \cdot \\ \cdot \\ \boldsymbol{v}'((k-1)h) \end{bmatrix} + \begin{bmatrix} \mathbf{0} \\ \mathbf{0} \\ \cdot \\ \cdot \\ \cdot \\ \mathbf{0} \\ \mathbf{I} \end{bmatrix} \boldsymbol{v}'(kh)$$

$$\boldsymbol{y}(kh) = \boldsymbol{C}\boldsymbol{x}(kh) \tag{9.37}$$

Exercise 9.12 Show that

$$z^{-l}\boldsymbol{C}(z\mathbf{I} - \boldsymbol{G})^{-1}(\boldsymbol{H}_1 z^{-1} + \boldsymbol{H}_0)$$

is the transfer matrix of (9.37).

9.3.2 Discretization of Transfer Functions

The purpose of this section is to compute the relation between the z-transforms of the input and output of a system composed of a zero-order hold and a plant with transfer function $W(s)$ as shown in Fig. 9.8.

When the plant contains no pure transport delays, we already know from exercise 9.9 that the ratio $Y(z)/U(z)$ is independent of u and y, therefore a characteristic of the system, and is known as the *discrete* or *pulse transfer function*. It was shown that $Y(z)/U(z)$ is the z-transform of y^* when u^* is a unit Dirac

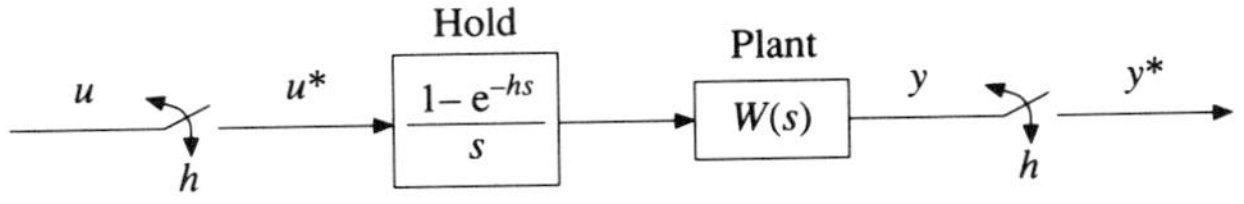

Figure 9.8 Zero-order hold sampling of a continuous-time plant.

impulse at $t=0$. Then we use the following algorithm to obtain the pulse transfer function:

1. Compute the impulse response of $W(s)/s$.
2. Compute its z-transform.
3. Multiply the above result by $(1-z^{-1})$.

Notice step (3) is an immediate consequence of the time-shift property of z-transforms.

Example 9.3 Let us compute the pulse transfer function in Fig. 9.8 when the plant is a pure integrator, i.e. $W(s)=1/s$. Since such a transfer function is the algebraic sum of the pulse transfer functions of $W(s)/s$ and $(W(s)e^{-hs})/s$, and the latter produces delayed versions of the former by h time units, all we need to do is to compute the impulse response of $W(s)/s$. In terms of z-transform such a delay will simply produce a multiplication by z^{-1} (cf. section 9.2). Therefore

$$\mathfrak{Z}\left\{\frac{1-e^{-hs}}{s}W(s)\right\} = (1-z^{-1})\,\mathfrak{Z}\left(\frac{W(s)}{s}\right)$$

The impulse response of $W(s)/s$ is $\mathcal{L}^{-1}(1/s^2)=t$, $t\geqslant 0$, whose z-transform is $(hz)/[(z-1)^2]$ (see Table 9.1). Multiplying by $(1-z^{-1})$ we get $h/(z-1)$ which is the desired result.

An extensive table of pulse transfer functions for the combination of a zero-order hold and plants with transfer functions of orders not greater than three without transport delays can be found in Neuman and Baradello (1979).

When the plant contains a pure delay, not necessarily a multiple of h, the above algorithm still holds, namely $Y(z)/U(z)$ is still the z-transform of the impulse response of the plant and hold combination as shown in the following.

Fact: The pulse transfer function of the system in Fig. 9.8 is the z-transform of the unit impulse response.

Proof: We have seen in Chapter 3 that the response v of a linear time-invariant system is related to the input m by the well-known convolution integral

$$v(t) = \int_0^t g(t-t')m(t')\,dt' \tag{9.38}$$

where g is the impulse response functions of the system. Then we can write for Fig. 9.8,

$$y(kh) = \sum_{j=0}^{k} g(kh-jh)\,u(jh) \tag{9.39}$$

because u^* is a sequence of impulses; $g(t)$ is the impulse response of the hold and plant combination, i.e.

$$g(t) = \mathcal{L}^{-1}\left(W(s)\frac{1-\exp(-hs)}{s}\right)$$

The z-transform of y^* can now be computed using (9.39) and the fact that $g(t)=0$, for t<0; thus

$$Y(z) = \sum_{l=0}^{\infty} y(lh)\ z^{-l} = \sum_{l=0}^{\infty} \sum_{j=0}^{\infty} g(lh - jh)\ u(jh)\ z^{-l}$$

If we introduce the variable i, where $l=i+j$, we can write

$$\begin{aligned} Y(z) &= \sum_{l=0}^{\infty} \sum_{j=0}^{\infty} g(ih)\ u(jh)\ z^{-(i+j)} \\ &= \sum_{i=0}^{\infty} g(ih)z^{-i} \sum_{j=0}^{\infty} u(jh)z^{-j} \\ &= \sum_{i=0}^{\infty} g(ih)z^{-i}\ U(z) \\ &= G(z)\ U(z) \end{aligned}$$

where $G(z) = \sum_{i=0}^{\infty} g(ih)z^{-i}$ is the z-transform of $g(t)$. Therefore

$$G(z) = Y(z)/U(z)$$

QED.

This result provides a method to compute the pulse transfer function when the plant delay is not a multiple of h. An alternative procedure is to find a state-space representation of the type (9.34) and then compute its pulse transfer function (cf. exercise 9.14).

Example 9.4 To illustrate the application of the above algorithm, let us compute the pulse transfer function in Fig. 9.8 when the plant contains a pure delay, e.g.

$$W(s) = \frac{e^{-0.5s}}{(s+1)(s+4)}$$

assuming $h=1$.

Step (i): Calculation of the impulse response of $W(s)/s$. We have that

$$\mathcal{L}^{-1}\left\{\frac{1}{s\,(s+1)(s+4)}\right\} = \frac{1}{4}\,(1 - 4/3\ e^{-t} + 1/3\ e^{-4t})$$

for $t \geqq 0$ and zero otherwise. Therefore

$$\mathcal{L}^{-1}\left\{\frac{e^{-0.5s}}{s(s+1)(s+4)}\right\} = \begin{cases} 1/4[1-4/3\ e^{-(t-0.5)}+1/3\ e^{-4(t-0.5)}];\ \text{for}\ t \geqq 0.5 \\ 0;\ \text{for}\ t<0.5 \end{cases}$$

Let g(t) denote this impulse response.

Step (ii): Computation of the z-transform of $g(t)$. If

$$v(t) = \exp(-a(t-\mathrm{d})) \text{ for } t \geqslant d$$

and

$$v(t) = 0 \text{ for } t<d,\ 0<d<h$$

we have

$$\begin{aligned}\mathcal{Z}(v(t)) &= \mathrm{e}^{ad}(0 + \mathrm{e}^{-ah} z^{-1} + \mathrm{e}^{-2ah} z^{-2} + \ldots) \\ &= \mathrm{e}^{ad} z^{-1} \mathrm{e}^{-ah} (1 + \mathrm{e}^{-ah} z^{-1} + \ldots) \\ &= z^{-1} \mathrm{e}^{-a(h-d)} z/(z - \mathrm{e}^{-ah}) \\ &= \mathrm{e}^{-a(h-d)}/(z - \mathrm{e}^{-ah})\end{aligned}$$

With this result in mind, the computation of the z-transform of $g(t)$ is straightforward, leading to

$$\mathcal{Z}(g(t)) = \frac{1}{4}\left[\frac{1}{z-1} - \frac{4}{3}\frac{\mathrm{e}^{-0.5}}{z-\mathrm{e}^{-1}} + \frac{1}{3}\frac{\mathrm{e}^{-2}}{z-\mathrm{e}^{-4}}\right] = G(z)$$

Step (iii):

$$\begin{aligned}Y(z)/U(z) &= (1 - z^{-1})\, G(z) \\ &= \frac{1 - \frac{4}{3}\frac{(z-1)\mathrm{e}^{-0.5}}{z - \mathrm{e}^{-1}} + \frac{1}{3}\frac{(z-1)\mathrm{e}^{-2}}{z-\mathrm{e}^{-4}}}{4z} \\ &= \frac{(3 - 4\mathrm{e}^{-0.5} + \mathrm{e}^{-2})z^2 + (-3\mathrm{e}^{-1} - 3\mathrm{e}^{-4} + 4\mathrm{e}^{-0.5} + 4\mathrm{e}^{-4.5} - \mathrm{e}^{-2} - \mathrm{e}^{-3})z + (3\mathrm{e}^{-5} - 4\mathrm{e}^{-4.5} + \mathrm{e}^{-3})}{12\, z\, (z^2 - z(\mathrm{e}^{-1} + \mathrm{e}^{-4}) + \mathrm{e}^{-5})} \\ &= \frac{0.7092\, z^2 + 1.1269\, z + 0.025\,56}{12\, z\, (z^2 - 0.3862\, z + 0.006\,738)}\end{aligned}$$

Exercise 9.13 Consider the discrete time system described by the difference equation (9.13), where $k \in \mathbb{N}_0$, and n is a fixed positive integer; using the properties of the z-transform, compute its pulse transfer function $Y(z)/U(z)$.
Answer: $(b_n z^n + b_{n-1} z^{n-1} + \ldots + b_1 z + b_0)/(z^n + a_{n-1} z^{n-1} + \ldots + a_1 z + a_0)$.

Exercise 9.14

(a) Compute the pulse transfer function for the set-up of Fig. 9.8 when:
  (i) $W(s) = \exp(-ds)$, $0<d<h$;
  (ii) $W(s) = \dfrac{\exp(-ds)}{s+a}$, $0 < d < h$

(b) Using the final value theorem, check the result in (ii) by comparing the steady-state responses of the continuous and pulse transfer functions when the input u is a unit step.

Answers:
(i) z^{-1}

$$\text{(ii)}\ \ Y(z)/U(z) = \frac{1}{a}\,\frac{z(1 - \exp^{(-a(h-d))}) + \exp^{(-ah)}\,(\exp^{(ad-1)})}{z(z - \exp^{(-ah)})}$$

9.3.2.1 Poles and zeros of sampled transfer functions

When a plant with a proper rational transfer function

$$W(s) = \frac{\beta_m s^m + \beta_{m-1} s^{m-1} + \ldots + \beta_1 s + \beta_0}{s^n + \alpha_{n-1} s^{n-1} + \ldots + \alpha_1 s + \alpha_0} \tag{9.40}$$

is sampled as shown in Fig. 9.8, the poles of $W(z)$ are $\exp(\lambda_1 h),\ldots, \exp(\lambda_n h)$, where $\lambda_1,\ldots,\lambda_n$ are the poles of $W(s)$; this is a straightforward consequence of (9.17) and a property of the exponential of a matrix.

Exercise 9.15 Prove the last mentioned statement.

At this point it is convenient to analyze in detail the transformation e^{hs} from the s-plane into the z-plane: e^{hs} is a function, with period $2\pi j/h$ (recall $j=\sqrt{-1}$); as a consequence a pair of s-plane points, s_1 and s_2, such that $s_1 - s_2 = \pm k\, 2\pi j/h$, $k \in \mathbb{N}$, will have the same z-plane image.

An s-plane line parallel to the imaginary axis with abscissa x, $x \in \mathbb{R}$, is mapped into a z-plane circle with centre at the origin and radius e^{hx}; in particular the imaginary axis is mapped into the unit circle.

A line in the left half s-plane, parallel to the real axis with ordinate jy is mapped into a line segment joining the origin of the z-plane with the point e^{jhy} on the unit circle.

Figure 9.9 illustrates what has been said above. Other loci of interest, in connection with s-plane specification, are the lines of constant damping ratio ζ (the z-plane image of such a locus, for $\zeta=0.259$, is the logarithmic spiral shown in Fig. 9.10), and the semicircles of constant ω_n: the undamped natural frequency. Because such lines are orthogonal to the ζ=constant lines, and the transformation e^{hs} is *conformal*, i.e. it preserves the s-plane angles, the z-plane image of these semicircles will be lines orthogonal to the logarithmic spirals (and the unit circle) as sketched in Fig. 9.11.

As far as the zeros of $W(z)$ are concerned, the problem is not quite so straightforward. $W(z)$ will always have zeros if $n>1$ in (9.40). Such zeros are non-linear functions of h and of the coefficients of $W(s)$. It can be shown (Aström *et al.*, 1984; Neuman and Baradello, 1979) that when $W(s)$ is a rational function, the resulting $W(z)$ in Fig. 9.8 has n poles and $(n-1)$ zeros if $m<n$, or n zeros if $m=n$.

The problem with zero-order hold sampling is that it is not true that a continuous-time plant with zeros only in the left half-plane will give rise to a pulse transfer function with zeros inside the unit disc (stable zeros) and vice versa (Aström *et al.*, 1984). For example $W(s)=1/(s+1)^3$ gives rise to a pulse transfer function with zeros outside the unit circle if the sampling period is less than 1.8399; on the other hand

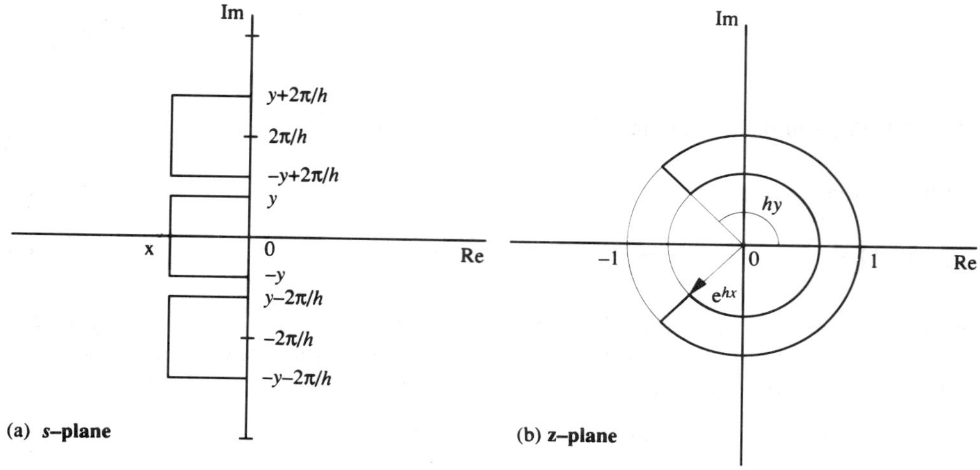

Figure 9.9 Illustrating the map $z = e^{hs}$.

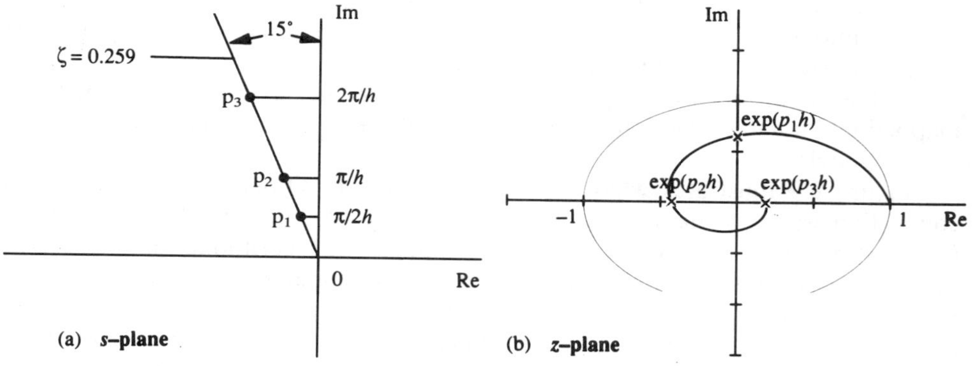

Figure 9.10 Illustrating the z-plane image of s-plane loci of constant damping ratio ζ.

$$W(s) = \frac{6(1 - s)}{(s + 2)(s + 3)}$$

which has an *unstable zero*, i.e. a zero in the right hand side of the s-plane, produces a pulse transfer function with no zeros outside the unit disc if the sampling period is greater than 1.2485, as shown in the above-mentioned paper.

Zeros outside the unit disc are undesirable for many reasons: for example, they rule out the possibility of feedforward compensation and the use of controller synthesis techniques based on zero cancellations. Loosely speaking, the

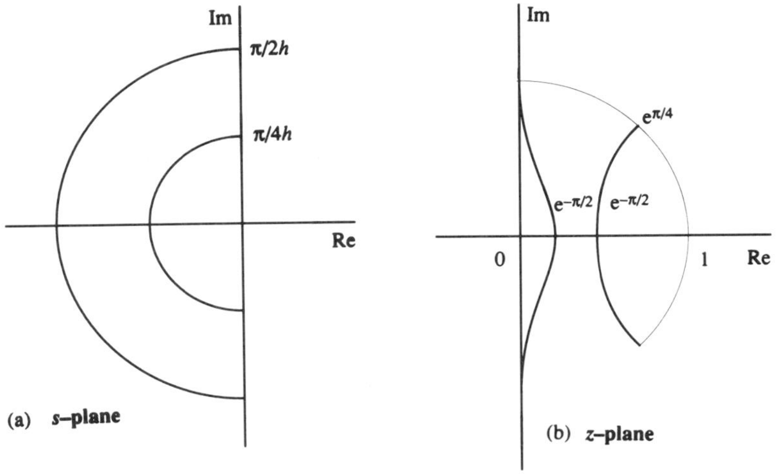

Figure 9.11 Image of s-plane loci of constant ω_n by $z=e^{hs}$.

effect of a zero outside the unit disc is an increase in the settling-time combined, sometimes, with an inverse type response as illustrated in Fig. 9.12. This explains why 'well-behaved' plants such as $W(s)=1/(s+1)^3$ may give rise to pulse transfer functions with zeros outside the unit circle. In fact the impulse response of Fig. 9.8, with the above $W(s)$, exhibits a settling time (in units of sampling period) that increases as the sampling period decreases. As depicted in Fig. 9.13 the sequence of samples takes longer to settle when h is small! When we are dealing with purely discrete-time signals what counts in terms of z-transform is the relative variation between consecutive terms in the sequence, and not the time elapsed between samples. For a rigorous treatment of the zeros of discrete-time systems obtained when sampling continuous-time systems, the reader is referred to Aström *et al.* (1984).

Exercise 9.16 Show that the $W(z)$ of Fig. 9.8 has a zero outside the unit circle when:
(a) $h>0$ and $W(s)=1-s/(1+s)$.
Hint: $W(s)=(2-\exp(-h)-z)/(z-\exp(-h))$.
(b) $W(s)=e^{-ds}/(s+1)$ and $0.62011<d<1$, assuming $h=1$.
Hint: cf. exercise 9.14.

9.3.3 State-space Representation of Difference Equations

Consider the linear nth-order difference equation

$$y(k+n) + a_{n-1}y(k+n-1) + \ldots + a_0y(k) = b_nu(k+n) + b_{n-1}u(k+n-1) + \ldots + b_0u(k) \tag{9.13}$$

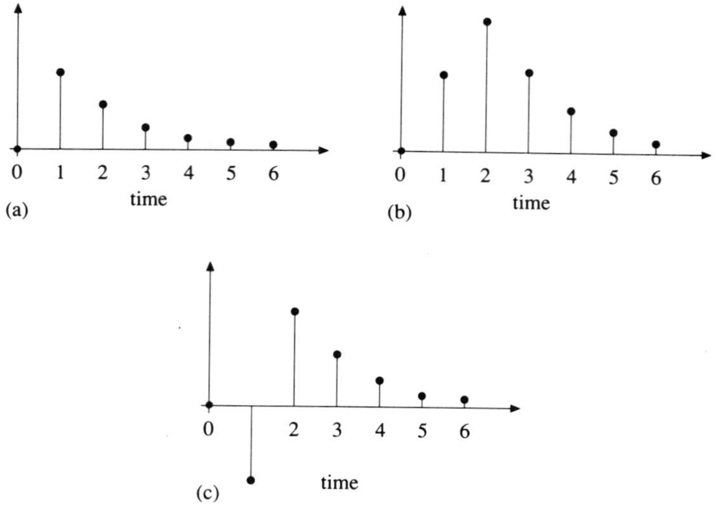

Figure 9.12 Impulse responses of: (a) $z/((z-e^{-1})(z-e^{-2}))$; (b) $z+2/((z-e^{-1})(z-e^{-2}))$; (c) $2-z/((z-e^{-1})(z-e^{-2}))$.

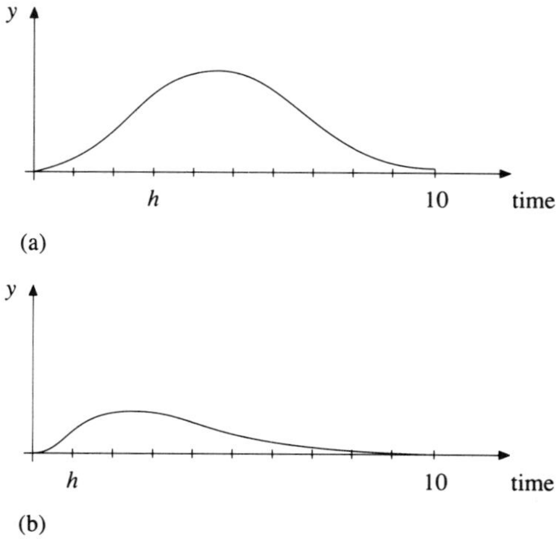

Figure 9.13 Impulse responses in Fig. 9.8, for $W(s)=1/(s+1)^3$ and h as indicated.

where $k=0,1,2,\ldots$. We start by deriving a state-space representation for a particular situation, namely when $b_1=b_2=\ldots=b_n=0$. In this hypothesis it is straightforward to recognize that (9.13) has the dynamical diagram shown in Fig. 9.14 where D denotes the delay operator, i.e. $D(y(k))=y(k-1)$, $\forall k\in\mathbb{N}$.

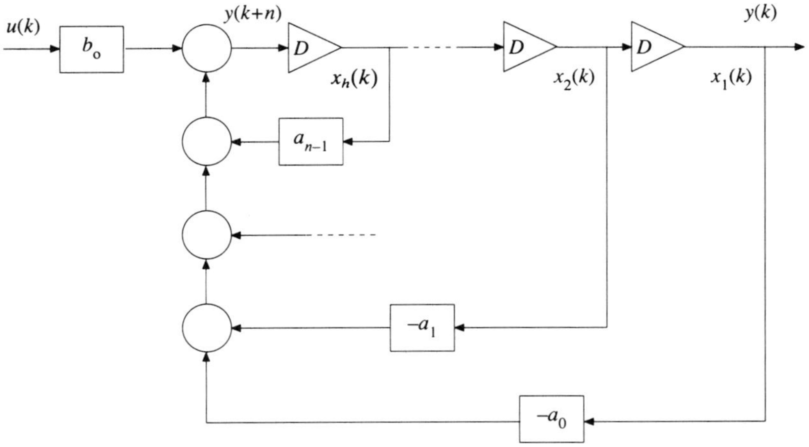

Figure 9.14 Dynamical diagram of equation (9.13) when $b_1=b_2=\ldots=b_n=0$.

Comparing this diagram with Fig. 2.1 we see the delay operator plays the same role as the integrator in the continuous-time case. If the outputs of the delay operators are chosen as state variables, then defining

$$x(k) = (x_1(k), x_2(k), \ldots, x_n(k))^{\mathrm{T}}$$

we can write

$$x(k+1) = \begin{bmatrix} 0 & 1 & 0 & \ldots & 0 \\ 0 & 0 & 1 & \ldots & 0 \\ \cdot & \cdot & \cdot & & \cdot \\ \cdot & \cdot & \cdot & & \cdot \\ \cdot & \cdot & \cdot & & \cdot \\ 0 & 0 & 0 & \ldots & 1 \\ -a_0 & -a_1 & -a_2 & \ldots & -a_{n-1} \end{bmatrix} x(k) + \begin{bmatrix} 0 \\ 0 \\ \cdot \\ \cdot \\ \cdot \\ 0 \\ b_0 \end{bmatrix} u_k$$

$$y(k) = (1,0,\ldots,0)\, x(k) \tag{9.41}$$

In general cases we can easily derive a state-space representation, noticing that the output of (9.13) is a linear combination of shifted versions of the output of the system with transfer function

$$\frac{1}{z^n + a_{n-1} z^{n-1} + \ldots + a_1 z + a_0} \tag{9.42}$$

as shown in exercise 9.13: thus the dynamical diagram of Fig. 9.15. With the

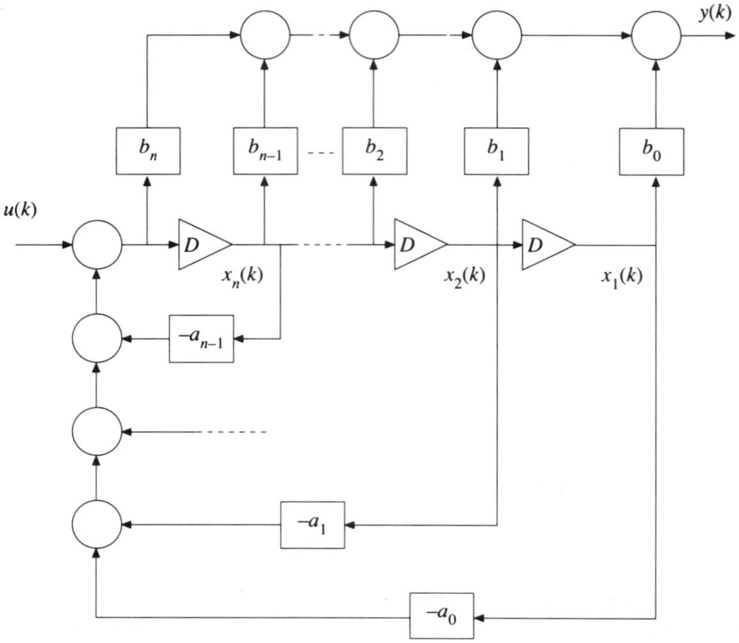

Figure 9.15 Dynamical diagram of equation 9.13.

choice of state variables as indicated we now have

$$x(k+1) = \begin{bmatrix} 0 & 1 & 0 & \dots & 0 \\ 0 & 0 & 1 & \dots & 0 \\ \cdot & \cdot & \cdot & & \cdot \\ \cdot & \cdot & \cdot & & \cdot \\ \cdot & \cdot & \cdot & & \cdot \\ 0 & 0 & 0 & \dots & 1 \\ -a_0 & -a_1 & -a_2 & \dots & -a_{n-1} \end{bmatrix} x(k) + \begin{bmatrix} 0 \\ 0 \\ \cdot \\ \cdot \\ \cdot \\ 0 \\ 1 \end{bmatrix} u(k) \qquad (9.43)$$

$$y(k) = ((b_0 - a_0 b_n); (b_1 - a_1 b_n); \dots; (b_{n-1} - a_{n-1} b_n))\, x(k) + b_n u(k)$$

Exercise 9.17 Write the difference equation, the pulse transfer function and a state-space representation for the system with the dynamical diagram of Fig. 9.16; compare the result with (9.37).

Answer:

$$y(k+1) - y(k) = u(k-1)$$

$$Y(z)/U(z) = 1/(z(z-1))$$

$$x(k+1) = \begin{bmatrix} 1\,1 \\ 0\,0 \end{bmatrix} x(k) + \begin{bmatrix} 0 \\ 1 \end{bmatrix} u(k)$$

$$y(k) = (1,0)\, x(k)$$

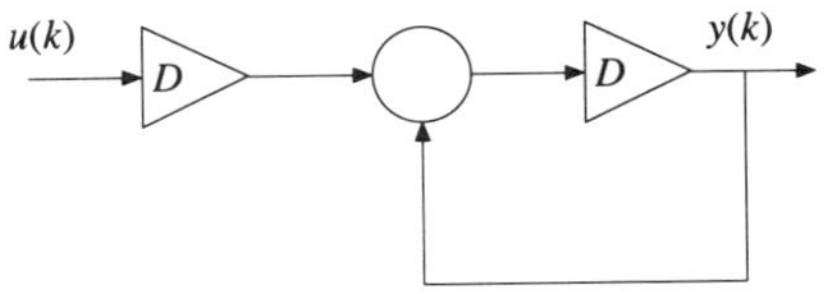

Figure 9.16 Dynamical diagram for exercise 9.17.

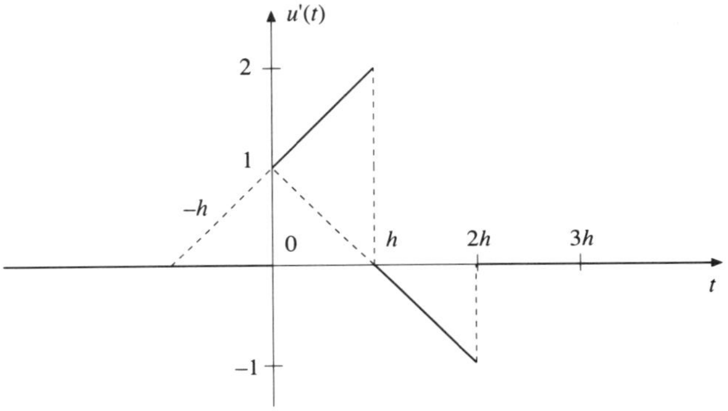

Figure 9.17 Unit impulse response of the first-order hold.

9.3.4 First-order Holds

The process of reconstructing continuous-time signals considered so far was the zero-order hold (ZOH), where the continuous-time signal was reconstructed by a piecewise-constant signal. Smaller errors can be obtained with higher-order approximations, namely first-order ones. In this case the continuous-time signal $u(t)$ is reconstructed by means of a piecewise-linear linear signal, $u'(t)$.

If

$$u'(t) = u(kh) + (t - kh)(u(kh) - u((k - 1)h))/h \tag{9.44}$$

$kh \leqslant t < (k+1)h$, with $k \in \mathbb{N}_0$, the process is known as first-order hold (FOH) because, between sampling instants, the signal is approximated by a polynomial of degree one, namely a line through the two most recent samples. The impulse response of the FOH is shown in Fig. 9.17; its Laplace transform is

$$(1 - e^{-sh})^2 (sh + 1)/(s^2 h) \tag{9.45}$$

Therefore (9.45) is also the transfer function of the FOH.

Exercise 9.18 Compute the Laplace transforms of the signal of Fig. 9.17. *Hint*: if $u_1(t)$ and $u_2(t)$ denote the unit step and the ramp of slope $1/h$ at t=0, respectively, then

$$u'(t) = u_1(t) + u_2(t) - 2\,[u_1(t - h) + u_2(t - h)] + u_1(t - 2h) + u_2(t - 2h)$$

When a plant is driven by the output of an FOH, as shown in Fig. 9.18, then the discrete transfer function of the plant and hold combination becomes

$$\frac{Y(z)}{U(z)} = \frac{1}{h}(1 - z^{-1})^2 \mathscr{Z}\left(\frac{sh+1}{s^2}G(s)\right) \tag{9.46}$$

where $G(s)$ is the plant transfer function.

The additional complexity of the FOH is not justified by an improved performance. In fact, there are no clear guidelines suggesting that FOH is preferred to ZOH in any situation. A better option is to use a forward difference to approximate the continuous-time signal $u(t)$, namely

$$u''(t) = u(kh) + \frac{(t-kh)}{h}(u((k+1)h) - u(kh)) \tag{9.47}$$

for $kh \leqslant t < (k+1)h$.

This process is known as predictive first-order hold (PFOH) reconstruction (Bernhardsson, 1990) because it requires the knowledge of $u((k+1)h)$ at time $t=kh$. Figure 9.19 depicts the reconstruction of signal $u(t)$ by both methods: the

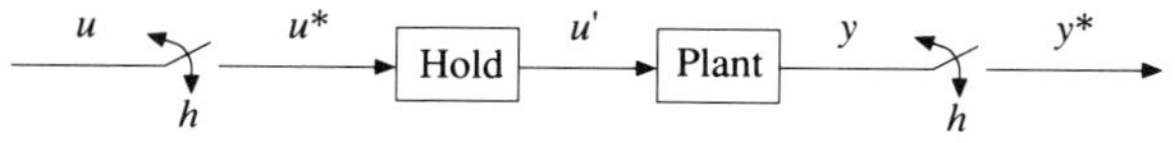

Figure 9.18 First-order hold sampling.

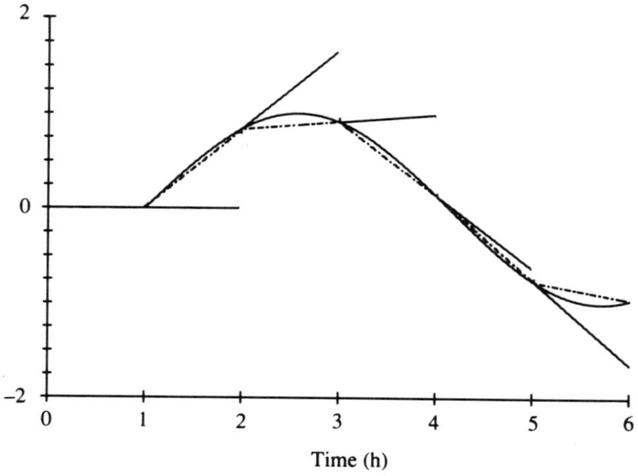

Figure 9.19 First order-hold reconstruction (solid line) and predictive first-order hold reconstruction (dot-dashed line) of a continuous-time signal $u(t)$.

PFOH reconstruction, besides being more accurate than ZOH and FOH reconstruction, also produces a smooth signal. Notice that when $u(t)$ is piecewise linear and continuous, its PFOH reconstruction is exact; the same does not apply to FOH reconstruction.

The PFOH is physically realizable when the signal to be reconstructed is the output of a control algorithm described by a strictly proper transfer function. For example if, in Fig. 9.1, the control algorithm is described by the transfer function

$$D(z) = \frac{M(z)}{E(s)} = \frac{b_{n-1}z^{n-1} + b_{n-2}z^{n-2} + \ldots + b_1 z + b_0}{z^n + a_{n-1}z^{n-1} + \ldots + a_1 z + a_0}$$

where $E(z)=R(z)-Y(z)$ and $M(z)$ is the transform of the input to the hold device, we have

$$\begin{aligned} m((k+1)h) = &- a_{n-1}m(kh) - \ldots - a_1 m(k-n+z) - a_0 m(k-n+1) \\ &+ b_{n-1}e(kh) + \ldots + b_0 e(k-n+1) \end{aligned}$$

showing that the use of PFOH reconstruction of $m(t)$ is feasible, since $m((k+1)h)$ only depends of information available at time $t=kh$.

The transfer function of the PFOH is

$$\frac{e^{hs} + e^{-hs} - 2}{hs^2} \tag{9.48}$$

with an impulse response as in Fig. 9.20. Therefore, when a plant, with transfer function $G(s)$, is driven by a PFOH (cf. Fig. 9.18), the discrete transfer function of the plant and hold combination becomes

$$\frac{Y(z)}{U(z)} = \frac{1}{h}\frac{(z-1)^2}{z}\mathcal{Z}\left\{\mathcal{L}^{-1}\frac{G(s)}{s^2}\right\} \tag{9.49}$$

Exercise 9.19 Show that the transfer function of the PFOH is

$$\frac{1}{h}\frac{e^{hs} + e^{-hs} - 2}{s^2}$$

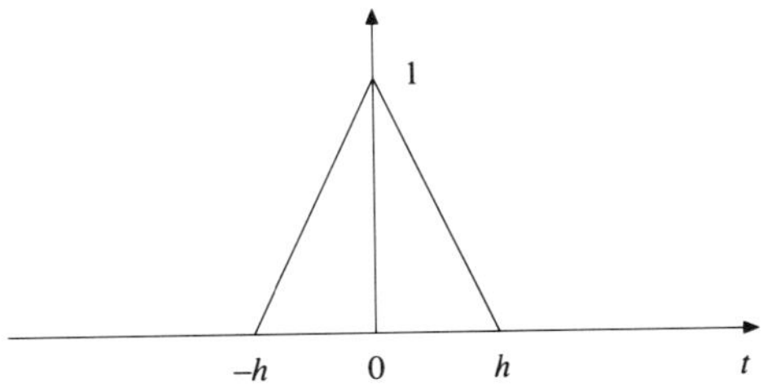

Figure 9.20 Impulse response of predictive first-order hold.

Exercise 9.20 Assuming in Fig. 9.18, $G(s)=1/s^2$ and $h=1$, compute the discrete transfer function $Y(z)/U(z)$ assuming the hold device is:
(a) ZOH
(b) PFOH
Answer:

(a) $\dfrac{z+1}{(z-1)^2} \times \dfrac{1}{2}$

(b) $\dfrac{z^2+4z+1}{(z-1)^2} \times \dfrac{1}{6}$

Exercise 9.21
(a) Compute and compare the FOH and PFOH reconstruction of the continuous-time unit ramp, at time $t=0$.
(b) Show that the PFOH reconstruction of continuous piecewise linear signals is error free.
Answer: (a) The FOH reconstruction between times $t=0$ and $t=h$ is zero.

If, in Fig. 9.18 we have a PFOH and a plant described by the state-space model (9.15), then the discrete-time state-space representation for the PFOH driven plant becomes:

$$\boldsymbol{x}((k+1)h) = \boldsymbol{G}\,\boldsymbol{x}(kh) + \frac{1}{h}\boldsymbol{H}_1\, u(kh+h) + (\boldsymbol{H} - \frac{1}{h}\boldsymbol{H}_1)u(kh) \tag{9.50}$$

with $\boldsymbol{G}$ and $\boldsymbol{H}$ as defined by (9.17) and (9.18), and

$$\boldsymbol{H}_1 = \left(\int_0^h \mathrm{e}^{\boldsymbol{A}t}\,(h-t)\mathrm{d}t\right)\boldsymbol{B} \tag{9.51}$$

This discrete-time representation is easily derived, since the signal is linear between consecutive sampling instants. In fact

$$\boldsymbol{x}(kh+h)=\mathrm{e}^{\boldsymbol{A}h}\,\boldsymbol{x}(kh) + \int_{kh}^{kh+h} \mathrm{e}^{\boldsymbol{A}(kh+h-t')}.\boldsymbol{B}\,\left[u(kh) + \frac{t'-kh}{h}(u(kh+h)-u(kh))\right]\mathrm{d}t' \tag{9.52}$$

which, after a few simple manipulations yields (9.50).

This discrete-time state-space representation can also be derived in an elegant manner, by noticing that the state-space plant model (9.15) and (9.47) can be combined as

$$\frac{\mathrm{d}}{\mathrm{d}t}\begin{bmatrix}\boldsymbol{x}\\ u\\ \dot{u}\end{bmatrix} = \begin{bmatrix}\boldsymbol{A} & \boldsymbol{B} & 0\\ 0 & 0 & \mathrm{I}\\ 0 & 0 & 0\end{bmatrix}\begin{bmatrix}\boldsymbol{x}\\ u\\ \dot{u}\end{bmatrix} \tag{9.53}$$

(cf. Bernhardsson, 1990). Therefore

$$\begin{bmatrix} \boldsymbol{x}((k+1)h) \\ \boldsymbol{u}((k+1)h) \\ \dot{\boldsymbol{u}}((k+1)h) \end{bmatrix} = \exp\left(\begin{bmatrix} \boldsymbol{A} & \boldsymbol{B} & \boldsymbol{0} \\ \boldsymbol{0} & \boldsymbol{0} & \mathbf{I} \\ \boldsymbol{0} & \boldsymbol{0} & \boldsymbol{0} \end{bmatrix} h\right) \begin{bmatrix} \boldsymbol{x}(kh) \\ \boldsymbol{u}(kh) \\ \frac{1}{h}(\boldsymbol{u}((kh+h)-\boldsymbol{u}(kh)) \end{bmatrix} \tag{9.54}$$

If we rewrite (9.50) as

$$\boldsymbol{x}((k+1)h) = \boldsymbol{G}\,\boldsymbol{x}(kh) + \boldsymbol{H}\,\boldsymbol{u}(kh) + \boldsymbol{H}_1 \frac{1}{h}(\boldsymbol{u}(kh+h) - \boldsymbol{u}(kh)) \tag{9.55}$$

we find that $[\boldsymbol{G}, \boldsymbol{H}, \boldsymbol{H}_1]$ is the first row block of exp(.) in (9.54). In other words, $\boldsymbol{G}$, $\boldsymbol{H}$, and $\boldsymbol{H}_1$ can be computed simultaneously by the formula

$$[\boldsymbol{G}\;\boldsymbol{H}\;\boldsymbol{H}_1] = [\mathbf{I}\;\boldsymbol{0}\;\boldsymbol{0}]\exp\left(\begin{bmatrix} \boldsymbol{A} & \boldsymbol{B} & \boldsymbol{0} \\ \boldsymbol{0} & \boldsymbol{0} & \mathbf{I} \\ \boldsymbol{0} & \boldsymbol{0} & \boldsymbol{0} \end{bmatrix} h\right) \tag{9.56}$$

Because of its simplicity of implementation the ZOH is mostly used in practical applications. In the text we always assume ZOH signal reconstruction, unless otherwise specified.

9.4 SIGNALS AND z-PLANE POLE LOCATIONS

We already know that most of our continuous-time signals can be expressed as linear combinations of exponentials, i.e. given $f(t)$, $t \in \mathbb{R}$, we can write

$$f(t) = c_1 \exp(p_1 t) + c_2 \exp(p_2 t) + \ldots + c_n \exp(p_n t)$$

where c_i and p_i, i=1,2,. . .,n, when complex, occur in conjugate pairs. Therefore their sampled versions will also appear as linear combinations of sampled exponentials. In section 9.2 we have seen that the z-transform of the (sampled) exponential e^{pt}, p real or complex, has a pole at $\exp(ph)$. Noting that the sampled exponential is the sequence

$$1,\ e^{ph},\ (e^{ph})^2,\ (e^{ph})^3,\ \ldots$$

we see that the z-plane pole $\exp(ph)=\exp((\sigma+j\omega)h)$ can be identified with a sequence whose terms are the positions at times kh, $k \in \mathbb{N}_0$, of a rotating phasor, with constant angular speed ω that changes its magnitude by a factor $e^{\sigma h}$ every sampling instant.

With this interpretation in mind it is now easy to relate z-plane poles and their time functions, as shown in Fig. 9.21. For example, $p_3=e^{j\pi}$ is the z-transform pole of the sequence generated by a phasor with unit magnitude, rotating an angle

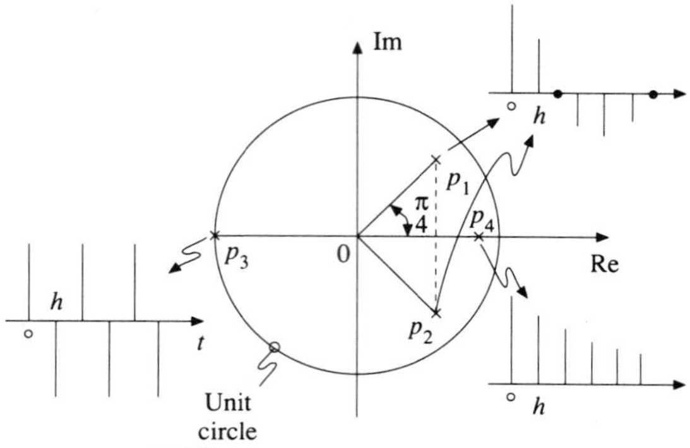

Figure 9.21 Relation between z-plane poles and time functions.

π every sampling interval, i.e. the sequence $\{(-1)^n\}_{n\in\mathbb{N}_0}$ as shown; on the other hand $p_4=e^{\sigma h}$ is the pole of the z-transform of a monotonically decreasing sequence because $\omega=0$ and $e^{\sigma h}<1$. p_1 and p_2 are related with phasors of identical magnitude, rotating in opposite directions with identical speeds; as a result the imaginary components cancel, producing a real sequence. Notice that in this case the number of sampling periods in a phasor cycle is $2\pi/(\pi/4)=8$. Summing up we have:

1. The interior of the unit circle corresponds to decaying sequences and the exterior to increasing sequences. The unit circle corresponds to constant amplitude sequences.
2. Positive real poles correspond to monotone sequences; negative real poles correspond to oscillating sequences with two sample intervals per cycle.
3. Each pair of complex conjugate poles corresponds to an oscillating sequence with $2\pi/(\omega h)$ sampling periods per cycle.

Exercise 9.22 Plot the time sequences corresponding to:
(a) a pole at -0.5.
(b) the conjugate pole pair $\pm$j.

9.5 CONTROLLABILITY, REACHABILITY AND OBSERVABILITY

The concepts of controllability and observability, presented in Chapter 7 for continuous-time systems, can be translated *mutatis mutandis* to the discrete-time case, when the system matrix $\boldsymbol{G}$, in model (9.14), is non-singular. The essential difference occurs in connection with controllability when $\boldsymbol{G}$ is singular because the system is no longer *reversible*, i.e. the possibility of transferring the system from any initial state to the zero state is not equivalent to the requirement of being transferable from the origin to an arbitrary state. A new concept must therefore be introduced, namely that of reachability.

Definition: The system (9.14) is said to be *controllable* if, given any initial state, it is possible to find a control sequence that drives the system to the zero state in a finite time.

For example the system

$$x(k+1) = \begin{bmatrix} 0 & 1 \\ 0 & 0 \end{bmatrix} x(k) + \begin{bmatrix} 1 \\ 0 \end{bmatrix} u(k) \tag{9.57}$$

is controllable; in fact any initial state is driven to zero, in two steps, with a zero input sequence. A stronger and related concept is that of reachability which is defined as follows.

Definition: The representation (9.14) is *reachable* if, given an arbitrary state $\boldsymbol{x}_f$, there exists a control sequence that drives the origin into $\boldsymbol{x}_f$ in a finite time.

In (9.57) the state $\boldsymbol{x}_f=(0,1)^T$ is not reachable; the only reachable states are those on the line supported by the vector $(1,0)^T$. In fact from (9.19) we have that the state of (9.14) at time k, $k=1,2,\ldots,$ is given by

$$\begin{aligned} \boldsymbol{x}(k) &= \boldsymbol{G}^k \boldsymbol{x}(0) + \boldsymbol{G}^{k-1}\boldsymbol{H}\boldsymbol{u}(0) + \boldsymbol{G}^{k-2}\boldsymbol{H}\boldsymbol{u}(1) + \ldots + \boldsymbol{G}\boldsymbol{H}\boldsymbol{u}(k-2) + \boldsymbol{H}\boldsymbol{u}(k-1) \\ &= \boldsymbol{G}^k \boldsymbol{x}(0) + [\boldsymbol{H} \mid \boldsymbol{G}\boldsymbol{H} \mid \ldots \mid \boldsymbol{G}^{k-1}\boldsymbol{H}]\,[\boldsymbol{u}^T(k-1), \boldsymbol{u}^T(k-2), \ldots, \boldsymbol{u}^T(0)]^T \\ &= \boldsymbol{G}^k \boldsymbol{x}(0) + \boldsymbol{M}_k \boldsymbol{U}_k \end{aligned} \tag{9.58}$$

where

$$\boldsymbol{M}_k = [\boldsymbol{H} \mid \boldsymbol{G}\boldsymbol{H} \mid \ldots \mid \boldsymbol{G}^{k-1}\boldsymbol{H}]$$

and

$$\boldsymbol{U}_k = [\boldsymbol{u}^T(k-1), \boldsymbol{u}^T(k-2), \ldots, \boldsymbol{u}^T(0)]^T$$

which shows that the set (subspace) of states that can be reached from the origin, in at most k steps, is the range of $\boldsymbol{M}_k$. Although the number of columns of $\boldsymbol{M}_k$ increases with k, its rank stabilizes for $k \geqslant n$, where n is the dimension of the state vector, as a consequence of the Cayley–Hamilton theorem (cf. section 3.7). Therefore we have.

Fact: The subspace of reachable states of (9.14) is the range of $\boldsymbol{M}_n$.

Theorem: The state-space representation (9.14) is reachable if and only if $\boldsymbol{M}_n$ has rank n.

We invite the reader to compare these results with their continuous-time counterparts, particularly the simplicity of the proofs in the discrete-time case.

Exercise 9.23 Check that the sub-space of reachable states in the system defined by (9.57) is span $\{(1,0)^T\}$ and compute a control sequence that steers the origin to $(-1,0)^T$.
Answer: it can be reached in one step with $u(0)=-1$.

Exercise 9.24 When (9.14) is reachable, then (9.58) gives a control sequence that steers the origin to any desired state in at most n steps. Show that if u is not a scalar, i.e. if we have at least two control inputs, then such a sequence is not unique. Verify the result assuming that in (9.14)

$$\boldsymbol{G} = \begin{bmatrix} 0.1 & 0.1 \\ 0.0 & 0.3 \end{bmatrix}, \boldsymbol{H} = \begin{bmatrix} 1 & 1 \\ 0 & 1 \end{bmatrix}, \boldsymbol{x}(0) = \begin{bmatrix} 0 \\ 0 \end{bmatrix} \text{ and } \boldsymbol{x}_f = \begin{bmatrix} 1 \\ 0 \end{bmatrix}$$

Answer: a possible set of solutions is

$$\begin{bmatrix} u_1(0) \\ u_2(0) \end{bmatrix} = \begin{bmatrix} 0 \\ 0 \end{bmatrix} \text{ and } \begin{bmatrix} u_1(1) \\ u_2(1) \end{bmatrix} = \begin{bmatrix} 10 \\ 0 \end{bmatrix}$$

or

$$\begin{bmatrix} u_1(0) \\ u_2(0) \end{bmatrix} = \begin{bmatrix} 1 \\ 0 \end{bmatrix} \text{ and } \begin{bmatrix} u_1(1) \\ u_2(1) \end{bmatrix} = \begin{bmatrix} 0 \\ 0 \end{bmatrix}$$

When $\boldsymbol{G}$ has full rank, which is the case of systems generated by sampling continuous-time ones (cf. exercise 9.7), reachability and controllability are equivalent. In fact, assuming controllability, given a point $\boldsymbol{x}_f$ in the state-space we know that there exists $\boldsymbol{x}_i \in \mathbb{R}^n$ such that $\boldsymbol{x}_f = \boldsymbol{G}^n \boldsymbol{x}_i$ because $\boldsymbol{G}^n$ is not singular. From (9.58) and the assumption of controllability we can see that there exists a control sequence $u(0), \ldots, u(n-1)$ that steers $-\boldsymbol{G}^n \boldsymbol{x}_i$ to the origin, i.e.

$$\boldsymbol{0} = -\boldsymbol{G}^n \boldsymbol{x}_i + \boldsymbol{M}_n \boldsymbol{U}_n$$

But such control also steers the origin to $\boldsymbol{x}_f = \boldsymbol{G}^n \boldsymbol{x}_i$; therefore controllability implies reachability if $\boldsymbol{G}$ is not singular. The reverse implication can be shown by similar arguments.

Exercise 9.25 Consider the discrete-time system described by (9.14). Then show:
(a) Reachability $\Rightarrow$ controllability.
(b) Reachability and controllability are equivalent if and only if $\boldsymbol{G}$ is a non-singular matrix.

Again, the treatment of observability is greatly simplified in comparison with the continuous-time case, as shown below.

Definition: The state $\boldsymbol{x}_0 \neq \boldsymbol{0}$ of (9.14) is said to be *unobservable* if there exists a finite integer l, $l \geq n$, such that $y(k)=0$ for $0 \leq k \leq l-1$, when $x(0)=x_0$ and $u(k)=0$ for $0 \leq k \leq l-1$.

The above definition implies that an unobservable state cannot be distinguished from the zero state on the basis of input and output observations.

Definition: The representation (9.14) is said to be *observable* if it has no unobservable states.

From (9.14), and assuming $u(k)=0$, $k=0,1,\ldots,l-1$, we can write:

$$y(0) = \boldsymbol{C}\,\boldsymbol{x}(0)$$
$$y(1) = \boldsymbol{C}\,\boldsymbol{G}\,\boldsymbol{x}(0)$$
$$y(2) = \boldsymbol{C}\,\boldsymbol{G}^2\,\boldsymbol{x}(0)$$
$$\vdots$$
$$y(l-1) = \boldsymbol{C}\,\boldsymbol{G}^{l-1}\,\boldsymbol{x}(0)$$

or in a more compact form

$$\boldsymbol{Y}_l = \boldsymbol{N}_l\,\boldsymbol{x}(0) \tag{9.59}$$

where

$$\boldsymbol{N}_l = \begin{bmatrix} \boldsymbol{C} \\ \boldsymbol{C}\,\boldsymbol{G} \\ \boldsymbol{C}\,\boldsymbol{G}^2 \\ \cdot \\ \cdot \\ \cdot \\ \boldsymbol{C}\,\boldsymbol{G}^{l-1} \end{bmatrix} \; ; \; \boldsymbol{Y}_l = \begin{bmatrix} y(0) \\ y(1) \\ \cdot \\ \cdot \\ \cdot \\ y^{l-1} \end{bmatrix} \tag{9.60}$$

$\boldsymbol{N}_l$ has n columns; therefore its rank is at most n; on the other hand we know from the Cayley–Hamilton theorem that there is no point in taking l greater than n to increase the rank of $\boldsymbol{N}_l$.

When $\boldsymbol{N}_n$ has not full rank then $\boldsymbol{N}_n\,\boldsymbol{x}(0)=\boldsymbol{0}$ does not imply $\boldsymbol{x}(0)=\boldsymbol{0}$: every point in the null space of $\boldsymbol{N}_n$, $N(\boldsymbol{N}_n)$, is a possible solution. This means that the elements of $N(\boldsymbol{N}_n)$ cannot be distinguished from the zero state on the basis of input and output observations; then from (9.59) we have

Theorem: The set (sub-space) of *unobservable* states is the null space of $\boldsymbol{N}_n$.

Theorem: Representation (9.14) is observable if and only if $\boldsymbol{N}_n$ has rank n.

Notice that the assumption $u(k)=0$, $k=0,1,\ldots,l-1$, represents no loss of generality because we can always subtract from $\boldsymbol{Y}_l$ the effect of a known input; in fact we have in general

$$\begin{bmatrix} y(0) \\ y(1) \\ y(2) \\ \cdot \\ \cdot \\ \cdot \\ y(l-1) \end{bmatrix} = \begin{bmatrix} \boldsymbol{C} \\ \boldsymbol{C}\boldsymbol{G} \\ \boldsymbol{C}\boldsymbol{G}^2 \\ \cdot \\ \cdot \\ \cdot \\ \boldsymbol{C}\boldsymbol{G}^{l-1} \end{bmatrix} \begin{bmatrix} x_1(0) \\ x_2(0) \\ x_3(0) \\ \cdot \\ \cdot \\ \cdot \\ x_l(0) \end{bmatrix}$$

$$+\begin{bmatrix} D & 0 & 0 & \dots 0 \\ CH & D & 0 & \dots 0 \\ CGH & CH & D & \dots 0 \\ \cdot & \cdot & \cdot & \cdot \\ \cdot & \cdot & \cdot & \cdot \\ \cdot & \cdot & \cdot & \cdot \\ CG^{l-2}H & CG^{l-3}H & CG^{l-4}H & \dots D \end{bmatrix} \begin{bmatrix} u(0) \\ u(1) \\ u(2) \\ \cdot \\ \cdot \\ \cdot \\ u(l-1) \end{bmatrix} \tag{9.61}$$

Exercise 9.26 Check the correctness of (9.61) by deriving the result directly from (9.14).

9.5.1 Reachability and Observability after Sampling

If the continuous-time plant in Fig. 9.5 is controllable, the resulting discrete-time system (9.14) is not necessarily reachable. The following example shows how reachability (and observability) are affected by the choice of the sampling interval h.

Example 9.5 Consider a single-input, single-output continuous-time plant described by the model

$$\begin{bmatrix} \dot{x}_1 \\ \dot{x}_2 \end{bmatrix} = \begin{bmatrix} 0 & -1 \\ 1 & 0 \end{bmatrix} \begin{bmatrix} x_1 \\ x_2 \end{bmatrix} + \begin{bmatrix} 1 \\ 0 \end{bmatrix} u' \tag{9.62}$$

$$y = (1,0) \begin{bmatrix} x_1 \\ x_2 \end{bmatrix}$$

which is illustrated in Fig. 9.22; this model is controllable and observable because the matrices

$$[B \mid AB] = \begin{bmatrix} 1 & 0 \\ 0 & 1 \end{bmatrix} \quad \text{and} \quad \begin{bmatrix} C \\ \overline{CA} \end{bmatrix} = \begin{bmatrix} 1 & 0 \\ 0 & -1 \end{bmatrix}$$

are not singular.

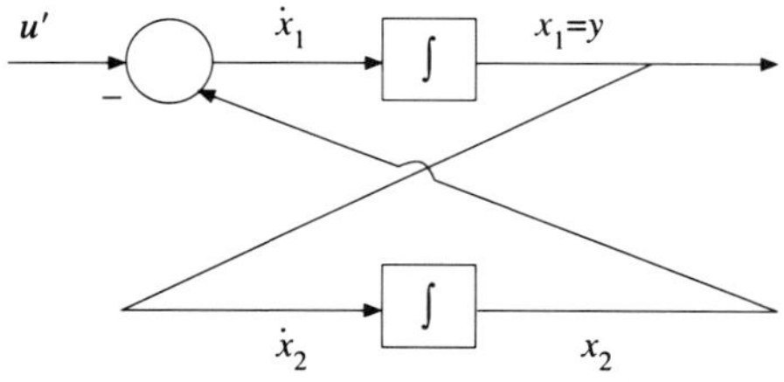

Figure 9.22 System for example 9.5.

Exercise 9.27 Compute the eigenvalues and eigenvectors of the system matrix in (9.62).
Answer: $\pm$j with eigenvectors $(\pm \mathrm{j}, 1)^{\mathrm{T}}$.
When (9.62) is sampled as shown in Fig. 9.5 we get

$$\boldsymbol{x}((k+1)h) = \boldsymbol{G}(h)\boldsymbol{x}(kh) + \boldsymbol{H}(h)u(kh) \tag{9.63}$$

with $\boldsymbol{G}(h)$ and $\boldsymbol{H}(h)$ computed by formulae (9.17) and (9.18).

Exercise 9.28 Compute $\boldsymbol{G}(h)$ and $\boldsymbol{H}(h)$ in (9.63). *Hint*: by formula (3.56) we can write $\exp(\boldsymbol{A}h)=\boldsymbol{M}\exp(\boldsymbol{\Lambda}h)\boldsymbol{M}^{-1}$ with

$$\boldsymbol{M} = \begin{bmatrix} \mathrm{j} & -\mathrm{j} \\ 1 & 1 \end{bmatrix} \text{ and } \boldsymbol{\Lambda} = \begin{bmatrix} \mathrm{j} & 0 \\ 0 & -\mathrm{j} \end{bmatrix}$$

Answer:

$$\boldsymbol{G}(h) = \begin{bmatrix} \cos h & -\sin h \\ \sin h & \cos h \end{bmatrix}; \ \boldsymbol{H}(h) = \begin{bmatrix} \sin h \\ 1-\cos h \end{bmatrix}$$

Figure 9.23 shows the vectors $\boldsymbol{H}$ and $\boldsymbol{GH}$ for $h \in \{\pi/4, \pi/2, 3\pi/4, 15\pi/16, \pi\}$ that characterize the reachability properties of model (9.63). When h equals π, $\boldsymbol{H}$ and $\boldsymbol{GH}$ have the same direction and the system is no longer reachable. Recall that the control sequence $\{u(0), u(\mathrm{h})\}$ that drives the origin to a given state $\boldsymbol{x}_f$ is made of the components of $\boldsymbol{x}_f$ along $\boldsymbol{H}$ and $\boldsymbol{GH}$; when these vectors approach the same direction at least one of the components goes to infinity. In practice, however, input amplitudes are always bounded; therefore an important consequence of the above analysis is that the sampling period determines the size of the set of reachable states in the case of amplitude-constrained inputs. The shaded area in Fig. 9.23 represents the set of states that can be reached from the origin in two steps, for $h=\pi/4$ and $|u(kh)| \leq 0.5$. Obviously such a set converges to a line segment as h approaches π.

The loss of reachability of model (9.63) for $h=\pi$ means that we can only steer the state along the x_2 axis (Fig. 9.23) and that x_1 will always be zero at the sampling instants. However, this does not imply that x_1 will be identically zero in the corresponding continuous-time system. Recall that when the plant of Fig. 9.22 is sampled, u' becomes a piecewise constant signal; for example, if u' is a unit step applied at time $t=0$ then $x_1(t)=y(t)=\sin t$, $t \in \mathbb{R}_0$, which shows that x_1 vanishes only at the sampling instants. Observability is also lost for $h=\pi$ because y being zero at the sampling instants can give no information about x_2. In fact the observability matrix of (9.63) is

$$\begin{bmatrix} \boldsymbol{C} \\ \boldsymbol{CG} \end{bmatrix} = \begin{bmatrix} 1 & 0 \\ \cos h & -\sin h \end{bmatrix}$$

For $h=\pi$ we have

$$\begin{bmatrix} \boldsymbol{C} \\ \boldsymbol{CG} \end{bmatrix} = \begin{bmatrix} 1 & 0 \\ -1 & 0 \end{bmatrix}$$

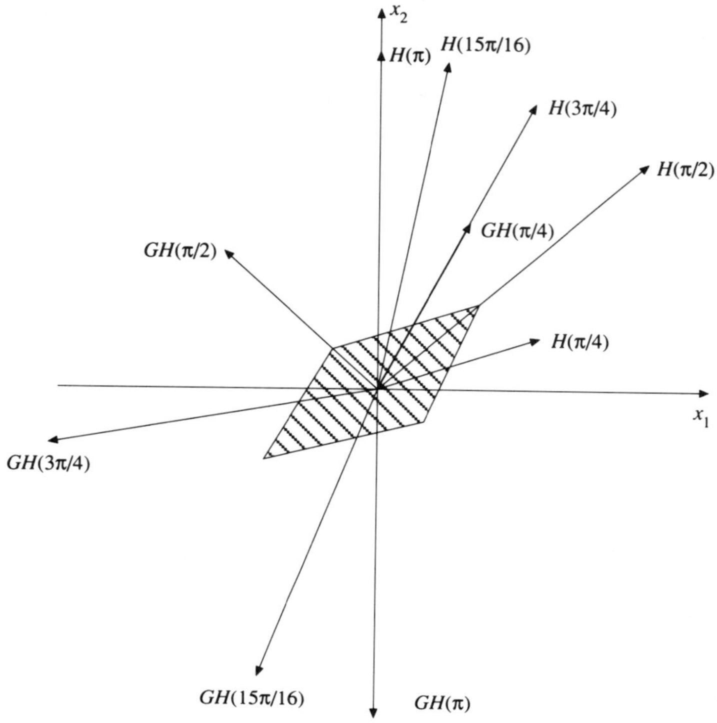

Figure 9.23 Evolution of the columns of the reachability matrix of model (9.63) as the sampling period varies from 0 to π. The shaded area represents the set of reachable states for $h=\pi/4$ and $|u(kh)|\leqslant 0.5$.

which shows that span $\{(0,1)^T\}$ is the sub-space of *unobservable* states, as expected; the component x_1 is directly observable.

This example suggests that the loss of reachability (and observability) of a continuous-time controllable plant, after sampling, may be the result of a choice of sampling frequency equal to a half of a system's natural frequency. The following theorem states this result in the case of a single-input, single-output plant.

Theorem: Assume that a controllable continuous-time single-input, single-output representation

$$\dot{x} = Ax + bu'$$

$$y = cx \tag{9.64}$$

is sampled as shown in Fig. 9.5. Then the resulting discrete-time state-space model is *reachable* if and only if for every pair $(\lambda_i,\bar{\lambda}_i)$ of conjugate eigenvalues of $\boldsymbol{A}$ we have $(\lambda_i-\bar{\lambda}_i)\neq j2\pi l/h$, $l=\pm1, \pm2,\ldots$; $j=\sqrt{-1}$, where h is the sampling period.

Proof: The proof will be given for the case of distinct eigenvalues only. Without any loss of generality, we can assume the continuous-time state-space representation in diagonal form, i.e.

$$\boldsymbol{A} = \begin{bmatrix} \lambda_1 & & \mathbf{0} \\ & \ddots & \\ \mathbf{0} & & \lambda_n \end{bmatrix} \quad \text{and } \boldsymbol{b} = \begin{bmatrix} b_1 \\ \vdots \\ b_n \end{bmatrix}$$

with $b_i\neq 0$, $i=1,\ldots,n$ (cf. exercise 7.19). Then

$$\mathbf{G} = \begin{bmatrix} \exp(\lambda_1 h) & & \mathbf{0} \\ & \ddots & \\ \mathbf{0} & & \exp(\lambda_n h) \end{bmatrix}, \quad \boldsymbol{H} = \begin{bmatrix} h_1 \\ \vdots \\ h_n \end{bmatrix} = \begin{bmatrix} \frac{b_1}{\lambda_1}(\exp(\lambda_1 h)-1) \\ \vdots \\ \frac{b_n}{\lambda_n}(\exp(\lambda_n h)-1) \end{bmatrix}$$

and

$$\boldsymbol{M}_n = \begin{bmatrix} h_1 & h_1\exp(\lambda_1 h) & \ldots & h_1(\exp(\lambda_1 h))^{n-1} \\ \vdots & \vdots & & \vdots \\ h_n & h_n\exp(\lambda_n h) & \ldots & h_n(\exp(\lambda_n h))^{n-1} \end{bmatrix}$$

$$= \begin{bmatrix} h_1 & & \mathbf{0} \\ & \ddots & \\ \mathbf{0} & & h_n \end{bmatrix} \begin{bmatrix} 1 & \exp(\lambda_1 h) & (\exp(\lambda_1 h))^2 & \ldots & (\exp(\lambda_1 h))^{n-1} \\ \vdots & \vdots & \vdots & & \vdots \\ 1 & \exp(\lambda_n h) & (\exp(\lambda_n h))^2 & \ldots & (\exp(\lambda_n h))^{n-1} \end{bmatrix}$$

$\boldsymbol{H}$ has non-zero elements because $h>0$. Then, by the same reasoning as in exercise 7.19, we see that $\boldsymbol{M}_n$ has full rank if and only if the rows of the second matrix on the right hand side are distinct. But this is equivalent to the requirement that

$$\lambda_l h - \lambda_k h \neq jn2\pi$$

for $l\neq k$ and $n=\pm1, \pm2, \ldots$

9.6 MODAL ANALYSIS

The modal analysis carried out in section 7.1 for continuous-time systems is also applicable to the discrete case. Therefore we shall make here only a brief analysis with the help of an example.

Example 9.6 Consider the Lanchester model of warfare described by equation (2.30):

$$\boldsymbol{n}(k+1) = \begin{bmatrix} 1 & -h_2 \\ -h_1 & 1 \end{bmatrix} \boldsymbol{n}(k)$$

The system eigenvalues are $1 \pm \surd(h_1 h_2)$ with associated eigenvectors $(1, -\surd(h_1 h_2)/h_2)$and $(1, \surd(h_1 h_2)/h_2)$ respectively. Defining

$$\boldsymbol{W} = \begin{bmatrix} 1 & 1 \\ \surd(h_1 h_2)/h_2 & -\surd(h_1 h_2)/h_2 \end{bmatrix} \tag{9.65}$$

and

$$\boldsymbol{n}(k) = \boldsymbol{W}\,\mathbf{d}(k) \tag{9.66}$$

we get the following diagonal representation for equation (2.30):

$$\boldsymbol{d}(k+1) = \begin{bmatrix} 1-\surd(h_1 h_2) & 0 \\ 0 & 1+\surd(h_1 h_2) \end{bmatrix} \boldsymbol{d}(k) \tag{9.67}$$

The vector $\boldsymbol{d}$ is made of the components of the state vector along the eigenvectors. If $h_1=0.4$ and $h_2=0.1$ we have the situation illustrated in Fig. 2.30 with

$$\boldsymbol{d}(k) = \begin{bmatrix} (0.8)^k & 0 \\ 0 & (1.2)^k \end{bmatrix} \boldsymbol{d}(0) \tag{9.68}$$

The 'locus of mutual annihilation' in this figure is the direction defined by the eigenvector $(1, \surd(h_1 h_2)/h_2)$: any initial condition on this line starts a trajectory towards the origin along the line. In general a trajectory will be a composition (linear combination) of straight line trajectories along the eigenvectors, as shown in Fig. 2.30 for $\boldsymbol{n}(0)=(50,90)^{\mathrm{T}}$. Because (9.68) has real positive eigenvalues, the components of the state vector evolve monotonically; however, the mode associated with eigenvalue 1.2 tends to infinity because its magnitude is greater than 1. This is in agreement with the discussion in section 9.4, about signals and z-plane pole locations, and the illustrations in Fig. 9.21.

9.7 STABILITY

Although stability is one of the most important requirements in control systems design, here we shall discuss it briefly because the stability analysis carried out in Chapter 3, for continuous-time systems, is also applicable to the discrete case. In particular the stability concepts defined there apply, word for word, to the discrete-time case. The only significant change is the stability boundary from the imaginary axis to the unit circle. For example, the plant with transfer function

$$H(z) = \frac{1}{(z-0.9)\,(z-0.5)}$$

is b.i.b.o. stable because it has all its poles inside the unit disc. On the other hand, the system described by equation (9.68) is neither asymptotically stable nor stable because it has one eigenvalue outside the unit disc. Also the well-known methods of stability analysis for continuous-time systems are immediately applicable to the discrete case with obvious modifications. The application of the root-locus method and the Nyquist criterion to the stability analysis of closed-loop discrete-time systems, with known open loop characteristics, is treated in the next chapter in connection with digital controller synthesis. Here we shall look for alternatives to the Routh–Hurwitz criterion.

In Chapter 3 we found that the Routh–Hurwitz criterion was a useful method for determining if a polynomial has all its zeros inside the left half of the complex plane. A possible way to apply the Routh–Hurwitz test to a discrete-time system is by means of the transformation

$$w = \frac{z+1}{z-1} \tag{9.69}$$

which maps the z-plane unit disc into the left half w-plane. Therefore if we wish to know if the polynomial

$$P(z) = a_n z^n + a_{n-1} z^{n-1} + \ldots + a_1 z + a_0 \tag{9.70}$$

has all its roots inside the unit disc all we need is to transform it into a polynomial in w by means of (9.69) and then apply the Routh–Hurwitz test to the latter. The following example illustrates this procedure.

Let $P(z)=z^2-1.4\ z+0.45$. Using the suggested transformation, $P(z)=0$ becomes

$$\left(\frac{w+1}{w-1}\right)^2 - 1.4\left(\frac{w+1}{w-1}\right) + 0.45 = 0$$

which is equivalent to

$$Q(w) = 0.05\ w^2 + 1.1w + 2.85 = 0$$

Now applying the Routh-Hurwitz test to this polynomial we get

$$\begin{array}{|ll} 0.05 & 2.85 \\ 1.1 & 0 \\ 2.85 & \end{array}$$

which shows that $Q(w)$ has all its roots inside the left half w-plane, or equivalently, that the roots of $P(z)$ lie inside the unit disc. In fact the roots of $P(z)$ are 0.9 and 0.5.

Exercise 9.29 Show that transformation (9.69) maps the z-plane unit disc into the left half of the w-plane. *Hint*: compute the real part of w for $z=x+jy$; then show that $x^2+y^2 \geq 1$ implies a non-negative real part for w.

A useful alternative is Jury's criterion described below (Kuo, 1980).

9.7.1 Jury's Stability Test

This is a stability test that determines whether all the roots of the polynomial (9.70) are inside the unit disc, and it has all the advantages of the Routh–Hurwitz test. First we need to construct the following table:

$$
\begin{array}{llll}
a_n & a_{n-1} \ldots & a_1 & a_0 \\
\underline{a_0} & \underline{a_1} \quad \ldots & \underline{a_{n-1}} & \underline{a_n} \\
a_{n-1}^{n-1} & a_{n-2}^{n-1} \ldots & a_0^{n-1} & \\
a_0^{n-1} & a_1^{n-1} \ldots & a_{n-1}^{n-1} & \\
\hline
\cdot & & & \\
\cdot & & & \\
\cdot & & & \\
\hline
a_0^0 & & &
\end{array}
$$

where a_0, a_1,. . ., a_n are the coefficients of $P(z)$ in (9.70); this table is formed by pairs of rows, computed recursively; in each row pair the second row is the first row in reverse order. For example, the third row is obtained by multiplying the second row by a_0/a_n and subtracting from the first row; therefore the last element becomes zero. Similarly, the fifth row is the difference between the third row and the fourth row multiplied by

$$a_0^{n-1}/a_{n-1}^{n-1}$$

This procedure is repeated until there are $2n+1$ rows, where the last one consists of a single element only; then we have *Jury's stability test*: $P(z)$ in (9.70) has all roots inside the unit disc if and only if a_0^0 and the first element of the first row, in each row pair, is not zero and has the same sign.

Before embarking on the construction of the above table, it is wise to check if the conditions

$$a_n\, P(1) > 0$$

$$a_n(-1)^n\, P(-1) > 0 \tag{9.71}$$

are satisfied because it can be shown that they constitute a set of necessary conditions for the roots of $P(z)$ to lie inside the unit circle. The following example illustrates what has been said so far.

Let $P(z)=z^2-1.4\,z+0.45$. Because $P(1)=0.05>0$ and $(-1)^n\,P(-1)=2.85>0$ we can proceed to the construction of Jury's table. The result is as follows:

$$
\begin{array}{lll}
1 & -1.4 & 0.45 \\
0.45 & -1.4 & 1 \\
\hline
0.80 & -0.77 & \\
-0.77 & 0.80 & \\
\hline
0.054 & &
\end{array}
$$

Because 1, 0.80 and 0.054 are of the same sign, this shows that $P(z)$ has all roots inside the unit disc.

9.8 FREQUENCY-DOMAIN ANALYSIS

9.8.1 Frequency Response Analysis of Discrete-time Systems

We have shown in Chapter 5 that the steady-state response of a continuous-time, linear, stable system to the sinusoidal input $\sin \omega t$, is also a sinusoid of the *same* frequency, namely

$$|G(j\omega)|\sin(\omega t + arg\ G(j\omega)) \qquad (9.72)$$

where $G(s)$ is the system transfer function.

The discrete-time case is also amenable to a similar interpretation, as shown in the following. Let us make precise, at the outset, what we mean by the 'frequency' of a discrete-time sinusoidal signal. Recall that when we are dealing with discrete-time signals, we have sequences of numbers and not sequences of Dirac impulses. The latter are only meaningful when the independent variable changes continuously. Signal $u_1{}^*(t)$, $t \in \mathbb{R}$, in Fig. 9.2, is such an example; furthermore if $u_1(t) = \sin(\omega t)$, $t \in \mathbb{R}$, then $u_1{}^*(t)$ contains, besides the frequency ω, an infinite number of frequencies, as shown in Chapter 10.

Recalling that

$$\sin(\alpha x) = \frac{e^{j\alpha x} - e^{-j\alpha x}}{2j} \qquad (9.73)$$

we may regard the discrete-time sinusoidal sequence

$$(\sin \alpha k)_{k \in \mathbb{N}_0} \qquad (9.74)$$

as being generated by the rotating phasors $e^{j\alpha}$ and $e^{-j\alpha}$. However such rotation is not done continuously but by 'quantum leaps' of magnitude α and $-\alpha$, respectively.

By analogy with the continuous-time case, we then define α as the 'frequency' of the sinusoidal sequence (9.74). Note that we only require

$$0 \leqslant |\alpha| \leqslant \pi$$

because

$$(\sin(\alpha + \pi)\ k) = -\sin(\alpha k)$$

α can be interpreted as the 'number of radians per leap'; as a result the sequence

$$(\sin \alpha k)_{k \in \mathbb{N}_0}$$

will not be periodic unless α/π is rational.

Consequently, the sampling of a continuous-time sinusoid may lead to a non-periodic or periodic sequence. Furthermore, in the latter case, the period of the resulting sequence will be a function of the sampling period. For example, the sequences

$$\left\{\sin\left(\frac{\pi}{4}k\right)\right\}_{k\in\mathbb{N}_0}$$

$$\left\{\sin\left(\frac{\pi}{3.5}k\right)\right\}_{k\in\mathbb{N}_0}$$

$$\left\{\sin\left(\frac{\pi}{3.1}k\right)\right\}_{k\in\mathbb{N}_0}$$

have periods 8, 7 and 31, respectively, and can be regarded as the result of sampling

$$(\sin t)_{t\in\mathbb{R}_0^+}$$

with sampling periods $h_1=\pi/4$, $h_2=\pi/3.5$ and $h_3=\pi/3.1$ respectively.

Consider now a linear stable, discrete-time system with transfer function $W(z)$, and input

$$(u_k)_{k\in\mathbb{N}_0} = (\sin(\alpha k))_{k\in\mathbb{N}_0} \tag{9.75}$$

The z-transform of the output is

$$Y(z) = W(z)\frac{z\sin\alpha}{(z-e^{+j\alpha})(z-e^{-j\alpha})} \tag{9.76}$$

A partial fraction expansion leads to

$$Y(z) = \frac{a\,z}{z-e^{j\alpha}} + \frac{\bar{a}\,z}{z-e^{-j\alpha}} + \text{contributions from the poles of } W(z) \tag{9.77}$$

Because we have assumed a stable $W(z)$, only the first two terms contribute to the steady-state response. Now

$$a\,z = \lim_{z\to\exp(j\alpha)} W(z)\frac{z\sin\alpha}{z-e^{-j\alpha}} \tag{9.78}$$

Therefore

$$a = \frac{W(e^{j\alpha})}{2j} \tag{9.79}$$

$$\bar{a} = \frac{W(e^{-j\alpha})}{-2j}$$

provided $\alpha \neq 0$ and $\alpha \neq \pi$. Otherwise $a = \bar{a} = 0$. Setting

$$W(e^{j\alpha}) = M\, e^{j\theta} \tag{9.80}$$

we can rewrite (9.77) as

$$Y(z) = \frac{M}{2j}\left(\frac{z\, e^{j\theta}}{z - e^{j\alpha}} - \frac{z\, e^{-j\theta}}{z - e^{-j\alpha}}\right) + \ldots \tag{9.81}$$

The steady-state response is then

$$(y(k))_{k \in \mathbb{N}_0} = \frac{M}{2j}(e^{j(k\alpha+\theta)} - e^{-j(k\alpha+\theta)}) = (M \sin(k\alpha + \theta))_{k \in \mathbb{N}_0} \tag{9.82}$$

where

$$\begin{cases} M = |W(e^{j\alpha})| \\ \theta = \arg W(e^{j\alpha}) \end{cases} \tag{9.83}$$

This constitutes the discrete-time equivalent of the continuous-time frequency domain interpretation. In order to prove this result also holds when $\alpha = 0$ or $\alpha = \pi$, we redefine the input sequence as

$$(u_k)_{k \in \mathbb{N}_0} = (\sin(k\alpha + \varphi))_{k \in \mathbb{N}_0} \tag{9.84}$$

For $\alpha = 0$ we have a constant signal. A trivial analysis reveals that

$$\lim_{\alpha \to 0} W(e^{j\alpha})$$

is in fact the 'steady-state' system gain in this case.

> **Exercise 9.30** Given a system with transfer function $W(z)$, prove that $W(1)$ is the steady-state gain for constant inputs. *Hint*: use the final value theorem of z-transforms.

When $\alpha = \pi$ we have the maximum frequency, i.e. the alternating sequence

$$(u_k)_{k \in \mathbb{N}_0} = (-1)^k \sin \varphi \tag{9.85}$$

In this case the output has the z-transform

$$Y(z) = W(z)\left(\frac{z}{z+1} \sin \varphi\right) \tag{9.86}$$

Because $W(z)$ is assumed stable, the output converges to the signal

$$\bar{Y}(z) = W(-1)\frac{z}{z+1} \sin \varphi \tag{9.87}$$

therefore the steady-state gain, at frequency π, is

$$\lim_{\alpha \to \pi} W(e^{j\alpha}) = \lim_{\omega \to -1} W(z) \tag{9.88}$$

as expected. This completes our proof.

9.8.2 Nyquist Stability Criterion

Let us address the stability problem of a closed-loop system as in Figure 9.24, when $G(e^{j\alpha})$, $0 \leqslant \alpha \leqslant 2\pi$ is known. In continuous-time systems the Nyquist criterion is used to detect the zeros of $(1+G(s))$ in the right half-plane. A similar criterion can be developed for discrete-time systems when the zeros of $(1+G(z))$ outside the unit circle need to be detected. A natural choice in the z-plane is the contour shown in Fig. 9.25: it consists of the unit circle indented as shown at the pole $z=1$, in order to exclude it (similarly for other poles on the unit circle, if any), together with a second circle expanding to infinity. Here the map $G(e^{j\alpha})$, $0 \leqslant \alpha \leqslant 2\pi$ is called the Nyquist curve of the system.

Exercise 9.31 Generalize the continuous time root-locus ideas to the discrete system of Fig. 9.24.

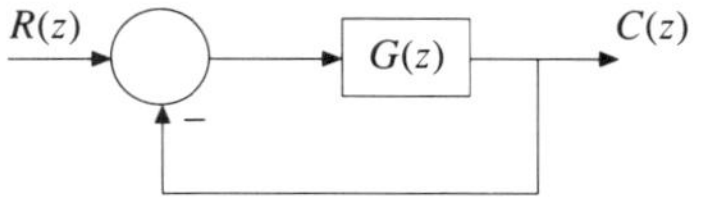

Figure 9.24 Discrete-time feedback system with transfer function $C(z)/Y(z) = G(z)/(1+G(z))$.

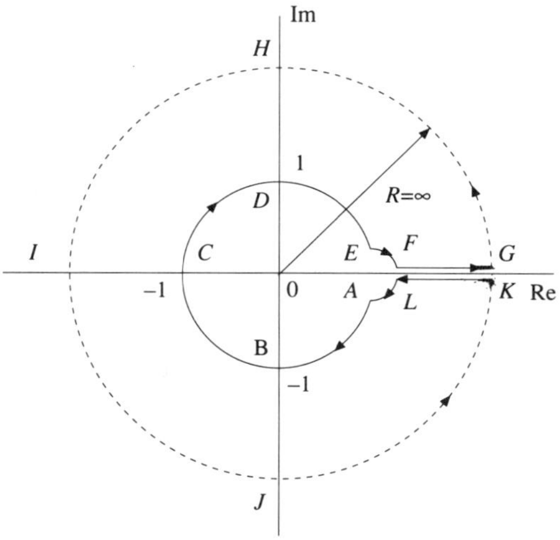

Figure 9.25 A contour enclosing the area outside the unit circle.

Example 9.7 Consider the closed-loop system of Figure 9.26 with $Gp(s)=K/s$, $K \geqslant 0$. Defining

$$G(z) = \mathcal{Z}\left(\mathcal{L}^{-1} \frac{(1 - e^{-hs})}{s} \frac{K}{s} \right) = \frac{K h}{z - 1}$$

we have

$$\frac{Y(z)}{R(z)} = \frac{G(z)}{1 + G(z)} = \frac{K h}{z - 1 + K h}$$

The root-locus is depicted in Fig. 9.27, showing the closed-loop pole outside the unit circle for $K > 2/h$. This is in sharp contrast with the continuous-time case where a first-order system is always stable.

A Nyquist analysis confirms the above result. The image of the z-plane contour of Fig. 9.25 is shown in Fig. 9.28, assuming a non-zero, but very small, radius for the semicircle *EFLA* and a very large radius for the outer circle *GHIJK*. In particular the image of the origin centred unit circle in the z-plane is very easy to draw since

$$\left. \frac{K h}{(z - 1)} \right|_{z = e^{j\alpha}} = -\frac{K h}{2} - j \frac{K h}{2} \frac{\sin \alpha}{(1 - \cos \alpha)}$$

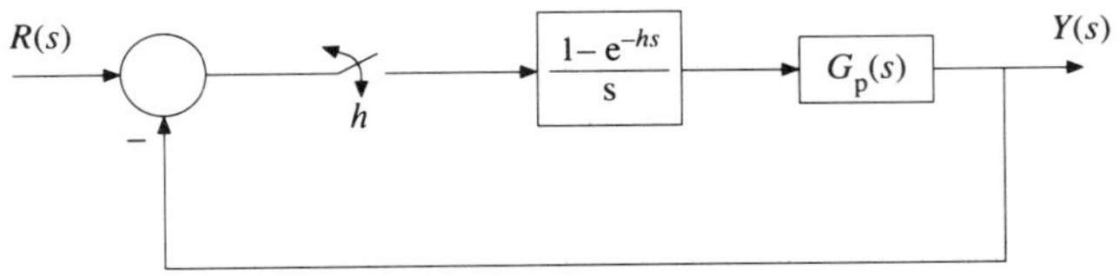

Figure 9.26 A computer-controlled system.

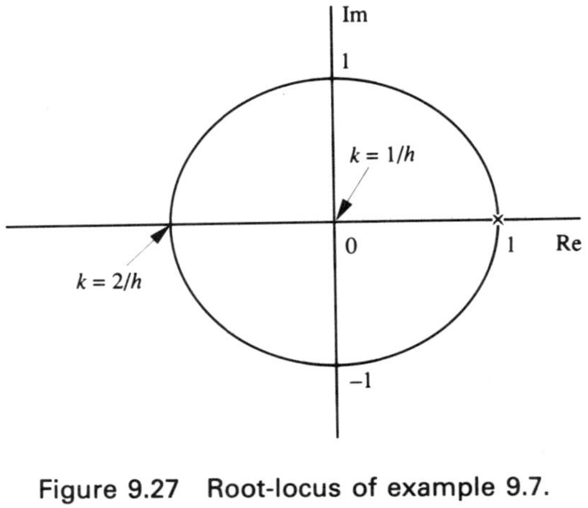

Figure 9.27 Root-locus of example 9.7.

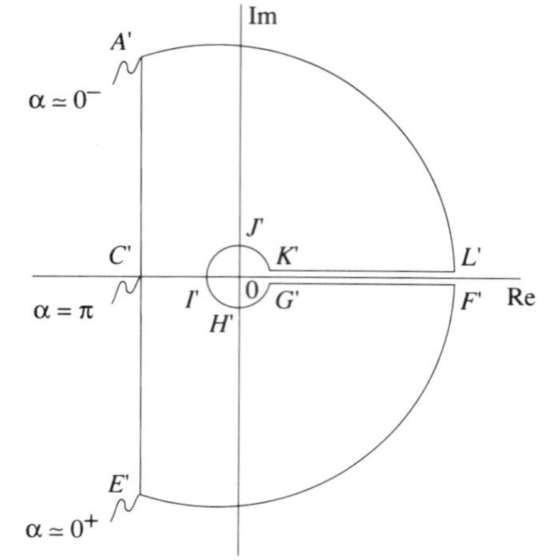

Figure 9.28 Image of the *z*-plane contour of Fig. 9.25 in example 9.7. The abscissa of line $\overline{A'C'E'}$ is $Kh/2$.

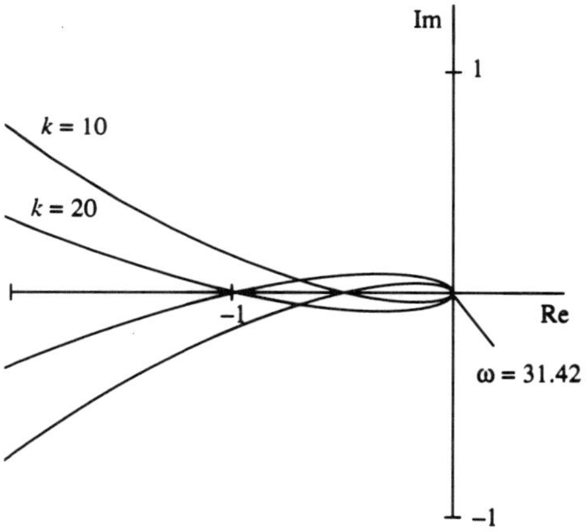

Figure 9.29 Nyquist plots of example 9.8 with sampling period h=0.1.

Example 9.8 Consider now the system of Fig. 9.26 with h=0.1 s and

$$G_p(s) = \frac{K}{s(s+1)}$$

The Nyquist plots of Fig. 9.29 reveal that the system becomes unstable for K slightly larger than 20 (K_c=21.14) when the frequency is about 4.5 rad/s.

The discrete transfer function of the plant and zero-hold combination (Neuman and Baradello, 1979) is

$$G(z) = \frac{K\{[-(1 - h - e^{-h})] z + [1 - e^{-h} - h\, e^{-h}]\}}{z^2 - (1 + e^{-h}) z + e^{-h}}$$

For $h=0.1$ s we get

$$G(z) = 10^{-3} K \frac{(4.837 z + 4.679)}{z^2 - 1.9048 z + 0.9048}$$

which has a zero at -0.9672 and poles at $+1$ and $+0.9048$.

The roots-loci are shown in Fig. 9.30, confirming the results of the Nyquist analysis: the loci leave the unit circle for $K=21.14$, at the point $e^{j0.45}$.

The Nyquist curves are also very useful to assess the sensitivity of our system to parameter variations. In Fig. 9.31 we have such curves for $h \in \{0.05, 0.1, 0.2\}$

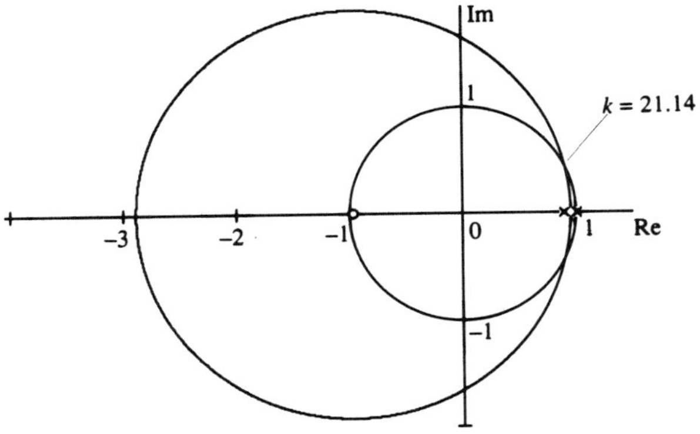

Figure 9.30 Root-loci for example 9.8 with $h=0.1$ s.

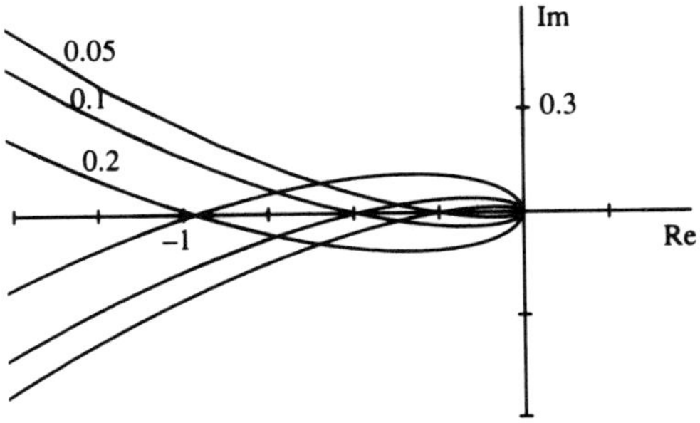

Figure 9.31 Nyquist plots for example 9.8, for sampling periods $h=0.05$, 0.1, 0.2 and gain $k=10$.

and $K=10$. As expected, an increase of the sampling period reduces the system phase margin: for $h=0.2$ the system is very near instability.

If we now replace our system, with $K=10$, by a continuous-time one, with the same $G_p(s)$ and a pure transport delay of $h/2=0.05$ s instead of the ZOH, i.e. a continuous-time, unit feedback, system with open-loop transfer function

$$e^{-0.05s} \times \frac{10}{s\,(s+1)}$$

as in Fig. 9.32, we find the Nyquist curves remain virtually unchanged within the working frequency range as depicted in Fig. 9.33. This is an important result because it allows the use of continuous-time design methods in computer-controlled systems. This point will be resumed later in the book.

Also of interest, in the Nyquist analysis, is the fact that the Nyquist plot, at half the sampling frequency, is real. Because, on the Nyquist path, $z=e^{j\alpha}$, the Nyquist plot becomes real when $\alpha=\pi$ which happens precisely at half the sampling frequency (cf. Figures 9.28, 9.29 and 9.33).

9.9 CONCLUSIONS

In this chapter we have developed the tools required to analyze computer-controlled systems as seen by the computer. We found that the mathematical

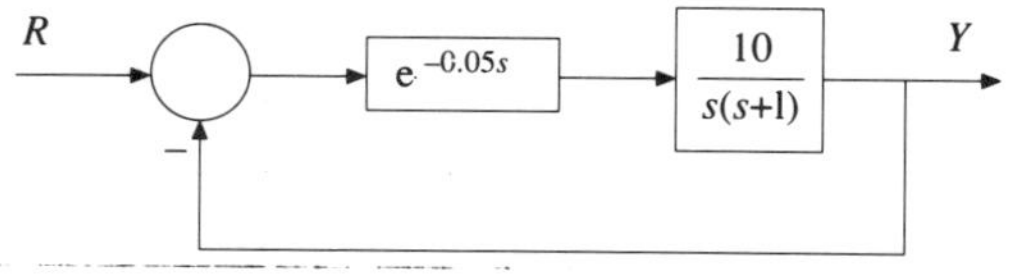

Figure 9.32 Continuous-time approximation of system of Fig. 9.26.

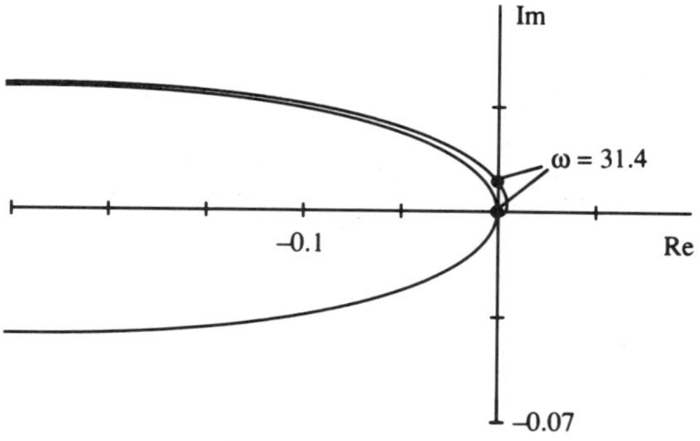

Figure 9.33 Nyquist plots of systems of Figs 9.26 and 9.32.

treatment was easier in comparison with the continuous-time case, mainly because discretization has made our models finite dimensional, even in the presence of plants with pure transport delays. Therefore it comes as no surprise to find that most of the continuous-time results are also applicable to the discrete-time case, with obvious alterations, such as the stability boundary from the imaginary axis to the unit circle.

However, new problems arose mainly in connection with the choice of the sampling interval. It was shown that continuous-time controllable or observable representations, with complex eigenvalues, may lead to discrete uncontrollable or unobservable representations for certain values of the sampling interval. It was also observed that, in general, the magnitude of the control input varies inversely with the sampling interval and, as a consequence, controllability may be lost in practice, owing to actuator saturation, for small sampling times. Last but not least, was the fact that a continuous-time all-pole plant, or a plant with zeros in the left half-plane only, may give rise to a pulse transfer function with zeros outside the unit circle (unstable zeros) for a small sampling period. On the other hand, continuous-time plants with zeros in the right half-plane (non-minimum phase plants) can lead to a discrete model with all zeros inside the unit circle provided a large enough sampling period is selected.

REFERENCES

Astrom, K.J., P. Hagander and J. Sternby (1984) 'Zeroes of sampled systems', *Automatica*, **20**(1), 31–8.

Bernhardsson, Bo (1990) 'The predictive first-order hold circuit', 29th IEEE Conference on Decision and Control, 5–7 December, Honolulu, Hawaii.

Kuo, B.C. (1980) *Digital Control Systems*, Holt, Rinehart and Winston.

Luenberger, D.G (1979) *Introduction to Dynamic Systems*, John Wiley & Sons.

Neuman, C.P. and C.S. Baradello (1979) 'Digital transfer functions for microcomputer control', *IEEE Trans. Syst. Man. Cybernetics*, **SMC–9**(12), pp. 856–60.

Smith, J.M. (1987) *Mathematical Modelling and Digital Simulation for Engineers and Scientists*, John Wiley & Sons.

10

Digital Controller Synthesis

10.1 INTRODUCTION

The structure of a computer-controlled system was introduced in Chapter 9, and is summarized in Fig. 9.1. In essence, the analog compensator is replaced by a digital computer program. Although the basic principles of controller design remain unaltered, new problems arise. Some of them are intrinsic to the nature of the digital computer, such as finite-precision and non-zero computation time. This contrasts with analog machines which allow a continuous range of variation for all the coefficients and variables, and perform the computations instantaneously. These facts must be taken into consideration when implementing a digital controller. An excellent survey of these and other relevant issues can be found in Hanselmann (1987).

On the other hand, infinite dimensional, continuous-time systems (e.g. time-delay systems) are seen by the digital computer as finite dimensional, which constitutes an advantage as far as controller design is concerned. However, new problems may arise as a consequence of the interaction of the discrete- and continuous-time. The most puzzling is the folding back of high frequency continuous-time signals or disturbances into the lower working frequency range. Such problems did not occur in continuous-time systems; for example, high frequency measurement noise was attenuated by the low pass characteristics of the loop. In a sampled-data system, such signals are likely to be interpreted by the computer as reference or output variations, and consequently may induce unnecessary actuator variations. This is why we will spend some time, in the next section, analyzing this phenomenon and learning how to avoid it, before entering the discussion of controller design methods.

10.2 SAMPLED-DATA SYSTEMS

10.2.1 The Alias or Frequency Foldback Phenomenon

Consider a continuous-time signal $u(t)$ with Laplace transform $U(s)$. If $u(t)$ is sampled as shown in Fig. 9.2, we can write

$$u^*(t) = \sum_{n=0}^{\infty} u(n\,h)\;\delta(t - n\,h) \tag{10.1}$$

It can be shown that the Laplace transform of $u^*(t)$ is given by

$$U^*(s) = \frac{1}{h}\sum_{k=-\infty}^{+\infty} U\left(s + j\frac{2\pi\,k}{h}\right) \tag{10.2}$$

Therefore the result of sampling is to repeat the pole-zero pattern of the original signal, at intervals j $2\pi/h$ in the s-plane, the overall result being the sum of the individual patterns as in Fig. 10.1. The restriction of the Laplace transform to the imaginary axis is known as the frequency spectrum of the signal. Hence the spectrum of the sampled signal, $U^*(j\omega)$, is the sum of the original spectrum $U(j\omega)$ displaced by intervals j $2\pi/h$ as in Fig. 10.2.

Consequently it is only possible to recover a band-limited signal perfectly from samples, when the sampling frequency ω_s,

$$\omega_s = \frac{2\pi}{h}$$

is at least twice as large as the highest frequency in the signal. This is in essence the well-known 'sampling theorem'.

Assume now a single frequency signal, namely $u(t)=\sin(\omega_c t)$. The spectra of $u(t)$ and $u^*(t)$ are as in Fig. 10.3, if $\omega_c<\omega_s/2$. When $\omega_c>\omega_s/2$, the spectrum of $u^*(t)$ will contain frequencies lower than ω_c. This is known as the 'alias' or 'frequency foldback' phenomenon for the following reason: since the highest frequency the digital computer (cf. Fig. 9.1) is able to discern, from the sampled values, is $\omega_s/2$, the computer will not see ω_c but its 'alias' in the interval $[0, \pi/h]$; or, alternatively, we can say that inside the computer, the frequency ω_c is folded back into the interval $[0, \pi/h]$. $\omega_s/2$ is known as the Nyquist frequency.

Exercise 10.1 Show that the sampled versions of $\sin(\omega_c t)$, $\sin((\omega_s-\omega_c)t)$, $\sin((\omega_s+\omega_c)t)$, $\sin((2\omega_s-\omega_c)t)$, . . . have identical spectra, when sampled at frequency ω_s (assume $\omega_c \leqslant \omega_s/2$).

The following examples illustrate familiar consequences of this phenomenon.

Example 10.1 Assume that a disc, rotating counter-clockwise at 3 revolutions per second, is illuminated by a stroboscope flashing at 5 Hz. In practice the disc looks

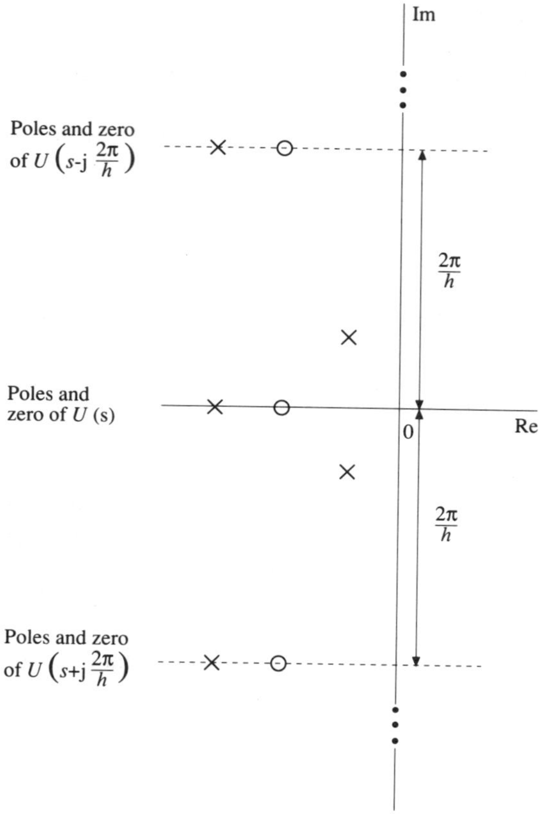

Figure 10.1 Pole-zero pattern for sampled signal. h is the sampling period.

to be rotating at 2 revolutions per second, in the opposite direction!

In fact, this is an immediate consequence of what has just been said: the frequency of the 'sampled signal', 3 Hz, is greater than half the sampling frequency, i.e. 2.5 Hz. Therefore it will have an alias in the interval [0, 2.5], namely (5–3) Hz=2 Hz. We can also reach the same conclusion by working out the position, at the flashing instants, of a fixed point on the disc.

Between two consecutive flashes, i.e. every 0.2 seconds, the disc makes $3\times0.2=0.6$ anti-clockwise turns, which correspond to $(180°+36°)$.

Assuming our point is on the x axis at the first flash, we have the situation depicted in Fig. 10.4: after five flashes, i.e. 1 second later, our point seems to have completed two clockwise revolutions, when in reality it has made three complete turns in the opposite direction.

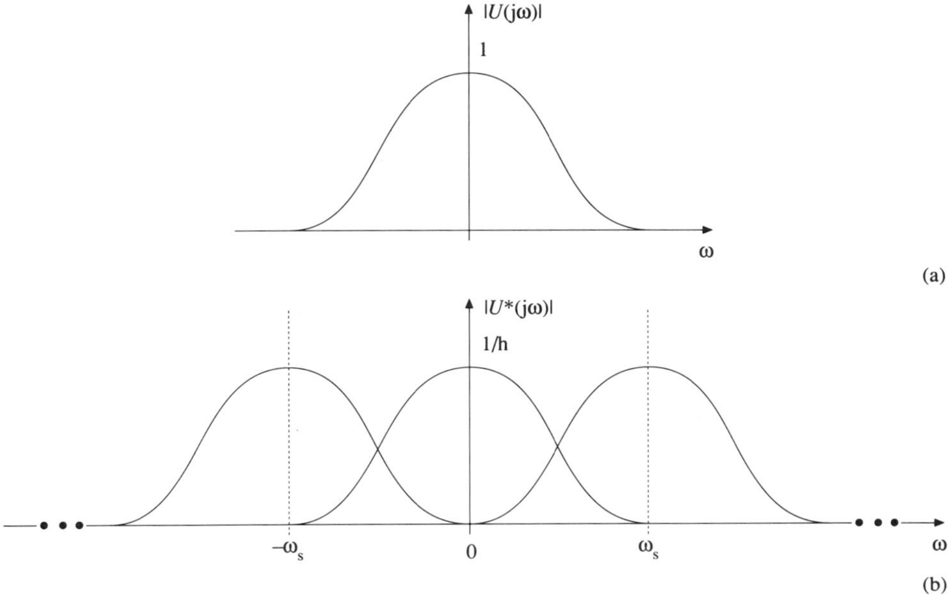

Figure 10.2 Frequency spectra of: (a) continuous-time signal $u(t)$; (b) sampled $u(t)$.

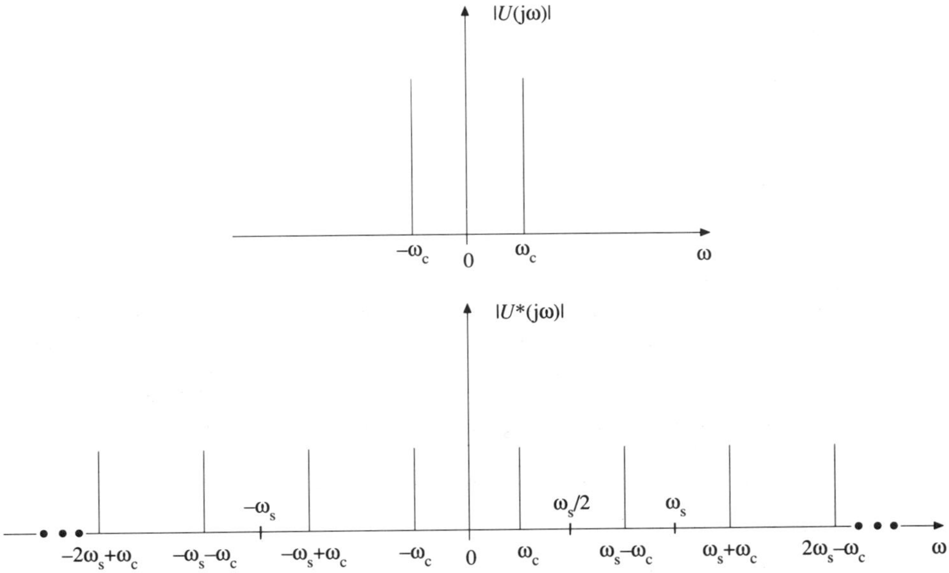

Figure 10.3 Spectra of $\sin(\omega_c t)$ and sampled $\sin(\omega_c t)$.

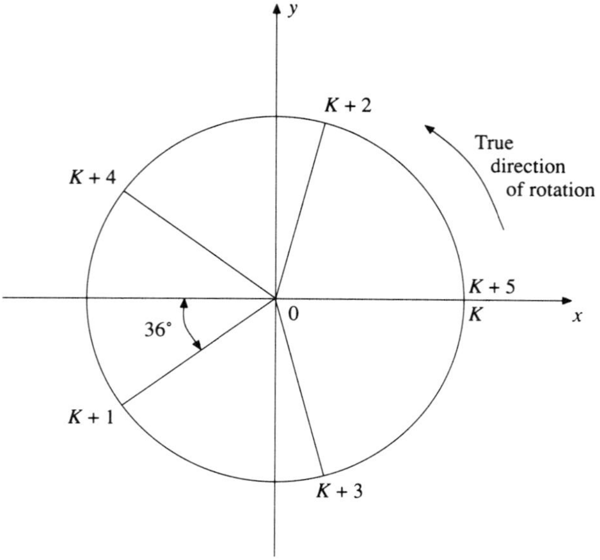

Figure 10.4 Consecutive positions, at the flashing times of a stroboscope, of a fixed point on a disc rotating counterclockwise, at 3 revolutions per second. The flashing frequency is 5 Hz.

Example 10.2 The sine waves

$$\sin(2t)$$

$$\sin\left((2\pi - 2)\left(t + \frac{\pi}{2\pi - 2}\right)\right)$$

$$\sin((2\pi + 2)\, t)$$

when sampled at 1 Hz, produce identical sequences.

In fact, the frequency 2 rad/s is the alias of the frequencies $(2\pi-2)$ rad/s and $(2\pi+2)$ rad/s, when these are sampled at $\omega_s = 2\pi$ rad/s. Figure 10.5 depicts $\sin(2t)$ and $\sin((2\pi+2)\,t)$: note the intersections at the sampling instants.

Example 10.3 Let us go back to our computer-controlled system of Fig. 9.1. What we have been discussing has a very intriguing consequence: in the absence of disturbances, the outputs, $y(t)$, produced by reference inputs of the form

$$r(t) = \sin(\omega t), \ t \in \mathbb{R}$$

$$\omega \in \{\omega_c, \omega_s - \omega_c, \omega_s + \omega_c, 2\omega_s - \omega_c, 2\omega_s + \omega_c, \ldots\}$$

$$0 \leq \omega_c \leq 2\pi/h$$

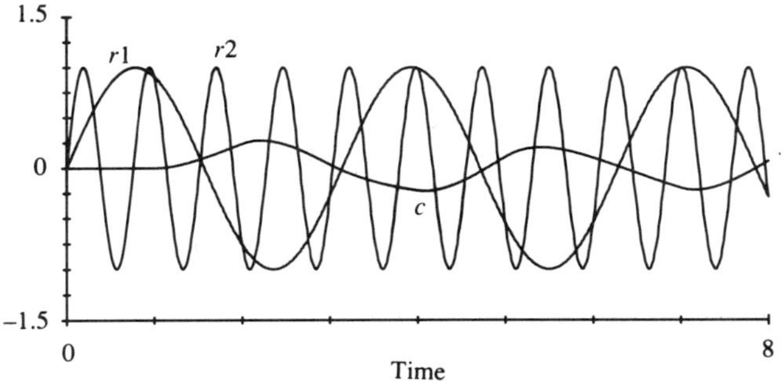

Figure 10.5 Response, $c(t)$, of the computer-controlled system of Fig. 9.1 to the reference inputs $r_1(t)=\sin(2t)$ and $r_2(t)=\sin((2\pi+2)t)$, when $h=1$. The plant transfer function is $1/(s+1)^2$ and the input to the D/A converter is (r^*-y^*).

are identical or shifted versions of each other! This is illustrated in Fig. 10.5 for two such inputs, namely $\sin(2t)$ and $\sin((2\pi+2)t)$, assuming $h=1$, the plant transfer function is $1/(s+1)^2$ and the input to the D/A converter is equal to $(r^*(t)-y^*(t))$, i.e. the action of the computer is simply a sample-and-hold operation. The explanation lies in that such input signals are seen by the computer as identical sequences, because they coincide at the sampling instants.

At this stage the perceptive reader may be asking about the significance of continuous-time frequency response tests on sampled-data systems as in Fig. 9.1. Even when the input is a single frequency signal, the output will always contain an infinite number of frequencies. Therefore, a sensible procedure is to define the frequency response, as the relation between the input sinusoid and the component of the output with the same frequency. Because of the alias phenomenon, such a test is of little practical value for frequencies above the Nyquist frequency. In example 10.4 we show how to compute the (continuous-time) frequency response for a sampled-data system.

The alias phenomenon is undesirable in a computer-controlled system, and the only way to prevent it is to ensure, by adequate filtering, that the signals to be sampled contain negligible power above the Nyquist frequency. If such precautions are not taken, the presence of high frequency measurement noise, unmodelled resonant modes of mechanical systems, etc., may produce low frequency aliases in the interval $[0, \omega_s/2]$. These will be interpreted by the computer as reference changes or errors in the controlled variable, giving rise to unnecessary actuator variations. Therefore we require analog prefiltering, the bandwidth of the filter being chosen equal to $\omega_s/2$.

For example, the second-order system

$$\frac{1}{1+2\zeta(j\omega/\omega_n)+(j\omega/\omega_n)^2}$$

discussed in Chapter 5 (5.21), becomes an anti-aliasing filter with bandwidth $\omega_s/2$, if we set

$$\omega_n = \frac{\omega_s}{2} \text{ and } \zeta = \sqrt{2}/2$$

For the design of analog filters the reader is referred to standard texts on networks and manufacturers' handbooks.

From what has just been said, we can also conclude that it is preferable to suppress low frequency disturbances, e.g. drifts, through digital filtering.

10.2.2 Block Diagram Algebra of Computer-controlled Systems

Let us begin by recalling that

$$\mathcal{Z}\,(\mathcal{L}^{-1}\,G(s)) = G(z)$$

is the z-transform of the unit-impulse response of the continuous-time system with transfer function $G(s)$, as in Fig. 10.6; this is equivalent to

$$G(z) = \frac{Y(z)}{U(z)}$$

as shown in exercise 9.9. Also note that $Y(s)/U(s)$ is no longer a transfer function because it is a function of the input signal.

The great majority of practical configurations in sampled-data systems can be reduced to the ones shown in Fig. 10.7. In part (d), the discrete transfer function can not be defined, since the system behaviour is determined by the values of the input between sampling instants.

Example 10.4 Consider the sampled-data system of Fig. 10.8 (a). In Chapter 9 we saw that

$$\frac{Y^*(s)}{R^*(s)} = \frac{Y(z)}{R(z)} = G(z)$$

and an interpretation of $G\,(e^{j\omega_c h})$, $0 \leq \omega_c \leq \pi/h$, was given.

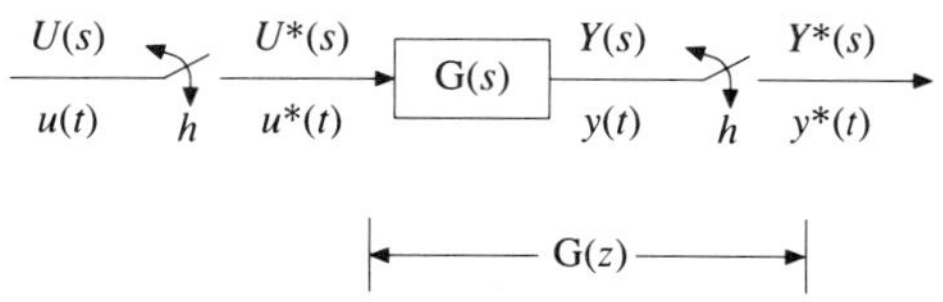

Figure 10.6 Computation of the discrete transfer function of a continuous-time system.

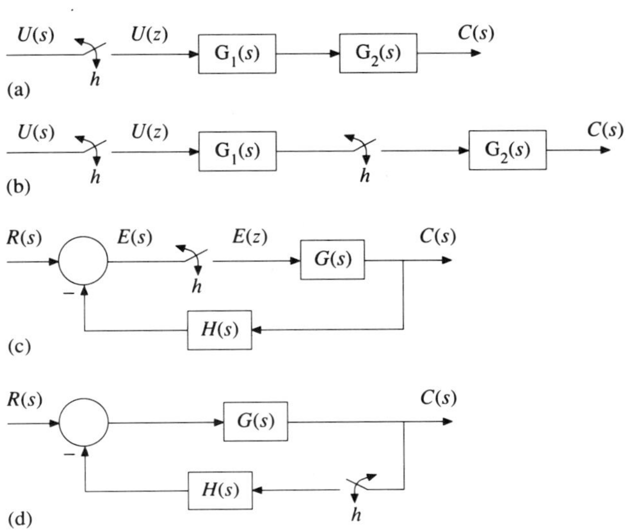

Figure 10.7 Block diagram algebra of sampled-data systems. Note that (d) has no discrete transfer function: (a) $\mathcal{Z}\{G_1(s)\ G_2(s)\}=C(z)/U(z)$; (b) $G_1(z)\ G_2(z)=C(z)/U(z)$; (c) $G(z)/(1+\mathcal{Z}\ (G(s)\ H(s)))=C(z)/R(z)$; (d) $\mathcal{Z}\ (R(s)\ G(s))/(1+\mathcal{Z}\ (G(s)\ H(s)))=C(z)$.

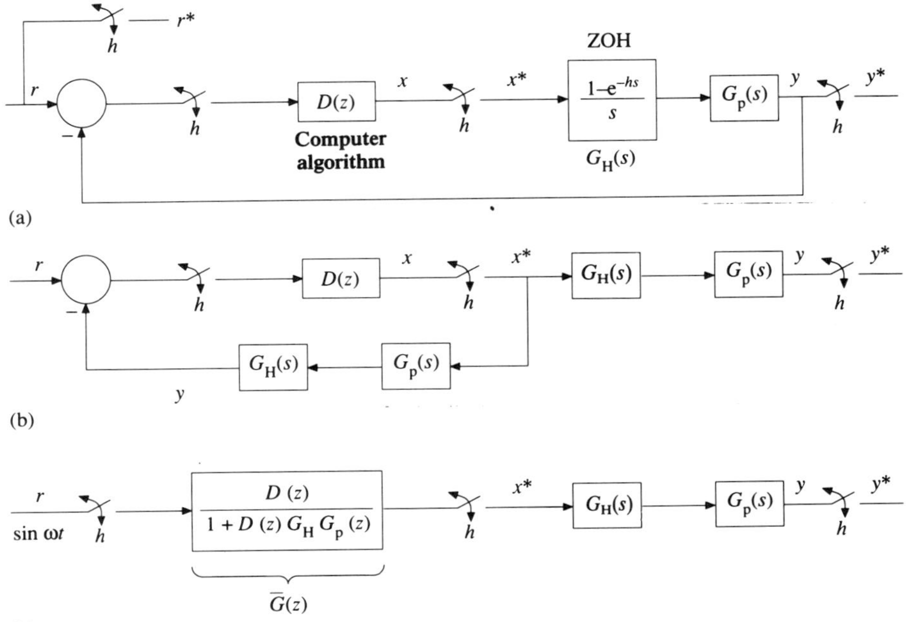

Figure 10.8 Figure of example 10.4: (b) and (c) are equivalent to (a).

We now show that the continuous frequency response of the sampled-data system can be computed from

$$\frac{1}{h} G_H(j\omega)\, G_p(j\omega) \frac{D(z)}{1 + D(z) G_H G_p(z)}\bigg|_{z = e^{j\omega_c h}} \tag{10.3}$$

for $0 \leq \omega < \infty$, where $0 \leq \omega_c \leq \pi/h$, and ω_c is such that $\omega = \pm\omega_c + 2\pi/h\, n_0$, for some $n_0 \in \{0,1,2,3,\ldots\}$. Also

$$G_H\, G_p\, (z) = \mathfrak{Z}\,(G_H\,(s)\, G_p\,(s)) \tag{10.4}$$

From the equivalent diagram shown in part (c) of the figure, the result can be easily obtained. Assume $r\,(t) = \sin\,(\omega\, t)$, ω as above. Then $x^*\,(t)$ is a sequence of Dirac impulses whose areas are

$$(x\,(k\,h))k \in \mathbb{N}_0 = |\bar{G}\,(e^{j\omega_c h})|\, \sin\,(\omega_c\, h\, k + \bar{G}\,(e^{j\omega_c h}))k \in \mathbb{N}_0 \tag{10.5}$$

$x^*(t)$ can also be regarded as the result of sampling the continuous-time sine wave

$$v(t) = A\, \sin\,(\omega_c\, t + \varphi),\ t \in \mathbb{R}$$
$$A = |\ \bar{G}(e^{j\omega_c h})\ |,\ \varphi = \bar{G}(e^{j\omega_c h})$$

From (10.2) we have

$$x^*\,(t) = \frac{A}{h}\,\{\sin\,(\omega_c\, t + \varphi) + \sum_{n=1}^{\infty}\,[\sin\left((n\frac{2\pi}{h} + \omega_c)\, t + \varphi\right)$$
$$- \sin\left((n\frac{2\pi}{h} - \omega_c)\, t - \varphi\right)]\} \tag{10.6}$$

Therefore either

$$\frac{A}{h}\,\sin\left((n_0\frac{2\pi}{h} + \omega_c)\, t + \varphi\right) \quad \text{or} \quad \frac{A}{h}\,\sin\left((n_0\frac{2\pi}{h} - \omega_c)\, t + \varphi\right)$$

is the component of x^* with the same frequency as $r(t)$. The 'frequency response' of the sampled-data system of Fig. 10.8(a), at $\omega = \pm\omega_c + (2\pi/h)\, n_0$, is then

$$\frac{1}{h} G_H(j\omega)\, G_p(j\omega)\, \bar{G}(e^{j\omega_c h}) \tag{10.7}$$

Exercise 10.2 Show that $\bar{G}(e^{j\omega_c h}) = \bar{G}(e^{j\omega h})$ if $\omega = \pm\omega_c + (2\pi/h)\, n_0$.
Answer: since e^{jah} is a periodic function of a, with period $2\pi/h$, the result follows.

In essence, the above discussion resides in the fact that the spectra of r^* and x^* contain exactly the same frequencies, with the amplitudes of the latter being $|\bar{G}(\exp(j\omega_c h)|$ times the amplitudes of the former.

Consequently if the frequency of the sinusoid at $r(t)$ is changed by an integer multiple of $2\pi/h$, $x^*(t)$ will remain unchanged (in steady-state) and so will $y(t)$. However, for continuous-time frequency response determination, we are interested in the relation between $r(t)$ and the component of $y(t)$ with the same frequency as $r(t)$.

10.3 CONTROLLER DESIGN METHODS

Digital controllers can be designed directly in discrete form or may be the result of discretization of a continuous design. Discrete design methods require an early decision about the sampling frequency; if such an estimate turns out to be inadequate, the control system has to be redesigned. However, discrete designs allow, in principle, lower sampling rates and performances not attainable with the discretization of continuous designs; 'dead-beat' behaviour is such an example.

We start this section by looking at some ways to approximate continuous-time controllers. With reference to discrete-time system design methods, they are essentially the same as in the analog case; they are briefly surveyed in the second part of this section.

10.3.1 Analog-to-digital Redesign

The techniques under this heading enable the replacement of an already existing analog controller by digital one. Following Hanselmann (1987) the available discretization methods can be classified into two main groups, namely open-loop and closed-loop methods. In the former the controller is discretized without taking into account the closed-loop use of the controller, and is recommended when the plant is poorly known.

10.3.1.1 Open-loop methods

These methods do not take into account the effect of the (almost) invariable use of the ZOH in the signal reconstruction. Therefore, if the analog controller is yet to be designed, it is advisable to add, to the continuous-time open-loop model, a pure delay of half a sampling interval before starting the design exercise (cf. example 9.8).

The methods under this heading can be grouped into three categories: transform ($s \to z$) substitution, simulation and invariance.

Amongst the substitution methods, we emphasize the bilinear (or Tustin) transformation

$$s \to \frac{2(z-1)}{h(z+1)} \tag{10.8}$$

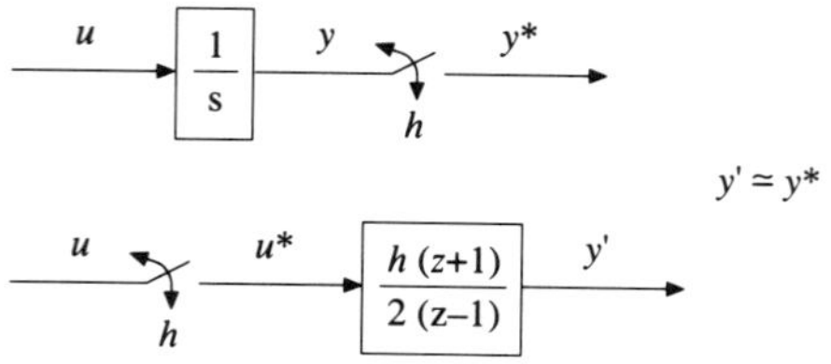

Figure 10.9 A discrete-time (trapezoidal) approximation of continuous-time integration.

This transformation corresponds to a simulation of the continuous-time system via implicit trapezoidal integration. In fact, if u and y are the input and output of an integrator (transfer function $1/s$), the above formula yields the approximating discrete transfer function

$$H(z) = \frac{Y'(z)}{U(z)} = \frac{h}{2}\frac{z+1}{z-1} \tag{10.9}$$

which is equivalent to

$$y'((k+1)\,h) = y'(kh) + \frac{h}{2}(u((k+1)\,h) + u(kh)) \tag{10.10}$$

But $(h/2)\,(u\,((k+1)\,h) + u(k\,h))$ is precisely the area under the piecewise linear approximation of $u(t)$, $t \in \mathbb{R}$, in the interval $[k, (k+1)h]$. Therefore y' is an approximation of y^* (sampled integral of u) as depicted in Fig. 10.9.

The bilinear transformation (10.8) maps the s-plane imaginary axis into the origin centred z-plane unit circle. Since it is a conformal mapping, we can also conclude that the interior of the disc is the image of the left half s-plane. This is an important property, since continuous-time stable systems will remain stable after discretization.

Let us look in detail at this mapping. When s is on the imaginary axis we have the situation depicted in Fig. 10.10 since

$$z = \frac{2 + h\,s}{2 - h\,s} \tag{10.11}$$

In particular, as ω increases from 0 to $+\infty$, its image travels along the unit circle, in the counter-clockwise direction, from $(+1\ +j0)$ to $(-1\ +j0)$, because

$$\frac{2 + h\,j\omega}{2 - h\,j\omega} = 2\,\alpha = 2\ \text{arc}\tan\frac{\omega h}{2} \tag{10.12}$$

If $G'(z)$ is the transfer function obtained from $G(s)$, by the Tustin transformation, we have another important property: the frequency response $G(j\omega)$ is transformed into the frequency response $G'(e^{j\omega h})$ of the discrete-time

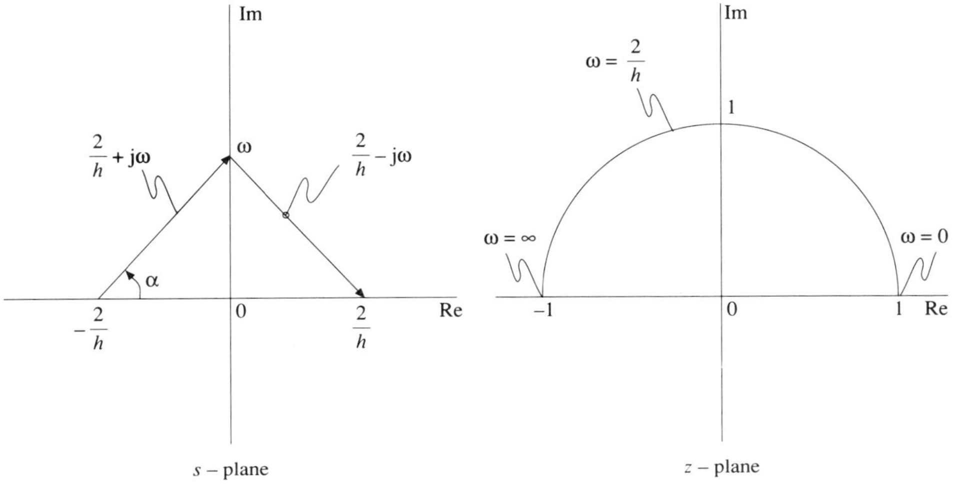

Figure 10.10 Illustration of the $s \rightarrow z$ bilinear transformation: $z=2+h\,s/(2-h\,s)$.

approximation. However, one has to be careful with this interpretation. Because of the non-linear relation (10.12), when we write $G'(e^{j\omega' h})$ we are not describing (approximately) the steady-state response of the continuous-time system to the input $\sin(\omega' t)$, but the response to the input

$$\sin\left(\frac{2}{h}\tan\left(\frac{\omega' h}{2}\right) t\right) \tag{10.13}$$

This effect is generally known as 'frequency warping'.

Example 10.5 Consider the Tustin approximation, with $h=0.1$ s,

$$H'(z) = \frac{0.1\,(z + 1)}{2.1z - 1.9} \tag{10.14}$$

of the transfer function

$$H(s) = \frac{1}{s + 1} \tag{10.15}$$

We now compare the steady-state sinusoidal gains of these transfer functions, to the inputs $\sin t$, $\sin 10\,t$, and $\sin 20\,t$, $t \in \mathbb{R}$. The result is shown is Table 10.1: as the frequency increases, the differences become apparent, which is not unexpected since the continuous-time frequency range $[0, \infty]$ was shrunk to $[0, \pi/h]$, by the bilinear transformation.

Table 10.1 Sinusoidal gains for transfer functions of example 10.5

ω (rad/s)	$\varphi = 2 \text{ arc tan} \frac{\omega h}{2}$ (rad)	$H(j\omega)$	$H'(e^{j\omega h})$
1	9.991 68 E^{-2}	0.707 1 $e^{-j45°}$	0.706 8 $e^{-j45.02°}$
10	0.927 295	0.099 5 $e^{-j84.29°}$	0.091 14 $e^{-j84.77°}$
20	1.570 80	0.049 94 $e^{-j87.14°}$	0.032 09 $e^{-j88.16°}$

If we want $H'(z)$ to produce a correct description of the continuous-time system at a given frequency, we need to first 'prewarp' the frequency scale by the transformation

$$\omega' = \frac{2}{h} \text{ arc tan} \left(\frac{\omega h}{2} \right) \tag{10.16}$$

which is plotted in Fig. 10.11, for h=0.1 s. The continuous-time frequencies ω_1=1, ω_2=10 and ω_3=20 are mapped into ω_1'=0.9992, ω_2'=9.273 and ω_3'=15.71,

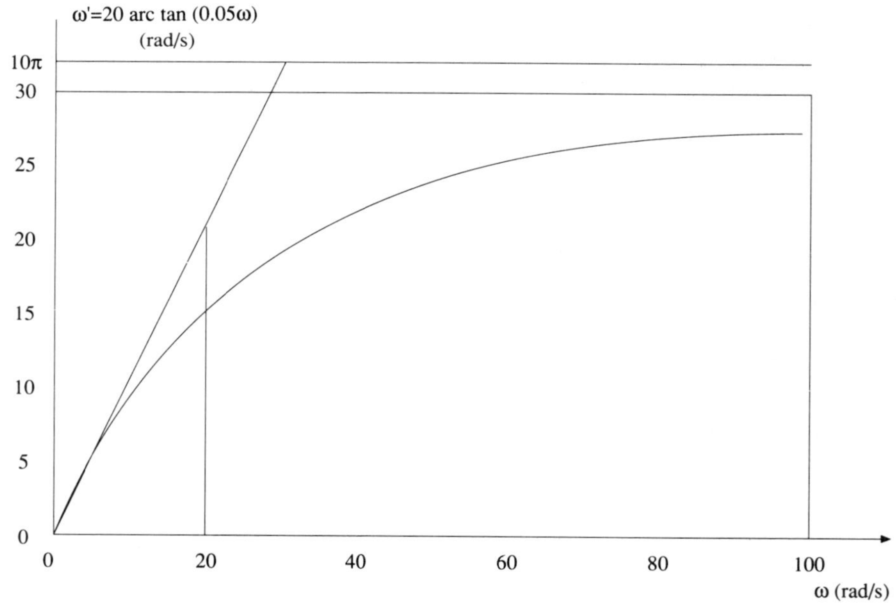

Figure 10.11 Distortion of the frequency scale by the bilinear transformation, when h=0.1 s.

respectively. The values of $H'(\exp(j\omega'_i h))$, $i=1, 2, 3$, are shown in Table 10.2, confirming that

$$H'(\exp(j\omega'_i h)) = H(j\omega_i)$$

Note the linear behaviour of the transformation in the low frequency range.

Table 10.2 Sinusoidal gains of the discrete-time approximation, in example 10.5, with 'prewarped' frequency scale

φ (rad)	$\omega' = \varphi/h$ (rad/s)	$H'(e^{j\omega' h})$
9.991 68 E^{-2}	0.999 168	$0.7071\ e^{-j45°}$
0.927 295	9.272 95	$0.0995\ e^{-j84.29°}$
1.570 80	15.708 0	$0.04994\ e^{-j87.14°}$

Another group of discretization methods, known as invariance methods, is based on agreement of the outputs of the continuous-time approximation, at the sampling instants, for predetermined inputs. In case of step and ramp inputs we have the zero-order (ZOH) and predictive first-order hold (PFOH) equivalence, respectively. These methods were already considered in Chapter 9 for transfer functions and state-space representations.

At this point, we would like to stress the advantages of the ramp-invariance discretization method (Hanselmann, 1987; Bernhardsson, 1990), when compared with the step-invariance method. Unlike the signal reconstruction problem, where hardware limitations dictate almost invariably the use of the ZOH, the discretization method has no such limitations since it is performed by software. In practice, the use of PFOH discretization allows lower sampling rates than ZOH discretization, for equivalent performances.

Table 10.3 shows several discretizations. Note that the Tustin and PFOH discretizations are equal for the pure integrator. Ramp and step invariant methods do not suffer from frequency distortion, since a continuous-time pole p is transformed into a discrete-time pole e^{ph}. Recall that the output, at times $0, h, 2h, 3h, \ldots$, of a continuous-time transfer function, driven by a piecewise linear continuous input, coincides with the output of its PFOH discretized transfer function.

Exercise 10.3 Can you explain why ZOH discretization of a strictly proper transfer function $G(s)$ always leads to a strictly proper $G(z)$? Why PFOH discretization does not? *Hint*: compare formula (9.46) with (9.49).

Table 10.3 Several discretizations of continuou-time transfer functions

$H(s)$	Tustin	ZOH sampling	PFOH sampling
$\frac{1}{s}$	$\frac{h}{2}\frac{z+1}{z-1}$	$\frac{h}{z-1}$	$\frac{h}{2}\frac{z+1}{z-1}$
$\frac{1}{s^2}$	$\left(\frac{h}{2}\right)^2\frac{(z+1)^2}{(z-1)^2}$	$\frac{h^2}{2}\frac{(z+1)}{(z-1)^2}$	$\frac{h^2}{6}\frac{z^2+4z+1}{(z-1)^2}$
$\frac{1}{s+a}$	$\frac{h(z+1)}{z(2+ah)+(ah-2)}$	$\frac{1}{a}\frac{1-e^{-ah}}{z-e^{-ah}}$	$\frac{1-e^{-ah}(ah+1)+z(ah-2+(1+e^{-ah}))}{ha^2(z-e^{-ah})}$

10.3.1.2 Closed-loop methods

When significant information about the plant is available we may expect a better analog-to-digital redesign, in the sense that lower sampling rates become possible, therefore reducing hardware complexity. Under this heading, a large number of methods have appeared, and will continue to appear since it constitutes a topic for ongoing research. We can group these methods as transfer function and state-space methods.

Amongst the transfer function methods we highlight frequency-response matching. Rattan (1984) presents a computer-aided method for synthesizing the transfer function of the digital controller, by matching the frequency response of the digital control system to that of the continuous system, with a minimum weighted mean-square error.

The advantage of frequency response matching lies in the fact that we are not restricted to the order of the analog controller. Since we no longer have hardware limitations, we are free to increase the order of the (digital) controller to improve performance.

With reference to state-feedback redesign, the situation is complex, as a result of the nature of the algorithm. The state-feedback controller is a multi-input, single-output (static) device, whose inputs are the state-vector components. The discretization of such a controller, by an open-loop method, naturally leaves it unaltered, since it has no dynamics. If we wish to improve matters, by considering the closed-loop system, we find that some improvement is possible; however, the method does not allow for large decreases in sampling rates. This is dictated by the conflict between the nature of the algorithm, state-feedback and the absence of feedback between sampling instants. Because the components of an n-dimensional state-vector contain linear combinations of derivatives, up to order $(n-1)$, the effectiveness of such procedure requires a large controller bandwidth. But this is incompatible with low sampling rates.

Exercise 10.4 Consider the continuous-time plant

$$\begin{cases} \dot{\boldsymbol{x}} = \boldsymbol{A}\boldsymbol{x} + \boldsymbol{B}u \\ y = \boldsymbol{C}\boldsymbol{x} \end{cases}$$

Assuming $u = \boldsymbol{M}\, u_c - \boldsymbol{K}\, \boldsymbol{x}$, and that u_c is constant in the interval $nh \leqslant t < (n+1)\, h$, $n \in_0$, the equation becomes

$$\boldsymbol{x}\,((n+1)\,h) = \boldsymbol{\Phi}_c\, \boldsymbol{x}(nh) + \boldsymbol{\Gamma}_c\, \boldsymbol{M}\, u_c\,(nh)$$

(a) Compute $\boldsymbol{\Phi}_c$ and $\boldsymbol{\Gamma}_c$
Answer: $\boldsymbol{\Phi}_c = \exp\,(\boldsymbol{A}_c h)$

$$\boldsymbol{\Gamma}_c = \int_0^h \exp\,(\boldsymbol{A}_c\, u)\, du\, \boldsymbol{B},\; \boldsymbol{A}_c = (\boldsymbol{A} - \boldsymbol{B}\,\boldsymbol{K})$$

(b) Let the plant input be the ZOH sampled version of $(\tilde{\boldsymbol{M}}\, u_c - \tilde{\boldsymbol{K}}\, \boldsymbol{x})$, i.e.

$$u(t) = \tilde{\boldsymbol{M}} u_c\,(nh) - \tilde{\boldsymbol{K}}\boldsymbol{x}\,(nh),\; nh \leqslant t < (n+1)\,h$$

Show that

$$\boldsymbol{x}\,((n+1)\,h) = (\boldsymbol{G} - \boldsymbol{H}\,\tilde{\boldsymbol{K}})\,\boldsymbol{x}\,(nh) + \mathbf{H}\,\tilde{\boldsymbol{M}}\,u_c\,(nh)$$

and compute $\boldsymbol{G}$ and $\boldsymbol{H}$.
Answer:

$$\boldsymbol{G} = \exp\,(\boldsymbol{A}h)$$

$$\boldsymbol{H} = \int_0^h \exp\,(\boldsymbol{A}\,t)\,\mathrm{d}t\,\boldsymbol{B}$$

The continuous-time plant

$$\dot{\boldsymbol{x}} = \boldsymbol{A}\boldsymbol{x} + \boldsymbol{B}u;\; y = \boldsymbol{c}\,\boldsymbol{x} \tag{10.17}$$

with (continuous) state-feedback $u(t) = -\boldsymbol{K}\,\boldsymbol{x}(t)$, can be described at instants of the form $t = nh$, $n \in_0$, by

$$\boldsymbol{x}\,((n+1)\,h) = \boldsymbol{\Phi}_c\,\boldsymbol{x}(nh) \tag{10.18}$$

where

$$\boldsymbol{\Phi}_c = \exp\,((\boldsymbol{A} - \boldsymbol{B}\,\boldsymbol{K})\,h) \tag{10.19}$$

If the control system is discretized in such a way that

$$u(t) = -\,\tilde{\boldsymbol{K}}\,\boldsymbol{x}(nh),\; nh \leqslant t < (n+1)\,h$$

the equation becomes

$$\boldsymbol{x}\,((n+1)\,h) = (\boldsymbol{G} - \boldsymbol{H}\,\tilde{\boldsymbol{K}})\,\boldsymbol{x}(nh) \tag{10.20}$$

with

$$G = \exp(Ah)$$
$$H = \int_0^h \exp(At)\, dt\, B$$

Quite naturally, we would like to have $\tilde{K}$ such that

$$G - H\tilde{K} = \Phi_c \tag{10.21}$$

in order to match the continuous- and discrete-time performances. However, a simple inspection shows that this is not possible in general. In fact (10.21) is equivalent to a system of n^2 scalar equations with n unknowns (the components of $\tilde{K}$).

Kuo (1980) discusses this problem in detail, and suggests several formulations in order to compute $\tilde{K}$. For example, one may choose $\tilde{K}$ in order to match linear combinations of the state components, at each sampling instant. Since $y = c\,x$, we may attempt to match in this case the outputs of the digital and of the analog system. In fact, if we multiply both sides of (10.21) by c, c as in (10.17), we get n scalar equations with n unknowns; if a solution exists, the outputs will coincide, provided the system remains stable!

Another formulation in Kuo (1980) consists of expanding $(G - H\tilde{K})$ and Φ_c into a power series about $h=0$, and equating terms of equal powers. By using up to three terms in the expansion we arrive at

$$\tilde{K} = K\left[I + \frac{h}{2}(A - B\,K)\right] \tag{10.22}$$

Exercise 10.5 Show that the first three terms of the series expansions of $\Phi_c = \exp(A_c h)$ and $(G - H\tilde{K})$ coincide, provided

$$\tilde{K} = K\left[I + \frac{h}{2}(A - B\,K)\right]$$

Answer:

$$G = I + Ah + \frac{h^2}{2}A^2 + \ldots$$

$$H = \left(h\,I + \frac{h^2}{2}A + \ldots\right)B$$

$$\tilde{K} = K_0 + \frac{h}{2}K_1$$

$$(G - H\tilde{K}) = I + h\,(A - B\,K_0) + \frac{h^2}{2}(A^2 - B\,K_1 - A\,B\,K_0) + \ldots$$

$$\exp(A_c h) = I + h\,(A - B\,K) + \frac{h^2}{2}(A - B\,K)^2 + \ldots$$

Setting $\boldsymbol{K}_0=\boldsymbol{K}$ and $\boldsymbol{K}_1=\boldsymbol{K}\ (\boldsymbol{A}-\boldsymbol{B}\ \boldsymbol{K})$ the result follows.

The following example compares the performance of the expansion and output matching methods for the plant of example 9.5.

Example 10.6 Given the (continuous-time) plant

$$\left|\begin{array}{l} \dot{\boldsymbol{x}} = \boldsymbol{A}\boldsymbol{x} + \boldsymbol{B}\mathrm{u} \\ y = \boldsymbol{C}\boldsymbol{x} \end{array}\right.$$

$\boldsymbol{A}$, $\boldsymbol{B}$ and $\boldsymbol{C}$ as in example 9.5, the state-feedback $u=-\boldsymbol{K}\ \boldsymbol{x}$, $\boldsymbol{K}=(4,3)$ produces a closed-loop system with a double eigenvalue at -2, i.e. det $(s\boldsymbol{I}-\boldsymbol{A}+\boldsymbol{B}\ \boldsymbol{K})=(s+2)^2$.

Assuming $\boldsymbol{x}(0)=(2,\ 0)$, this system evolves as (2), Fig. 10.13. A sampler and ZOH device is now placed before the plant input, as in Fig. 10.12. If we set $h=0.4$ s and $\tilde{\boldsymbol{K}}=\boldsymbol{K}=(4,3)$ we get the response (1) as in Fig. 10.13: the discretization has caused a severe performance degradation. The alternative is either to reduce the sampling interval or to redesign the controller.

The series expansion method yields

$$\tilde{\boldsymbol{K}} = \boldsymbol{K}\left[\mathrm{I} + \frac{h}{2}(\boldsymbol{A} - \boldsymbol{B}\ \boldsymbol{K})\right] = \left(4 - \frac{13}{2}h,\ -8h + 3\right) \tag{10.23}$$

Note that $\tilde{\boldsymbol{K}}$ equals the continuous-time gain for $h=0$, as expected. Also $\tilde{\boldsymbol{K}}=(1.4,\ -0.2)$ for $h=0.4$.

The matching of outputs, at the sampling instants, occurs if

$$\boldsymbol{C}\ (\boldsymbol{G} - \boldsymbol{H}\ \tilde{\boldsymbol{K}}) = \boldsymbol{C}\ \boldsymbol{\Phi}_{\mathrm{c}} \tag{10.24}$$

Since $\boldsymbol{G}$ and $\boldsymbol{H}$ were already computed in exercise 9.28, let us compute $\boldsymbol{\Phi}_{\mathrm{c}}$. From (10.19)

$$\boldsymbol{\Phi}_{\mathrm{c}} = \exp\ ((\boldsymbol{A} - \boldsymbol{B}\ \boldsymbol{K})\ h) = \mathrm{e}^{-2h}\begin{bmatrix} -2h+1 & -4h \\ h & 2h+1 \end{bmatrix} \tag{10.25}$$

Also

$$(\boldsymbol{G} - \boldsymbol{H}\ \tilde{\boldsymbol{K}}) = \begin{bmatrix} \cos h-\tilde{K}_1 \sin h & -\sin h\ (1+\tilde{K}_2) \\ \sin h-\tilde{K}_1\ (1-\cos h) & \cos h-(1-\cos h)\ K_2 \end{bmatrix} \tag{10.26}$$

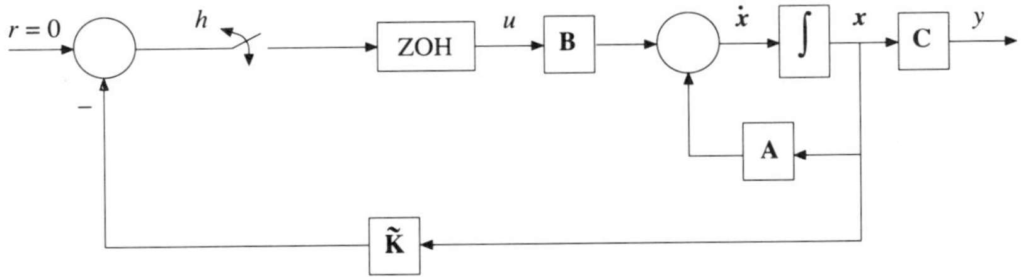

Figure 10.12 Plant of example 10.6.

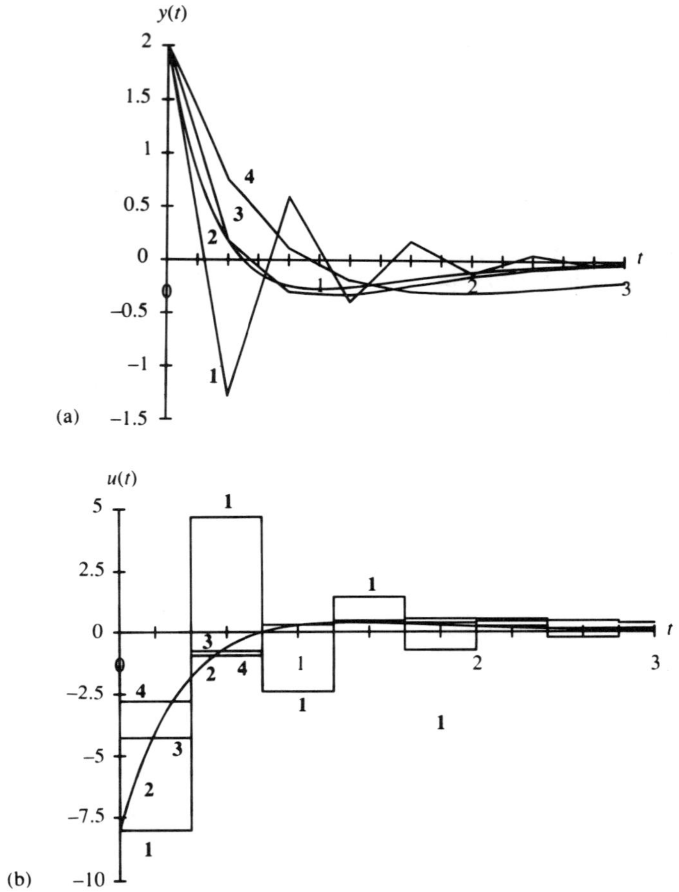

Figure 10.13 Time plots in Fig. 10.12, assuming $\boldsymbol{x}(0)=(2,0)$: 1, $\tilde{\boldsymbol{K}}=(4,3)$, $h=0.4$; 2, $\tilde{\boldsymbol{K}}=(4,3)$, $h=0$; 3, $\tilde{\boldsymbol{K}}=(2.134;\ 0.8462)$, $h=0.4$; 4, $\tilde{\boldsymbol{K}}=(1.4;\ -0.2)$; $h=0.4$; for (a) $y(t)$ (b) $u(t)$.

Now substituting the above matrices into (10.24) gives the desired value of $\tilde{\boldsymbol{K}}=(\tilde{k}_1\ \tilde{k}_2)$, namely

$$\begin{cases} \tilde{k}_1 = \dfrac{\cos h - e^{-2h}(-2h+1)}{\sin h} \\ \tilde{k}_2 = (4h\,e^{-2h} - \sin h)/\sin h \end{cases} \qquad (10.27)$$

Again, note $\tilde{\boldsymbol{K}}=(4,3)$ for $h=0$. For $h=0.4$ we get $\tilde{\boldsymbol{K}}=(2.134,\ 0.8462)$.

The performance of these two redesigns is depicted in Fig. 10.13. Note that the outputs of the continuous system, and of the discretized system by the output matching method, quite naturally are almost coincident at sampling instants; however, that is achieved at the expenses of a slightly larger control effort, as shown in part (b) of the figure. On the other hand, we found the output matching method was still exhibiting a satisfactory performance, for h=0.5. This is in contrast with the series expansion method which, for h=0.5, produced an unstable design.

The system of Fig. 10.12, with h=0.5 and $\tilde{\boldsymbol{K}}=(4,3)$, is also unstable.

10.3.1.3 PID controller discretization

Although we could include this topic under the open-loop discretization methods, the ubiquity of the PID controller justifies a separate treatment.

As shown in Chapters 4 and 8, the continuous-time PID controller is made of three (cf. equation 8.87) terms, namely proportional, integral and derivative, as follows:

$$m(t) = p(t) + i(t) + d(t) \tag{10.28}$$

The discretization of the proportional term is straightforward, since it has no dynamics, leading to

$$p(kh) = K\,(f\,r(kh) - y(kh)) \tag{10.29}$$

For the integral term, it is frequently discretized by means of the rectangular integration rule. Therefore

$$i((k+1)\,h) = i\,(kh) + \frac{Kh}{T_i}(r\,(kh) - y\,(kh)) \tag{10.30}$$

With reference to the derivative term, the usual approach is the so-called backward differences method, which approximates the derivative of a continuous-time signal $x(t)$, at time $t=kh$, by

$$\frac{\mathrm{d}\,x(t)}{\mathrm{d}\,t} \simeq \frac{x(kh) - x\,((k-1)\,h)}{h} \tag{10.31}$$

Consequently

$$\mathrm{d}(kh) = \frac{1}{(Nh + T_\mathrm{d})}\,[T_\mathrm{d}\,\mathrm{d}\,((k-1)\,h) - T_\mathrm{d}\,K\,N\,(y(kh) - y\,((k-1)\,h))] \tag{10.32}$$

Exercise 10.6 Discretize the derivative term of the PID controller by the following methods:
(a) forward difference.
(b) Tustin.
Recall that

$$D(s) = -\frac{s\,T_d}{1 + s\,T_d/N} K\,Y(s)$$

Answer:

(a) $$d\,(kh + h) = \left[1 - \frac{h\,N}{T_d}\right] d\,(kh) - K\,N\,[y(kh + h) - y(kh)]$$

(b) $$d\,(kh) = \frac{2T_d - h\,N}{2T_d + h\,N} d\,(kh - h) - \frac{2K\,N\,T_d}{2T_d + h\,N} [y(kh) - y(kh - h)]$$

The sampling rates of today's PID controllers are sufficiently large to allow simple discretization methods such as the backward differences. On the other hand, the discretization of the derivative term, by Tustin's method for example, creates a pole at

$$\frac{2T_d - h\,N}{2T_d + h\,N}$$

This pole becomes negative for $T_d < (hN)/2$, which is undesirable. Real negative poles correspond to the fastest possible rate of variation, causing unnecessary actuator wear.

For a professional treatment of PID controllers the reader is referred to Aström and Hägglund (1988).

10.3.2 Discrete-time System Design Methods

All the methods presented in earlier chapters for analog controller design are applicable in the discrete-time case, with obvious modifications. The only exception is the frequency domain, as shown shortly.

The following example illustrates the application of root-locus ideas to discrete-time systems.

Example 10.7 Consider a position control system, as in Fig.10.8(a), with $h=0.2$, $D(z)=1$, and

$$G_p(s) = \frac{K}{s(s + 1)}$$

The open-loop transfer function is (cf. with example 9.8):

$$G(z) = \mathcal{Z}\left\{\mathcal{L}^{-1}\left(\frac{1 - e^{-hs}}{s} \times \frac{K}{s(s + 1)}\right)\right\}$$

$$= K\frac{(0.018\,73\,z + 0.017\,52)}{(z - 1)\,(z - e^{-0.2})}$$

with a zero at -0.9355. The root-locus is sketched in Fig. 10.14 showing rather unsatisfactory closed-loop poles.

Assume we wish a pair of closed-loop poles (p_d, $\bar{p}_d$) at ($-0.5 \pm j0.5$). Since the phase of $G(z)$ at $z=p_d$ is $-237.6°$, the compensator $D(z)$ must contribute with $(237.6°-180°)=57.6°$ at $z=p_d$.

Let the compensator have the form

$$D(z) = \frac{z - z_c}{z - p_c}$$

If we cancel the plant pole at $e^{-0.2}$ by setting $z_c=e^{-0.2}$, then the computer pole must be $p_c=0.266$. Therefore

$$D(z) = \frac{z - 0.8187}{z - 0.266}$$

The root-locus, for the compensated system, is also depicted in Fig. 10.14, showing the desired closed-loop poles for $K=13.76$. The closed-loop step responses, before and after compensation, for the indicated gain values, are compared in Fig. 10.15. Notice the same overshoot in both situations and the poor performances before compensation, namely in terms of rise-time and settling-time.

Given a computer-controlled system, as in Fig. 9.1, we may wish to design $D(z)$ using frequency-domain techniques as in the analog case. The fact that $D(e^{j\omega h})$, $0 \leq \omega \leq \pi/h$, is not a rational function in $(j\omega)$, precludes such use. However we can transform $D(z)$ into a rational function of a variable $\bar{s}$, by means of our well-known transformation (10.11), namely

$$\bar{s} = \frac{2}{h}\frac{z - 1}{z + 1}$$

For $z=e^{j\omega h}$

$$\bar{s} = j\frac{2 \sin \omega h}{h(1 + \cos \omega h)} = j\frac{2}{h}\tan\left(\frac{\omega h}{2}\right) = j\,\bar{\omega} \tag{10.33}$$

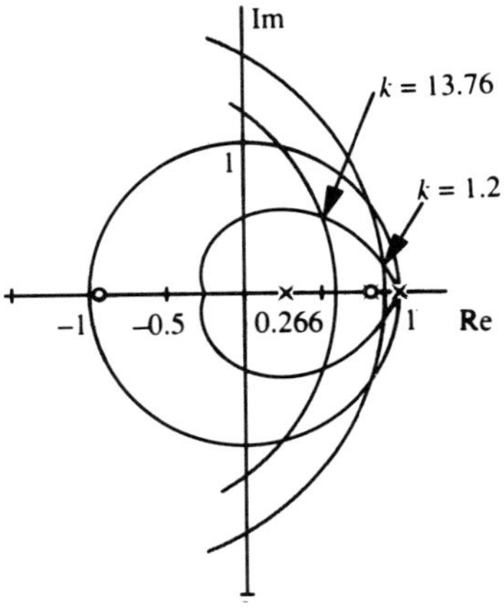

Figure 10.14 Root-loci for example 10.7. Superimposed is the locus of constant damping ratio $\zeta=0.4$.

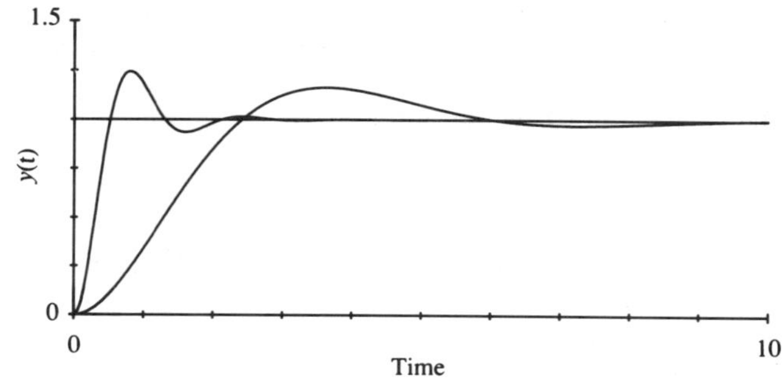

Figure 10.15 Step responses of the uncompensated system with $K=1.2$, and of the compensated system with $K=13.76$, in example 10.7.

where $0 \leqslant \bar{\omega} < \infty$. The compensator can then be designed, in the $\bar{\omega}$ domain, by conventional frequency domain methods. Naturally the pulse transfer function of the plant and ZOH combination, $G(z)$, must be also converted to the $\bar{\omega}$ domain, by replacing z by

$$z = \frac{2 + \mathrm{j}\,\bar{\omega}h}{2 - \mathrm{j}\,\bar{\omega}h} \tag{10.34}$$

The last step is to convert the transfer function of the resulting compensator, $D(\bar{s})$, back to the z-plane, by substituting

$$\bar{s} = \frac{2}{h}\frac{z - 1}{z + 1}$$

in $D(\bar{s})$. Note that the design specifications in the $\bar{\omega}$ domain must take into account the warping of the frequency scale, as discussed in example 10.5, section 10.3.1.

The discrete-time design allows a new type of behaviour, which was neither possible in continuous-time nor with discretized designs. It is the so-called 'deadbeat' or 'minimal response time' transfer function. With such a transfer function, the output y equals the reference signal r, at the earliest possible sampling instant, after an error is detected, and at every sampling instant after that. Consequently

$$\frac{Y(z)}{R(z)} = \frac{1}{z^n}$$

for some non-negative integer n.

The following example illustrates the design of such a controller.

Example 10.8 Consider the control system of Fig. 10.8(a), with plant transfer function

$$G_{\mathrm{p}}(s) = \frac{1}{s\,(s + 1)} \tag{10.35}$$

The pulse transfer function of the ZOH plant combination is

$$G_H\ G_p(z) = \frac{(h - 1 + e^{-h})\ z + (1 - e^{-h} - h\ e^{-h})}{z^2 - (1 + e^{-h})\ z + e^{-h}} \tag{10.36}$$

In this case, $Y/R=1/z^2$ is a feasible closed-loop transfer function; it satisfies all the implementability conditions discussed in section 8.2. In fact, this transfer function is achieved with the controller

$$D(z) = \frac{z^2 - (1 + e^{-h})\ z + e^{-h}}{(h - 1 + e^{-h})\ z^3 + (1 - e^{-h}\ (1 + h))\ z^2 - (h - 1 + e^{-h})\ z - (1 - e^{-h}\ (1 + h))} \tag{10.37}$$

Figure 10.16 shows the response $y(t)$, and the control effort (ZOH output), of this system to a unit-step reference change, for two sampling periods, namely $h=0.5$ and $h=2$. Note the dramatic reduction in control effort, when the sampling

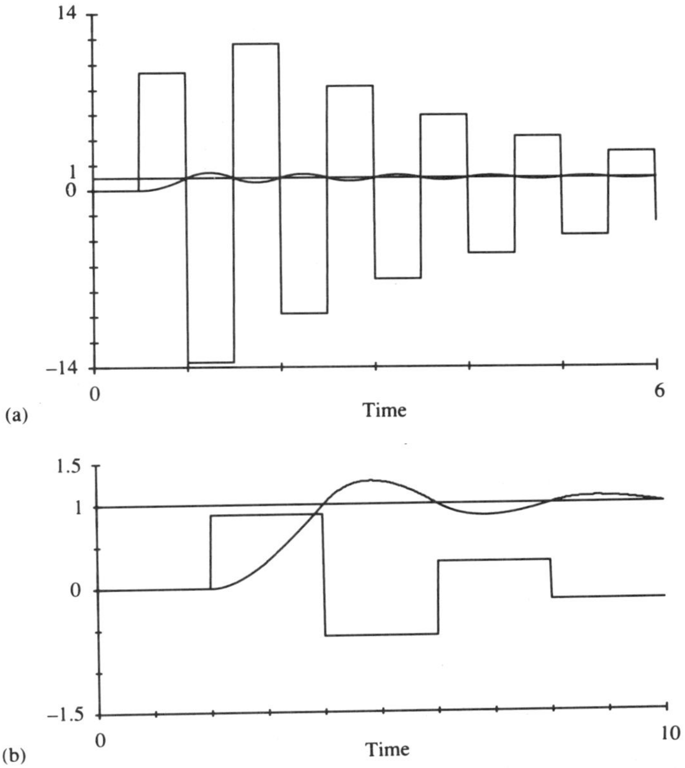

Figure 10.16 Response $y(t)$, and control effort of system of Fig. 10.8(a), to a unit-step reference change, when the sampling interval is: (a) $h=0.5$; (b) $h=2$. $D(z)$ is a deadbeat controller, computed in example 10.8, and the plant transfer function $G_p(s)=1/(s\ (s+1))$.

period is increased from 0.5 to 2, and the coincidence of output and reference input, after the second sampling instant, as desired.

Also of interest to note is the oscillation 'ringing' of the controller output, at each sample, around its final value. This is general undesirable, since it causes excessive actuator activity. Its cause is the pole of

$$\frac{X(z)}{R(z)} = \frac{z^2 - (1 + e^{-h})\, z + e^{-h}}{z^2\, [(h - 1 + e^{-h})\, z + (1 - e^{-h}(1 + \mathrm{h}))]} \tag{10.38}$$

on the negative real axis.

Exercise 10.7 The closed-loop system, of example 10.8 presents an oscillation which cannot be seen at the sampling instants. Is there any possible loss of observability caused by the cancellation of the plant poles?
Answer: yes; there are cancellations in the closed-loop transfer function $Y(z)/R(z)$. These will cause loss of either reachability or observability.

Exercise 10.8 Show that in transfer function (10.38), the amplitude of $(x(kh))_{k\in\mathbb{N}_0}$, at time k=1, goes to infinity as h approaches 0, when $r(t)$ is a step at t=0.
Answer:

$$X(z) = \frac{z^2 - (1 + e^{-h})\, z + e^{-h}}{z\, [(h - 1 + e^{-h})\, z + (1 - e^{-h}(1 + h))](z - 1)}$$

because

$$R(z) = \frac{z}{z - 1}$$

Since $x(0)$=0,

$$X(z) = x(h)\, z^{-1} + x(2h)\, z^{-2} + \dots$$

$$z\, X(z) = x(h) + x(2h)\, z^{-1} + \dots$$

then

$$\lim_{z\to\infty} z\, X(z) = x(h)$$

But

$$\lim_{z\to\infty} z\, X(z) = \frac{1}{h - 1 + e^{-h}}$$

Since the right hand side goes to infinity, as h goes to zero, the result follows.

The deadbeat algorithm is very sensitive to modelling inaccuracies. This is shown in Fig.10.17, where we have used the controller, as in Fig. 10.16, with h=2, and two different plants, namely

$$G_p(s) = \frac{1.1}{s\,(s + 1.1)} \tag{i}$$

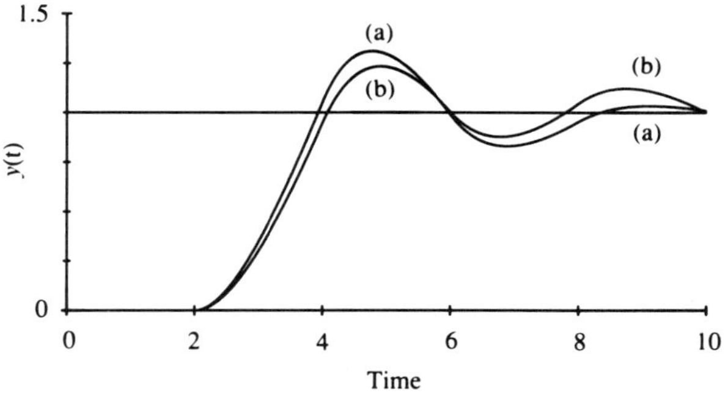

Figure 10.17 As in Fig. 10.16, with $h=2$, except (a) $G_p(s)=1/(s\ (s+1.1))$; (b) $G_p(s)=0.9/(s\ (s+0.9))$.

$$G_p(s) = \frac{0.9}{s\ (s + 0.9)} \tag{ii}$$

Note the output y and reference r no longer coincide, at the first and third sampling instants; a small modelling error has destroyed the 'deadbeat' behaviour.

10.4 CHOICE OF SAMPLING FREQUENCY

Sampling frequency selection is a difficult problem. The present state-of-art still requires our decisions to be based on heuristic arguments, and practical experience, which are translated, whenever possible, into 'rules-of-thumb'.

We already know that amplitude constraints, on the control effort, create a lower bound for the sampling period, h. What we need now is to find 'reasonable' upper bounds for h. With respect to discretization of analog designs, the popular approach is to approximate the (zero-order) hold block by a pure transport delay, of half a sampling period. Such an approach is very accurate, provided h is small. Then we select h on the basis of how much phase margin we are prepared to lose, at crossover frequency ω_c, as a consequence of discretization. If $\bar{\varphi}$ denotes such phase loss, we can write

$$h \leqslant \frac{2\bar{\varphi}}{\omega_c} \tag{10.39}$$

since the hold device, at the frequency ω_c, contributes with a phase of $-\omega_c h/2$, approximately. This rule is rather conservative. In practice we find that it is often possible to use larger sampling periods, and still maintain a good approximation of the continuous-time behaviour.

When the controller is designed in the discrete domain, the choice of h depends on the nature of the specifications. If, for example, the desired closed-loop bandwidth, ω_B, is specified, the usual procedure is to select the sampling frequency 10–30 times ω_B. Since the important part of the closed-loop frequency response goes as far as, say, 10 times the bandwidth, the sampling rate is chosen accordingly. When the specifications are given in the time-domain, e.g. by means of the closed-loop dominant poles, or by the shape of the desired closed-loop step response, we can use the closed-loop rise time, T_r, as a guideline and choose h such that

$$\frac{T_r}{10} \leqslant h \leqslant \frac{T_r}{4}$$

i.e. 4 to 10 samples per rise time.

With reference to PID controllers, many of them have a fixed sampling period, of a few tenths of a second, and pose no problems in the majority of industrial processes. When we have the choice of selecting h, we must bear in mind that the presence of pure transport delays dictates much smaller sampling periods for PID than for PI controllers. As we have seen in Chapter 4, long pure delays and derivative action are somewhat conflicting situations, which can be aggravated by long sampling intervals.

REFERENCES

Bernhardsson, B. (1990) 'The predictive first-order hold circuit', 29th IEEE Conference on Decision and Control, 5–7 December, Honolulu, Hawaii.

Hanselmann, H. (1987) 'Implementation of digital controllers – a survey' *Automatica*, **23**(1), 7–32.

Kuo, B. (1980) *Digital Control Systems*, Holt, Rinehart and Winston.

Rattan, K.S. (1984) 'Digitalization of existing continuous control systems', *IEEE Trans. Auto. Control*, **AC–29**(3), pp. 282–5.

Appendix A

The Laplace Transform

Given a real function of a real variable $f(t)$, such that $f(t)=0$ for $t<0$, the Laplace transform (LT) of $f(t)$, $\mathcal{L}[f(t)]$, is a function of a complex variable, $F(s)$, defined by

$$\mathcal{L}[f(t)] = F(s) = \int_{0^-}^{\infty} f(t)e^{-st}\,dt \tag{A.1}$$

when such integral exists.

It can be shown that, when

$$|f(t)| \leqslant k \exp(\sigma^+ t) \tag{A.2}$$

the convergence of integral (A.1) is ensured for values of s, $s=\sigma+j\omega$, such that $\sigma>\sigma^+$. Recall $j=\sqrt{-1}$.

For example, given the time function

$$f(t) = \begin{cases} 0; \ t < 0 \\ e^{at}; \ t \geqslant 0 \end{cases}$$

we can write

$$F(s) = \int_{0^-}^{\infty} e^{at}\, e^{-st}\, dt = \int_{0^-}^{\infty} e^{-(s-a)t}\, dt = \frac{1}{s-a}$$

provided $\mathrm{Re}(s)>a$.

There exists a one-to-one correspondence between time functions and Laplace transforms. To be more precise, if $f_1(t)\neq f_2(t)$ on a subset of $\overset{+}{0}$ of non-zero measure, then $F_1(s)\neq F(s)$. Therefore, if $f_1(t)$ and $f_2(t)$ are equal, except at a finite set of instants, then $F_1(s)=F(s)$.

The computation of $f(t)$, given $F(s)$, is called the LT inversion problem, and is defined by means of the integral

$$f(t) = \frac{1}{2\pi j}\int_{\sigma_1-j\infty}^{\sigma_1+j\infty} F(s) \tag{A.3}$$

for suitable values of σ_1.

From the LT definition and its inverse, we immediately conclude that these are linear transformations. For example

$$\mathcal{L}[a\,f(t) + b\,g(t)] = a\,F(s) + b\,G(s) \tag{A.4}$$

Other important properties of the LT are as follows:

$$\mathcal{L}\left[\frac{\mathrm{d}f}{\mathrm{d}t}\right] = s\,F(s) - f(0^-) \tag{A.5}$$

$$\mathcal{L}\left[\int_{0^-}^{t} f(t')\,\mathrm{d}t'\right] = \frac{F(s)}{s} \tag{A.6}$$

$$\lim_{s\to\infty} s\,F(s) = \lim_{t\to 0^+} f(t) = f(0^+) \tag{A.7}$$

This property is known as the *initial-value theorem*.

$$\lim_{s\to 0} s\,F(s) = \lim_{t\to\infty} f(t) = f(\infty) \tag{A.8}$$

when such a limit exists. This property is known as the *final-value theorem*.

$$\mathcal{L}\,[f(t-T)] = \mathrm{e}^{-sT}\,F(s) \tag{A.9}$$

where $f(t-T)$ is $f(t)$ delayed by T time units.

$$\mathcal{L}^{-1}[W(s)\,U(s)] = \int_{0^-}^{t} \omega(t-t')\,u(t')\,\mathrm{d}t' \tag{A.10}$$

This property is known as the *convolution theorem*.

Property (A.5) is the gist of Laplace transform theory, since it transforms linear, constant coefficient, differential equations into algebraic ones. Its proof is straightforward, as shown below.

By definition

$$\mathcal{L}\left[\frac{\mathrm{d}f}{\mathrm{d}t}\right] = \int_{0^-}^{\infty} \frac{\mathrm{d}f}{dt}\mathrm{e}^{-st}\,dt \tag{A.11}$$

Recalling the integration by parts formula

$$\int u'v = uv - \int uv'$$

we can write

$$\mathcal{L}\left[\frac{\mathrm{d}f}{\mathrm{d}t}\right] = f(t)\mathrm{e}^{-st}\Big|_{0^-}^{\infty} - \int_{0^-}^{\infty} f(t)\,(-\,s\mathrm{e}^{-st})\mathrm{d}t = -f(0^-) + sF(s)$$

q.e.d.

Table A.1 Some time functions and corresponding Laplace transforms. Time functions are zero for negative values of t

Function	*Laplace transforms*
Unit impulse $\delta(t)$	1
Unit step (1)	$\frac{1}{s}$
Unit ramp (t)	$\frac{1}{s^2}$
e^{-at}	$\frac{1}{s+a}$
$\frac{t^{n-1}\,e^{-at}}{(n-1)!}$	$\frac{1}{(s+a)^n}$
$\sin\omega t$	$\frac{\omega}{s^2+\omega^2}$
$\cos\omega t$	$\frac{s}{s^2+\omega^2}$
$\sin(\omega t+\phi)$	$\frac{s\sin\phi+\omega\cos\phi}{s^2+\omega^2}$
$\frac{1}{\omega}e^{-at}\sin\omega t$	$\frac{1}{[s+(a+j\omega)]\,[s+(a-j\omega)]}$

Laplace transforms which are a quotient of two polynomials, can be easily inverted by means of a partial fraction decomposition and the help of Table A.1. For example:

$$\mathcal{L}^{-1}\left[\frac{1}{s(s+1)^2}\right]=\mathcal{L}^{-1}\left[\frac{1}{s}-\frac{1}{(s+1)^2}-\frac{1}{s+1}\right]$$

$$=1-\frac{t\,e^{-t}}{0!}-e^{-t}$$

$$=1-e^{-t}(t+1)$$

for $t \geqslant 0$. For $t<0$, $\mathcal{L}^{-1}(.)=0$.

Appendix B

Matrices and Determinants

Consider the following set of linear algebraic equations

$$
\begin{aligned}
y_1 &= a_{11}\, x_1 + a_{12}\, x_2 + \ . \ . \ . + a_{11n}\, x_n \\
y_2 &= a_{21}\, x_1 + a_{22}\, x_2 + \ . \ . \ . + a_{2n}\, x_n \\
&\cdot \\
&\cdot \\
&\cdot \\
y_m &= a_{m1}\, x_1 + a_{m2}\, x_2 + \ . \ . \ . + a_{mn}\, x_n
\end{aligned} \tag{B.1}
$$

in the unknowns $x_1, x_1, \ldots, x_n$.

Defining the vectors

$$
\boldsymbol{y} = \begin{bmatrix} y_1 \\ y_2 \\ \cdot \\ \cdot \\ \cdot \\ y_m \end{bmatrix}, \quad \boldsymbol{x} = \begin{bmatrix} x_1 \\ x_2 \\ \cdot \\ \cdot \\ \cdot \\ x_n \end{bmatrix}
$$

and the array

$$
\boldsymbol{A} = [a_{ij}] = \begin{bmatrix} a_{11} & a_{12} & \ldots & a_{1n} \\ a_{21} & a_{22} & \ldots & a_{2n} \\ \cdot & \cdot & & \cdot \\ \cdot & \cdot & & \cdot \\ \cdot & \cdot & & \cdot \\ a_{m1} & a_{m2} & \ldots & a_{mn} \end{bmatrix} \tag{B.3}
$$

(B.1) can be expressed in the more compact form

$$
\boldsymbol{y} = \boldsymbol{A}\boldsymbol{x} \tag{B.4}
$$

provided we define the multiplication of array $\boldsymbol{A}$ by vector $\boldsymbol{x}$ as in (B.1). Array (B.3), also denoted by $[a_{ij}]$ is what we call an $m \times n$ *matrix*, because it has *m rows* and *n columns*.

a_{ij} denotes the element in row i and column j in matrix A. Consequently

$$y_j = \sum_{i=1}^{n} a_{ji} x_i, j = 1, 2, \ldots, m$$

It can be shown that any linear operation between two finite dimensional vector spaces can be described as in (B.4). In particular we see that **y** belongs to the linear space generated by the columns of $\boldsymbol{A}$, known as the *range* of $\boldsymbol{A}$. In fact we can write (B.1) in the following equivalent form:

$$\boldsymbol{y} = x_1 \begin{bmatrix} a_{11} \\ a_{21} \\ \cdot \\ \cdot \\ \cdot \\ a_{m1} \end{bmatrix} + x_2 \begin{bmatrix} a_{12} \\ a_{22} \\ \cdot \\ \cdot \\ \cdot \\ a_{m2} \end{bmatrix} + \ldots + x_n \begin{bmatrix} a_{1n} \\ a_{2n} \\ \cdot \\ \cdot \\ \cdot \\ a_{mn} \end{bmatrix} \tag{B.5}$$

which shows that $\boldsymbol{y}$ is a linear combination of the columns of matrix $\boldsymbol{A}$.

If the columns of matrix $\boldsymbol{A}$ are linearly dependent, then there exists a vector $\boldsymbol{x}$ such that

$$\mathbf{O} = \boldsymbol{Ax}$$

The set of such vectors also constitutes a linear space and is called the *null-space* of $\boldsymbol{A}$.

We define the *rank* of a matrix as the number of its linearly independent rows or columns. It can be shown that row rank = column rank.

Given the $m \times n$ matrices $\boldsymbol{A}=[a_{ij}]$ and $\boldsymbol{B}=[b_{ij}]$ we define the sum of $\boldsymbol{A}$ and $\boldsymbol{B}$,

$$(\mathrm{A} + \mathrm{B}) \tag{B.6}$$

as the matrix $\boldsymbol{C}=[c_{ij}]$ such that

$$c_{ij} = a_{ij} + b_{ij} \tag{B.7}$$

Assume we have the following situation

$$\boldsymbol{y} = \boldsymbol{A}\ \boldsymbol{x} \text{ and } \boldsymbol{z} = \boldsymbol{By}$$

$$\boldsymbol{x} \in \mathbb{R}^n,\ \boldsymbol{z} = \in \mathbb{R}^l$$

Consequently

$$\boldsymbol{z} = \boldsymbol{B}(\boldsymbol{Ax}) \tag{B.8}$$

Using form (B.5) we can write

$$\boldsymbol{z} = x_1\boldsymbol{B} \begin{bmatrix} a_{11} \\ a_{21} \\ \cdot \\ \cdot \\ \cdot \\ a_{m1} \end{bmatrix} + x_1\boldsymbol{B} \begin{bmatrix} a_{12} \\ a_{22} \\ \cdot \\ \cdot \\ \cdot \\ a_{m2} \end{bmatrix} + \ldots + x_n\boldsymbol{B} \begin{bmatrix} a_{1n} \\ a_{2n} \\ \cdot \\ \cdot \\ \cdot \\ a_{mn} \end{bmatrix} \tag{B.9}$$

or equivalently,

$$z = \left[B \begin{bmatrix} a_{11} \\ a_{21} \\ \cdot \\ \cdot \\ \cdot \\ a_{m1} \end{bmatrix} B \begin{bmatrix} a_{12} \\ a_{22} \\ \cdot \\ \cdot \\ \cdot \\ a_{m2} \end{bmatrix} \ldots B \begin{bmatrix} a_{1n} \\ a_{2n} \\ \cdot \\ \cdot \\ \cdot \\ a_{mn} \end{bmatrix} \right] x \tag{B.10}$$

In a more compact form

$$z = Px \tag{B.11}$$

The $l \times n$ matrix P in the above equation is what we call the *product* of matrices B and A, i.e.

$$P = BA \tag{B.12}$$

Notice that the (i,j)th element of P, p_{ij}, is obtained by multiplying componentwise the ith row of B by the jth column of A, i.e.

$$p_{ij} = \sum_{r=1}^{m} b_{ir}\, a_{rj} \tag{B.13}$$

Consequently we have in general $BA \neq AB$.

Given matrices A and B, we say B is the *transpose of* A, if the columns of A equal the rows of B. We denote transposition by the superscript T, i.e. A^{T}.

When a matrix equals its transpose we say the matrix is *symmetric*. Note that symmetric matrices have the same number of rows and columns.

It is easy to show that

$$(A^{T})^{T} = A$$

$$(AB)^{T} = \mathbf{B}^{T}\mathbf{A}^{T}$$

Matrices with equal number of rows and columns are known as *square* matrices.

Given an $n \times m$ matrix A we define the *trace* of A, $\mathrm{Tr}(A)$, as the sum of its diagonal elements. Therefore

$$\mathrm{Tr}(A) = \sum_{i=1}^{n} a_{ii} \tag{B.14}$$

An $m \times n$ matrix A defines a (linear) mapping between vector spaces ^{n} and ^{m}. Such a mapping is invertible if it is one-to-one and on to. Therefore we need, in order to define the inverse of A, A^{-1}, the following conditions:

1. $m=n$.
2. The columns of A are linearly independent. (B.15)

If the columns of A are not linearly independent, there are *non-zero* elements $\mathbf{x}$ in

$\mathbb{R}^n$ such that $\boldsymbol{Ax}=\boldsymbol{0}$; since $\bar{\boldsymbol{A}}\boldsymbol{0}=\boldsymbol{0}$, $\boldsymbol{A}$ cannot be inverted, because the transformation is not one-to-one.

Naturally, if $\boldsymbol{A}$ has an inverse, we can write

$$\boldsymbol{AA}^{-1} = \boldsymbol{A}^{-1}\boldsymbol{A} = \boldsymbol{I} \tag{B.16}$$

where $\boldsymbol{I}$ is the *identity matrix*:

$$\boldsymbol{I} = [i_{kl}] \tag{B.17}$$

where

$$i_{kl} = 0, \text{ if } k \neq l, \text{ and } i_{kk} = 1$$

Therefore the identity matrix is made of 1s along the main diagonal and 0s everywhere else. For example

$$\boldsymbol{I} = \begin{bmatrix} 1\,0\,0 \\ 0\,1\,0 \\ 0\,0\,1 \end{bmatrix}$$

is the 3×3 identity matrix.

Besides the identity matrix, there are matrices with special structures of interest to us.

The *diagonal* matrix:

$$[a_{ij}] = \begin{bmatrix} a_{11} & & & \boldsymbol{0} \\ & a_{22} & & \\ & & \ddots & \\ \boldsymbol{0} & & & a_{nn} \end{bmatrix} \tag{B.18}$$

is characterized by the property $a_{ij}=0$, if $i \neq j$.

The *lower triangular* matrix:

$$[a_{ij}] = \begin{bmatrix} a_{11} & & & \boldsymbol{0} \\ a_{21} & a_{22} & & \\ & & \ddots & \\ a_{n1} & a_{n2} & \cdots & a_{nn} \end{bmatrix} \tag{B.19}$$

has zero elements above the main diagonal, i.e.

$$[a_{ij}] = 0 \text{ if } (i - j) < 0$$

The *Toeplitz* matrix, which is constant along the diagonals; for example

$$\boldsymbol{A} = \begin{bmatrix} a & b & c & d \\ e & a & b & c \\ f & e & a & b \\ g & f & e & a \end{bmatrix} \tag{B.20}$$

is a 4×4 *Toeplitz* matrix.

The *companion* matrix, which has the structure

$$A = \begin{bmatrix} 0 & 1 & 0 & \dots & 0 & 0 \\ 0 & 0 & 1 & & \cdot & \cdot \\ \cdot & \cdot & \cdot & & \cdot & \cdot \\ \cdot & \cdot & \cdot & & \cdot & \cdot \\ \cdot & \cdot & \cdot & & 1 & 0 \\ 0 & 0 & 0 & & 0 & 1 \\ -a_n & -a_{n-1} & -a_{n-2} & \dots & -a_2 & -a_1 \end{bmatrix} \tag{B.21}$$

To every $n \times m$ square matrix $\boldsymbol{A}$ we can assign a number, the *determinant* of $\boldsymbol{A}$, denoted by 'det $\boldsymbol{A}$', which is equal to the absolute value of the volume, in $\mathbb{R}^n$, enclosed by the parallelepiped generated by the columns in the matrix.

The determinant of $\boldsymbol{A}$ can be evaluated by using Laplace's expansion

$$\det \boldsymbol{A} = \sum_{j=1}^{n} a_{ij}\, \gamma_{ij} \tag{B.22}$$

for any $i=1, 2, \dots, n$. γ_{ij} denotes the *cofactor* of the element a_{ij} and is equal to $(-1)^{i+j} \det \boldsymbol{M}_{ij}$. $\boldsymbol{M}_{ij}$ is the submatrix obtained from $\boldsymbol{A}$ by deleting row i and column j. det $\boldsymbol{M}_{ij}$ is called the *ijth minor* of the matrix.

An important consequence of expansion (B.22) is that the determinant of a triangular matrix is equal to the product of the diagonal elements. Another obvious consequence is

$$\det \boldsymbol{A} = \det \boldsymbol{A}^{\mathrm{T}} \tag{B.23}$$

Furthermore, if row (or column) i of $\boldsymbol{A}$ is multiplied by a scalar α, the determinant of the resulting matrix is $\alpha \det \boldsymbol{A}$.

Another very useful property of the determinant is

$$\det (\boldsymbol{AB}) = \det \boldsymbol{A} \det \mathbf{B} \tag{B.24}$$

The *inverse* of the (square) matrix **A** can be computed by means of Cramer's rule:

$$\boldsymbol{A}^{-1} = \frac{1}{\det \boldsymbol{A}} \operatorname{Adj} \boldsymbol{A} \tag{B.25}$$

where Adj $\boldsymbol{A}$ is the *adjugate* matrix of $\boldsymbol{A}$, defined by

$$\operatorname{Adj} \boldsymbol{A} = [\gamma_{ij}]^{\mathrm{T}} \tag{B.26}$$

where γ_{ij} is the cofactor of a_{ij} as defined above.

There are three types of *elementary row (column) operations* that we can perform on a square matrix $\boldsymbol{A}$ without changing its rank:

1. Multiplication of a row (column) by a non-zero scalar.

2. Add to row (column) i the product of row (column) j, $i \neq j$, by a non-zero scalar α.
3. Interchange of two rows (columns).

Since the effect of any elementary row (column) operation is the same as multiplying $\boldsymbol{A}$ on the left (right) by a non-singular matrix $\boldsymbol{E}$, det $\boldsymbol{E} \neq 0$, the result follows. Matrix $\boldsymbol{E}$ is obtained from the identity matrix I by performing on I the desired row (column) operation, e.g. let

$$\boldsymbol{A} = \begin{bmatrix} a_{11} & a_{12} \\ a_{21} & a_{22} \end{bmatrix}$$

Then

$$\bar{\boldsymbol{A}} = \begin{bmatrix} a_{21} & a_{22} \\ a_{11} & a_{12} \end{bmatrix} = \begin{bmatrix} 0 & 1 \\ 1 & 0 \end{bmatrix} \boldsymbol{A}$$

Therefore det $\bar{\boldsymbol{A}}$=det $\boldsymbol{A}$.

If

$$\bar{\boldsymbol{A}} = \begin{bmatrix} \alpha a_{11} & \alpha a_{12} \\ a_{21} & a_{22} \end{bmatrix}$$

then

$$\bar{\boldsymbol{A}} = \begin{bmatrix} \alpha & 0 \\ 0 & 1 \end{bmatrix} \boldsymbol{A}$$

therefore det $\bar{\boldsymbol{A}}=\alpha$ det $\boldsymbol{A}$, as noted earlier.

On the other hand, if

$$\bar{\boldsymbol{A}} = \begin{bmatrix} a_{11} & a_{12} \\ \alpha a_{11}+a_{21} & \alpha a_{12}+a_{22} \end{bmatrix}$$

then

$$\bar{\boldsymbol{A}} = \begin{bmatrix} 1 & 0 \\ \alpha & 1 \end{bmatrix} \boldsymbol{A}$$

Consequently det $\bar{\boldsymbol{A}}$=det $\boldsymbol{A}$.

Index